Developmental Exposure to Environmental Contaminants

Developmental Exposure to Environmental Contaminants

Editors

Kimberly Keil Stietz
Tracie Baker
Jessica Plavicki

MDPI • Basel • Beijing • Wuhan • Barcelona • Belgrade • Manchester • Tokyo • Cluj • Tianjin

Editors

Kimberly Keil Stietz
Comparative Biosciences
University of
Wisconsin-Madison
Madison
United States

Tracie Baker
Aquatic Pathobiology
Laboratory Department of
Environmental Global Health
University of Florida
Gainesville
United States

Jessica Plavicki
Pathology and Laboratory
Medicine
Brown University
Providence
United States

Editorial Office
MDPI
St. Alban-Anlage 66
4052 Basel, Switzerland

This is a reprint of articles from the Special Issue published online in the open access journal *Toxics* (ISSN 2305-6304) (available at: www.mdpi.com/journal/toxics/special_issues/developmental_exposure_contaminants).

For citation purposes, cite each article independently as indicated on the article page online and as indicated below:

LastName, A.A.; LastName, B.B.; LastName, C.C. Article Title. *Journal Name* **Year**, *Volume Number*, Page Range.

ISBN 978-3-0365-5886-8 (Hbk)
ISBN 978-3-0365-5885-1 (PDF)

© 2022 by the authors. Articles in this book are Open Access and distributed under the Creative Commons Attribution (CC BY) license, which allows users to download, copy and build upon published articles, as long as the author and publisher are properly credited, which ensures maximum dissemination and a wider impact of our publications.

The book as a whole is distributed by MDPI under the terms and conditions of the Creative Commons license CC BY-NC-ND.

Contents

Korawin Triyasakorn, Ubah Dominic Babah Ubah, Brandon Roan, Minsyusheen Conlin, Ken Aho and Prabha S. Awale
The Antiepileptic Drug and Toxic Teratogen Valproic Acid Alters Microglia in an Environmental Mouse Model of Autism
Reprinted from: *Toxics* 2022, 10, 379, doi:10.3390/toxics10070379 1

Alex Haimbaugh, Chia-Chen Wu, Camille Akemann, Danielle N. Meyer, Mackenzie Connell and Mohammad Abdi et al.
Multi- and Transgenerational Effects of Developmental Exposure to Environmental Levels of PFAS and PFAS Mixture in Zebrafish (*Danio rerio*)
Reprinted from: *Toxics* 2022, 10, 334, doi:10.3390/toxics10060334 11

Kimberly A. Jarema, Deborah L. Hunter, Bridgett N. Hill, Jeanene K. Olin, Katy N. Britton and Matthew R. Waalkes et al.
Developmental Neurotoxicity and Behavioral Screening in Larval Zebrafish with a Comparison to Other Published Results
Reprinted from: *Toxics* 2022, 10, 256, doi:10.3390/toxics10050256 37

Narimane Djekkoun, Flore Depeint, Marion Guibourdenche, Hiba El Khayat El Sabbouri, Aurélie Corona and Larbi Rhazi et al.
Chronic Perigestational Exposure to Chlorpyrifos Induces Perturbations in Gut Bacteria and Glucose and Lipid Markers in Female Rats and Their Offspring
Reprinted from: *Toxics* 2022, 10, 138, doi:10.3390/toxics10030138 73

Hao Chen, Rhianna K. Carty, Adrienne C. Bautista, Keri A. Hayakawa and Pamela J. Lein
Triiodothyronine or Antioxidants Block the Inhibitory Effects of BDE-47 and BDE-49 on Axonal Growth in Rat Hippocampal Neuron-Glia Co-Cultures
Reprinted from: *Toxics* 2022, 10, 92, doi:10.3390/toxics10020092 91

Nelson T. Peterson and Chad M. Vezina
Male Lower Urinary Tract Dysfunction: An Underrepresented Endpoint in Toxicology Research
Reprinted from: *Toxics* 2022, 10, 89, doi:10.3390/toxics10020089 105

Jessica Phillips, Alex S. Haimbaugh, Camille Akemann, Jeremiah N. Shields, Chia-Chen Wu and Danielle N. Meyer et al.
Developmental Phenotypic and Transcriptomic Effects of Exposure to Nanomolar Levels of 4-Nonylphenol, Triclosan, and Triclocarban in Zebrafish (*Danio rerio*)
Reprinted from: *Toxics* 2022, 10, 53, doi:10.3390/toxics10020053 129

Bailey A. Kermath, Lindsay M. Thompson, Justin R. Jefferson, Mary H. B. Ward and Andrea C. Gore
Transgenerational Effects of Prenatal Endocrine Disruption on Reproductive and Sociosexual Behaviors in Sprague Dawley Male and Female Rats
Reprinted from: *Toxics* 2022, 10, 47, doi:10.3390/toxics10020047 151

Ross Gillette, Michelle Dias, Michael P. Reilly, Lindsay M. Thompson, Norma J. Castillo and Erin L. Vasquez et al.
Two Hits of EDCs Three Generations Apart: Effects on Social Behaviors in Rats, and Analysis by Machine Learning
Reprinted from: *Toxics* 2022, 10, 30, doi:10.3390/toxics10010030 171

Rafael R. Domingues, Hannah P. Fricke, Celeste M. Sheftel, Autumn M. Bell, Luma C. Sartori and Robbie S. J. Manuel et al.
Effect of Low and High Doses of Two Selective Serotonin Reuptake Inhibitors on Pregnancy Outcomes and Neonatal Mortality
Reprinted from: *Toxics* **2022**, *10*, 11, doi:10.3390/toxics10010011 . . . . **191**

Jelonia T. Rumph, Kayla J. Rayford, Victoria R. Stephens, Sharareh Ameli, Pius N. Nde and Kevin G. Osteen et al.
A Preconception Paternal Fish Oil Diet Prevents Toxicant-Driven New Bronchopulmonary Dysplasia in Neonatal Mice
Reprinted from: *Toxics* **2021**, *10*, 7, doi:10.3390/toxics10010007 . . . . **203**

Conner L. Kennedy, Audrey Spiegelhoff, Kathy Wang, Thomas Lavery, Alexandra Nunez and Robbie Manuel et al.
The Bladder Is a Novel Target of Developmental Polychlorinated Biphenyl Exposure Linked to Increased Inflammatory Cells in the Bladder of Young Mice
Reprinted from: *Toxics* **2021**, *9*, 214, doi:10.3390/toxics9090214 . . . . **221**

Karl F. W. Foley, Daniel Barnett, Deborah A. Cory-Slechta and Houhui Xia
Early Low-Level Arsenic Exposure Impacts Post-Synaptic Hippocampal Function in Juvenile Mice
Reprinted from: *Toxics* **2021**, *9*, 206, doi:10.3390/toxics9090206 . . . . **239**

Communication

The Antiepileptic Drug and Toxic Teratogen Valproic Acid Alters Microglia in an Environmental Mouse Model of Autism

Korawin Triyasakorn [1], Ubah Dominic Babah Ubah [1], Brandon Roan [2], Minsyusheen Conlin [3], Ken Aho [3] and Prabha S. Awale [1,*]

1. Department of Biomedical and Pharmaceutical Sciences, College of Pharmacy, Idaho State University, 921 S 8th Avenue, Mail Stop 8288, Pocatello, ID 83209, USA; korawintriyasakor@isu.edu (K.T.); dominicbabahubah@isu.edu (U.D.B.U.)
2. Division of Health Sciences, Idaho State University, 921 S 8th Avenue, Mail Stop 8288, Pocatello, ID 83209, USA; brandonroan@isu.edu
3. Department of Biological Sciences, Idaho State University, 921 S 8th Avenue, Mail Stop 8288, Pocatello, ID 83209, USA; sheenconlin@isu.edu (M.C.); kenaho1@gmail.com (K.A.)
* Correspondence: awalprab@isu.edu

Abstract: Autism spectrum disorder (ASD), a neurodevelopmental condition affecting approximately 1 in 44 children in North America, is thought to be a connectivity disorder. Valproic acid (VPA) is a multi-target drug widely used to treat epilepsy. It is also a toxic teratogen as well as a histone deacetylase inhibitor, and fetal exposure to VPA increases the risk of ASD. While the VPA model has been well-characterized for behavioral and neuronal deficits including hyperconnectivity, microglia, the principal immune cells of CNS that regulate dendrite and synapse formation during early brain development, have not been well-characterized and may provide potential hints regarding the etiology of this disorder. Therefore, in this study, we determined the effect of prenatal exposure to VPA on microglial numbers during early postnatal brain development. We found that prenatal exposure to VPA causes a significant reduction in the number of microglia in the primary motor cortex (PMC) during early postnatal brain development, particularly at postnatal day 6 (P6) and postnatal day 10 (P10) in male mice. The early microglial reduction in the VPA model coincides with active cortical synaptogenesis and is significant because it may potentially play a role in mediating impaired connectivity in ASD.

Keywords: microglia; valproic acid; primary motor cortex

1. Introduction

Microglia are the principal resident immune cells of the CNS that colonize the brain during early prenatal development and comprise 5–15% of brain cells [1]. Recent advances in microglial and neuronal development have revealed they are implicated in all processes of neurogenesis and synaptogenesis including neuronal proliferation, migration, differentiation, as well as the formation and maturation of synaptic networks [2]. In early development, neurons make far more synaptic connections than are maintained in the adult brain. The large number of synapses that form in early development are eliminated by synaptic pruning, a developmental program that eliminates a large excess of synapses while also maintaining and strengthening a subset of them [3]. The precise colonization of microglia around areas undergoing active synaptogenesis during postnatal development strongly suggests that microglia and neuronal synapse formation may be influenced by each other. Indeed, several studies have shown that microglia interact with synapses, and this interaction is critical for synaptic maturation and synaptic connectivity. Microglia–synapse interaction occurs through several different pathways depending on the brain region. For example, the fractalkine receptor (CX3CR1) exclusively expressed by microglia is one mechanism involved in synaptic pruning. Genetically knocking out

CX3CR1 in mouse hippocampus during postnatal development led to delayed synaptic pruning demonstrated by transient excess of dendritic spines that was also associated with decreased microglial numbers [4]. Another mechanism of microglia-mediated synaptic pruning in an activity and complement-dependent mechanism (CR3/CD11b-CD18/Mac-1) is found in the retinogeniculate system. Genetic (CR3 and C3 KO) and pharmacological perturbations specific to microglia resulted in sustained deficits in synaptic wiring in the postnatal dorso lateral geniculate nucleus (dLGN) of the thalamus [5]. These and other observations underscore the indispensable contribution and function of microglia in synaptic pruning and connectivity. Consequently, then, any chemicals, drugs or environmental agents that may potentially be toxic to microglia during the prenatal or postnatal period can adversely affect brain development, resulting in impaired synaptic connectivity and neurodevelopmental disorders such as autism.

Autism spectrum disorder (ASD), a neurodevelopmental disorder affecting nearly 1 in 54 children, is thought to result from aberrant brain connectivity [6] https://paperpile.com/c/hbo56Y/v8bG (accessed on 15 March 2022). Multiple studies suggest that brain connectivity in adults with ASD differs from young autistic children. In general, adults exhibit underconnectivity between brain areas engaged in cognitive tasks, while children exhibit functional hyperconnectivity at the whole brain and subsystems level, including sensory and association cortices [6,7]. Hyperconnectivity is also observed in animal models of autism including the valproic acid (VPA) model of autism. A model for idiopathic autism, the VPA model, has been well-characterized for behavioral deficits and neural deficits including neural hyperconnectivity in different brain regions [8–10]. The cause of this hyperconnectivity is largely unknown and may be due to the direct effect of VPA on neurons or as a consequence of the toxic effect of VPA on microglia or a combination of the two. The impact of prenatal VPA exposure on microglia in regions undergoing synaptogenesis in early postnatal development is also unknown and may provide some clues to the etiology of neural hyperconnectivity. Therefore, the current study seeks to first investigate the putative toxic effects of VPA on microglial changes in early postnatal brain development in a mouse model of autism. We hypothesize that prenatal exposure to VPA induces changes in the microglial population in this model. We demonstrate that prenatal exposure to VPA significantly reduces the microglial number in the primary motor cortex (PMC) at postnatal day 6 (P6) and P10 in male mice, a critical period coinciding with active synaptogenesis when microglia are essential for synaptic pruning. Together, these data suggest that environmental agents such as VPA induce changes in microglia, the crucial cellular machinery needed for all aspects of normal brain development. Such disruptions may profoundly affect synaptogenesis, synaptic pruning, synaptic maturation, and synaptic connectivity resulting in neurodevelopmental disorders, including autism.

2. Materials and Methods

2.1. Animals

Adult male and female BALB/c mice were paired for breeding. Female breeders were visually examined daily before 8:00 a.m. for the presence of a vaginal plug, which was recorded as day 0 of embryonic development (E0). Pregnant females were treated with either sterile saline or VPA (sodium valproate: Sigma, St. Louis, MO, USA) 600 mg/kg dissolved in sterile saline (filtered using a sterile Millex syringe filter Millipore Corporation, Bedford, MA, USA), injected subcutaneously (*s.c*) on E13 and administered at a volume of 10 mL/kg [11]. Day of birth was recorded as day 0 (P0). All animals were housed in the Northeast Ohio Medical University vivarium in temperature and humidity-controlled rooms under a 12 h/12 h light/dark cycle. Water and laboratory chow were available ad libitum. All NIH guidelines were strictly adhered to, and treatment of the mice was approved by the institutional animal care and use committee at Northeast Ohio Medical University. All efforts were made to minimize animal suffering, reduce the number of animals used, and to utilize alternatives to in vivo techniques, if available.

2.2. Tissue Collection

Male brains were collected at P6 and P10, (n = 6 brains/group; number of sections/brains used = 5). At the time of tissue collection, the mice were anesthetized (ketamine 100 mg/kg and xylazine 7.5 mg/kg) and perfused transcardially with cold 0.9% saline followed by 4% paraformaldehyde. The animals were decapitated, and the brains were removed and placed in cold 4% paraformaldehyde for 4 h, after which they were transferred into 20% sucrose for 24 h.

2.3. Immuno-Histochemical Detection of Iba1 Protein

Brains were sectioned (20 µm) coronally on a Leica cryostat at −20 °C, mounted directly onto Super frost ++ Micro slides (VWR), and allowed to dry before being stored at 4 °C. The ionized calcium-binding adaptor (Iba)-1 protein was used because its expression is constitutive in both activated (amoeboid) and quiescent microglia (ramified). The staining of all sections from all the groups was carried out under identical conditions. For staining, slides were removed from the freezer and allowed to thaw at RT for 10 min. The slides were washed with PBS and incubated for 1 h in PBS with 5% normal donkey serum (NDS) and 0.1% Triton X-100 to block and permeabilize the tissue. Following blocking, slides were incubated with primary antibody (rabbit anti-Iba1 at 1:500 dilution Wako Chemicals Richmond, VA, USA in 3% NDS and 0.1% Triton X-100) overnight at 4 °C. On day 2, the slides were washed and incubated with fluorescently labeled secondary antibody (Alexa Fluor 488 Donkey anti rabbit IgG H+L, at 1:500 dilution, Molecular Probes, USA in 1% NDS and 0.1% Triton X-100) for 2 h at RT in the dark. The slides were washed with PBS and incubated with the nuclear stain Red Dot 2 (far red nuclear dye: Biotium, CA, USA) for 20 min. The slides were washed and cover slipped with Fluoromount G (Southern Biotech, Birmingham, AL, USA) and dried overnight at RT.

2.4. Cell Counting

Iba1-labeled microglia were examined under a confocal microscope (Olympus IX 70) at 200× and at 600× magnification. The multi-laser feature on the confocal microscope and Flouview V 5.0 software were used to simultaneously acquire images of both microglia stained with Iba1 and the nuclei stained with Red Dot 2. This feature covers a wide spectrum of wavelengths ranging from 405 nm to 644 nm. An excitation wavelength of 488 nm with emission at 519 nm was used to acquire images of microglia fluorescently labeled with Alexa Fluor 488 nm. Red Dot 2 was excited at 634 nm with an emission at 695 nm. A systematic sampling of PMC area in each section was carried out. After the acquisition of images, the cells were counted using the cell counter feature of Image J (rsbweb.nih.gov/ij/ accessed on 24 April 2022) throughout each 20 µm section. Counting areas were set at of 800 × 600 µm, with a final magnification of 200×. To avoid counting cell fragments, cells were only counted as positive if the cell body was visible as yellow (merger of green and red) and the stain appeared uniform throughout the cell. For each animal, sections taken every 300 µm throughout the PMC region were analyzed. The number of representative sections analyzed for the PMC for the different groups was 5 sections/animal. The total number of Iba1-positive cells was obtained by an experimenter blind to experimental conditions, across all representative sections for the PMC visualized. The Atlas of the Developing Mouse Brain at E17.5, P0 and P10 (George Paxinos, Glenda Halliday, Charles Watson, Yuri Koutcherov, Hong Qin Wang) served as a reference guide to delineate the PMC of the sections used.

2.5. Statistical Analysis

The data obtained for the number of Iba1-positive microglia were averaged from 5 consecutive sections for each animal, and there were 6 animals (or 6 independent experiments from 6 different mothers) in each group, providing a sample size of 6. We tested for treatment effects using pooled variance t-test. Day 10 data required log transformation to meet linear model assumptions. Treatment effects were highly significant on both day 6

($p = 3.7 \times 10^{-5}$) and particularly day 10 ($p = 2.3 \times 10^{-9}$). GraphPad Prism version 7.0 was used to draw the data figures and calculate statistical significance. The microglial number on the Y axis of the graphs was obtained by analyzing the number of microglia/area/μm^3 (area = length × width × height) and converting the value obtained to mm^3, which rendered the number of microglia/mm^3.

3. Results

3.1. VPA Treatment Reduces Microglial Number in the PMC of Male Mice at P6

The photomicrographs of Iba1-positive microglia (green) in the PMC (right side of the brain) from saline-treated and VPA-treated male mice at P6 are represented in Figure 1. Two times points were chosen for this study, namely P10 and P6. The P10 time point was selected based on the fact that in mice, P8–P16 constitutes the critical period of neuronal remodeling in the cerebral cortex [12]. We also wanted to determine the effect of VPA on microglia just before neuronal remodeling, and therefore, P6 was chosen. Because ASD disproportionally affects males compared to females with a 3:1 ratio, we focused our microglial studies on male animals. The primary motor cortex (PMC) was selected based on current literature that demonstrates complex arborizations of dendrites in PMC in the VPA rat model of autism [13]. In addition, children with autism also suffer from motor abnormalities as comorbid symptoms [14]. A summary of the data analyses of the number of microglia in the PMC is presented in the bar graphs shown in Figure 1e.

The number of microglia at P6 in VPA-treated mice (Mean ± SEM: 124 ± 25) was significantly lower than those found in saline-treated mice (Mean ± SEM: 372 ± 24). The microglia in the saline-treated mice appear to be healthy with a distinct cell body and nuclear staining (Figure 1a). At higher magnification (600×), the microglia appear to have characteristics of either amoeboid or intermediate morphology with short processes (Figure 1c). However, the majority of microglia in the VPA-treated mice upon close examination at higher magnification (600×) appear to be fragmented with no distinct cell body (Figure 1b,d).

3.2. VPA Treatment Depletes Microglia in PMC of Male Mice at P10

The photomicrographs of Iba1 positive microglia (green) in the PMC from saline-treated and VPA-treated male mice at P10 are represented in Figure 2. A summary of the data analyses of the number of microglia in the PMC is presented in the bar graphs of Figure 2e.

The number of microglia at P10 in VPA-treated mice (Mean ± SEM: 26 ± 3) was significantly lower than saline-treated mice (Mean ± SD: 824 ± 54). The microglia in the saline-treated mice appear to be healthy with a distinct cell body and nuclear staining (Figure 2a). At higher magnification (600×), the microglia appear to have characteristics of either an intermediate morphology with short processes or ramified morphology with thin processes and a distinct cell body (Figure 2c). In contrast, the microglia in the VPA-treated mice appear to have been completely depleted by the treatment (Figure 2b). Even upon close examination at higher magnification (600×), very few fragments are observed with no distinct cell body (Figure 2d).

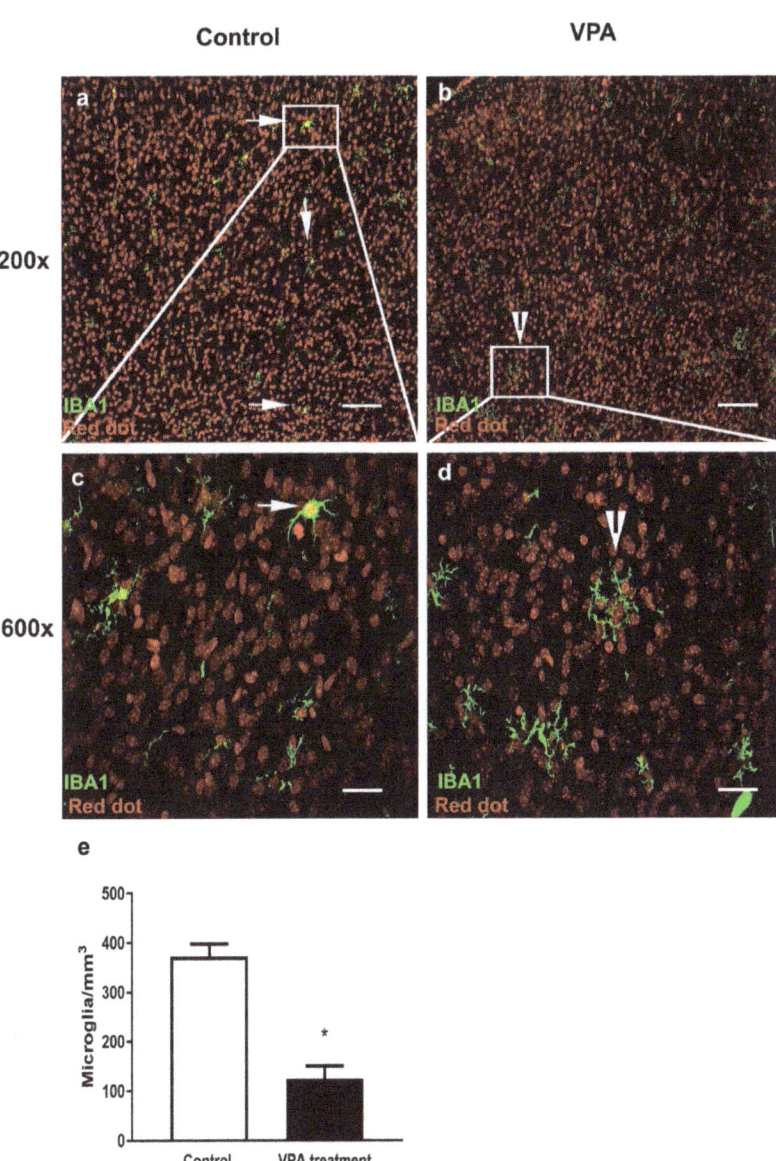

Figure 1. VPA-treated male mice show decreases in microglial number and fragmented morphology in primary motor cortex at P6. Confocal images showing labeled microglia in the PMC of saline control and age-matched brains of VPA mouse model of autism. (**a,b**) Microglia labeled with Iba1 (green) in control and VPA-treated mice at low magnification (200×). (**c,d**) Microglia labeled with Iba1 (green) at high magnification (600×). White square region (insert) in a and b is magnified in c and d. White arrows and arrowheads indicate differences in morphology of microglia in control and VPA model. Scale bar (**a,b**) 200 μm and in (**c,d**) 50 μm. (**e**) Analysis of number of microglia as summarized within the bar graphs (Mean ± SEM) reveals a statistically significant decrease in microglial number in VPA-treated mice (* $p = 3.7 \times 10^{-5}$).

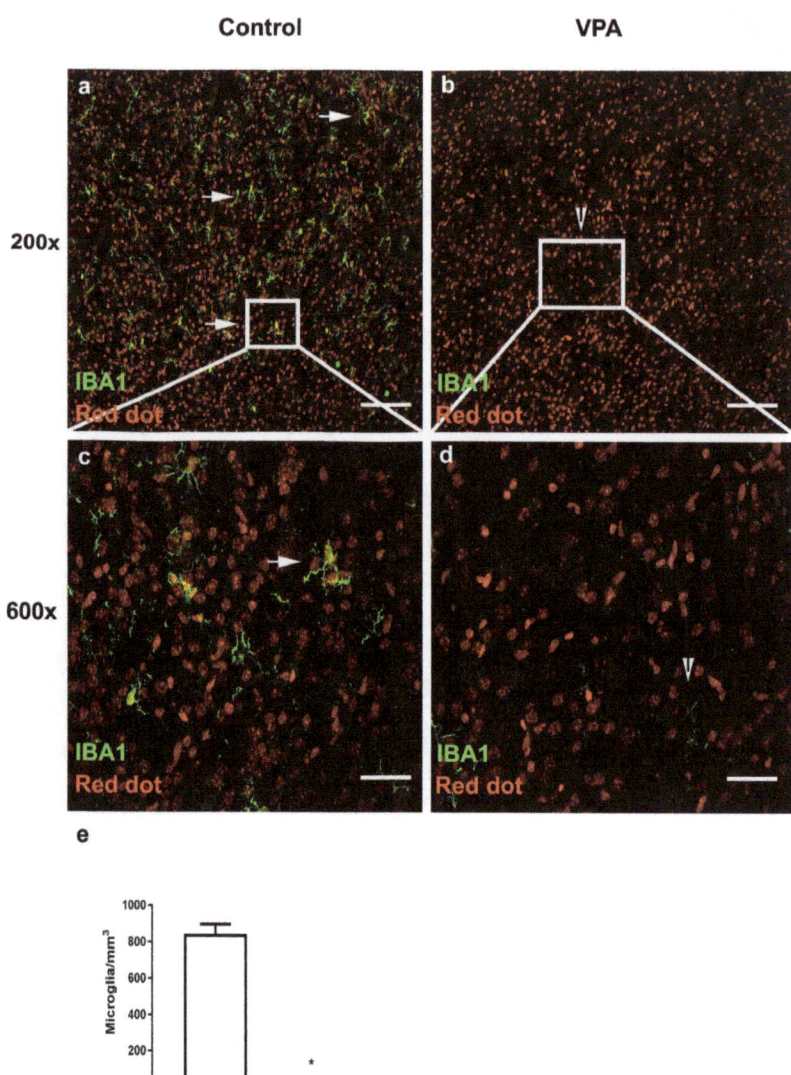

Figure 2. VPA-treated male mice show significant decrease in microglial number and fragmented morphology in primary motor cortex at P10. Confocal images showing labeled microglia in cortex of saline control and age matched brains in VPA mouse model of autism. (**a**,**b**) Microglia labeled with Iba1 (green) in control mice and VPA-treated mice at low magnification (200×). (**c**,**d**) Microglia labeled with Iba1 (green) at high magnification (600×). White square region (insert) in a and b is magnified in c and d. White arrows and arrowheads indicate differences in morphology of microglia in control and VPA model. Scale bar (**a**,**b**) 200 µm and in (**c**,**d**) 50 µm. (**e**) Analysis of number of microglia as summarized within the bar graphs (Mean ± SEM) reveals a statistically significant decrease in microglial number in VPA-treated mice (* $p = 2.3 \times 10^{-9}$).

4. Discussion

To our knowledge, this study marks the first attempt to investigate the effect of embryonic exposure to VPA on microglial number particularly during early postnatal development in a male VPA mouse model of autism. We demonstrate that embryonic exposure to VPA has distinct effects on the microglial number during early postnatal period of brain development. Specifically, we demonstrate reduced microglial number and fragmented morphology at P6 and significantly lower microglial numbers at P10, particularly in male offspring.

The current study analyzed the early postnatal period of brain development to more closely match the period of early symptoms seen in humans with ASD. Clinically, autism is associated with impairments in social behavior, verbal and nonverbal communication and stereotypical repetitive behaviors and narrowed interest [15,16]. One of the hallmarks of autism is hyperconnectivity, found to be higher in children than in adults, inferring neurodevelopmental origin, the cause of which is elusive, but it is hypothesized that a lack of pruning of synapses may be responsible for impairments in connectivity [6,7]. Microglia are key players in pruning extranumerary synapses in development [3]. VPA-exposed animals exhibit impaired communication, reduced exploratory behavior, and anxiety-like and repetitive behaviors, a phenotype that models behavioral impairments in children with autism [17]. Multiple reports in the literature have demonstrated either an increase, decrease or no change in microglial density in different neuroanatomical regions, at later time points in postnatal development in both male and female offspring exposed to VPA prenatally. Of particular interest is the observation that no changes in microglial density were observed in the Dentate Gyrus (DG) (molecular layer, granular cell layer, Hilus), CA1 region of the hippocampus (stratum oriens, pyramidal cell layer, stratum radiatum) and cerebellum (molecular layer, granular cell layer) at P7 in a female VPA model of autism [18]. This is in contrast to our data that show fragmented microglia at P6 and near complete depletion of microglia during peak period of synaptogenesis and synaptic maturation at P10 in the PMC in a male VPA model, highlighting age-specific differences, region-specific differences and sexual differences to drug response. It is also important to note that the mouse strains used in these studies are different from what we used. It is possible that microglia in BALBc mice may be particularly susceptible to the deleterious effect of VPA.

The reduction in microglia number close to the time of insult we observe in our experiments is also in contrast to other studies in young adult animals at a time point after remodeling occurs in the brain. For example, increased microglial density has been reported in the molecular layer of the dentate gyrus in female BALBc mice at 5 months [19]. This is in contrast to studies by Kazlauskas et al., who showed no change in microglial density in the same molecular layer of the dentate gyrus; however, this was carried out in early neonatal (PD-7) development and in a different strain Cr1Fcen:CF1 lineage showing age-specific and strain specific effects of the drug [18].

In yet another interesting report, Gassowska-Dobrowolska et al. studied changes induced by VPA on protein expression of IBA-1 a microglial marker and mRNA levels of pro and anti-inflammatory markers in the cerebral cortex and hippocampus at PD58 in rats. Prenatal exposure to VPA leads to a significant increase in protein levels of IBA1 in the cerebral cortex, which may reflect a higher microglial number. Additionally, the mRNA levels of proinflammatory cytokines (IL-1β, IL-6. TNF-α), and anti-inflammatory neuroprotective phenotypes (Arg1, Chi3L1, Mrc1, CD86, Fcgr1a, TGFβ1, and Sphk1) were increased. On the other hand, the same VPA treatment resulted in no change in the above parameters, except for the anti-inflammatory and neuroprotective phenotype that showed an increase in the hippocampus, indicating ongoing immune system impairments, which may be more robust in the cortex compared to the hippocampus [17]. An anti-inflammatory phenotype in the hippocampus may reflect the possible resolution of immune activation that may have started earlier than in the cerebral cortex. These observations also correlate well with observations seen in the clinical symptoms of autism. For example, in high-functioning males with ASD, there is an elevated plasma level of several cytokines IL-1β, IL-1RA, IL-5, IL-8, IL-12, (p70), IL-13, and IL-17 [20]. In addition to increases in the proinflammatory cytokine Il-1β, another proinflammatory cytokine TNF-α is also found to be elevated in the cerebrospinal fluid of ASD [21]. There are frequent reports of an increase in IL-6 both centrally and peripherally in ASD patients [22–25]. Interestingly, researchers have noticed an association between peripheral cytokine levels and the severity of behavioral impairments. For example, elevated IL-1β and IL-6 are associated with increased stereotypical behavior [25], and the dysregulation of IL-1β is associated with impairments in memory and learning [26]. Reduced levels of the regulatory cytokine TGFβ are associated with reduced adaptive behavior and worsening behavioral symptoms [27]. Elevated levels of IL-8 and IL-12p40 are also associated with greater impairment of aberrant behavior including lethargy and stereotypy. Now, as the expression of IL-8 decreases, cognitive and adaptive ability improves [25]. All these studies indicate that ASD patients are in a state of continuous immune dysregulation, which impacts behavior, although many of these studies are carried out in either adult animals or humans. The immune response of microglia in different brain regions to VPA in early development in both inbred and outbred strains of mice remains to be fully elucidated and may be useful in determining the pathology of ASD.

It is unclear how prenatal exposure to VPA reduces microglia during early postnatal development, which remains to be fully elucidated. One possibility is that VPA induces microglial apoptosis, and it is likely the fragments at P6 are apoptotic bodies. Interestingly, at P10, we observe a near complete absence of microglia including fragments. There is some evidence from cell culture studies where VPA selectively killed cultured murine BV-2 microglia by a caspase 3-mediated mechanism, sparing the neurons and astrocytes, suggesting that this might be one possible mechanism [28]. A second possibility is that VPA induces microglial necrosis. Alternatively, a third possibility is that VPA reduces microglial cell proliferation. Regardless of the mechanism, VPA's deleterious and toxic impact on microglial number particularly during the critical window of synaptogenesis raises the possibility that this might potentially have an impact on neuronal synaptogenesis, so the further exploration of cause and effect would be of interest. Contrary to adulthood where either an increase, decrease or no changes in microglial density are observed, our results demonstrate microglial reduction may be an early effect of VPA. It is noteworthy that impaired microglial function in mice has been shown to induce behavioral deficits related to clinical symptoms of ASD, obsessive compulsive disorder and schizophrenia [29].

ASD is thought to result from aberrant hyperconnectivity. There is also evidence for general hyperconnectivity in the somatosensory cortex, medial prefrontal cortex (mPFC), amygdala and auditory cortex in the rodent VPA model of ASD [8–10]. Although strains used in these studies are different, a clear picture of hyperconnectivity is emerging, the underlying cause of which is not clear, but dysfunctional/depleted microglia may be a key player in hyperconnectivity in ASD. Future experiments should aim to test the hypothesis that microglial dysfunction alters synaptic connectivity in the VPA model of autism. One limitation of our study is that we performed these experiments in male mice only. In the future, it is equally important to study the effect of VPA on female mice as well.

In summary, the present experiments are the first to show that, in mammals, VPA reduces the microglial number particularly during critical periods of synaptogenesis in PMC. The data reported in this study are important and relevant to pathological states associated with autism. Clearly, this is an interesting area that merits further investigation.

Author Contributions: Methodology, validation, writing—original draft preparation K.T., U.D.B.U., B.R. and M.C.; Statistical analysis K.A.; Conceptualization, supervision, project administration, writing review and editing of manuscript, P.S.A. All authors have read and agreed to the published version of the manuscript.

Funding: We would like to acknowledge National Institute of General Medical Sciences (NIGMS) of the National Institutes of Health (P20GM103408), Idaho INBRE for their support towards undergraduate student research and the cost of this publication.

Institutional Review Board Statement: The animal study protocol was approved by the Institutional animal care and use committee at Northeast Ohio Medical University (protocol code 37 and 10 October

2010). All efforts were made to minimize animal suffering, to reduce the number of animals used and to utilize alternatives to in vivo techniques, if available.

Informed Consent Statement: Not applicable.

Data Availability Statement: Not applicable.

Conflicts of Interest: The authors declare no conflict of interest.

References

1. Monier, A.; Adle-Biassette, H.; Delezoide, A.-L.; Evrard, P.; Gressens, P.; Verney, C. Entry and distribution of microglial cells in human embryonic and fetal cerebral cortex. *J. Neuropathol. Exp. Neurol.* **2007**, *66*, 372–382. [CrossRef] [PubMed]
2. Mosser, C.-A.; Baptista, S.; Arnoux, I.; Audinat, E. Microglia in CNS development: Shaping the brain for the future. *Prog. Neurobiol.* **2017**, *149–150*, 1–20. [CrossRef] [PubMed]
3. Hua, J.Y.; Smith, S.J. Neural activity and the dynamics of central nervous system development. *Nat. Neurosci.* **2004**, *7*, 327–332. [CrossRef] [PubMed]
4. Paolicelli, R.C.; Bolasco, G.; Pagani, F.; Maggi, L.; Scianni, M.; Panzanelli, P.; Giustetto, M.; Ferreira, T.A.; Guiducci, E.; Dumas, L.; et al. Synaptic pruning by microglia is necessary for normal brain development. *Science* **2011**, *333*, 1456–1458. [CrossRef] [PubMed]
5. Schafer, D.P.; Lehrman, E.K.; Kautzman, A.G.; Koyama, R.; Mardinly, A.R.; Yamasaki, R.; Ransohoff, R.M.; Greenberg, M.E.; Barres, B.A.; Stevens, B. Microglia sculpt postnatal neural circuits in an activity and complement-dependent manner. *Neuron* **2012**, *74*, 691–705. [CrossRef]
6. Supekar, K.; Uddin, L.Q.; Khouzam, A.; Phillips, J.; Gaillard, W.D.; Kenworthy, L.E.; Yerys, B.E.; Vaidya, C.J.; Menon, V. Brain hyperconnectivity in children with autism and its links to social deficits. *Cell Rep.* **2013**, *5*, 738–747. [CrossRef]
7. Kleinhans, N.M.; Richards, T.; Sterling, L.; Stegbauer, K.C.; Mahurin, R.; Johnson, L.C.; Greenson, J.; Dawson, G.; Aylward, E. Abnormal functional connectivity in autism spectrum disorders during face processing. *Brain* **2008**, *131 Pt 4*, 1000–1012. [CrossRef]
8. Rinaldi, T.; Perrodin, C.; Markram, H. Hyper-connectivity and hyper-plasticity in the medial prefrontal cortex in the valproic Acid animal model of autism. *Front. Neural Circuits* **2008**, *2*, 4. [CrossRef]
9. Rinaldi, T.; Silberberg, G.; Markram, H. Hyperconnectivity of local neocortical microcircuitry induced by prenatal exposure to valproic acid. *Cereb. Cortex* **2008**, *18*, 763–770. [CrossRef]
10. Nagode, D.A.; Meng, X.; Winkowski, D.E.; Smith, E.; Khan-Tareen, H.; Kareddy, V.; Kao, J.P.; Kanold, P.O. Abnormal Development of the Earliest Cortical Circuits in a Mouse Model of Autism Spectrum Disorder. *Cell Rep.* **2017**, *18*, 1100–1108. [CrossRef]
11. Wagner, G.C.; Reuhl, K.R.; Cheh, M.; McRae, P.; Halladay, A.K. A new neurobehavioral model of autism in mice: Pre- and postnatal exposure to sodium valproate. *J. Autism Dev. Disord.* **2006**, *36*, 779–793. [CrossRef] [PubMed]
12. Litzinger, M.J.; Mouritsen, C.L.; Grover, B.B.; Esplin, M.S.; Abbott, J.R. Regional differences in the critical period neurodevelopment in the mouse: Implications for neonatal seizures. *J. Child Neurol.* **1994**, *9*, 77–80. [CrossRef] [PubMed]
13. Snow, W.M.; Hartle, K.; Ivanco, T.L. Altered morphology of motor cortex neurons in the VPA rat model of autism. *Dev. Psychobiol.* **2008**, *50*, 633–639. [CrossRef] [PubMed]
14. Ming, X.; Brimacombe, M.; Wagner, G.C. Prevalence of motor impairment in autism spectrum disorders. *Brain Dev.* **2007**, *29*, 565–570. [CrossRef]
15. Zoghbi, H.Y.; Bear, M.F. Synaptic dysfunction in neurodevelopmental disorders associated with autism and intellectual disabilities. *Cold Spring Harb. Perspect. Biol.* **2012**, *4*, a009886. [CrossRef]
16. Ebrahimi-Fakhari, D.; Sahin, M. Autism and the synapse: Emerging mechanisms and mechanism-based therapies. *Curr. Opin. Neurol.* **2015**, *28*, 91–102. [CrossRef]
17. Gassowska-Dobrowolska, M.; Cieslik, M.; Czapski, G.A.; Jesko, H.; Frontczak-Baniewicz, M.; Gewartowska, M.; Dominiak, A.; Polowy, R.; Filipkowski, R.K.; Babiec, L.; et al. Prenatal Exposure to Valproic Acid Affects Microglia and Synaptic Ultrastructure in a Brain-Region-Specific Manner in Young-Adult Male Rats: Relevance to Autism Spectrum Disorders. *Int. J. Mol. Sci.* **2020**, *21*, 3576. [CrossRef]
18. Kazlauskas, N.; Campolongo, M.; Lucchina, L.; Zappala, C.; Depino, A.M. Postnatal behavioral and inflammatory alterations in female pups prenatally exposed to valproic acid. *Psychoneuroendocrinology* **2016**, *72*, 11–21. [CrossRef]
19. Dos Santos, A.L.G.; de Leao, E.; de Almeida Miranda, D.; de Souza, D.N.C.; Picanco Diniz, C.W.; Diniz, D.G. BALB/c female subjected to valproic acid during gestational period exhibited greater microglial and behavioral changes than male mice: A significant contra intuitive result. *Int. J. Dev. Neurosci.* **2021**, *81*, 37–50. [CrossRef]
20. Suzuki, K.; Matsuzaki, H.; Iwata, K.; Kameno, Y.; Shimmura, C.; Kawai, S.; Yoshihara, Y.; Wakuda, T.; Takebayashi, K.; Takagai, S.; et al. Plasma cytokine profiles in subjects with high-functioning autism spectrum disorders. *PLoS ONE* **2011**, *6*, e20470. [CrossRef]
21. Chez, M.G.; Dowling, T.; Patel, P.B.; Khanna, P.; Kominsky, M. Elevation of tumor necrosis factor-alpha in cerebrospinal fluid of autistic children. *Pediatr Neurol.* **2007**, *36*, 361–365. [CrossRef] [PubMed]
22. Vargas, D.L.; Nascimbene, C.; Krishnan, C.; Zimmerman, A.W.; Pardo, C.A. Neuroglial activation and neuroinflammation in the brain of patients with autism. *Ann. Neurol.* **2005**, *57*, 67–81. [CrossRef] [PubMed]

23. Li, X.; Chauhan, A.; Sheikh, A.M.; Patil, S.; Chauhan, V.; Li, X.M.; Ji, L.; Brown, T.; Malik, M. Elevated immune response in the brain of autistic patients. *J. Neuroimmunol.* **2009**, *207*, 111–116. [CrossRef]
24. Wei, H.; Zou, H.; Sheikh, A.M.; Malik, M.; Dobkin, C.; Brown, W.T.; Li, X. IL-6 is increased in the cerebellum of autistic brain and alters neural cell adhesion, migration and synaptic formation. *J. Neuroinflammation* **2011**, *8*, 52. [CrossRef]
25. Ashwood, P.; Krakowiak, P.; Hertz-Picciotto, I.; Hansen, R.; Pessah, I.; Van de Water, J. Elevated plasma cytokines in autism spectrum disorders provide evidence of immune dysfunction and are associated with impaired behavioral outcome. *Brain Behav. Immun.* **2011**, *25*, 40–45. [CrossRef]
26. Goines, P.E.; Ashwood, P. Cytokine dysregulation in autism spectrum disorders (ASD): Possible role of the environment. *Neurotoxicol. Teratol.* **2013**, *36*, 67–81. [CrossRef] [PubMed]
27. Ashwood, P.; Enstrom, A.; Krakowiak, P.; Hertz-Picciotto, I.; Hansen, R.L.; Croen, L.A.; Ozonoff, S.; Pessah, I.N.; DeWater, J. Decreased transforming growth factor beta1 in autism: A potential link between immune dysregulation and impairment in clinical behavioral outcomes. *J. Neuroimmunol.* **2008**, *204*, 149–153. [CrossRef] [PubMed]
28. Dragunow, M.; Greenwood, J.M.; Cameron, R.E.; Narayan, P.J.; O'Carroll, S.J.; Pearson, A.G.; Gibbons, H. M Valproic acid induces caspase 3-mediated apoptosis in microglial cells. *Neuroscience* **2006**, *140*, 1149–1156. [CrossRef]
29. Zhan, Y.; Paolicelli, R.C.; Sforazzini, F.; Weinhard, L.; Bolasco, G.; Pagani, F.; Vyssotski, A.L.; Bifone, A.; Gozzi, A.; Ragozzino, D.; et al. Deficient neuron-microglia signaling results in impaired functional brain connectivity and social behavior. *Nat. Neurosci.* **2014**, *17*, 400–406. [CrossRef]

Article

Multi- and Transgenerational Effects of Developmental Exposure to Environmental Levels of PFAS and PFAS Mixture in Zebrafish (*Danio rerio*)

Alex Haimbaugh [1], Chia-Chen Wu [1], Camille Akemann [1], Danielle N. Meyer [1], Mackenzie Connell [2], Mohammad Abdi [2], Aicha Khalaf [2], Destiny Johnson [2] and Tracie R. Baker [1,2,3,*]

[1] Department of Pharmacology, Wayne State University, Detroit, MI 48202, USA; alexhaim@wayne.edu (A.H.); chiachenwu@ufl.edu (C.-C.W.); gi2263@wayne.edu (C.A.); danielle.meyer@ufl.edu (D.N.M.)
[2] Institute of Environmental Health Sciences, Wayne State University, Detroit, MI 48202, USA; gg8277@wayne.edu (M.C.); mohammed.abdi@wayne.edu (M.A.); aichakhalaf@wayne.edu (A.K.); destinyjohnson@wayne.edu (D.J.)
[3] Department of Environmental and Global Health, University of Florida, Gainesville, FL 32610, USA
* Correspondence: tracie.baker@ufl.edu

Abstract: Per- and polyfluoroalkyl substances (PFASs) are ubiquitous in the environment and are tied to myriad health effects. Despite the phasing out of the manufacturing of two types of PFASs (perfluorosulfonic acid (PFOS) and perfluorooctanoic acid (PFOA)), chemical composition renders them effectively indestructible by ambient environmental processes, where they thus remain in water. Exposure via water can affect both human and aquatic wildlife. PFASs easily cross the placenta, exposing the fetus at critical windows of development. Little is known about the effects of low-level exposure during this period; even less is known about the potential for multi- and transgenerational effects. We examined the effects of ultra-low, very low, and low-level PFAS exposure (7, 70, and 700 ng/L PFOA; 24, 240, 2400 ng/L PFOS; and stepwise mixtures) from 0–5 days post-fertilization (dpf) on larval zebrafish (*Danio rerio*) mortality, morphology, behavior and gene expression and fecundity in adult F0 and F1 fish. As expected, environmentally relevant PFAS levels did not affect survival. Morphological abnormalities were not observed until the F1 and F2 generations. Behavior was affected differentially by each chemical and generation. Gene expression was increasingly perturbed in each generation but consistently showed lipid pathway disruption across all generations. Dysregulation of behavior and gene expression is heritable, even in larvae with no direct or indirect exposure. This is the first report of the transgenerational effects of PFOA, PFOS, and their mixture in terms of zebrafish behavior and untargeted gene expression.

Keywords: PFAS; PFAS mixtures; epigenetics; zebrafish; transgenerational

1. Introduction

Per- and polyfluoroalkyl substances (PFASs) are a class of chemicals constituted by a polar head group attached to a chain of C-F bonds. The unique chemistry of these compounds renders them effectively indestructible and, thus, a prime candidate for high-heat industrial processes and long-lasting consumer goods such as non-stick cookware and waterproofed outerwear. The utility of PFASs is offset by their bioaccumulation and toxic health effects. PFAS are detected virtually everywhere—in diverse wildlife, multiple environmental matrices, and in >99% of the general public [1–3]. Drinking water is a significant source of exposure in humans [4], and drinking water treatment plants are not designed to remove these contaminants from source water. Likewise, wastewater treatment plants do not intentionally filter out PFASs. PFASs are commonly found to be in the parts per trillion (ppt; ng/L) range in both untreated and treated drinking water [5] and wastewater [6]. The widespread low-level exposure warrants investigation into the health effects on wildlife and humans.

PFASs readily cross the placental barrier, potentially exposing a fetus during sensitive time periods during development [7]. Chemical assault during critical windows in development can have effects later in life; this thinking stems from the developmental origins of health and disease (DOHaD) hypothesis [8]. DOHaD posits that the timing of the exposure is crucial in determining the result. Placental transfer of PFASs necessitates the study of early-life exposure and the heritable effects of exposure. As most people have small amounts of many types of PFASs in their bodies, it is of great general interest to study the effects of low-level exposure, including to mixtures, on developing organisms.

The two most common PFASs carried by the general population are perfluorooctanoic acid (PFOA) and perfluorooctane sulfonic acid (PFOS), and they are usually detected at higher levels than other types of PFASs. PFOS is present at approximately three times the levels of PFOA in humans [9]. Choosing a relevant exposure concentration is of importance when planning translational experiments to realistically inform public health. The general population carries serum PFAS levels in the µg/L range (>999 ng/L), with an average of 1.42 µg/L PFOA and 4.25 µg/L PFOS [9]. The Environmental Protection Agency (EPA) health advisory limit for drinking water of PFOA, PFOS, or their combined concentration in mixture is 70 ppt, or 70 ng/L. Much early work characterizing PFAS toxicity used, understandably, high dose experiments to define outcomes, such as the concentration at which 50% of exposed organisms die (LC_{50}). We now know that PFAS levels, while ubiquitous in all environmental compartments, are typically at ng/L or µg/L levels in water. A study of treated water from 25 drinking water treatment plants across the United States found a median concentration of 19.5 ng/L for 17 PFASs combined, with a maximum sum of 1.1 µg/L (1100 ng/L) [5]. Our group has previously shown that mean concentrations of PFOA and PFOS in a waterway that provides drinking water in a major metropolitan area in Michigan were 2.2 ng/L and 2.9 ng/L, respectively [10]. In order to advance public health knowledge of exposures at both environmentally relevant levels and levels encompassing the EPA health advisory, we chose exposure concentrations for PFOA of 7, 70, and 700 ng/L and for PFOS at approximately $3\times$ higher concentrations of 24, 240, and 2.4 µg/L (2400 ng/L), a ratio similar to reported human levels. The mixture concentrations contain half of each exposure level per chemical (e.g., the ultra-low mixture concentration contains 3.5 ng/L PFOA and 12 ng/L PFOS). Throughout this report, we will refer to these nominal concentrations of 7 ng/L PFOA exposure and 24 ng/L PFOS exposure as the "ultra-low" exposure level, the 70 ng/L PFOA and 240 ng/LPFOS exposure as "very low", and the 700 ng/L PFOA and 2.4 µg/L PFOS exposure as "low". Exposures in other studies within the ng/L range are referred to as "low", µg/L range as "moderate", mg/L range as "high", and g/L range as "very high".

As individual PFASs are seldom discovered in the environment or treated drinking water alone, it is critical to study mixtures at environmental levels. PFAS mixtures are increasingly studied, but their effects are still unclear and often unpredictable, especially at different concentrations. Ding et al. [11] characterized the 1:1 mixture of PFOA and PFOS at high concentrations to be synergistic towards early-life lethality in zebrafish, while increasing the PFOA:PFOS ratio resulted in antagonism, then additivity. In another study, individual PFASs alone significantly changed swim behavior in exposed fish at moderate levels, but a mixture of nine PFAS had no effect at environmental levels [12]. We sought to address this gap by characterizing a low-level mixture of the two most commonly detected PFAS.

There is emerging evidence that PFAS exposure confers heritable effects on later generations via epigenetic mechanisms [13] rather than direct genotoxicity. Epigenetic modifications to DNA or chromatin serve as a "biological memory" of environmental history that modulate gene regulatory networks in current and future generations [14]. Toxicoepigenetic initialization in the directly exposed organism can be perpetuated across multiple generations. When an individual is directly exposed, the exposure indirectly affects germ cells residing in the individual. "Multigenerational" (F1) effects are seen in the generation following the directly exposed (F0) generation. Even if the exposure ceases, indirect germ cell exposure has occurred and can present phenotypically in this next generation's life. Effects are considered "transgenerational" when observed in the subsequent (F2) generation, which has never been directly or indirectly exposed. Zebrafish exposures with explicit epigenetic outcomes through multiple generations have not been conducted for PFAS. However, in the F0 generation, Bouwmeester et al. [15] found that moderate-range PFOA exposure increased methylation associated with *vtg1*, a gene involved in fertility. Limited epigenetic studies have been done in rodents. Tian et al. (2019) found that non-specific methylation therapy administered with PFOS to F0 females resulted in better birth outcomes in F1 pups than F0 PFOS exposure without methylation therapy [16]. The potential heritability of PFAS exposure effects is pertinent, as measures taken now to prevent or reduce exposure could magnify public health benefits to the next generation(s) at scale. The results of the current study suggest that epigenetic mechanisms mediate each generation's response to exposure in terms of behavior and gene expression.

The zebrafish is an ideal model system for conducting early-life research on waterborne contaminants over multiple lifetimes. Zebrafish have been a useful, popular model in developmental toxicology due to their easy visibility, high n-values, quick generation time, and high homology with the human genome [17]. Additionally, the EPA plans to eliminate funding for mammalian vertebrate research completely by 2035 [18], positioning the zebrafish as a pertinent alternative model organism. From the outset, zebrafish eggs have a transparent chorion through which development can be observed from the single-cell stage to free-swimming larvae at five days post-fertilization (5 dpf). Zebrafish are prolific breeders, producing > 300 eggs per week, and are sexually mature in ~3 months [19], meaning transgenerational effects can be observed in about one year. They have been utilized as an ideal transgenerational model due to all of the above-mentioned benefits and the external fertilization of eggs, which reduces the number of generations compared to mammalian models [20].

This study aims to advance understanding of the short- and long-term health effects of developmental exposure to environmentally relevant levels of PFASs using the zebrafish model organism. After exposing embryonic zebrafish to environmental levels of two prevalent PFASs, PFOS and PFOA, and a mixture of the two chemicals (referred to throughout simply as "mixture") for the first 5 days of life, we found that swimming behavior and gene expression at 5 days post-fertilization (dpf) was affected by at least one concentration of all chemicals in all three generations (F0–F2). Pathway analysis of gene expression revealed upregulated pathways of immunotoxicity, movement disorders, and endocrine disruption. Adult fecundity (eggs produced per female) was statistically increased in the PFOA-exposed F0 generation and decreased in the F1 generation. Morphological abnormalities at 5 dpf were not observed until the F1 and F2 generations. As expected at these low doses, survival was uniformly unaffected by exposure.

It is the authors' aim that these results inform decision-making regarding safe contaminant limits in drinking water and in aquatic habitats. The federal health advisory limit set by the EPA for PFOS, PFOA, and their mixture is currently 70 ng/L [21], while some states legislate much lower levels. This study provides the first report of multigenerational effects of PFOA exposure on behavior and of mixture exposure on behavior and gene expression, supporting findings in other PFOS studies showing these endpoints are affected multigenerationally. Further, we show novel transgenerational effects on behavior and gene expression following low-level exposure to any PFAS during early life. Future efforts should include complex mixtures, and PFAS replacements, including "short-chain" alternatives, will be critical to study as well.

2. Materials and Methods

2.1. Animal Husbandry of Adult Fish

Adult AB strain zebrafish were maintained on a 14:10 h light:dark cycle, as previously described [22], on a recirculating system of RO water buffered to a neutral pH with Instant Ocean© salts (Spectrum Brands, Blacksburg, VA, USA) at 27–30 °C. Ammonia and nitrite levels remained at 0 ppm. Fish were fed twice daily (Aquatox Fish Diet, Zeigler Bros Inc., Gardners, PA, USA) and supplemented with brine shrimp (Artemia International, Fairview, TX, USA). All zebrafish use protocols were approved by the Institutional Animal Care and Use Committee at Wayne State University, according to the National Institutes Health Guide to the Care and Use of Laboratory Animals (Protocol 16-03-054; approved 4 August 2016).

2.2. PFAS Exposures

2.2.1. Spawning Procedure

To obtain F0 embryos, adult stock zebrafish were spawned in a 2:1 female:male ratio (at least 4 trios per concentration) (Figure S1) in the environmental conditions described above. Sexes were separated overnight by a plastic divider in a spawning tank and were allowed to spawn at 08:00 the next morning. Spawning tanks contained a slotted insert through which eggs fell to the bottom, away from the adults. Embryos were harvested after 2 h of spawning activity.

2.2.2. Egg Cleaning

Eggs were incubated at 27 °C in 58 ppm bleach for 10 min, rinsed with RO water, and then placed back in their normal environment of a weak salt solution (600 mg/L salt in RO water) containing Instant Ocean© salts (Spectrum Brands, Blacksburg, VA, USA).

2.2.3. Exposure Protocol

Perfluorooctanoic acid (PFOA) (CAS# 335-67-1, Sigma, St. Louis, MO, USA, 95% purity) and perfluorooctane sulfonic acid (PFOS) (CAS# 1763-23-1, Sigma, 99.4% purity) were used for stock solutions. From these stock solutions, serial dilutions in RO water buffered with Instant Ocean© salts were carried out each day of the exposure to reach the nominal concentrations of 7, 70, and 700 ng/L for PFOA; 24, 240, and 2400 ng/L PFOS; and a mixture with half of the individual concentrations and 1:1 volume ratios (e.g., the ultra-low mixture concentration would contain 3.5 ng/L PFOA and 12 ng/L PFOS). The control was exposed to RO water buffered with Instant Ocean salts; 30 embryos ($\leq$4 hpf) were placed into a well of a 6-well Falcon plate with 8.5 mL of their respective chemical concentration or buffered water (controls). Solutions were replenished daily with approximately 90% fresh solution. Larvae were maintained in an incubator at 27 °C. On day 5, all larvae were rinsed three times in buffered water solution to end the exposure before proceeding with further assays.

2.3. Survival and Abnormality Screening

Survival was recorded on day 5 post-fertilization. Embryos or larvae were considered dead if the heart was stopped. On day 5, all hatched survivors were screened via light microscope for cardiac edema, yolk sac edema, presence of swim bladder, and bent spine. Student's *t*-test was used to determine the statistical significance of each concentration compared to control in terms of the percent total abnormalities. Assays were repeated a minimum of 5 times, with at least 150 larvae per concentration (Table S1). Each repetition was performed on a different day with different larvae.

2.4. Behavioral Analysis

The behavioral assay measuring swim distance in light and dark cycles was performed and analyzed as previously reported [20]. Briefly, healthy (no morphological abnormalities) 5 dpf larvae from control and exposed groups were acclimated to a well plate for $\geq$1 h, then loaded into a DanioVision Chamber (Noldus Information Technology, Wageningen, The Netherlands), which alternated four light and dark cycles for three min each following a chamber acclimation period. Raw data were exported to Noldus EthoVisionXT14, and average distance moved (cm) was analyzed using ANOVA and Tukey's HSD tests in custom R scripts (File S1). The assay was replicated at least three times for each chemical or mixture, with at least 68 fish per concentration in each replicate (Table S1). Each repetition was performed on a different day with different larvae. Larvae were euthanized after the behavioral assay and not used for any further endpoints.

2.5. RNA-Seq and Pathway Analysis

At 5 dpf, five larvae were euthanized and pooled to create one sample, and at least 3 samples per concentration were analyzed for gene expression (Table S1). Each repetition was performed on a different day with a different cohort of larvae. Larvae were pooled to represent the ratio of healthy:abnormal larvae observed during the morphological abnormality assay. For example, if 20% of all low-level PFOA larvae presented abnormalities, 1 of each 5 pooled larvae would present an abnormality, while the other 4 were healthy. Larvae, once euthanized in 16.7 mg/mL tricaine methanesulfonate, were placed in RNALater. This was drained according to the manufacturer's instructions (i.e., between 1–7 days later) and then stored at -80 °C. Storage of larvae, RNA isolation, cDNA library preparation, sequencing, differential expression analysis, and pathway analysis were performed as previously reported [22]. Briefly, RNA isolation was performed with the Qiagen RNeasy Lipid Mini Kit (Qiagen, Hilden, Germany). cDNA libraries were prepared using the Quantseq™ 3' mRNA-seq kit (Lexogen, Vienna, Austria). RNA and cDNA concentrations were measured with a Qubit™ 2.0 fluorometer (Invitrogen, Carlsbad, CA, USA), and cDNA quality was also assessed with an Agilent TapeStation 2200 (Agilent Technologies, Santa Clara, CA, USA). F0 samples were sequenced on Illumina® MiSeq™ (Illumina, CA, USA) and F1–F2 were sequenced on Illumina® HiSeq 2500™ (Illumina, CA, USA) using the Lexogen Bluebee® Genomics Platform (Bluebee, Rijswijk, The Netherlands). F0 reads were aligned to *Danio rerio* genome Build GRCz10, and F1–F2 reads were aligned to *Danio rerio* genome Build GRCz11; differential expression analysis was determined via DESeq2. Differentially expressed genes (DEGs) with log2 fold changes ≥ 0.75 or ≤ -0.75, p-values <0.01, and ≥ 50 analysis-ready molecules were analyzed with Ingenuity Pathway Analysis (IPA®) software (Qiagen Bioinformatics, Redwood City, CA, USA).

2.6. Fecundity Assay

At sexual maturity and dimorphism (4–6 weeks), fish were spawned in a 1:1 male:female ratio in order to attribute the number of eggs produced to each individual female (Figure S1). Fish were not spawned more than once per week. Four randomly-chosen pairs per concentration and control were used per experiment (16 total spawning tanks) (Figure S1, Table S1). After two acclimation sessions of spawning, experiments were replicated a minimum of three times, and a minimum of 6 clutches per concentration were analyzed. Males and females were separated overnight by a plastic divider. At 08:00, dividers were removed and spawning allowed for 2 h. Then, each clutch was cleaned (as described in Section 2.2.2), and eggs were imaged for later quantification. Student's two-tailed t-test was used to determine the average number of eggs per female for each concentration and chemical.

2.7. Sex Ratio

At maturity, fish were visually assessed for female or male secondary sex characteristics. Chi-squared tests were used to determine the statistical significance of any concentration compared to control (Table S1). In F0 fish, dissection was performed for validation. Fish were euthanized in 1.67 mg/mL tricaine methanesulfonate (Syndel, Ferndale, WA, USA) for 10 min.

3. Results
3.1. F0 Generation

Table 1 shows significant endpoints in all chemicals and concentrations.

Table 1. Endpoints of PFAS exposure in zebrafish (*Danio rerio*) across all chemicals, concentrations, and generations. Survival, morphological abnormalities, swim distance, fecundity, sex ratio: percent change. DEGs: number. Blue: decreased endpoint. Orange: increased endpoint. Grey: both increased and decreased endpoints.

F0 Generation Endpoint	Concentration	PFOA	PFOS	Mixture
Survival	Ultra-low			
	Very low			
	Low			
Morphological abnormalities	Ultra-low			
	Very low			
	Low			
Swim distance (dark)	Ultra-low	−10.5%		+3.7%
	Very low	−10.2%		+3.6%
	Low	−4.2%		+12.1%
Swim distance (light)	Ultra-low	−11.6%		+9%
	Very low	−18.8%	−8.16%	+9.7%
	Low		−5.4%	+16%
Differentially-expressed genes	Ultra-low		1	6
	Very low	1	54	
	Low	14		2
Fecundity	Ultra-low	+85%		
	Very low	+42.7%		
	Low			
Sex ratio (% males)	Ultra-low			
	Very low			
	Low			
F1 Generation Endpoint	**Concentration**	**PFOA**	**PFOS**	**Mixture**
Survival	Ultra-low			
	Very low			
	Low	+26.7		
Morphological abnormalities	Ultra-low			
	Very low			
	Low	−7.4%		
Swim distance (dark)	Ultra-low		+15.4%	−12.2%
	Very low	+4.6%	+10.2%	−9.9%
	Low	+9%		−12%
Swim distance (light)	Ultra-low			−15.5%
	Very low	+9.6%		
	Low	+10.6%	−10.6%	
Differentially-expressed genes	Ultra-low	17	5	35
	Very low	106	2	12
	Low	49	149	7
Fecundity	Ultra-low			
	Very low			
	Low		−28.2%	
Sex ratio (% males)	Ultra-low	+30.2%		
	Very low	+55.4%		+28.9%
	Low	+57.1%		
F2 Generation Endpoint	**Concentration**	**PFOA**	**PFOS**	**Mixture**
Survival	Ultra-low			
	Very low			
	Low			
Morphological abnormalities	Ultra-low			
	Very low			
	Low			

Table 1. Cont.

F0 Generation Endpoint	Concentration	PFOA	PFOS	Mixture
Swim distance (dark)	Ultra-low	−8.8%		−3.8%
	Very low		+9.4%	+11.2%
	Low	−7.3%		
Swim distance (light)	Ultra-low			−14.5%
	Very low	−4.7%	+8.8%	
	Low	−14.8%		
Differentially-expressed genes	Ultra-low	112	484	69
	Very low	106	23	1
	Low	302	7	9

3.1.1. F0 Survival and Abnormalities

No statistically significant larval abnormalities or mortality were observed in any concentration of any chemical or mixture. Ultra-low PFOS exposure approached significance with a slightly higher rate of abnormalities ($p = 6.3 \times 10^{-2}$) (Table S1).

3.1.2. F0 Behavior

PFOA

Direct PFOA exposure significantly decreased larval swimming distance in both dark and light cycles at every concentration, with the exception of the low concentration in the light (Figures 1 and 2) ($p < 1 \times 10^{-8}$; $p < 1 \times 10^{-8}$; $p = 2.1 \times 10^{-2}$; ultra-low, very low, low exposure in the dark, respectively) ($p = 3.4 \times 10^{-4}$; $p < 1 \times 10^{-8}$; ultra-low and very low exposure in the light, respectively).

PFOS

Direct PFOS exposure had no effect on larvae from any concentration in the dark. In the light, very low and low exposure groups were significantly hypoactive (Figures 1 and 2) ($p = 4 \times 10^{-7}$, 1.2×10^{-3}, respectively).

Mixture

Direct exposure to the mixture of PFOA and PFOS resulted in increased swimming distance in larval zebrafish (*Danio rerio*), regardless of light/dark setting (Figures 1 and 2) (dark: $p = 1.6 \times 10^{-3}$; $p = 8.1 \times 10^{-4}$; $p < 1 \times 10^{-8}$; ultra-low, very low, low exposure, respectively) (light: $p = 6.7 \times 10^{-4}$; $p = 6.4 \times 10^{-5}$; $p < 1 \times 10^{-8}$; ultra-low, very low, low exposure, respectively).

3.1.3. F0 Transcriptomics

The full lists of DEGs for all chemicals and concentrations in the F0 generation can be found in Table S3; the top five up- and downregulated DEGs are shown in Table 2. Venn diagrams illustrating the overlap of generation-specific DEGs (all concentrations combined) are in Figure 2. Venn diagrams illustrating the overlap of F0 DEGs for each chemical (all concentrations combined) are in Figure 3. DEGs are considered significant at $p < 0.01$ and log2FC of ≥ 0.75 or ≤ -0.75. Pathway analysis could not be performed due to an insufficient number of DEGs.

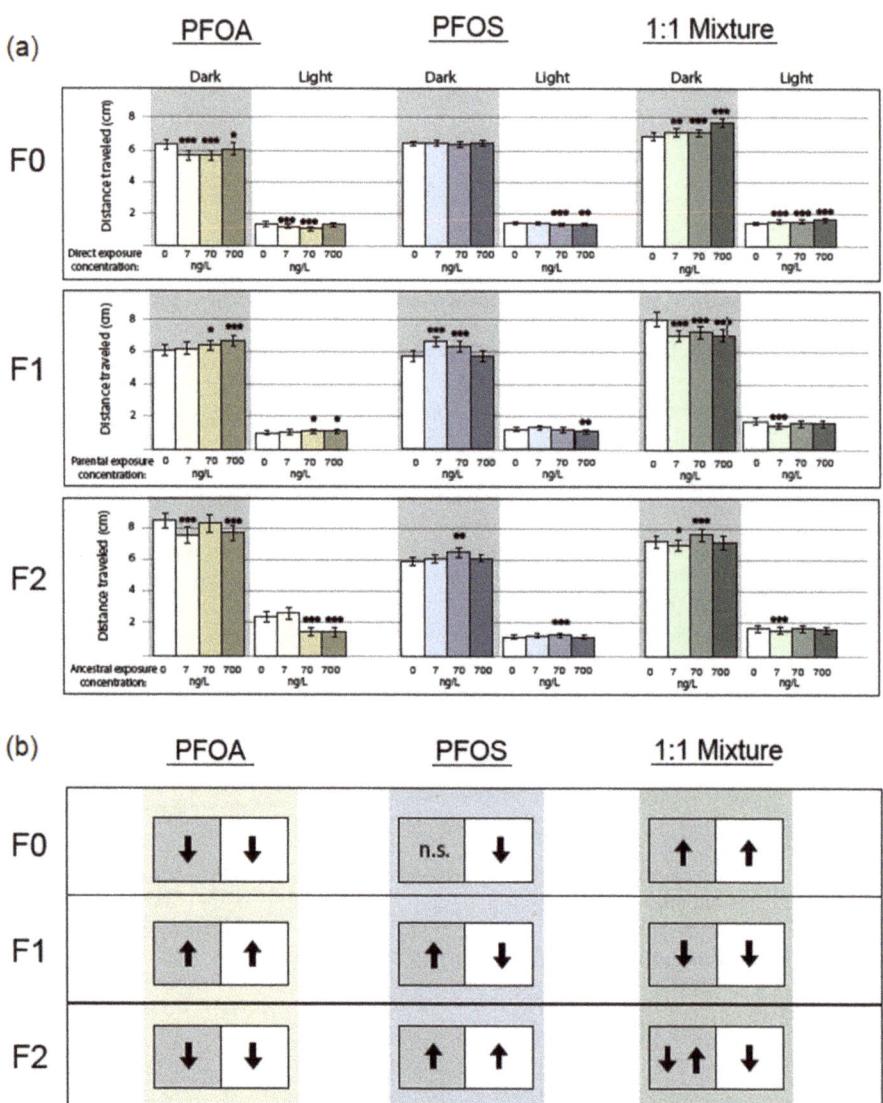

Figure 1. Locomotion following PFAS exposure in dark and light. Yellow: PFOA. Blue: PFOS. Green: mixture. (**a**) Top panel: F0 generation. Middle panel: F1 generation. Lower panel: F2 generation. * $p < 0.05$, ** $p < 0.01$, *** $p < 0.001$; ANOVA with Tukey pairwise test. 0: no exposure. UL: ultra-low exposure. VL: very low exposure. L: low exposure. (**b**) Simplified representation of significant behavioral direction. Upwards arrow: hyperactivity. Downwards arrow: hypoactivity. Two arrows: discordance between one or more concentrations on hyper- vs. hypoactivity. n.s.: not significant.

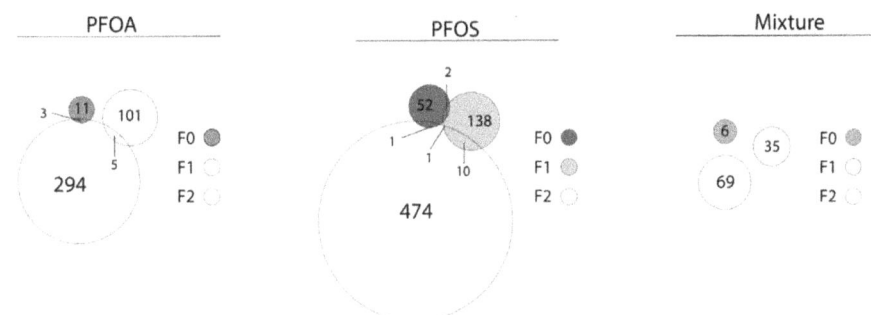

Figure 2. Number of DEGs in each generation for each chemical.

Table 2. Top 5 up- and downregulated DEGs in each chemical, concentration, and generation of zebrafish (*Danio rerio*) and the pathways affected.

Gen.	Chemical	Conc.	Upregulated	Downregulated	Pathways
F0	PFOA	Ultra-low	NA	NA	
		Very low	NA	rpe65a	
		Low	ENSDARG00000075180, gabarapl2, capzb, npc2.1, dusp1	atp6v0e1, trim36, phtf2,	
	PFOS	Ultra-low	irs2a	NA	
		Very low	irg1l, si:ch211-153b23.4, psma5, si:dkeyp-1h4.8, si:ch211-153b23.5	kif3c, ENSDARG00000087345, map4k3b, prodha, ca4a	
		Low	NA	NA	NA
	Mixture	Ultra-low	NA	fkbp9, si:ch211-251f6.6, hbbe1.3, ENSDARG00000092364, ENSDARG00000088687	
		Very low	NA	NA	
		Low	slc6a19a.1, entpd8	NA	
F1	PFOA	Ultra-low	dusp16, pycr1b, tmigd1, wbp2nl, serpina7	zgc:136410, lgals1l1, pkhd1l1.2, pcnx3, si:ch211-125e6.5	NA
		Very low	dusp27, gadd45ba, lims2, asb2b, cuzd1.2	c3a.2, c4b, mthfd1l, lgals1l1, si:ch211-125e6.5,	Xenobiotic metabolism, estrogen receptor signaling
		Low	tmigd1, npas4a, dusp16, gadd45ba, dusp27	trmt1, pitrm1, ercc6l, ifi44d, cdk16	NA
	PFOS	Ultra-low	satb1a, si:ch211-103n10.5, zgc:172051, spint1b, akap17a	NA	NA
		Very low	npas4a	slc43a2a	NA
		Low	zmat5, ENSDARG00000082716, slc9a2, dusp19b, gadd45bb	mfsd14ba, ggt5b, ppp6r2b, dennd5a, nkx3.3	Lipid metabolism, cell death

Table 2. Cont.

Gen.	Chemical	Conc.	Upregulated	Downregulated	Pathways
F2	Mixture	Ultra-low	zgc:92590, smyhc2, amy2a, calcoco1b, si:dkey-14d8.7	smtnl, fh, panx1a, trak2, g6pc1a.1	NA
		Very low	cela1.3, si:dkey-14d8.7, amy2a, si:ch211-240l19.8, pla2g1b	panx1a	NA
		Low	cpa4, zgc:92590, hsd11b2, pla2g1b, si:dkey-14d8.7	ms4a17a.8, rlbp1b	NA
	PFOA	Ultra-low	amy2al2, glg1a, slc17a6a, actl6a, ENSDARG00000096135	ms4a17a.8, tfdp2, prss59.2, srsf5b, LOC100538179	Mitochondrial membrane potential, organismal injury
		Very low	amy2al2, crp2, LOC103910030, eef2k, pcnp	scn2b, cela1.3, cela1.5, tmem97, LOC101882496	Cholesterol and other sterol synthesis
		Low	amy2al2, LOC103910030, irg1l, eef2k, si:ch211-260e23.9	lhx2b, smc1a, LOC110439320, rlbp1b, ms4a17a.8	Immune cell function and trafficking, cell death, glucose homeostasis
	PFOS	Ultra-low	cela1.5, haao, ENSDARG00000115830, atp9b, ENSDARG00000097916	pgk1, si:ch211-260e23.9, crtac1a, cyp8b1, rrm2	Steroid synthesis, bone mineral density, connective tissue
		Very low	cela1.5, lhx2b, cela1.3, mafb, zmp:0000001048	b3gntl1, si:ch211-196h16.5, arpc5a, rbm4.1, bnip4	NA
		Low	ENSDARG00000115830, LOC100536187, pcdh1b, smdt1a	cfp, ddx47, LOC108179091	NA
	Mixture	Ultra-low	fzd6, gatm, bub3, fgfbp2b, il20ra	tcap, mmp9, bnip4, pfkfb3, si:dkey-85k7.7	NA
		Very low	purab	hbae5, c4b, hbae1.3, cebpa	NA
		Low	fgfbp2b, si:dkey-102c8.3	si:ch211-281l24.3, anxa1c, si:ch211-240l19.8, calcoco1b, c4b	NA

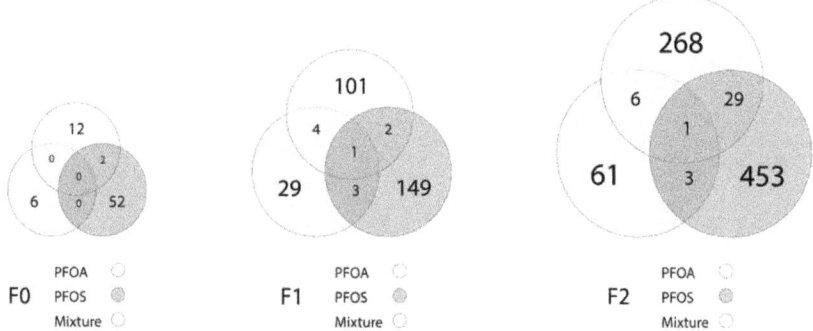

Figure 3. Number of DEGs from each chemical in every generation. Size of the circle indicates the proportion of genes expressed; color of the circle indicates the chemical.

PFOA

Exposure to the ultra-low level of PFOA had no effect on differential gene expression. At very low exposure, only *rpe65a* was significantly changed (LFC -0.89, $p = 9.7 \times 10^{-10}$). Basic cellular functions were impacted by low exposure. Of the 14 genes that were differentially expressed (DEGs) (11 up, 3 down), *tmem14c* was the most upregulated (LFC 0.96) and *atp6v0e1* the most downregulated (LFC -0.85).

PFOS

Exposure to the ultra-low level of PFOS significantly increased the expression of the insulin receptor substrate *irs2a* (LFC 0.77). Of 54 DEGs, the inflammatory response gene *irg1l* was the most upregulated following very low exposure (LFC 1.72); the most downregulated was the kinesin *kif3c* (LFC −1.24). Low exposure to PFOS did not elicit gene expression changes.

Mixture

Exposure to ultra-low levels of PFAS mixture significantly downregulated six genes. The most downregulated was the isomerase *fkbp9* (LFC −0.92). Very low exposure had no effect on gene expression. Low exposure induced the upregulation of two genes, the amino acid transporter *slc6a19a.1* (LFC 0.82) and the nucleoside biosynthesis gene *entpd8* (LFC 0.78).

3.1.4. F0 Fecundity

PFOA

Early-life PFOA exposure did not significantly affect adult female egg production at any concentration (Figure S2). Full fecundity data are shown in Table S2.

PFOS

Fecundity trended downwards with increasing concentrations of PFOS but did not reach statistical significance (low exposure: $p = 7 \times 10^{-2}$) (Figure S2). Full fecundity data are shown in Table S2.

Mixture

Early-life mixture exposure did not significantly affect fecundity at any concentration (Figure S2). Full fecundity data are shown in Table S2.

3.1.5. F0 Adult Body Weight/Length

Adult body weight or length was unaffected by any concentration of any chemical significantly. Females exposed to PFOA at very low and low levels trended towards being significantly heavier ($p = 5.1 \times 10^{-2}$, 6.5×10^{-2}, respectively); 15–17 fish were evaluated per concentration, with one replicate per exposure group.

3.1.6. F0 Sex Ratio

No chemical or concentration affected the sex ratio in the F0. Low mixture exposure approached a significant decrease in the male:female ratio (0.73, $p = 5.3 \times 10^{-2}$, $n = 22$–23). The control male:female ratio for PFOA was 1.11 ($n = 19$–25), for PFOS was 2.09 ($n = 29$–34), and for the mixture was 2.67; 19–34 fish were evaluated per concentration, with one replicate per exposure group.

3.2. F1 Generation

Table 1 shows significant endpoints for all chemicals and concentrations.

3.2.1. F1 Abnormalities and Survival

No statistically significant abnormalities or mortality were observed in any concentration of PFOS or the mixture. At low PFOA exposure, a significant decrease was observed in abnormalities ($p = 2.9 \times 10^{-2}$) (Table S1), and at ultra-low exposure, a significant increase was observed for survival ($p = 3.6 \times 10^{-4}$).

3.2.2. F1 Behavior

PFOA

Parental PFOA exposure was associated with increased swimming activity in both light and dark at the very low and low concentrations (Figures 1 and 2) (dark: $p = 1.3 \times 10^{-2}$; $p < 1 \times 10^{-8}$; 70 and 700 ng/L, respectively) (light: $p = 4 \times 10^{-2}, 1.3 \times 10^{-2}$, respectively).

PFOS

Parental PFOS exposure was associated with increased larval activity in the dark at the very low and ultra-low concentrations ($p < 1 \times 10^{-8}$ for both), yet activity decreased in the light at the low concentration ($p = 5.9 \times 10^{-3}$) (Figures 1 and 2).

Mixture

Parental exposure to the PFAS mixture strongly decreased swimming behavior in the dark at all concentrations ($p < 1 \times 10^{-8}$ for all) and in the light as well only at the ultra-low level ($p = 3.4 \times 10^{-6}$); $n = 72$ per concentration (Figures 1 and 2).

3.2.3. F1 Transcriptomics and Pathway Analysis

The full lists of DEGs for all chemicals and concentrations in the F1 generation can be found in Table S4; the top five up- and downregulated DEGs and affected pathways are shown in Table 2. The full lists of pathways (where applicable) for all chemicals, concentrations, and generations can be found in Table S5. Venn diagrams illustrating the overlap of generation-specific DEGs (all concentrations combined) are in Figure 2. Venn diagrams illustrating the overlap of F1 DEGs for each chemical (all concentrations combined) are in Figure 3. DEGs are considered significant at $p < 0.01$ and log2FC of ≥ 0.75 or ≤ -0.75.

PFOA

Parental PFOA exposure at the ultra-low level caused the significant upregulation of 12 genes and the downregulation of 5 genes (log2FC < −0.75). The most upregulated gene (log2FC = 1.03) was *dusp16* and the most downregulated *si:ch211-125e6.5* (log2FC: −0.84). At very low exposure, 106 genes were significantly differentially expressed (64 upregulated, 42 downregulated). The most upregulated gene was *dusp27* (log2FC: 1.24), and the most downregulated was the innate immunity-related *c3a.2* (log2FC: −1.44). The genes were involved in pathways of xenobiotic metabolism via the CAR pathway and estrogen receptor signaling. Other xenobiotic pathways involved were LXR, RXR, AhR, and FXR. The kinase *dusp27* was the most upregulated molecule. With low parental exposure, there were 49 DEGs, with the most highly upregulated gene of 37 genes being *tmigd1* (log2FC: 1.26) and the most downregulated *trmt1* (log2FC: −0.87), out of 12.

PFOS

Parental PFOS exposure at the ultra-low level caused the significant upregulation of five genes. The most upregulated gene (log2FC = 0.96) was *satb1a*. Very low parental exposure resulted in only two significant DEGs: *npas4a* (log2FC: 0.75) and *slc43a2a* (log2FC: −0.94). Low parental exposure resulted in 149 significant DEGs. The most highly upregulated of 118 genes was *zmat5* (log2FC: 1.29); the most downregulated gene of 80 genes was *mfsd14ba* (log2FC: −1.14). Pathway analysis indicated increased lipid metabolism and decreased cell death pathways.

Mixture

Parental exposure to the ultra-low level of PFAS mixture induced 35 significant DEGs. Of the 21 upregulated genes, *zgc:92590* was the highest (log2FC: 1.35). Of the 13 downregulated genes, *smtnl* was the most downregulated (log2FC: −1.17). Very low parental exposure was associated with the upregulation of 11 genes, with *cela1.3* being the most upregulated (log2FC: 1.19), and only 1 downregulated gene (*panx1a*, log2FC: −0.85). At low parental exposure, five genes were upregulated, with *cpa4* the most upregulated (log2FC: 0.88), and two genes were downregulated: *rlbp1b* and *ms4a17a.8* (log2FC: −0.79, −0.89, respectively).

3.2.4. F1 Fecundity

PFOA

The F1 generation of very low-level PFOA exposure lineage produced significantly fewer eggs than controls (−28.2%, $p < 0.01$). Ultra-low and low concentrations were not affected (Figure S2). Full fecundity data are shown in Table S2.

PFOS

Parental PFOS exposure had no effect on F1 fecundity (Figure S2). Full fecundity data are shown in Table S2.

Mixture

Parental mixture exposure had no effect on F1 fecundity (Figure S2). Full fecundity data are shown in Table S2.

3.2.5. F1 Sex Ratio

PFOA

At every concentration (ultra-low, very low, low), there was a significant increase in the male:female ratio of adult fish ($p = 7.4 \times 10^{-4}$, 2.6×10^{-9}, 3.2×10^{-9}, respectively). The authors note the abnormal lack of males in the control group, which may have led to a false-positive result (PFOA control male ratio: 0.17, PFOS control male ratio: 1.53, mixture control male ratio: 1.12). PFOA F1 contained significantly fewer males than PFOS and mixture F1 ($p = 4.1 \times 10^{-4}$, 4.24×10^{-5}, respectively (chi-square test)). There was no difference in control PFOS and mixture male ratio ($p = 0.59$). Additionally, due to a lack of access to research animals during the SARS-CoV-2-related institutional shutdown, only one cohort of fish ($n = 56$–66) could be observed.

PFOS

No sex ratio shift was observed ($n = 36$–86).

Mixture

At the very low exposure level, there was a significant increase in the male:female ratio (4.45 compared to 1.12 in controls, $p = 1.8 \times 10^{-3}$, $n = 55$–63). The authors note that due to a lack of access to research animals during the SARS-CoV-2-related institutional shutdown, only two cohorts of fish could be observed.

3.3. F2 Generation

Table 1 shows significant endpoints for all chemicals and concentrations.

3.3.1. F2 Abnormalities and Survival

No statistically significant abnormalities or mortality were observed in any concentration of any chemical or mixture.

3.3.2. F2 Behavior

PFOA

Transgenerational behavioral effects of legacy PFOA exposure manifested as hypoactivity at each concentration (Figures 1 and 2) (ultra-low: $p < 1 \times 10^{-8}$ (dark); very low: $p < 1 \times 10^{-8}$ (light); low: $p = 1 \times 10^{-2}$ (dark); $p < 1 \times 10^{-8}$ (light)).

PFOS

Transgenerational behavioral effects of legacy PFOS exposure manifested at only the very low exposure concentration, where hyperactivity was observed (Figures 1 and 2) (dark: $p = 7.6 \times 10^{-3}$; light: $p < 1 \times 10^{-8}$).

Mixture

Transgenerational behavioral effects of legacy exposure to the PFAS mixture presented as hypoactivity following ultra-low exposure (Figures 1 and 2) (dark: $p = 4.2 \times 10^{-2}$; light $p = 3.2 \times 10^{-4}$) and hyperactivity only at the very low concentration and only in the dark ($p < 1 \times 10^{-8}$).

3.3.3. F2 Transcriptomics and Pathway Analysis

The full lists of DEGs for all chemicals and concentrations in the F2 generation can be found in Table S5; the top five up- and downregulated DEGs and affected pathways are shown in Table 2. The full lists of pathways (where applicable) for all chemicals, concentrations, and generations can be found in Table S6. Venn diagrams illustrating the overlap of generation-specific DEGs (all concentrations combined) are in Figure 2. Venn diagrams illustrating the overlap of F2 DEGs for each chemical (all concentrations combined) are in Figure 3. DEGs are considered significant at $p < 0.01$ and log2FC of ≥ 0.75 or ≤ -0.75.

PFOA

Ultra-low-level ancestral exposure to PFOA resulted in 112 significant DEGs in the F2 generation (30 upregulated, 82 downregulated). The most upregulated gene was *amy2al2* (log2FC: 4.41) and the most downregulated was *ms4a17a.8* (log2FC: −2.21). Pathway analysis revealed dysregulation of mitochondrial membrane potential and increased organismal injury, including cancer. In total, 106 DEGs resulted from ancestral PFOA exposure at the very low level (38 upregulated, 68 downregulated). As with the ultra-low concentration, the most upregulated gene here was the carbohydrate-metabolism-related *amy2al2* (log2FC: 3.58); the most downregulated was the sodium channel gene *scn2b* (log2FC: −2.62). Pathway analysis implicated cholesterol synthesis via CYP51A1 and other canonical pathways of sterol synthesis. Ancestral low PFOA exposure resulted in 302 significant DEGs in the F2 generation (124 upregulated, 178 downregulated). The most upregulated gene was again *amy2al2* (log2FC: 4.22), and the most downregulated was *lhx2b* (log2FC: −2.10). Pathways of immune function were upregulated; the top five most upregulated pathways all regard the trafficking of various immune cell types; 9 of the top 20 most upregulated pathways also feature cellular movement, 7 of these in immune cells specifically. The 20 most downregulated pathways feature 9 cell-death-related functions and 5 involved in the dysregulation of glucose homeostasis. Comparisons with GSEA datasets concerning epigenetic and/or chromatin regulation returned multiple DEGs (Table 3).

PFOS

Ultra-low-level ancestral exposure to PFOS resulted in 484 significant DEGs in the F2 generation (209 upregulated, 275 downregulated). The most upregulated gene was *cela1.5* (log2FC: 2.07), and the most downregulated was *pgk1* (log2FC: −2.34). Pathway analysis shows increased lipid metabolism, with 8 of the top 20 most upregulated pathways having to do with the synthesis or metabolism of steroids and terpenoids. The most downregulated pathway was bone mineral density (bias-corrected z-score: −2.75); other connective tissue pathways were overrepresented in the 20 most downregulated pathways. In total, 23 DEGs resulted from ancestral PFOS exposure at the very low level (16 upregulated, 7 downregulated). As with the ultra-low concentration, the most upregulated gene here was *cela1.5* (log2FC: 1.52); however, the most downregulated was *b3gntl1* (log2FC: −0.87). Ancestral low PFOS exposure resulted in seven significant DEGs in the F2 generation (four upregulated, three downregulated). The most upregulated gene was an unnamed/unannotated gene on chromosome 3 (*ENSDARG00000115830*) (log2FC: 1.38), and the most downregulated was *cfp* (log2FC: −1.00). Comparisons with GSEA datasets concerning epigenetic and/or chromatin regulation returned multiple DEGs (Table 3).

Table 3. Significant DEGs involved in epigenetic processes in the F2 generation of zebrafish (*Danio rerio*).

Chemical	Gene Symbol	log2FC	p-Value	Function
PFOA	actl6a	1.7	0.0018	Chromatin modifying
	foxa3	1.15	0.0067	HAT recruitment
	glyr1	0.94	0.0085	Nucleosome activity
	kdm3b	0.99	0.0080	Histone lysine demethylase
	mat1a	−1.6356	0.0035	Methionine adenosyltransferase
	max	−1.24	0.0043	HMT interaction
	sap30l	−1.18	0.0001	HDAC subunit
	smc1a	−1.93	0.0000	Chromatid tethering
	ybx1	−1.33 (ultra-low); 1.06 (low)	<0.004	DNA binding
PFOS	chmp2a	−0.91	0.0079	Chromatin modifying
	h1-0	0.91	0.0028	H1.0 linker histone
	hbp1	0.97	0.0017	DNMT1 repressor
	hmg20a	0.89	0.0085	HMT recruitment
	hmgn2	−1.06	0.0046	Chromatin modifying
	hnrnpk	−1.19	0.0019	ssDNA binding
	kdm1a	−0.87	0.0071	Lysine demethylase 1A
	meaf6	−1.4	0.0013	HAT interactor
	prdm9	−1.41	0.0053	HMT recruitment
	riox2	1.48	0.0004	HDMT
	setd5	0.96	0.0020	KMT2E paralog
	tox2	1.26	0.0001	Chromatin modifying
	usf1	−1.01	0.0071	Chromatin modifying
Mixture	h2bc1	−0.95	0.0006	H2B clustered histone 1

Mixture

Ultra-low-level ancestral mixture exposure was associated with 69 significant DEGs in the F2 generation. Of the 27 upregulated genes, *fzd6* had the highest log2FC (1.31). Of the 42 downregulated genes, *tcap* had the lowest log2FC (−1.19).

At the very low concentration, F2 larvae exhibited only one significantly upregulated DEG (*purab*, log2FC: 0.76) and four downregulated. The most downregulated gene was *hbae5* (log2FC: −1.36).

Similarly, the low level exposure had few DEGs. Two were upregulated (*fgfbp2b* and *si:dkey-102c8.3*, log2FC: 0.77, 0.76, respectively), and seven were downregulated (most downregulated: *si:ch211-281l24.3*, log2FC: −0.93). Comparisons with GSEA datasets concerning epigenetic and/or chromatin regulation returned one DEG, *h2bc1* (Table 3). No generation alone produced a sufficient number DEGs for pathway analysis; however, when all generations and concentrations were collated, pathways of cell death and immune dysfunction emerged (Table 4).

Table 4. Pathway analysis (IPA) of all DEGs from each concentration and generation of mixture-exposed zebrafish (*Danio rerio*) larvae combined.

Rank	Diseases or Functions Annotation	p-Value	Bias-Corrected z-Score	# Molecules
1	Organismal death	3.59×10^{-3}	1.714	22
3	Morbidity or mortality	1.81×10^{-3}	1.429	23
10	Quantity of cytokine	3.53×10^{-3}	0.834	5
11	Infiltration by neutrophils	1.30×10^{-3}	0.793	5
12	Cell movement of neutrophils	3.56×10^{-3}	0.751	6
18	Necrosis	6.16×10^{-3}	0.603	23
24	Chemotaxis of leukocytes	9.40×10^{-4}	0.307	7
27	Quantity of myeloid cells	1.41×10^{-3}	0.301	9
41	Cellular infiltration by phagocytes	2.65×10^{-3}	−0.026	6
42	Cellular infiltration by myeloid cells	4.04×10^{-3}	−0.028	6
46	Cellular infiltration by leukocytes	1.09×10^{-3}	−0.144	8
57	Accumulation of leukocytes	5.69×10^{-3}	−0.402	5
87	Inflammatory response	5.20×10^{-3}	−1.872	10

4. Discussion

In this study, numerous endpoints were examined across three generations of zebrafish exposed to environmentally relevant PFAS concentrations. Locomotion, gene expression, and fecundity were significantly altered across all generations by at least one concentration of PFOA, PFOS, and/or their mixture.

Environmental levels of PFAS exposure, as expected, did not cause significant mortality in any generation. Mortality with PFAS exposure is typically not observed in zebrafish under 10 mg/L (10^7 ng/L) [23,24]. Jantzen et al. [25] also found no significant death or abnormalities using similar exposure methods to PFOS and PFOA. Gross morphological abnormalities were not increased by exposure, agreeing with literature noting abnormalities following ≥1 mg/L (10^6 ng/L) PFAS exposure [26]. In the F1 generation of PFOA larvae, decreased abnormalities and increased survival were observed in the low and ultra-low groups, respectively. It is possible these unexpected outcomes may be due to exposure solutions, which were carefully derived from commercially available certified stock solutions but were not analytically verified, which may lead to variability in dosing; additionally, neither stock solution was available at 100% purity, ranging from 95% (PFOA) to 99.4% (PFOS) purity. Impurities of unknown origin, constituting up to 5% of PFOA exposure (0.35, 3.5, and 35 ng/L of the ultra-low, very low, and low concentrations, respectively) and 0.6% of PFOS exposure (0.14, 1.4, and 14 ng/L), could potentially have influenced the results. Overall, these results do not point to a severe risk of bodily harm from environmental-level exposure, though analytical verification of our exposure doses would support a higher-confidence assessment.

A persistent endpoint across all chemicals and generations was alterations in the behavioral response to light and dark stimuli. Larval swimming behavior is used as an indicator of neurotoxicity [27]. By 5 dpf, all major organ systems, including the brain, are functional [27]. Larvae are naturally more inclined to reserve swim bouts for dark periods, where they are less susceptible to predators than in the light [28]. Exposure-induced excitability or lethargy may be modulated by CNS function, which could translate to negative health implications in humans, and erratic behavior could have ecological consequences in aquatic wildlife consistently exposed to PFASs. We report multigenerational behavioral effects in the F1 generation and report for the first time transgenerational PFAS-associated behavioral changes in the completely unexposed F2 generation. The presence of a behavioral phenotype in the F2 generation suggests epigenetic changes induced by F0 exposure. More research is needed to plot the mechanism of this phenotypic inheritance, as well as how animal and human health and ecology are affected by continued PFAS exposure over multiple generations.

PFOA-exposed F0 larvae were hypoactive in both light and dark; this pattern was reversed in the indirectly exposed F1 generation, then returned to the F0 pattern in the unexposed F2 generation. Hyperactivity is typically seen in moderately to highly exposed F0 larvae [12,29,30]; however, exposure sometimes has no behavioral effect [31,32]. It is possible that PFOA exerts a non-monotonic response, wherein ng/L concentrations produce the observed hypoactivity, while higher doses produce hyperactivity. More research at low doses in the F0 generation will be required to draw conclusions. To our knowledge, this is the first study to examine behavior in the F1 and F2 generations of PFOA-exposed F0. The reversal in each generation of the direction of behavior (hypoactivity in F0; hyperactivity in F1; hypoactivity in F2) also suggests the possibility of a neuromodulatory compensation mechanism overcorrecting for the previous generation's propensity for erratic behavior.

PFOS-exposed F0 larvae were hypoactive in the light only; this persisted in the F1 generation. Additionally, F1 larvae were hyperactive in the dark, and F2 larvae were hyperactive under both conditions. In F0 larvae, hyperactivity is generally observed at moderate to high doses [25,29,31–35]. One study at 2 mg/L (20^6 ng/L) found hypoactivity [12], but to our knowledge, this is the first study within a ng/L range, which may account for the diverging effect. Few studies have examined F1, and none at low doses. At moderate doses, Chen et al. observed the exact pattern that we observed of hyperactivity in the dark and hypoactivity in the light [33]; hyperactivity was also observed in other studies [36]. In contrast to PFOA-lineage F2, transgenerational PFOS effects present as totally different from the F0 pattern. The differences in the structure of sulfonic and carboxylic acids are known to exert different effects [31]; this phenomenon appears to continue into the F2 generation.

Mixture-exposed F0 larvae were hyperactive, F1 were hypoactive, and F2 possessed a variable response to dark and light stimuli. Though PFOS was present in a higher concentration than PFOA, PFOS did not appear to overpower PFOA's presence or drive the mixture results as the mixture endpoints were quite different from the PFOS endpoints. Mixtures are generally understudied. Despite PFOS and PFOA being two of the most thoroughly investigated individual PFAS chemicals, their mixture at human levels has not been well-studied for locomotor behavior. However, a complex mixture including both chemicals induced hyperactivity at putative human serum levels [32], though the presence of other chemicals likely influenced the outcome. Very high exposure to a >1 g/L (10^9 ng/L) mixture of nine PFASs in equal amounts was associated with hypoactivity [12]. However, this concentration may have caused lethargy-inducing toxicity as the LC50 of a 1:1 ratio of PFOA:PFOS has been demonstrated at ~37 mg/L (37^6 ng/L) in zebrafish [11]. Given that we are never exposed to a single PFAS alone, and PFAS mixtures have been measured in amniotic fluid [37,38], the lack of knowledge on mixtures across generations necessitates more research, especially at environmentally relevant levels.

Gene expression dysregulation was the most sensitive and persistent endpoint observed across all chemicals and generations. While PFASs are not directly genotoxic, they are known to cause transcriptomic changes [25,29,39]. As exposure to low concentrations of PFASs is understudied, we chose to explore the full transcriptome using RNA-seq rather than targeted expression analysis. This study may provide genes of interest for future biomarkers of effect and for targeted analysis in low-exposure schemes. We report multi- and transgenerational effects in gene expression. In fact, for every chemical, more genes were differentially expressed as the generations progressed. The F1 generation showed 2–4× more DEGs than F0. The F2 generation showed 2–3× more DEGs than F1 (and 8–9× more than F0) even in the absence of exposure in the F1 and F2 generations, suggesting epigenetic regulation of the transcriptome. The affected pathways ranged from the immune system, xenobiotic metabolism, and steroid metabolism and synthesis (PFOA) to movement disorders and bone mineral density (PFOS). Surprisingly, the mixture caused relatively few DEGs compared to the single chemicals alone. In general, each chemical was associated with a unique set of DEGs in each generation. However, common to all chemicals in the F1 generation was the dysregulation of *si:dkey-14d8*, and to the F2 generation, *wbp2nl*. Little is known about the *Danio rerio* gene *si:dkey-14d8*; it is predicted to be involved in collagen fibril organization [40]. The F2 gene *wbp2nl* encodes a sperm protein that promotes oocyte fertilization [41]. This gene was upregulated by PFOS and mixture exposure (log2 fold change 1.20, 0.85, respectively) but downregulated by PFOA exposure (log2 fold change −1.23). In line with our null findings of changes in fecundity, WBP2NL expression was not associated with reproductive outcomes in a human study seeking prognostic fertilization factors [42]. *Wbp2nl* is silenced during early development [43]; thus, its activation in the F2 larvae by PFOS and the mixture may indicate aberrant epigenetic programming. While *wbp2nl* is mainly expressed by sperm, it is also found in the breast and kidney [44]—areas known to be PFAS targets [45,46]. It will be interesting for future studies to further phenotypically anchor the diverse transcriptomic pathways of each chemical and the mixture and establish biomarkers of effect for PFAS exposure.

In PFOA-exposed F0 larvae, *rpe65a* and *atp6v0e1* were downregulated. Downregulation of *rpe65a* is associated with retinal degeneration in zebrafish [47], and loss of RPE65 function leads to blindness in humans [48]. *Atp6v0e1* is involved in visual–motor behavior [49]. The downregulation of optical-related *rpe65a* and *atp6v0e1* may have contributed to the hypoactivity we observed in PFOA-exposed F0 larvae. In the F1 generation, xenobiotic pathways predominated, driven by the upregulation of *cyp3a7*. PFASs are known xenobiotic inducers of PXR and CAR pathways in humans and rodents; in zebrafish, these receptors have been shown to be unresponsive to PFOA [50]. However, as these data seek to inform human health, the change in *cyp3a7* expression implicating PXR and CAR activation is still meaningful. Additionally, CYP3A7 in humans is enriched in fetal liver [51], underscoring the relevance of the embryonic zebrafish exposure model to human developmental health. Another upregulated molecule in xenobiotic pathways was *dusp16*, which has a role in immune function [52]. Though immune dysregulation does not feature prominently in the PFOA F1 pathway profile, PFASs are a demonstrated immunotoxicant [53], and the *dusp* genes appeared in DEGs of the F0 generation and additionally in the pathway analysis of the F2 generation. In addition to immune system pathway disruption in the F2 generation, steroid synthesis was affected, and the glucose-homeostasis-related pancreatic gene *amy2al2* was the most upregulated DEG at every concentration. PFOA has been shown to increase steroid hormone levels in zebrafish larvae [54], and links between PFOA serum levels and diabetes risk have been established in humans [55–57]. Immune dysfunction appears to be a significant outcome of low-level PFOA exposure, though effects may not be seen until later generations.

In PFOS-exposed F0 larvae, *irs2a* and inflammatory response gene *irg1l* were upregulated, and the kinesin gene *kif3c* was downregulated. Besides its known glucose metabolism function [56–58], *irs2a* has an emerging function in hypoxia protection [59,60]. Hypoxia and inflammation, in combination with decreased expression of the photoreceptor *kif3c* gene [61], may have contributed to the hypoactivity we observed in PFOS-exposed larvae. Others have found downregulation of the histamine H1 receptor [32] and steroidogenic enzymes [29,62] at moderate to high exposure. Relatively few genes were differentially expressed in the F0 generation as compared to the F1 and F2 generations. In the F1 generation, DNA-binding genes *satb1a* and *npas4a* were two of the most upregulated genes. *Satb1a* and *npsa4a* expression is localized to the CNS in larvae [63,64]. In line with the present behavioral results of hyperactivity from very low exposure, *npas4* expression is increased in response to neuronal activity [64]. Less is known about *satb1a* in zebrafish. In humans, SATB1 remodels chromatin in thymocyte differentiation into T-cells [65,66]. Pathway analysis results included lowered chemotaxis of immune cells, increased steroid synthesis, survival of neuronal cell types, and movement disorders. F1 larvae were the only group across all chemicals and generations where the direction of behavior (hyper- or hypoactivity) had no agreement between light and dark conditions. The upregulation of neuronal activity gene *npas4* may have contributed to hyperactivity in the dark; the movement disorder pathway is one molecular indication of the contrasting responses to light and dark. Other studies in zebrafish have not examined and compared for gene expression with

F1-lineage behavior. As in F1, F2 larvae showed increased pathways involving steroids and, additionally, other lipids. The pancreatic gene *cela1.5* was the most upregulated DEG at ultra-low and very low exposures and was moderately upregulated at low exposure. Lipid metabolism disruption has been previously linked to PFOS exposure at moderate levels in F0 fish [67], but we did not observe this effect until the F2 generation. Bone mineral density and other connective tissue pathway disruptions were also a PFOS F2-specific occurrence. Increased lipid pathways and decreased connective tissue pathways do not seem to explain the observed hyperactivity in the F2 generation; however, a non-specific movement disorder pathway was also increased. More research is required in F2 larvae to fully understand the scope of the ancestral effects of PFOS exposure. In most generations, PFOS exposure was associated with pathways of increased lipid synthesis, which complements the thoroughly-studied PFOS-associated high cholesterol in humans [68–70].

Mixtures are a rapidly expanding field of research, and low levels are highly relevant to human health. In mixture-exposed larvae, few DEGs were expressed in all generations compared to individual PFASs (Figure 2). The F0 generation exhibited dysregulation of genes involved in basic cellular processes, with no obvious influence on the observed hyperactivity. Similarly, in their assessment of behavior and gene expression in a complex mixture including PFOS and PFOA, Khezri et al. [30] could not rationalize a clear link between exposure-associated hyperactivity and DEGs. More research is certainly needed to elucidate the complex transcriptomic dynamics underpinning behavioral outcomes in mixtures. The F1 generation showed dysfunction in pancreatic genes *zgc:92590*, *cela1.5*, and *cpa4*. F1 downregulation in optic-related gene *rlbp1b* [71] could have contributed to the observed hypoactivity of larvae in light and dark. F2 larvae downregulated the muscular gene *tcap*, yet upregulated growth genes *fzd6* and *fgfbp2*. When significant DEGs from all generations were collated, pathway analysis revealed immune dysfunction and developmental deficits predicting organismal and cell death; however, mortality was unaffected in any generation. The present results suggest mixture exposure does not cause overt harm in any generation; however, the transcriptome of developmentally exposed fish may be an early indicator of latent embodied effects. Perhaps a longer experiment with aged fish would reveal mixture-associated latent mortality.

Egg production in females was measured to estimate fecundity in the F0 and F1 generations. Changes in fecundity may have implications for reproductive health, the offspring, as well as the ecosystem. In humans, there is no consensus on fecundity and PFAS exposure, possibly owing to the multiple ways of defining fecundity in humans. Multiple epidemiological studies have found a decrease in fecundity with PFOA or PFOS [72–75], while some have found no effect in either [76,77]. There was no effect on the fecundity of F0 exposure to PFAS. Of note, controls in the PFOA F0 group produced significantly fewer eggs than controls in the PFOS ($p = 4 \times 10^{-3}$) but not mixture ($p = 0.15$) groups (one-tailed t-test) (PFOS and mixture controls were not significantly different ($p = 0.22$, two-tailed t-test)). When egg production in PFOA-exposed larvae was compared to control data from the PFOS and mixture larvae, there was actually a significant decrease in egg production at low exposure ($p = 2.1 \times 10^{-2}$) (data not shown). Decreased zebrafish egg production was observed in another study on low-level exposure to PFOA [37] and at moderate exposure in the crustacean *Daphnia magna* [78]; however, in wild-caught fish, hepatic levels of PFOA had no association with fecundity [79]. In the F1 generation, very low exposure lineage fish produced significantly fewer eggs. In the only other transgenerational study of fecundity, Marziali et al. [80] found no effect in F0–F2 in harlequin flies. In PFOS studies, fecundity is found to be either decreased at moderate doses [78,81] or to have no effect [34,77], including no effect in F1 and/or F2 [34,78]. In an F0–F2 study of a moderate dose PFAS mixture containing low doses of PFOA and PFOS on Japanese medaka, Lee et al. [82] reported no significant effect. Overall, the present results suggest little to no effect on fecundity in F0–F1 zebrafish exposed to low levels of PFAS.

Sex determination in laboratory zebrafish is polygenic and is thought to be influenced by their environment, which can include exposure to contaminants [83]. Alterations in the male:female ratio are thus a common endpoint in endocrine disruption studies, with a shift in either direction indicating disruption. In the F0 generation, there was no significant change in sex ratio following exposure. In the F1 generation, the very low level of the mixture and every concentration of PFOA caused an increase in the ratio (significantly more males). The authors note the abnormal lack of males in the PFOA control group, which may have led to a false-positive result. Additionally, due to a lack of access to research animals during the SARS-CoV-2-related institutional shutdown, only one cohort of PFOA F1 and two cohorts of mixture F1 fish could be observed. No changes were observed in the PFOS-exposed sex ratio in either generation. Other studies have observed a decreased ratio following F0 PFOS exposure [34]. Exposing the F0 and F1 generations to a mixture of four PFAS, including PFOA and PFOS, did not result in any shift [82]. The scarcity of replicates for all groups and abnormal PFOA control fish do not allow meaningful conclusions to be drawn from this endpoint in the present study.

As each generation was differentially affected by each chemical, it is pertinent to summarize the similarities and differences of each chemical's discussed effects in the F0, F1, and F2 generations separately. Additionally, human health research and policies are mainly concerned with the directly exposed subject. In the F0 generation, PFOA exposure was associated with hypoactivity, with a potential visual–motor impact occurring via the down-regulation of vision-related genes $rpe65a$ and $atp6v0e1$ [47,49]. Similarly, PFOS-exposed larvae were also hypoactive in the light and had a downregulated photoreceptor gene ($kif3c$) [61]. Ophthalmic health should be observed more closely in future studies examining behavior response to visual cues, especially as zebrafish eyes are in constant contact with the exposure solution. The mixture-exposed larvae were, in contrast, hyperactive and showed no clear disruption of a particular pathway, dysregulating instead the genes involved in basic cellular processes. No chemical was associated with a significant change in adult fecundity, body weight, length, or sex ratio. In the F1 generation, PFOA exposure was associated with hyperactivity and xenobiotic response. PFOS-related behavior varied by light or dark status in the only locomotor disagreement in the study; upregulated CNS-related genes could account for the hyperactivity in the dark. The mixture larvae displayed hypoactivity and dysfunctional pancreatic genes. Additionally, the downregulation of optic-related gene $rlbp1b$ [71] in the mixture larvae could complicate behavior results, as in the F0 generation. Overall, each chemical was associated with disparate pathways in the F1 generation, in line with different behavioral patterns across the chemicals. No chemical was associated with a reliable change in adult fecundity or sex ratio. In the F2 generation of PFOA- and PFOS-exposed larvae, pancreatic genes were most affected, likely leading to the observed alterations in hormone-related pathways. Additionally, immune pathways were affected in the PFOA and mixture groups. Each chemical in the F2 generation was associated with a different behavioral pattern (PFOA: hypoactivity; PFOS: hyperactivity), with the mixture showing both hyper- and hypoactivity. In sum, the evidence points to varying effects of PFASs depending on both the specific chemical and degree of exposure.

This study provides the first report on the multigenerational effects of environmental-level PFOA exposure on zebrafish behavior and of a mixture of the two chemicals on behavior and gene expression. Further, it is the first report of the transgenerational effects of PFOA, PFOS, and a 1:1 mixture in terms of behavior and transcriptomics. The next steps in this line of research will be to examine the epigenetic influences set in motion by these PFASs. Effects onto the F2 generation have been reported in PFBS [84] and PFOS-alternative F-53B exposure [85] at moderate levels. Intriguingly, low-level PFAS exposure in the present study continued to exert effects generations after exposure cessation. Gene expression dysregulation increased as the generations progressed, with F2 exhibiting far more DEGs than F0, suggesting epigenetic regulation of expression in the absence of a chemical stressor. In general, the DEGs in each generation and in each chemical had little overlap. Interestingly, the mixture had a relatively small influence on the number of DEGs compared to the individual PFASs. The unique suites of DEGs underscore the differential effects of different functional groups of PFASs and individual PFASs versus a mixture and suggest different mechanisms of action in the production of the observed transcriptomic signatures and behavioral phenotypes.

It is the authors' aim that these results inform decision-making regarding safe contaminant limits in drinking water, food sources, and aquatic habitats. Future studies into the mechanisms of epigenetic dysregulation under exposure will be of great interest. PFAS replacements, including "short-chain" alternatives to PFOA and PFOS, will be critical to study as well, both individually and in environmentally relevant mixtures.

Supplementary Materials: The following supporting information can be downloaded at: https://www.mdpi.com/article/10.3390/toxics10060334/s1. Figure S1: Schematic of experimental design for obtaining embryos and for fecundity assay. Figure S2: Box plots of F0 and F1 fecundity data. Table S1: Replicates and n-values for all experiments. Table S2: Endpoint data and significance reporting for survival, abnormalities, behavior, fecundity, and sex ratio. Table S3: F0 DEGs. Table S4: F1 DEGs. Table S5: F2 DEGs. Table S6: IPA summaries. File S1: Custom R code for analysis of behavioral data.

Author Contributions: Conceptualization, T.R.B. and A.H.; methodology, A.H., C.A., D.N.M., M.A., M.C., A.K. and D.J.; formal analysis, A.H. and C.-C.W.; resources/funding acquisition, T.R.B.; writing: A.H.; editing: T.R.B. and A.H. All authors have read and agreed to the published version of the manuscript.

Funding: Funding was provided by Great Lakes Water Authority; the National Center for Advancing Translational Sciences (K01 OD01462 to T.R.B.), the National Institute of Environmental Health Sciences (R01 ES030722 to T.R.B., A.H., and D.N.M.; P30 ES020957 to D.N.M. and T.R.B.; F31 ES030278 to D.N.M.), the National Institute of General Medicine Sciences (R25 GM 058905 to D.J.), and the WSU reBUILD program (to M.C. and M.A.).

Institutional Review Board Statement: All zebrafish use protocols were approved by the Institutional Animal Care and Use Committee at Wayne State University, according to the National Institutes Health Guide to the Care and Use of Laboratory Animals (Protocol 16-03-054; approved 4 August 2016).

Acknowledgments: We acknowledge Emily Crofts, Kim Bauman, and all members of the Warrior Aquatic, Translational, and Environmental Research (WATER) lab at Wayne State University for help with zebrafish care and husbandry. We would like to acknowledge the Wayne State University Applied Genomics Technology Center, especially Katherine Gurdziel, for providing sequencing services and the use of Ingenuity Pathway Analysis software.

Conflicts of Interest: The authors declare no conflict of interest.

References

1. Giesy, J.; Kannan, K. Global distribution of perfluorooctane sulfonate in wildlife. *Environ. Sci. Technol.* **2001**, *35*, 1339–1342. [CrossRef] [PubMed]
2. Nakayama, S.; Yoshikane, M.; Onoda, Y.; Nishihama, Y.; Iwai-Shimada, M.; Takagi, M.; Kobayashi, Y.; Isobe, T. Worldwide trends in tracing poly- and perfluoroalkyl substances (PFAS) in the environment. *TrAC Trends Anal. Chem.* **2019**, *121*, 115410. [CrossRef]
3. Calafat, A.; Kato, K.; Hubbard, K.; Jia, T.; Botelho, J.; Wong, L. Legacy and alternative per- and polyfluoroalkyl substances in the U.S. general population: Paired serum-urine data from the 2013–2014 National Health and Nutrition Examination Survey. *Environ. Int.* **2019**, *131*, 105048. [CrossRef] [PubMed]
4. Domingo, J.; Nadal, M. Human exposure to per- and polyfluoroalkyl substances (PFAS) through drinking water: A review of the recent scientific literature. *Environ. Res.* **2019**, *177*, 108648. [CrossRef]
5. Boone, J.; Vigo, C.; Boone, T.; Byrne, C.; Ferrario, J.; Benson, R.; Donohue, J.; Simmons, J.; Kolpin, D.; Furlong, E.; et al. Per- and polyfluoroalkyl substances in source and treated drinking waters of the United States. *Sci. Total Environ.* **2019**, *653*, 359–369. [CrossRef]
6. Schultz, M.; Higgins, C.; Huset, C.; Luthy, R.; Barofsky, D.; Field, J. Fluorochemical mass flows in a municipal wastewater treatment facility. *Environ. Sci. Technol.* **2006**, *40*, 7350–7357. [CrossRef]
7. Gützkow, K.; Haug, L.; Thomsen, C.; Sabaredzovic, A.; Becher, G.; Brunborg, G. Placental transfer of perfluorinated com-pounds is selective—A Norwegian mother and child sub-cohort study. *Int. J. Hyg. Environ. Health* **2012**, *215*, 216–219. [CrossRef]
8. Barker, D.J.P.; Osmond, C. Infant mortality, childhood nutrition, and ischaemic heart disease in England and Wales. *Lancet* **1986**, *327*, 1077–1081. [CrossRef]
9. Biomonitoring Data Tables for Environmental Chemicals. Available online: https://www.cdc.gov/exposurereport/data_tables.html (accessed on 6 June 2022).

10. Baker, B.; Haimbaugh, A.; Sperone, F.; Johnson, D.; Baker, T. Persistent contaminants of emerging concern in a Great Lakes urban-dominant watershed. *J. Great Lakes Res.* **2022**, *48*, 171–182. [CrossRef]
11. Ding, G.; Zhang, J.; Chen, Y.; Wang, L.; Wang, M.; Xiong, D.; Sun, Y. Combined effects of PFOS and PFOA on zebrafish (*Danio rerio*) embryos. *Arch. Environ. Contam. Toxicol.* **2013**, *64*, 668–675. [CrossRef]
12. Menger, F.; Pohl, J.; Ahrens, L.; Carlsson, G.; Örn, S. Behavioural effects and bioconcentration of per- and polyfluoroalkyl substances (PFASs) in zebrafish (*Danio rerio*) embryos. *Chemosphere* **2020**, *245*, 125573. [CrossRef] [PubMed]
13. Kim, S.; Thapar, I.; Brooks, B. Epigenetic changes by per- and polyfluoroalkyl substances (PFAS). *Environ. Pollut.* **2021**, *279*, 116929. [CrossRef] [PubMed]
14. Bowers, E.; McCullough, S. Linking the epigenome with exposure effects and susceptibility: The epigenetic seed and soil model. *Toxicol. Sci.* **2016**, *155*, 302–314. [CrossRef] [PubMed]
15. Bouwmeester, M.; Ruiter, S.; Lommelaars, T.; Sippel, J.; Hodemaekers, H.; van den Brandhof, E.; Pennings, J.; Kamstra, J.; Jelinek, J.; Issa, J.; et al. Zebrafish embryos as a screen for DNA methylation modifications after compound exposure. *Toxicol. Appl. Pharmacol.* **2016**, *291*, 84–96. [CrossRef]
16. Tian, J.; Xu, H.; Zhang, Y.; Shi, X.; Wang, W.; Gao, H.; Bi, Y. SAM targeting methylation by the methyl donor, a novel therapeutic strategy for antagonize PFOS transgenerational fertility toxicity. *Ecotoxicol. Environ. Saf.* **2019**, *184*, 109579. [CrossRef]
17. Howe, K.; Clark, M.; Torroja, C.; Torrance, J.; Berthelot, C.; Muffato, M.; Collins, J.; Humphray, S.; McLaren, K.; Matthews, L.; et al. The zebrafish reference genome sequence and its relationship to the human genome. *Nature* **2013**, *496*, 498–503. [CrossRef]
18. Wheeler, A. *Memorandum: Directive to Prioritize Efforts to Reduce Animal Testing*; Environmental Protection Agency: Washington, DC, USA, 2019.
19. Hill, A.; Teraoka, H.; Heideman, W.; Peterson, R. Zebrafish as a model vertebrate for investigating chemical toxicity. *Toxicol. Sci.* **2005**, *86*, 6–19. [CrossRef]
20. Baker, T.; King-Heiden, T.; Peterson, R.; Heideman, W. Dioxin induction of transgenerational inheritance of disease in zebrafish. *Mol. Cell* **2014**, *398*, 36–41. [CrossRef]
21. Fact Sheet: PFOA & PFOS Drinking Water Health Advisories. Available online: https://www.epa.gov/sites/default/files/2016-06/documents/drinkingwaterhealthadvisories_pfoa_pfos_updated_5.31.16.pdf (accessed on 26 February 2022).
22. Phillips, J.; Haimbaugh, A.; Akemann, C.; Shields, J.; Wu, C.; Meyer, D.; Baker, B.; Siddiqua, Z.; Pitts, D.; Baker, T. Devel-opmental phenotypic and transcriptomic effects of exposure to nanomolar levels of 4-nonylphenol, triclosan, and triclocarban in zebrafish (*Danio rerio*). *Toxics* **2022**, *10*, 53. [CrossRef]
23. Ulhaq, M.; Carlsson, G.; Örn, S.; Norrgren, L. Comparison of developmental toxicity of seven perfluoroalkyl acids to zebrafish embryos. *Environ. Toxicol. Pharmacol.* **2013**, *36*, 423–426. [CrossRef]
24. Sharpe, R.; Benskin, J.; Laarman, A.; MacLeod, S.; Martin, J.; Wong, C.; Goss, G. Perfluorooctane sulfonate toxicity, iso-mer-specific accumulation, and maternal transfer in zebrafish (*Danio rerio*) and rainbow trout (*Oncorhynchus mykiss*). *Environ. Toxicol. Chem.* **2010**, *29*, 1957–1966. [CrossRef] [PubMed]
25. Jantzen, C.; Annunziato, K.; Bugel, S.; Cooper, K. PFOS, PFNA, And PFOA sub-lethal exposure to embryonic zebrafish have different toxicity profiles in terms of morphometrics, behavior and gene expression. *Aquat. Toxicol.* **2016**, *175*, 160–170. [CrossRef] [PubMed]
26. Shi, X.; Du, Y.; Lam, P.; Wu, R.; Zhou, B. Developmental toxicity and alteration of gene expression in zebrafish embryos exposed to PFOS. *Toxicol. Appl. Pharmacol.* **2008**, *230*, 23–32. [CrossRef] [PubMed]
27. Orger, M.; de Polavieja, G. Zebrafish behavior: Opportunities and challenges. *Annu. Rev. Neurosci.* **2017**, *40*, 125–147. [CrossRef]
28. Kalueff, A.; Gebhardt, M.; Stewart, A.; Cachat, J.; Brimmer, M.; Chawla, J.; Craddock, C.; Kyzar, E.; Roth, A.; Landsman, S.; et al. Towards a comprehensive catalog of zebrafish behavior 1.0 and beyond. *Zebrafish* **2013**, *10*, 70–86. [CrossRef]
29. Jantzen, C.; Annunziato, K.; Cooper, K. Behavioral, morphometric, and gene expression effects in adult zebrafish (*Danio rerio*) embryonically exposed to PFOA, PFOS, and PFNA. *Aquat. Toxicol.* **2016**, *180*, 123–130. [CrossRef]
30. Ulhaq, M.; Örn, S.; Carlsson, G.; Morrison, D.; Norrgren, L. Locomotor behavior in zebrafish (*Danio rerio*) larvae exposed to perfluoroalkyl acids. *Aquat. Toxicol.* **2013**, *144–145*, 332–340. [CrossRef]
31. Gaballah, S.; Swank, A.; Sobus, J.; Howey, X.; Schmid, J.; Catron, T.; McCord, J.; Hines, E.; Strynar, M.; Tal, T. Evaluation of developmental toxicity, developmental neurotoxicity, and tissue dose in zebrafish exposed to GenX and other PFAS. *Environ. Health Perspect.* **2020**, *128*, 047005. [CrossRef]
32. Khezri, A.; Fraser, T.; Nourizadeh-Lillabadi, R.; Kamstra, J.; Berg, V.; Zimmer, K.; Ropstad, E. A mixture of persistent organic pollutants and perfluorooctanesulfonic acid induces similar behavioural responses, but different gene expression profiles in zebrafish. *Int. J. Mol. Sci.* **2017**, *18*, 291. [CrossRef]
33. Chen, J.; Das, S.; La Du, J.; Corvi, M.; Bai, C.; Chen, Y.; Liu, X.; Zhu, G.; Tanguay, R.; Dong, Q.; et al. Chronic PFOS exposures induce life stage-specific behavioral deficits in adult zebrafish and produce malformation and behavioral deficits in F1 off-spring. *Environ. Toxicol. Chem.* **2012**, *32*, 201–206. [CrossRef]
34. Spulber, S.; Kilian, P.; Wan Ibrahim, W.; Onishchenko, N.; Ulhaq, M.; Norrgren, L.; Negri, S.; Di Tuccio, M.; Ceccatelli, S. PFOS induces behavioral alterations, including spontaneous hyperactivity that is corrected by dexamfetamine in zebrafish larvae. *PLoS ONE* **2014**, *9*, e94227. [CrossRef] [PubMed]

35. Huang, H.; Huang, C.; Wang, L.; Ye, X.; Bai, C.; Simonich, M.; Tanguay, R.; Dong, Q. Toxicity, uptake kinetics and behavior assessment in zebrafish embryos following exposure to perfluorooctanesulphonicacid (PFOS). *Aquat. Toxicol.* **2010**, *98*, 139–147. [CrossRef] [PubMed]
36. Wang, M.; Chen, J.; Lin, K.; Chen, Y.; Hu, W.; Tanguay, R.; Huang, C.; Dong, Q. Chronic zebrafish PFOS exposure alters sex ratio and maternal related effects in F1 offspring. *Environ. Toxicol. Chem.* **2011**, *30*, 2073–2080. [CrossRef] [PubMed]
37. Stein, C.; Wolff, M.; Calafat, A.; Kato, K.; Engel, S. Comparison of polyfluoroalkyl compound concentrations in maternal serum and amniotic fluid: A pilot study. *Reprod. Toxicol.* **2012**, *34*, 312–316. [CrossRef] [PubMed]
38. Long, M.; Ghisari, M.; Kjeldsen, L.; Wielsøe, M.; Nørgaard-Pedersen, B.; Mortensen, E.; Abdallah, M.; Bonefeld-Jørgensen, E. Autism spectrum disorders, endocrine disrupting compounds, and heavy metals in amniotic fluid: A case-control study. *Mol. Autism* **2019**, *10*, 1–19. [CrossRef] [PubMed]
39. Jantzen, C.; Toor, F.; Annunziato, K.; Cooper, K. Effects of chronic perfluorooctanoic acid (PFOA) at low concentration on morphometrics, gene expression, and fecundity in zebrafish (*Danio rerio*). *Reprod. Toxicol.* **2017**, *69*, 34–42. [CrossRef] [PubMed]
40. ZFIN Gene: Si:dkey-14d8.7.2022. Available online: https://zfin.org/ZDB-GENE-041210-143#summary (accessed on 20 April 2022).
41. Marchak, A.; Grant, P.A.; Neilson, K.M.; Datta Majumdar, H.; Yaklichkin, S.; Johnson, D.; Moody, S.A. Wbp2nl has a developmental role in establishing neural and non-neural ectodermal fates. *Dev. Biol.* **2017**, *429*, 213–224. [CrossRef]
42. Freour, T.; Barragan, M.; Ferrer-Vaquer, A.; Rodríguez, A.; Vassena, R. WBP2NL/PAWP mRNA and protein expression in sperm cells are not related to semen parameters, fertilization rate, or reproductive outcome. *J. Assist. Reprod. Genet.* **2017**, *34*, 803–810. [CrossRef]
43. Dahlet, T.; Truss, M.; Frede, U.; Al Adhami, H.; Bardet, A.F.; Dumas, M.; Vallet, J.; Chicher, J.; Hammann, P.; Kottnik, S.; et al. E2F6 initiates stable epigenetic silencing of germline genes during embryonic development. *Nat. Commun.* **2021**, *12*, 3582. [CrossRef]
44. Tissue Cell Type-WBP2NL-The Human Protein Atlas. Proteinatlas.org. 2022. Available online: https://www.proteinatlas.org/ENSG00000183066-WBP2NL/tissue+cell+type (accessed on 15 April 2022).
45. Pierozan, P.; Jerneren, F.; Karlsson, O. Perfluorooctanoic acid (PFOA) exposure promotes proliferation, migration and invasion potential in human breast epithelial cells. *Arch. Toxicol.* **2018**, *92*, 1729–1739. [CrossRef]
46. Stanifer, J.; Stapleton, H.; Souma, T.; Wittmer, A.; Zhao, X.; Boulware, L. Perfluorinated chemicals as emerging envi-ronmental threats to kidney health. *Clin. J. Am. Soc. Nephrol.* **2018**, *13*, 1479–1492. [CrossRef] [PubMed]
47. Magnuson, J.; Bautista, N.; Lucero, J.; Lund, A.; Xu, E.; Schlenk, D.; Burggren, W.; Roberts, A. Exposure to crude oil induces retinal apoptosis and impairs visual function in fish. *Environ. Sci. Technol.* **2020**, *54*, 2843–2850. [CrossRef] [PubMed]
48. Lopez-Rodriguez, R.; Lantero, E.; Blanco-Kelly, F.; Avila-Fernandez, A.; Martin Merida, I.; del Pozo-Valero, M.; Perea-Romero, I.; Zurita, O.; Jiménez-Rolando, B.; Swafiri, S.; et al. RPE65-related retinal dystrophy: Mutational and phenotypic spectrum in 45 affected patients. *Exp. Eye Res.* **2021**, *212*, 108761. [CrossRef] [PubMed]
49. Daly, C.; Shine, L.; Heffernan, T.; Deeti, S.; Reynolds, A.; O'Connor, J.; Dillon, E.; Duffy, D.; Kolch, W.; Cagney, G.; et al. A brain-derived neurotrophic factor mimetic is sufficient to restore cone photoreceptor visual function in an inherited blindness model. *Sci. Rep.* **2017**, *7*, 11320. [CrossRef]
50. Ren, H.; Vallanat, B.; Nelson, D.; Yeung, L.; Guruge, K.; Lam, P.; Lehman-McKeeman, L.; Corton, J. Evidence for the in-volvement of xenobiotic-responsive nuclear receptors in transcriptional effects upon perfluoroalkyl acid exposure in diverse species. *Reprod. Toxicol.* **2009**, *27*, 266–277. [CrossRef]
51. Leeder, J.; Gaedigk, R.; Marcucci, K.; Gaedigk, A.; Vyhlidal, C.; Schindel, B.; Pearce, R. Variability of CYP3A7 expression in human fetal liver. *J. Pharmacol. Exp. Ther.* **2005**, *314*, 626–635. [CrossRef]
52. Lang, R.; Raffi, F. Dual-specificity phosphatases in immunity and infection: An update. *Int. J. Mol. Sci.* **2019**, *20*, 2710. [CrossRef]
53. U.S. Department of Health and Human Services. NTP monograph on immunotoxicity associated with exposure to per-fluorooctanoic acid (PFOA) or perfluorooctane sulfonate (PFOS). *Natl. Toxicol. Prog.* **2016**, 22–80.
54. Xin, Y.; Ren, X.; Wan, B.; Guo, L. Comparative in vitro and in vivo evaluation of the estrogenic effect of hexafluoropropylene oxide homologues. *Environ. Sci. Technol.* **2019**, *53*, 8371–8380. [CrossRef]
55. Cardenas, A.; Gold, D.; Hauser, R.; Kleinman, K.; Hivert, M.; Calafat, A.; Ye, X.; Webster, T.; Horton, E.; Oken, E. Plasma concentrations of per- and polyfluoroalkyl substances at baseline and associations with glycemic indicators and diabetes incidence among high-risk adults in the diabetes prevention program trial. *Environ. Health Perspect.* **2017**, *125*, 107001. [CrossRef]
56. Lin, P.; Cardenas, A.; Hauser, R.; Gold, D.; Kleinman, K.; Hivert, M.; Fleisch, A.; Calafat, A.; Webster, T.; Horton, E.; et al. Per- and polyfluoroalkyl substances and blood lipid levels in pre-diabetic adults—longitudinal analysis of the diabetes pre-vention program outcomes study. *Environ. Int.* **2019**, *129*, 343–353. [CrossRef] [PubMed]
57. Mancini, F.; Rajaobelina, K.; Praud, D.; Dow, C.; Antignac, J.; Kvaskoff, M.; Severi, G.; Bonnet, F.; Boutron-Ruault, M.; Fagherazzi, G. Nonlinear associations between dietary exposures to perfluorooctanoic acid (PFOA) or perfluorooctane sulfonate (PFOS) and type 2 diabetes risk in women: Findings from the E3N cohort study. *Int. J. Hyg. Environ. Health* **2018**, *221*, 1054–1060. [CrossRef] [PubMed]
58. Withers, D.; Burks, D.; Towery, H.; Altamuro, S.; Flint, C.; White, M. Irs-2 coordinates igf-1 receptor-mediated β-cell de-velopment and peripheral insulin signalling. *Nat. Genet.* **1999**, *23*, 32–40. [CrossRef] [PubMed]
59. Yang, B.; Zhai, G.; Gong, Y.; Su, J.; Peng, X.; Shang, G.; Han, D.; Jin, J.; Liu, H.; Du, Z.; et al. Different physiological roles of insulin receptors in mediating nutrient metabolism in zebrafish. *Am. J. Physiol. Endocrinol. Metab.* **2018**, *315*, E38–E51. [CrossRef]

60. Manchenkov, T.; Pasillas, M.; Haddad, G.; Imam, F. Novel genes critical for hypoxic preconditioning in zebrafish are reg-ulators of insulin and glucose metabolism. *G3 Genes Genomes Genet.* **2015**, *5*, 1107–1116. [CrossRef]
61. Raghupathy, R.; Zhang, X.; Alhasani, R.; Zhou, X.; Mullin, M.; Reilly, J.; Li, W.; Liu, M.; Shu, X. Abnormal photoreceptor outer segment development and early retinal degeneration inkif3a mutant zebrafish. *Cell Biochem. Funct.* **2016**, *34*, 429–440. [CrossRef]
62. Du, G.; Huang, H.; Hu, J.; Qin, Y.; Wu, D.; Song, L.; Xia, Y.; Wang, X. Endocrine-related effects of perfluorooctanoic acid (PFOA) in zebrafish, H295R steroidogenesis and receptor reporter gene assays. *Chemosphere* **2013**, *91*, 1099–1106. [CrossRef]
63. Thisse, B.; Pflumio, S.; Fürthauer, M.; Loppin, B.; Heyer, V.; Degrave, A.; Woehl, R.; Lux, A.; Steffan, T.; Charbonnier, X.Q.; et al. Expression of the Zebrafish Genome during Embryogenesis (NIH R01 RR15402). ZFIN Direct Data Submission. 2001. Available online: https://www.scienceopen.com/document?vid=f98e6bdd-d74f-4d33-8aa4-4656a744c451 (accessed on 26 February 2022).
64. Klarić, T.; Lardelli, M.; Key, B.; Koblar, S.; Lewis, M. Activity-dependent expression of neuronal pas domain-containing protein 4 (npas4a) in the developing zebrafish brain. *Front. Neuroanat.* **2014**, *8*, 148. [CrossRef]
65. Cai, S.; Han, H.; Kohwi-Shigematsu, T. Tissue-specific nuclear architecture and gene expression regulated by SATB1. *Nat. Genet.* **2003**, *34*, 42–51. [CrossRef]
66. Yasui, D.; Miyano, M.; Cai, S.; Varga-Weisz, P.; Kohwi-Shigematsu, T. SATB1 targets chromatin remodelling to regulate genes over long distances. *Nature* **2002**, *419*, 641–645. [CrossRef]
67. Martínez, R.; Navarro-Martín, L.; Luccarelli, C.; Codina, A.; Raldúa, D.; Barata, C.; Tauler, R.; Piña, B. Unravelling the mechanisms of PFOS toxicity by combining morphological and transcriptomic analyses in zebrafish embryos. *Sci. Total Environ.* **2019**, *674*, 462–471. [CrossRef] [PubMed]
68. Dong, Z.; Wang, H.; Yu, Y.; Li, Y.; Naidu, R.; Liu, Y. Using 2003–2014 U.S. NHANES data to determine the associations between per- and polyfluoroalkyl substances and cholesterol: Trend and implications. *Ecotoxicol. Environ. Saf.* **2019**, *173*, 461–468. [CrossRef] [PubMed]
69. Nelson, J.; Hatch, E.; Webster, T. Exposure to polyfluoroalkyl chemicals and cholesterol, body weight, and insulin resistance in the general U.S. population. *Environ. Health Perspect.* **2010**, *118*, 197–202. [CrossRef] [PubMed]
70. Eriksen, K.; Raaschou-Nielsen, O.; McLaughlin, J.; Lipworth, L.; Tjønneland, A.; Overvad, K.; Sørensen, M. Association between plasma PFOA and PFOS levels and total cholesterol in a middle-aged Danish population. *PLoS ONE* **2013**, *8*, e56969. [CrossRef] [PubMed]
71. Thisse, B.; Thisse, C. Fast Release Clones: A High Throughput Expression Analysis. ZFIN Direct Data Submission. 2004. Available online: https://zfin.org/ZDB-PUB-040907-1 (accessed on 26 February 2022).
72. Lum, K.; Sundaram, R.; Barr, D.; Louis, T.; Buck Louis, G. Perfluoroalkyl chemicals, menstrual cycle length, and fecundity. *Epidemiology* **2017**, *28*, 90–98. [CrossRef] [PubMed]
73. Fei, C.; McLaughlin, J.; Lipworth, L.; Olsen, J. Maternal levels of perfluorinated chemicals and subfecundity. *Hum. Reprod.* **2009**, *24*, 1200–1205. [CrossRef] [PubMed]
74. Whitworth, K.; Haug, L.; Baird, D.; Becher, G.; Hoppin, J.; Skjaerven, R.; Thomsen, C.; Eggesbo, M.; Travlos, G.; Wilson, R.; et al. Perfluorinated compounds in relation to birth weight in the Norwegian mother and child cohort study. *Am. J. Epidemiol.* **2012**, *175*, 1209–1216. [CrossRef]
75. Velez, M.; Arbuckle, T.; Fraser, W. Maternal exposure to perfluorinated chemicals and reduced fecundity: The MIREC study. *Hum. Reprod.* **2015**, *30*, 701–709. [CrossRef]
76. Vestergaard, S.; Nielsen, F.; Andersson, A.; Hjollund, N.; Grandjean, P.; Andersen, H.; Jensen, T. Association between perfluori-nated compounds and time to pregnancy in a prospective cohort of Danish couples attempting to conceive. *Hum. Reprod.* **2012**, *27*, 873–880. [CrossRef]
77. Jørgensen, K.; Specht, I.; Lenters, V.; Bach, C.; Rylander, L.; Jönsson, B.; Lindh, C.; Giwercman, A.; Heederik, D.; Toft, G.; et al. Perfluoroalkyl substances and time to pregnancy in couples from Greenland, Poland and Ukraine. *Environ. Health* **2014**, *13*, 116. [CrossRef]
78. Seyoum, A.; Pradhan, A.; Jass, J.; Olsson, P. Perfluorinated alkyl substances impede growth, reproduction, lipid metabolism and lifespan in Daphnia magna. *Sci. Total Environ.* **2020**, *737*, 139682. [CrossRef] [PubMed]
79. Bangma, J.; Reiner, J.; Lowers, R.; Cantu, T.; Scott, J.; Korte, J.; Scheidt, D.; McDonough, C.; Tucker, J.; Back, B.; et al. Per-fluorinated alkyl acids and fecundity assessment in striped mullet (*Mugil cephalus*) at Merritt Island National Wildlife Refuge. *Sci. Total Environ.* **2018**, *619–620*, 740–747. [CrossRef] [PubMed]
80. Marziali, L.; Rosignoli, F.; Valsecchi, S.; Polesello, S.; Stefani, F. Effects of perfluoralkyl substances on a multigenerational scale: A case study with Chironomus riparius (Diptera, Chironomidae). *Environ. Toxicol. Chem.* **2019**, *38*, 988–999. [CrossRef] [PubMed]
81. Suski, J.; Salice, C.; Chanov, M.; Ayers, J.; Rewerts, J.; Field, J. Sensitivity and accumulation of perfluorooctanesulfonate and perfluorohexanesulfonic acid in fathead minnows (*Pimephales promelas*) exposed over critical life stages of reproduction and development. *Environ. Toxicol. Chem.* **2021**, *40*, 811–819. [CrossRef]
82. Lee, J.; Lee, J.; Shin, Y.; Kim, J.; Ryu, T.; Ryu, J.; Lee, J.; Kim, P.; Choi, K.; Park, K. Multi-generational xenoestrogenic effects of perfluoroalkyl acids (pfaas) mixture on Oryzias latipes using a flow-through exposure system. *Chemosphere* **2017**, *169*, 212–223. [CrossRef]
83. Nagabhushana, A.; Mishra, R. Finding clues to the riddle of sex determination in zebrafish. *J. Biosci.* **2016**, *41*, 145–155. [CrossRef]

84. Chen, L.; Lam, J.; Hu, C.; Tsui, M.; Lam, P.; Zhou, B. Perfluorobutanesulfonate exposure skews sex ratio in fish and transgenerationally impairs reproduction. *Environ. Sci. Technol.* **2019**, *53*, 8389–8397. [CrossRef]
85. Shi, G.; Guo, H.; Sheng, N.; Cui, Q.; Pan, Y.; Wang, J.; Guo, Y.; Dai, J. Two-generational reproductive toxicity assessment of 6:2 chlorinated polyfluorinated ether sulfonate (F-53B, a novel alternative to perfluorooctane sulfonate) in zebrafish. *Environ. Pollut.* **2018**, *243*, 1517–1527. [CrossRef]

Article

Developmental Neurotoxicity and Behavioral Screening in Larval Zebrafish with a Comparison to Other Published Results

Kimberly A. Jarema [1,*], Deborah L. Hunter [2], Bridgett N. Hill [3], Jeanene K. Olin [2], Katy N. Britton [4], Matthew R. Waalkes [5,†] and Stephanie Padilla [2,*]

[1] Center for Public Health and Environmental Assessment, Immediate Office, Program Operations Staff, U.S. Environmental Protection Agency, Research Triangle Park, NC 27711, USA
[2] Center for Computational Toxicology and Exposure, Biomolecular and Computational Toxicology Division, Rapid Assay Development Branch, U.S. Environmental Protection Agency, Research Triangle Park, NC 27711, USA; hunter.deborah@epa.gov (D.L.H.); olin.jeanene@epa.gov (J.K.O.)
[3] ORISE Research Participation Program Hosted by EPA, Center for Computational Toxicology and Exposure, Biomolecular and Computational Toxicology Division, Rapid Assay Development Branch, U.S. Environmental Protection Agency, Research Triangle Park, NC 27711, USA; hill.bridgett@epa.gov
[4] ORAU Research Participation Program Hosted by EPA, Center for Computational Toxicology and Exposure, Biomolecular and Computational Toxicology Division, Rapid Assay Development Branch, U.S. Environmental Protection Agency, Research Triangle Park, NC 27711, USA; britton.katy@epa.gov
[5] ORISE Research Participation Program Hosted by EPA, National Health and Environmental Effects Research Laboratory, Integrated Systems Toxicology Division, Genetic and Cellular Toxicology Branch, U.S. Environmental Protection Agency, Research Triangle Park, NC 27711, USA; mrw0051@mix.wvu.edu
* Correspondence: jarema.kimberly@epa.gov (K.A.J.); padilla.stephanie@epa.gov (S.P.); Tel.: +1-919-541-2299 (K.A.J.); +1-919-541-3650 (S.P.)
† Present Address: West Virginia Department of Biology, Morgantown, WV 26505, USA.

Citation: Jarema, K.A.; Hunter, D.L.; Hill, B.N.; Olin, J.K.; Britton, K.N.; Waalkes, M.R.; Padilla, S. Developmental Neurotoxicity and Behavioral Screening in Larval Zebrafish with a Comparison to Other Published Results. *Toxics* **2022**, *10*, 256. https://doi.org/10.3390/toxics10050256

Academic Editors: Kimberly Keil Stietz and Tracie Baker

Received: 1 April 2022
Accepted: 7 May 2022
Published: 17 May 2022

Publisher's Note: MDPI stays neutral with regard to jurisdictional claims in published maps and institutional affiliations.

Copyright: © 2022 by the authors. Licensee MDPI, Basel, Switzerland. This article is an open access article distributed under the terms and conditions of the Creative Commons Attribution (CC BY) license (https:// creativecommons.org/licenses/by/ 4.0/).

Abstract: With the abundance of chemicals in the environment that could potentially cause neurodevelopmental deficits, there is a need for rapid testing and chemical screening assays. This study evaluated the developmental toxicity and behavioral effects of 61 chemicals in zebrafish (*Danio rerio*) larvae using a behavioral Light/Dark assay. Larvae (n = 16–24 per concentration) were exposed to each chemical (0.0001–120 µM) during development and locomotor activity was assessed. Approximately half of the chemicals (n = 30) did not show any gross developmental toxicity (i.e., mortality, dysmorphology or non-hatching) at the highest concentration tested. Twelve of the 31 chemicals that did elicit developmental toxicity were toxic at the highest concentration only, and thirteen chemicals were developmentally toxic at concentrations of 10 µM or lower. Eleven chemicals caused behavioral effects; four chemicals (6-aminonicotinamide, cyclophosphamide, paraquat, phenobarbital) altered behavior in the absence of developmental toxicity. In addition to screening a library of chemicals for developmental neurotoxicity, we also compared our findings with previously published results for those chemicals. Our comparison revealed a general lack of standardized reporting of experimental details, and it also helped identify some chemicals that appear to be consistent positives and negatives across multiple laboratories.

Keywords: behavior; chemical screening; literature comparison; developmental toxicity; developmental neurotoxicity; negative control; positive control; rapid testing; zebrafish

1. Introduction

The incidence of neurodevelopmental deficits in children is steadily increasing (reviewed in [1,2]), accompanied by warnings from many scientific fronts regarding the possible adverse effects of environmental chemicals on nervous system development [3–5]. The evidence that chemicals may alter the trajectory of brain development has led to heightened awareness of the need for rapid testing of environmental chemicals for developmental neurotoxicity potential. An experimental model that appears to hold promise is a small, hardy, aquarium fish: zebrafish (*Danio rerio*). Elegant work in zebrafish has been published on the development of the nervous system, neuronal pathfinding, myelination, and the genetic or structural basis of nervous system function (e.g., [6,7]). Because of the concordance between zebrafish and human developmental and neurodevelopmental pathways, zebrafish are now used to discover mechanisms, and possibly treatments, of neurological diseases [8–11]. Many in vitro tests have been developed to assess specific aspects of brain development, but because brain development is complicated, with pre-described windows of migration and connectivity orchestrated by endocrine crosstalk and feedback, a whole animal model is often part of many developmental neurotoxicity screening batteries.

Behavioral assessment, regarded as a functional endpoint, is an integrative signal representing nervous system status or fitness [12–14]. Not only are larval zebrafish able to exhibit many different behaviors [15,16] but, analogous to mammalian neurodevelopment, the development of the zebrafish nervous system is guided and influenced by the interplay among brain development and endocrine systems such as the hypothalamus-pituitary-thyroid (HPT) axis [17–19] and the hypothalamo-pituitary-adrenal (inter-renal in zebrafish) (HPA/HPI) axis (reviewed in [20]). Moreover, zebrafish at all developmental stages metabolize toxic chemicals using pathways similar to mammals [21,22]. Because of these attributes, zebrafish are often proffered as a model for developmental neurotoxicity screening (reviewed in [23–25]).

Using the zebrafish model, we had two main goals for this study: to screen a library of chemicals for developmental neurotoxicity, and to compare our findings with previously published results for those chemicals. The specific chemical library was chosen because (1) some of the chemicals have been associated with developmental neurotoxicity in mammals [26]; (2) many of the chemicals were tested by other investigators within the U.S. Environmental Protection Agency (EPA) using in vitro assays for developmental neurotoxicity potential [27–29]; and/or (3) some of the chemicals have been tested by investigators external to EPA using zebrafish assays [30–63]. The first aspect of this study was to screen the library of 61 chemicals in zebrafish embryos/larvae to determine if the chemical (maximum nominal concentration = 120 µM) produced developmental toxicity (lethality, non-hatching or malformations) and/or neurotoxicity (changes in larval locomotor activity). The second aspect of this study was to compare our results with the results from other laboratories performing similar behavioral assays with larval zebrafish treated with the same chemicals during development.

2. Materials and Methods
2.1. Chemicals

Table 1 lists information about the chemicals used in this study. The chemical name, CAS number, DTXSID, molecular weight, and solvent (vehicles) are included. Also included are the predicted median and range of water solubility, as well as the predicted median and range of the octanol/water partition coefficient, all of which were obtained from the EPA's Chemicals Dashboard (https://comptox.epa.gov/dashboard/; last accessed on 31 January 2022). For the creation of stock plates, stock solutions of each chemical were prepared in their respective vehicles, either dimethyl sulfoxide (DMSO; Anhydrous (>99.9% pure) from Sigma-Aldrich] or deionized water, which were then used for subsequent serial dilutions for dosing of the experimental plates. Chlorpyrifos [ethyl; CAS# 39475-55-3] served as the positive control for behavioral alterations [64]. The highest nominal concentration tested of any chemical was 120 µM because human plasma rarely exceeds micromolar levels of most environmental chemicals.

Table 1. List of Chemicals Tested. Each chemical is listed by row. The columns (from left to right) contain the following information on a given chemical: name; CAS number; DTXSID; molecular weight; solvent in which the chemical was dissolved; predicted median and range of water solubility; and predicted median and range of the octanol/water partition coefficient. Values obtained from https://comptox.epa.gov/dashboard/ Last Accessed: 25 January 2022.

Chemical	Cas #	DTXSID	Molecular Weight	Solvent	Water Solubility (µmol/L) Predicted Median	Water Solubility (µmol/L) Predicted Range	Octanol Water Coeff (LogKow) Predicted Median	Octanol Water Coeff (LogKow) Predicted Range
5,5-Diphenylhydantoin	57-41-0	DTXSID8020541	252.3	DMSO	5.67×10^4	1.07×10^2 to 6.00×10^6	2.39	2.16 to 2.52
5-Fluorouracil	51-21-8	DTXSID2020634	130.1	DMSO	1.44×10^5	3.07×10^4 to 3.69×10^6	−0.906	−1.37 to −0.810
6-Aminonicotinamide	329-89-5	DTXSID5051446	137.1	DMSO	6.68×10^4	6.41×10^4 to 1.28×10^5	0.027	−0.730 to 0.698
6-Propyl-2-thiouracil	51-52-5	DTXSID5021209	170.2	DMSO	3.00×10^4	6.93×10^3 to 4.98×10^6	0.523	−0.386 to 1.37
Acetaminophen	103-90-2	DTXSID2020006	151.1	DMSO	1.47×10^5	3.95×10^4 to 5.70×10^6	0.372	0.270 to 0.462
Acrylamide	79-06-1	DTXSID5020027	71.1	DMSO	7.05×10^6	2.66×10^6 to 8.99×10^6	−0.726	−0.810 to −0.670
Aldicarb	116-06-3	DTXSID0039223	190.3	DMSO	2.79×10^4	2.55×10^4 to 3.03×10^4	1.13	1.13 to 1.36
Amoxicillin	26787-78-0	DTXSID3037044	365.4	DMSO	9.36×10^3	5.58×10^3 to 4.93×10^6	0.742	0.48 to 0.97
Amphetamine	51-63-8	DTXSID2057865	184.3	DMSO	5.70×10^6	1.33×10^5 to 1.13×10^7	1.81	0.602 to 1.82
Arsenic	7784-46-5	DTXSID5020104	129.9	H_2O	–	–	−3.28	−3.28
Bisphenol A (BPA)	80-05-7	DTXSID7020182	228.3	DMSO	1.00×10^3	7.45×10^2 to 6.76×10^6	3.53	3.32 to 3.64
Bis(tributyltin) Oxide	56-35-9	DTXSID9020166	596.1	DMSO	1.5×10^{-1}	1.5×10^{-1}	4.05	4.05
Cadmium chloride	654054-66-7	–	183.3	DMSO	–	–	–	–
Caffeine	58-08-2	DTXSID0020232	194.2	DMSO	8.30×10^4	1.36×10^4 to 7.14×10^6	0.045	−0.131 to 0.283
Captopril	62571-86-2	DTXSID1037197	217.2	DMSO	9.47×10^4	3.98×10^5 to 2.46×10^6	0.481	0.272 to 0.840
Carbamazepine	298-46-4	DTXSID4022731	236.3	DMSO	2.83×10^2	2.55×10^1 to 7.00×10^6	2.37	2.25 to 2.67
Chloramben	133-90-4	DTXSID2020262	206.0	DMSO	3.40×10^3	2.92×10^3 to 4.68×10^3	2.15	0.912 to 2.52
Chlorpyrifos (ethyl)	2921-88-2	DTXSID4020458	350.6	DMSO	2.83	1.02 to 7.00×10^6	4.78	4.66 to 4.96
Chlorpyrifos (ethyl) oxon	5598-15-2	DTXSID1038666	334.5	DMSO	2.10×10^2	7.76×10^1 to 2.26×10^2	3.32	2.89 to 3.73
Cocaine base	50-36-2	DTXSID2038443	184.3	H_2O	4.93×10^6	5.73×10^3 to 9.85×10^6	2.79	2.3 to 3.08
Colchicine	64-86-8	DTXSID2024845	399.4	DMSO	5.65×10^4	5.25×10^2 to 7.00×10^6	1.2	0.920 to 1.86
Cotinine	486-56-6	DTXSID1047576	176.2	DMSO	2.99×10^6	3.70×10^4 to 9.02×10^6	0.119	−0.228 to 0.340
Cyclophosphamide	6055-19-2	DTXSID6024888	279.1	H_2O	1.52×10^5	5.58×10^4 to 8.02×10^6	0.526	0.230 to 1.30
Cytosine arabinoside	147-94-4	DTXSID3022877	243.2	DMSO	4.54×10^5	4.39×10^4 to 8.32×10^6	−2.32	−2.51 to −1.94
Deltamethrin	52918-63-5	DTXSID8020381	505.2	DMSO	1.96×10^{-2}	1.86×10^{-3} to 7.00×10^6	6.19	6.12 to 6.20
Dexamethasone	50-02-2	DTXSID3020384	392.4	DMSO	1.95×10^2	1.05×10^2 to 7.00×10^6	1.89	1.72 to 1.92
Di(2-ethylhexyl)phthalate (DEHP)	117-81-7	DTXSID5020607	390.6	DMSO	4.23×10^{-1}	2.90×10^{-3} to 7.00×10^6	8.15	7.52 to 8.71
Diazepam	439-14-5	DTXSID4020406	284.7	DMSO	1.91×10^2	1.07×10^2 to 7.07×10^6	2.91	2.70 to 2.92
Dieldrin	60-57-1	DTXSID9020453	380.9	DMSO	1.57	5.42×10^{-1} to 2.60	4.94	4.88 to 5.12
Diethylene Glycol	111-46-6	DTXSID8020462	106.1	DMSO	6.51×10^6	5.40×10^6 to 9.42×10^6	−1.28	−1.51 to −1.09

39

Table 1. Cont.

Chemical	Cas #	DTXSID	Molecular Weight	Solvent	Water Solubility (μmol/L) Predicted Median	Water Solubility (μmol/L) Predicted Range	Octanol Water Coeff (LogKow) Predicted Median	Octanol Water Coeff (LogKow) Predicted Range
Diethyl-stilbesterol	56-53-1	DTXSID3020465	268.4	DMSO	4.37×10^1	1.24×10^1 to 6.88×10^6	5.35	4.80 to 5.93
D-sorbitol	50-70-4	DTXSID5023588	182.2	DMSO	3.31×10^6	1.72×10^6 to 6.07×10^6	−3.15	−4.67 to −2.38
Fluconazole	86386-73-4	DTXSID3020627	306.2	DMSO	9.68×10^3	1.35×10^3 to 7.15×10^6	0.501	0.250 to 0.698
Fluoxetine	56296-78-7	DTXSID7020635	345.8	DMSO	1.94×10^2	2.37×10^1 to 1.02×10^7	4.09	0.768 to 4.23
Glyphosate	1071-83-6	DTXSID1024122	169.1	H$_2$O	1.99×10^6	6.56×10^4 to 8.41×10^6	−2.88	−4.47 to −2.26
Haloperidol	52-86-8	DTXSID4034150	375.9	DMSO	3.10×10^1	2.34×10^1 to 9.11×10^6	3.84	3.01 to 4.29
Heptachlor	76-44-8	DTXSID3020679	373.3	DMSO	9.25×10^{-2}	7.39×10^{-2} to 3.82×10^{-1}	5.7	5.46 to 6.10
Heptachlor epoxide	1024-57-3	DTXSID1024126	389.3	DMSO	5.68×10^{-1}	5.68×10^{-1}	5.29	4.98 to 5.47
Hexachlorophene	70-30-4	DTXSID6020690	406.9	DMSO	1.73×10^2	9.43×10^{-3} to 6.66×10^6	7.23	6.92 to 7.54
Hydoxy-urea	127-07-1	DTXSID6025438	76.1	DMSO	5.42×10^6	2.95×10^6 to 1.32×10^7	−1.74	−1.80 to −1.54
Isoniazid	54-85-3	DTXSID8020755	137.1	DMSO	7.66×10^5	1.22×10^6 to 7.32×10^6	0.754	−0.887 to −0.635
Lead acetate	6080-56-4	DTXSID3031521	379.3	H$_2$O	7.77×10^6	2.14×10^6 to 1.34×10^7	−0.285	−2.21 to −7.10 × 10^{-2}
Loperamide	34552-83-5	DTXSID00880006	513.5	DMSO	4.46×10^6	2.21×10^1 to 8.91×10^6	4.26	1.32 to 4.47
Maneb	12427-38-2	DTXSID9020794	265.3	DMSO	1.01×10^6	7.72×10^5 to 1.25×10^6	1.4	−2.70 to 1.66
Manganese	7773-01-5	DTXSID9040681	126.0	H$_2$O				
Methotrexate	59-05-2	DTXSID4020822	454.4	DMSO	3.2×10^3	1.89×10^2 to 5.37×10^6	−0.922	−1.85 to −0.241
Naloxon	51481-60-8	DTXSID90199452	399.9	H$_2$O	4.06×10^3	2.74×10^3 to 7.99×10^6	1.45	0.243 to 1.53
Nicotine	54-11-5	DTXSID1020930	162.2	DMSO	6.15×10^6	8.00×10^4 to 1.10×10^7	0.91	0.720 to 1.17
Paraquat	1910-42-5	DTXSID7024243	257.1	H$_2$O	4.88×10^6	2.76×10^6 to 7.00×10^6	−4.58	−5.11 to −4.50
Permethrin	52645-53-1	DTXSID8022292	391.2	DMSO	1.32×10^{-1}	2.49×10^{-2} to 7.00×10^6	6.82	6.47 to 7.43
Phenobarbital	57-30-7	DTXSID0021123	254.2	DMSO	1.68×10^4	1.02×10^4 to 3.89×10^5	−0.285	−2.29 to 1.13
Phenol	108-95-2	DTXSID5021124	94.1	DMSO	6.04×10^5	2.78×10^5 to 4.91×10^6	1.5	1.46 to 1.63
Polybrominated diphenyl ether (PBDE)-47	5436-43-1	DTXSID3030056	485.8	DMSO	5.61×10^{-3}	3.01×10^{-3} to 1.23×10^{-1}	6.79	6.59 to 7.39
Saccharin	82385-42-0	DTXSID7021992	205.1	DMSO	1.91×10^4	9.43×10^3 to 1.85×10^6	0.705	−2.01 to 0.910
Sodium benzoate	532-32-1	DTXSID1020140	144.1	H$_2$O	3.32×10^5	6.44×10^4 to 2.84×10^6	0.158	−2.27 to 1.90
Sodium fluoride	7681-49-4	DTXSID2020630	42.0	H$_2$O	1.42×10^7	1.42×10^7	−0.77	−0.77
Tebuconazole	107534-96-3	DTXSID9032113	307.8	DMSO	1.03×10^2	8.04×10^1 to 7.09×10^5	3.72	3.58 to 3.89
Terbutaline	23031-32-5	DTXSID3045437	274.3	DMSO	4.71×10^6	4.63×10^4 to 9.37×10^6	0.477	0.439 to 0.523
Thalidomide	50-35-1	DTXSID9022524	258.2	DMSO	1.74×10^3	6.49×10^2 to 6.42×10^6	0.405	−0.240 to 0.541
Triethyltin	2767-54-6	DTXSID9040712	285.8	DMSO	1.38×10^3	1.38×10^3	1.84	1.84
Valproate	99-66-1	DTXSID6023733	144.2	DMSO	1.99×10^4	6.20×10^3 to 3.33×10^6	2.73	2.65 to 2.96

The primary medium for rearing the embryos was 10% Hanks' Balanced Salt Solution (13.7 mM NaCl, 0.54 mM KCl, 25 µM Na2HPO4, 44 µM KH2PO4, 130 µM CaCl2, 100 µM MgSO4, and 420 µM NaHCO3; pH = 7.6 ± 0.2; all salts obtained from Sigma-Aldrich, St. Louis, MO; hereafter referred to as 10% Hanks'). The lead (Pb) exposed larvae were not exposed in 10% Hanks' solution because of concerns about possible precipitation of the lead in that solution. Rather, larvae exposed to lead were reared in 1X EPA Moderately Hard Reconstituted Water (MHW: 54 µM KCl, 0.5 mM MgSO4·7H2O, 1.1 mM NaHCO3, 350 µM CaSO4; hereafter referred to as MHW). We have previously shown that control animals reared in either Hanks' solution or MHW do not differ in their locomotor activity [65].

2.2. DMSO Evaluation

Some publications [66–68] have noted that exposure to DMSO at very low concentrations can affect larval zebrafish behavior. Therefore, we determined if the vehicle concentration (0.4% DMSO) in our developmental exposure regimen caused any behavioral changes in 6 days post fertilization (dpf) larvae tested using our behavioral protocol. The experiment was conducted under the same experimental conditions described below with both DMSO exposed and non-DMSO exposed animals on the same microtiter plate. For non-DMSO exposed animals, water was added in place of the DMSO. The results presented in Supplemental Figure S1 show no effect of DMSO exposure during development on the behavior of the zebrafish larvae.

2.3. Experimental Animals

All studies were carried out in accordance with the guidelines of, and approved by, the Office of Research and Development's Institutional Animal Care and Use Committee (IACUC) at the U.S. Environmental Protection Agency (EPA) in Research Triangle Park, NC, USA.

In-house, wild type adult zebrafish (*Danio rerio*) descended from undefined, outbred stock originally obtained from Aquatic Research Organisms (Hampton, NH, USA) and EkkWill Waterlife Resources (Ruskin, FL, USA) were used. Each year, as replacement breeders are reared, embryos of a new strain are mixed with the in-house strain to maintain the outbred status of the colony. Animals were housed in an American Association for Accreditation of Laboratory Animal Care (AAALAC) approved animal facility with a 14:10 h light/dark cycle (lights on at 0700 h). Adult fish were kept in flow-through colony tanks (Tecniplast USA, West Chester, PA or Aquaneering Inc., San Diego, CA, USA) with a water temperature of 28 °C. The system water is composed of Durham, NC city tap water that is purified via reverse osmosis and buffered with sea salt (Instant Ocean, Spectrum Brands, Blacksburg, VA, USA) and sodium bicarbonate (Church & Dwight Co., Ewing, NJ, USA). This water is maintained at pH 7.4, conductivity of 1000 µS/cm, with negligible ammonia and nitrate/nitrite present. For egg collection, adults from colony tanks were placed in a 2-L (static) breeding tank (Aquatic Habitats, Apopka, FL, USA) the night prior to embryo collection. At 0730 h the following morning, approximately 30 min after the light illumination, eggs were collected.

2.4. Experimental Procedure

The Experimental Procedure is outlined in Figure 1 and explained in detail below. In the conduct and analysis of our behavioral assay, it was important that developmental neurotoxicity rather than the pharmacological effects of each chemical was assessed. To accomplish this, our experimental procedure included removal of the chemical from the dosing solution 24 h before testing and replacing the test chemical with a vehicle solution. We have previously shown that this removal of the test chemical markedly alters the behavioral profile, separating neuroactive from neurodevelopmental effects [64], although it is possible that this depuration time interval may not be long enough for all chemicals. We also wanted to limit the possibility that morphological changes alter the swimming behavior of the larvae, as this would seriously confound the interpretation of behavioral changes. We are assuming that any changes in swimming activity during the behavioral assessment is due to nervous system function and not changes in physical locomotor ability precipitated by teratological changes. To accomplish this, each animal was carefully assessed for any morphological changes, including swim bladder inflation as swim bladder inflation status has been shown to affect behavioral endpoints [69,70].

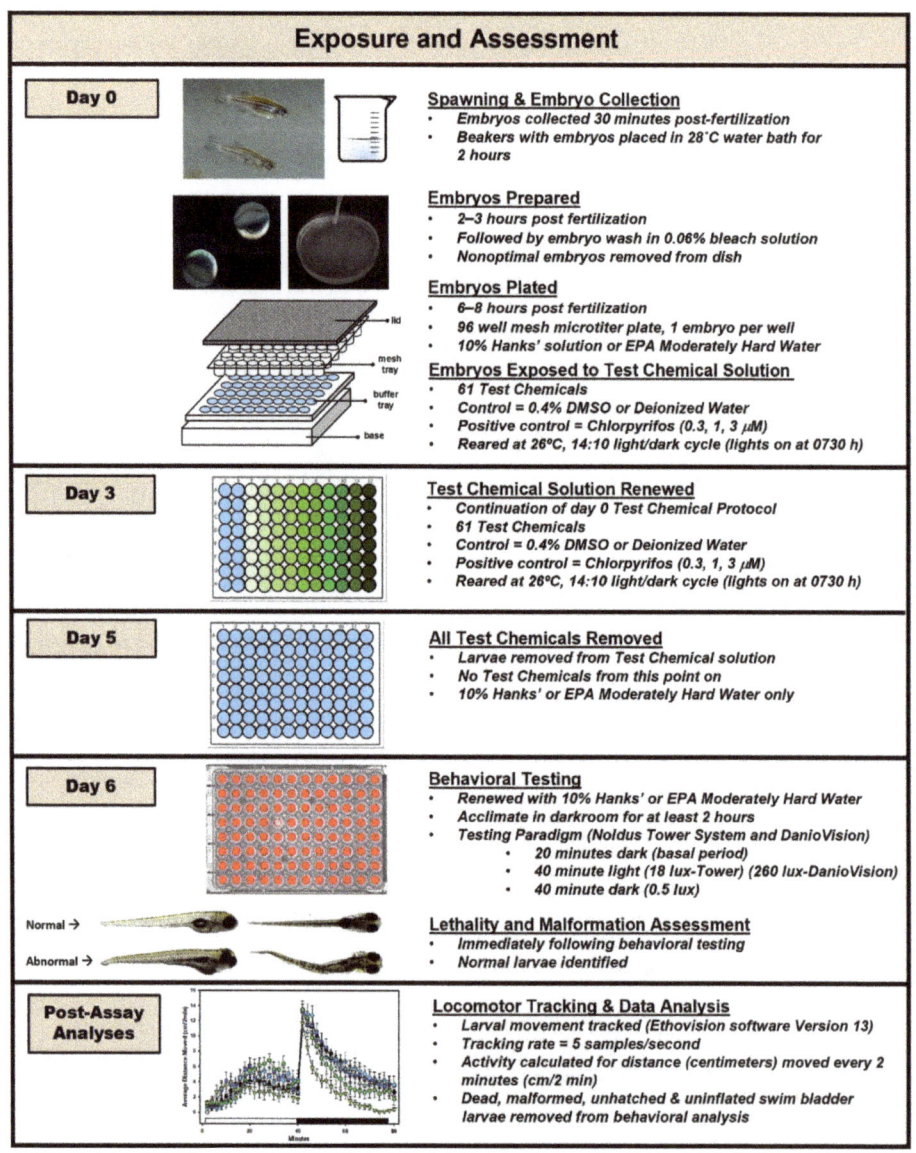

Figure 1. Experimental Design. Detailed timeline of the experimental process from spawning to analyses. The diagram is divided into sections for each critical time period, then further subdivided for event.

2.5. General Embryo Rearing

Newly collected embryos were washed with a dilute bleach solution shortly after collection. This process consisted of submerging the embryos in 0.06% bleach (v/v) in 10% Hanks' two times, for five minutes each, and then briefly rinsing in 10% Hanks' 3 times after each bleach wash [71]. Healthy, normal appearing embryos were individually placed, with their intact chorion (i.e., embryos were not dechorionated), into the upper mesh insert of a 96-well microtiter plate (Multiscreen™, Millipore Sigma, Burlington, MA, USA), which was submerged in a receiver plate containing 10% Hanks' solution.

2.6. Chemical Exposure

After plating (6–8 h post fertilization (hpf)), embryos were immersed in the appropriate chemical solution. To accomplish this, the upper mesh insert containing the embryos was blotted on glass fiber filter paper (Whatman GF/B paper (fired) Brandel, Gaithersburg, MD) and placed in the new 96-well receiving plate, which contained the appropriate chemical concentration. To dilute the chemicals, 1 µL from the stock plate was added per well to the receiving plate containing 150 µL of 10% Hanks', followed by an additional 100 µL of 10% Hanks' solution after the transfer of mesh insert. All concentrations of each chemical, along with vehicle controls, were included on every plate. Each plate was sealed with a non-adhesive material (Microseal® A, BioRad, Hercules, CA, USA), covered with a lid, and wrapped in Parafilm™ to secure the lid to the plate. The treated embryo plates were placed in a secondary container in the incubator (Lab-Line Imperial III, Barnstead International, Dubuque, IA, USA) and reared for 6 days at 26 °C under a 14:10 light:dark cycle (lights on at 0730 h). In addition to day 0, the 250 µL of 10% Hanks' solution, along with the appropriate chemical and concentration, in each well was completely renewed on 3 dpf (as described above). On 5 dpf, larvae were transferred to 10% Hanks' solution only (i.e., did not contain experimental chemical). On the morning of 6 dpf, the larvae were transferred again to 10% Hanks' without chemical and placed in the pre-warmed behavioral testing darkroom. Zebrafish larvae at 6 dpf, reared at 26 °C (5 dpf if reared at 28.5 °C), are at an optimal age for behavioral testing since their locomotor activity and response to visual stimuli are well developed in preparation for independent feeding behaviors that begin at 7 dpf.

The chlorpyrifos (0.3, 1.0 or 3 µM) positive control plates followed the same chemical exposure procedure described above. These positive control plates were tested throughout the study at intervals of about 60 days to ensure that the system was working properly.

2.7. Behavioral Testing Systems

These experiments utilized two larval zebrafish behavior systems for recording fish locomotion: a Noldus Tower System and a Noldus DanioVision System (model DVOC-0030), both manufactured by Noldus Information Technology, Leesburg, VA, USA. These systems are hereafter referred to as "Tower" or "DanioVision". Each system was equipped with a light box that provided both infrared and visible light. The luminance of the Light portion of the testing paradigm was 260 lux (DanioVision) or 18 lux (Tower), and that of the Dark portion was 0.5 lux on both systems. Luminance measures were taken at the level of the recording platform using a photometer (Sper Scientific, model # 840022, Scottsdale, AZ, USA).

Due to unavoidable circumstances, it was necessary to switch recording systems while the experiments were underway. Using two different systems for behavioral assessment is not ideal; however, data from each system indicated they were comparable. For this comparison, larval zebrafish (6 dpf) treated with the positive control, chlorpyrifos (0.3, 1.0 or 3 µM), following identical exposure configuration and behavior testing protocols, were tested on each system. This comparison (Figure 2) shows that the animals tested on the two systems exhibited different levels of baseline activity, but when animals exposed to chlorpyrifos during development (our positive control) were tested on both systems, the results did not differ. Panel A in Figure 2 shows the activity of the larvae in the Light or

Dark period in either the DanioVision (left panel) system or the Tower (right panel) system. Note that the animals appear to be more active in the DanioVision system (overall effect of system: $p < 0.0001$): about 40% more activity in the Light and about twice as much activity in the Dark period. There is also an overall effect of chlorpyrifos ($p < 0.0001$), but there is no interaction of the chlorpyrifos effect with system used, meaning that the pattern of the chlorpyrifos effect is not dependent on whether the DanioVision or Tower system was used. Figure 2 (Panel B) shows the effect of chlorpyrifos when the data from both systems are combined as percent of control to correct for the differences in baseline activity. In the Light period chlorpyrifos exposure during development depressed locomotor activity at all three concentrations, while in the Dark only the animals exposed to the highest concentration (3 µM) of chlorpyrifos during development showed hypoactivity.

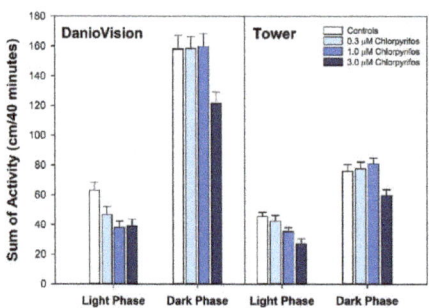

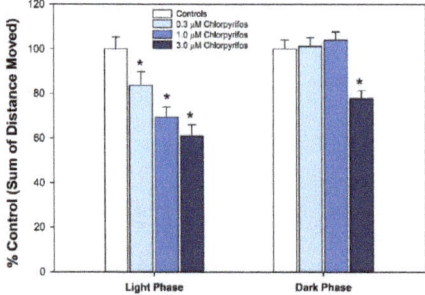

Figure 2. Comparison of the Developmental Chlorpyrifos Effect When Tested on Either the DanioVision or Tower System. Upper (**A**) shows the results when larvae treated with chlorpyrifos during development were tested on either the DanioVision system (**left** panel) or Tower system (**right** panel). Using an ANOVA with chlorpyrifos treatment and the system tested as independent variables, and locomotor activity as the dependent variable, it was found that there was an overall effect of chlorpyrifos ($p < 0.0001$) and of the system used ($p < 0.0001$), but that there was no interaction between those two variables ($p = 0.22$). Because the effect of the chlorpyrifos did not depend on the system that was used for testing, the data from both systems were combined, expressed as a percent of control and analyzed to delineate the effect of chlorpyrifos (**B**). In this case the data were analyzed using an ANOVA (chlorpyrifos or Light/Dark period were independent variables and locomotor activity was the dependent variable). This analysis showed that there was an overall effect of chlorpyrifos ($p < 0.0001$), Light/Dark period ($p < 0.0001$), and that there was an interaction

between the two ($p = 0.0003$), meaning that the effect of chlorpyrifos was different depending on whether the animals were tested in the Light or Dark period. Using an ANOVA and testing each period separately, it was first determined whether there was an overall effect of chlorpyrifos concentration ($p < 0.0001$ in either the Light or Dark) and then a Fisher's PLSD *post hoc* test was conducted to determine which chlorpyrifos concentration was different from control in either the Light or Dark period. Those concentrations that were different from control are indicated by an asterisk. In the Light period the 0.3 µM ($p = 0.03$), 1.0 µM ($p < 0.0001$) and the 3.0 µM ($p < 0.0001$) chlorpyrifos were all different from control, while in the Dark period, only the highest concentration 3.0 µM ($p < 0.0001$) was different from control. For the DanioVision system testing, the sample sizes were 63 controls, 65 at 0.3 µM, 63 at 1.0 µM, and 61 at 3.0 µM, and for the Tower system, the sample sizes were 69 controls, 62 at 0.3 µM, 69 at 1.0 µM and 54 at 3.0 µM. The sample sizes for (**B**) were a combination of each of those sample sizes for each system at each concentration.

2.8. Behavioral Testing

All testing was performed on 6 dpf larvae in the same 96-well mesh plate in which they had been exposed and reared. On the morning of testing (6 dpf) the rearing solution was totally renewed, and the plates were moved to a light-tight drawer in the behavioral testing darkroom where the ambient temperature was the same as the rearing incubator (26 °C). For all experiments, testing occurred between 1200 and 1630 h. After acclimating in the behavioral testing room for at least 2 h, the plates were transferred to either the Tower or DanioVision recording platform light box to begin behavioral testing. The testing paradigm consisted of a 20-min acclimation period in the dark (Basal period), followed by 40 min of light (Light) followed by 40 min of dark (Dark). Prior research in this laboratory, and several others [72–76], have demonstrated that zebrafish larvae exposed to light drastically increase locomotor activity when transitioned to darkness. The Basal period serves to minimize any behavioral disruption due to transfer of the plate and larvae to the recording platform. Data were collected during this acclimation period but were not analyzed further because of a lack of specification and stimulus control.

For both the Tower and DanioVision systems, fish movement (locomotion) was recorded using Media Recorder software (Noldus Information Technology, Leesburg, VA, USA) and saved as MPEG2 files, a process initially described by MacPhail [77].

2.9. Lethality and Malformation Assessment and Inclusion Criteria

Immediately following behavioral testing, larvae were assessed by observers (blinded to treatment conditions) for death and malformations using an Olympus SZH10 stereo microscope. Morphological assessments focused on the following: craniofacial (abnormal eyes or head), spinal (stunted, curved, or kinked tail), abdominal region (edema or emaciation), thoracic region (distention or heart malformations), swim bladder inflation, and position in the water column (floating or lying on side). All dead, unhatched, malformed larvae, and those with uninflated swim bladders, were eliminated from any behavioral analysis; malformed 6 dpf zebrafish larvae, as well as normal appearing larvae with uninflated swim bladders, do not behave normally in our behavioral paradigm [65,69]. Following the assessments, larvae were anaesthetized using cold shock and then euthanized with 20% (v/v) bleach solution.

There were multiple levels of embryo quality acceptance for inclusion in the behavioral data. First, at the plate level, if more than 15% of the control larvae were abnormal, then no data from that plate were used; the plate was discarded and repeated. Next, at the concentration level, if more than 25% of the larvae from any concentration group were abnormal, then that entire concentration was removed from further behavior analyses, though the data were still used for developmental toxicity evaluation. The 75% concentration group threshold was established because it was thought that if any more than 25% of the animals were abnormal at a given concentration, then the developmental toxicity of that chemical concentration outweighed the neurodevelopmental toxicity. Lastly, each individual embryo included in the behavioral analyses must have appeared normal (i.e., no obvious malformations).

The statistical results and number of larvae in every concentration group are noted in Figure 3, which also notes the concentration groups for each chemical that were excluded from behavioral analyses.

Figure 3. *Cont.*

Colchicine			
Overall effect		0.4705	0.8094
Control	68		
0.12	17	0.4006	0.3282
0.40	17	0.3282	0.9605
1.20	18	0.1401	0.5381
4.00	18	0.0728	0.5593
12.00	18	0.4969	0.5451
40.00	13	⊘	⊘
120.00	1	⊘	⊘

Cotinine			
Overall effect		0.4177	0.4604
Control	69		
0.03	17	0.5188	0.7947
0.10	18	0.1938	0.6373
0.30	18	0.2163	0.9332
1.00	17	0.4843	0.4007
3.00	16	0.1702	0.8661
10.00	17	0.4381	0.2225
30.00	18	0.6599	0.0699

Cyclophosphamide			
Overall effect		0.0166	0.6936
Control	65		
0.12	18	0.6035	0.6426
0.40	16	0.2309	0.5611
1.20	17	0.0039	0.8146
4.00	17	0.0062	0.9772
12.00	17	0.0133	0.4333
40.00	18	0.0443	0.0997
120.00	15	0.2247	0.4054

Cytosine arabinoside			
Overall effect		0.0442	0.8829
Control	70		
0.12	18	0.7173	0.4440
0.40	16	0.6332	0.6571
1.20	17	0.4159	0.9829
4.00	17	0.1083	0.7591
12.00	17	0.0108	0.9829
40.00	18	0.0063	0.2342
120.00	17	0.3573	0.3685

Deltamethrin			
Overall effect		0.0819	0.6453
Control	64		
0.001	16	0.6476	0.5966
0.004	16	0.1165	0.1741
0.012	17	0.1121	0.6933
0.040	18	0.0085	0.2532
0.120	13	⊘	⊘
0.400	0	⊘	⊘
1.200	0	⊘	⊘

Dexamethazone			
Overall effect		0.1989	0.3845
Control	70		
0.12	18	0.1889	0.5213
0.40	18	0.0518	0.6416
1.20	15	0.7295	0.2174
4.00	15	0.9540	0.1496
12.00	10	⊘	⊘
40.00	9	⊘	⊘
120.00	5	⊘	⊘

Di(2-ethylhexyl)phthalate (DEHP)			
Overall effect		0.1836	0.8413
Control	67		
0.12	17	0.5514	0.8718
0.40	17	0.0103	0.8718
1.20	18	0.7469	0.7551
4.00	16	0.2041	0.9448
12.00	17	0.6763	0.2979
40.00	17	0.3383	0.6125
120.00	17	0.3031	0.1231

Diazepam			
Overall effect		<0.0001	0.0326
Control	186		
0.004	24	0.1750	0.2701
0.012	23	0.0258	0.7812
0.040	23	0.2572	0.7953
0.120	23	<0.0001	0.6504
0.400	24	0.0791	0.0496
1.200	24	0.0001	0.0300
4.000	24	<0.0001	0.0034
12.000	21	<0.0001	0.0359
40.000	0	⊘	⊘
120.000	0	⊘	⊘

Dieldrin			
Overall effect		0.5950	0.4642
Control	64		
0.001	18	0.3438	0.8842
0.004	18	0.8842	0.9955
0.012	14	0.3902	0.0478
0.040	17	0.4103	0.3657
0.120	14	0.2916	0.3037
0.400	1	⊘	⊘
1.200	0	⊘	⊘

Diethylene glycol			
Overall effect		0.0376	0.3197
Control	81		
0.12	21	0.8007	0.3648
0.40	20	0.4847	0.1547
1.20	21	0.7250	0.1582
4.00	21	0.3561	0.3067
12.00	20	0.4847	0.0735
40.00	21	0.0107	0.9110
120.00	19	0.0583	0.8502

Diethylstilbesterol			
Overall effect		0.0175	0.5189
Control	69		
0.03	17	0.0218	0.7408
0.10	17	0.4708	0.3237
0.30	18	0.0134	0.5574
1.00	17	0.0918	0.4446
3.00	16	0.1439	0.2805
10.00	0	⊘	⊘
30.00	0	⊘	⊘

D-sorbitol			
Overall effect		0.1176	0.2255
Control	63		
0.12	17	0.3319	0.1033
0.40	18	0.6333	0.0727
1.20	18	0.0578	0.2805
4.00	17	0.3746	0.2019
12.00	16	0.7144	0.7790
40.00	18	0.0863	0.8114
120.00	17	0.2372	0.1743

Fluconazole			
Overall effect		0.6665	0.8804
Control	81		
0.12	21	0.8914	0.6760
0.40	21	0.4150	0.7943
1.20	21	0.2069	0.7034
4.00	21	0.6283	0.5708
12.00	21	0.1487	0.7250
40.00	20	0.1754	0.3507
120.00	20	0.9592	0.2072

Fluoxetine			
Overall effect		<0.0001	0.0060
Control	126		
0.004	16	0.7324	0.8566
0.012	14	0.5928	0.7758
0.040	16	0.2616	0.9794
0.120	15	0.0533	0.0833
0.400	15	0.0659	0.0938
1.200	15	0.0009	0.0026
4.000	15	<0.0001	0.0052
12.000	10	⊘	⊘
40.000	0	⊘	⊘
120.000	0	⊘	⊘

Glyphosate			
Overall effect		0.0728	0.7909
Control	65		
0.12	18	0.0467	0.7994
0.40	18	0.6347	0.6112
1.20	17	0.0368	0.1267
4.00	17	0.4537	0.6473
12.00	18	0.6989	0.1885
40.00	18	0.0080	0.9383
120.00	16	0.6867	0.7850

Haloperidol			
Overall effect		0.1754	0.2430
Control	138		
0.004	16	0.5941	0.2682
0.012	16	0.4556	0.1907
0.040	15	0.7778	0.7105
0.120	15	0.5194	0.0548
0.400	15	0.8061	0.4127
1.200	16	0.0060	0.6442
4.000	1	⊘	⊘
12.000	0	⊘	⊘
40.000	0	⊘	⊘
120.000	0	⊘	⊘

Heptachlor			
Overall effect		0.9711	0.0089
Control	186		
0.004	24	0.7158	0.4036
0.012	22	0.7388	0.5365
0.040	24	0.5440	0.2907
0.120	23	0.8939	0.5538
0.400	21	0.5932	0.2458
1.200	23	0.6557	0.0003
4.000	4	⊘	⊘
12.000	0	⊘	⊘
40.000	0	⊘	⊘
120.000	0	⊘	⊘

Heptachlor epoxide			
Overall effect		0.0356	0.2997
Control	23		
0.0004	24	0.6396	0.4693
0.0012	24	0.2251	0.2970
0.0040	24	0.0969	0.4628
0.0120	23	0.2227	0.7006
0.0400	23	0.0272	0.1766
0.1200	22	0.0016	0.7679
0.4000	12	⊘	⊘

Hexachlorophene			
Overall effect		0.0846	0.6570
Control	81		
0.01	21	0.1057	0.5485
0.04	21	0.3871	0.4245
0.12	20	0.0682	0.5623
0.40	21	0.8523	0.4197
1.20	21	0.1534	0.2517
4.00	1	⊘	⊘
12.00	0	⊘	⊘

Hydoxyurea			
Overall effect		0.4549	0.6041
Control	67		
0.12	17	0.2680	0.6521
0.40	15	0.1811	0.2520
1.20	18	0.2005	0.6056
4.00	15	0.6442	0.7965
12.00	16	0.1921	0.7729
40.00	15	0.6878	0.1623
120.00	16	0.4393	0.7509

Figure 3. Cont.

Isoniazid			
Overall effect		0.1723	0.8182
Control	69		
0.03	18	0.5159	0.8669
0.10	18	0.0106	0.5295
0.30	17	0.8325	0.5655
1.00	17	0.0437	0.1755
3.00	17	0.5188	0.5840
10.00	16	0.8572	0.9016
30.00	17	0.8836	0.5618

Lead acetate			
Overall effect		0.4467	0.1957
Control	28		
0.01	15	0.8986	0.2622
0.04	17	0.5742	0.3512
0.12	17	0.1114	0.4538
0.40	16	0.9708	0.1184
1.20	3	⊘	⊘
4.00	0	⊘	⊘
12.00	1	⊘	⊘
40.00	0	⊘	⊘

Loperamide			
Overall effect		0.0008	0.0077
Control	104		
0.004	16	0.7959	0.0121
0.012	16	0.0771	0.3193
0.040	16	0.5678	0.2000
0.120	16	0.5067	0.1623
0.400	16	0.7167	0.0851
1.200	16	0.6599	0.0171
4.000	13	0.2314	0.0288
12.000	13	0.0575	0.1652
40.000	14	<0.0001	0.0763
120.000	6	⊘	⊘

Maneb			
Overall effect		0.6906	0.3610
Control	138		
0.004	14	0.1610	0.1216
0.012	14	0.6788	0.7890
0.040	16	0.3712	0.5657
0.120	15	0.4323	0.0722
0.400	15	0.3385	0.8061
1.200	14	0.7792	0.5837
4.000	12	0.6878	0.7870
12.000	16	0.4292	0.8917
40.000	13	0.3101	0.0652

Manganese			
Overall effect		0.0400	0.1314
Control	65		
0.12	15	0.6796	0.3985
0.40	16	0.0048	0.4766
1.20	17	0.2320	0.8503
4.00	17	0.0187	0.0151
12.00	17	0.4266	0.8057
40.00	16	0.0337	0.2402
120.00	11	⊘	⊘

Methotrexate			
Overall effect		0.9155	0.8952
Control	61		
0.12	17	0.9085	0.8184
0.40	17	0.8893	0.7576
1.20	18	0.5241	0.9627
4.00	16	0.3337	0.3152
12.00	17	0.6240	0.3065
40.00	15	0.8193	0.9948
120.00	10	⊘	⊘

Naloxon			
Overall effect		0.4105	0.2753
Control	68		
0.12	17	0.1470	0.2866
0.40	17	0.5098	0.3017
1.20	16	0.6817	0.9274
4.00	18	0.1089	0.0508
12.00	18	0.9577	0.2303
40.00	18	0.9408	0.7589
120.00	17	0.1837	0.3765

Nicotine			
Overall effect		0.3491	0.4239
Control	69		
0.03	18	0.9791	0.2086
0.10	16	0.6129	0.8661
0.30	17	0.0289	0.6909
1.00	16	0.4788	0.4379
3.00	18	0.6637	0.1541
10.00	17	0.7738	0.5401
30.00	2	⊘	⊘

Paraquat			
Overall effect		0.3553	<0.0001
Control	68		
0.12	17	0.8864	0.1986
0.40	18	0.1237	0.7744
1.20	18	0.0385	0.0471
4.00	18	0.1401	<0.0001
12.00	16	0.9546	<0.0001
40.00	18	0.9070	<0.0001
120.00	15	0.2320	<0.0001

Permethrin			
Overall effect		0.4518	0.2471
Control	63		
0.001	21	0.6681	0.2713
0.004	23	0.9184	0.7660
0.013	20	0.8731	0.6702
0.040	24	0.0482	0.0668
0.126	24	0.1965	0.5495
0.393	22	0.2008	0.3354
1.229	21	0.9876	0.4792
3.840	19	0.7046	0.1123
12.000	1	⊘	⊘

Phenobarbital			
Overall effect		0.0120	0.0239
Control	72		
0.12	18	0.0203	0.0050
0.40	18	0.0245	0.0220
1.20	17	0.0255	0.1440
4.00	17	0.3979	0.0359
12.00	18	0.0620	0.1252
40.00	16	0.1520	0.5860
120.00	18	0.4615	0.2110

Phenol			
Overall effect		0.2495	0.1764
Control	71		
0.12	17	0.7794	0.5089
0.40	18	0.7131	0.7671
1.20	17	0.0261	0.0157
4.00	18	0.0880	0.9349
12.00	17	0.5790	0.3078
40.00	18	0.2401	0.2203
120.00	17	0.6009	0.7957

Polybrominated diphenyl ether (PBDE)-47			
Overall effect		0.2502	<0.0001
Control	80		
0.04	20	0.1627	0.5991
0.13	20	0.3588	0.8903
0.40	21	0.3401	0.2282
1.32	20	0.1052	0.0151
4.00	19	0.5398	<0.0001
13.00	15	⊘	⊘
40.00	2	⊘	⊘

Saccharin			
Overall effect		0.9077	0.5658
Control	71		
0.12	17	0.3442	0.2559
0.40	18	1.0000	0.5468
1.20	18	0.6385	0.8302
4.00	18	0.2569	0.0630
12.00	17	0.4756	0.9978
40.00	18	0.9349	0.3370
120.00	17	0.6609	0.8367

Sodium benzoate			
Overall effect		0.2920	0.0689
Control	30		
0.04	15	0.1016	0.6647
0.12	17	0.4516	0.4925
0.40	20	0.8122	0.0569
1.20	19	0.5180	0.0072
4.00	19	0.7582	0.4726
12.00	18	0.3943	0.2774
40.00	17	0.1266	0.0514
120.00	15	0.4701	0.0736

Sodium fluoride			
Overall effect		0.0768	0.7213
Control	65		
0.12	17	0.0412	0.5791
0.40	16	0.2263	0.4952
1.20	17	0.0311	0.7618
4.00	18	0.3476	0.1246
12.00	17	0.0045	0.4818
40.00	15	0.8581	0.8581
120.00	17	0.0525	0.4333

Tebuconazole			
Overall effect		0.5517	0.3863
Control	63		
0.12	17	0.8369	0.7110
0.40	18	0.8646	0.3099
1.20	18	0.8916	0.1764
4.00	17	0.1058	0.3348
12.00	18	0.4199	0.2512
40.00	3	⊘	⊘
120.00	0	⊘	⊘

Terbutaline			
Overall effect		0.0970	0.4565
Control	70		
0.12	18	0.6565	0.2222
0.40	14	0.0906	0.8010
1.20	17	0.1232	0.4799
4.00	17	0.1015	0.5069
12.00	17	0.0260	0.6607
40.00	16	0.3073	0.1904
120.00	18	0.6416	0.3412

Thalidomide			
Overall effect		0.1078	0.1959
Control	81		
0.12	20	0.2772	0.0549
0.40	20	0.2430	0.4696
1.20	20	0.1018	0.2829
4.00	20	0.4305	0.0305
12.00	21	0.7437	0.7250
40.00	21	0.0668	0.5213
120.00	21	0.6342	0.9242

Triethyltin			
Overall effect		0.6951	0.8830
Control	33		
0.0004	21	0.8383	0.9505
0.0012	19	0.4303	0.9621
0.0040	21	0.2310	0.5884
0.0120	21	0.1338	0.9788
0.0400	21	0.2752	0.3247
0.1200	22	0.5028	0.3491
0.4000	0	⊘	⊘

Valproate			
Overall effect		0.4110	0.3518
Control	104		
0.004	15	0.4187	0.3656
0.012	15	0.9585	0.3407
0.040	16	0.5731	0.2894
0.120	16	0.5471	0.0499
0.400	16	0.1555	0.4210
1.200	16	0.1207	0.3872
4.000	16	0.8712	0.8228
12.000	15	0.9521	0.4280
40.000	12	0.1189	0.0671
120.000	0	⊘	⊘

Kruskal-Wallis Non-Parametric test w/Bonferroni correction (α= 0.025) for overall effect and Wilcoxon-Mann-Whitney post-hoc test (α= 0.05) for step down comparisons

⊘ = Developmental Toxicity. Concentration removed from behavioral analyses due to the number of Dead, Malformed and Uninflated Swim Bladders exceeding 25%

Figure 3. Behavioral Nonparametric Statistics Results. Results of the Kruskal-Wallis Nonparametric test for each chemical. A Bonferroni correction was applied to the overall effect to account for the Light and Dark periods, resulting in α = 0.025. The Wilcoxon-Mann-Whitney post-hoc test (α = 0.05) compared each concentration to the control for that chemical. The circle with the slash symbol (⊘) indicates developmental toxicity: that concentration was not included in behavioral analyses due to the number of dead, malformed and uninflated swim bladders exceeding 25%. Overall effect is listed under the chemical name followed by the sample size and results for each concentration, with the Dark period shaded gray. Statistically significant results are highlighted with the light-yellow shading in the Light period and dark-yellow shading in the Dark period.

2.10. Analysis of Fish Movement

The videos recorded during the behavioral testing session were later analyzed using Ethovision XT (Noldus Information Technology) software Version 13 to quantify the distance moved by each larva. Tracking rate was 5 samples/sec (i.e., an image was captured every 200 ms). A dynamic subtraction method was used to detect objects that were darker than the background, with a minimum object size of 10 pixels. Tracks were analyzed for total distance moved (cm). An input filter of 0.135 cm (minimum distance moved) was used to remove system noise. All locomotion data is expressed as distance moved per segment of testing, from which total activity was calculated for each larva in both Light and Dark periods.

2.11. Data Analysis and Statistics

Under control conditions, the distributions of locomotion data were not normally distributed, but were markedly skewed (Figure 4). In the Light, there was a preponderance of low values and increasingly fewer instances of higher distance-moved values. Positive skew was also noted in the control values of distance moved during the Dark. Therefore, no "outliers" were removed, and nonparametric statistical analyses were conducted on concentration-response data (all data for each animal for the Light period or the Dark period were summed) using SAS software (v.9.4). Data were first analyzed using the Kruskal-Wallis Test assessing if there was an overall dose-response relationship between the activity in the Light or the Dark and the concentration of the test chemical. If the results of this test were significant ($\alpha \leq 0.025$ (Bonferroni corrected for the repeated measures aspect of the Light and Dark periods)) it was followed by Wilcoxon-Mann-Whitney post-hoc tests ($\alpha \leq 0.05$) that compared data for each concentration group to the vehicle-control group.

The Kruskal-Wallis nonparametric test was also used for total activity in the Light or Dark periods to analyze the effect of developmental DMSO exposure on activity (Supplemental Figure S1). A repeated measures ANOVA was used to compare the Tower and DanioVision systems with activity as the dependent variable and system, chlorpyrifos concentration and Light/Dark as independent variables. In addition to statistical analyses, the percent change between each concentration and control was also calculated.

2.12. Comparison of Results with Previously Published Data

One of the goals of this study was to compare these present results to those reported in the literature. A systematic literature review was conducted (latest publication date was 30 November 2020) by gathering abstracts using the Abstract Sifter [78], searching by chemical name and/or CAS number in combination with "zebrafish" or "zebrafish and behavior" as search terms. After publications were gathered, each was further screened for methodological relevance by targeting publications that (1) specified a developmental window during chemical exposure (0–3 dpf); (2) had at least 24 h of chemical exposure; (3) included an acclimation period prior to behavioral testing; (4) conducted the behavioral test sometime between 5–7 dpf; and (5) the behavioral paradigm had at least one transition from Light to Dark. These methodological aspects were selected to focus on assays similar to our protocol. This decision was made due to the proposed influence of methodological variables on zebrafish behavior and toxicity outcome [79,80]. Information on how behavioral changes were reported, concentrations included in the dose response, and concentrations that were noted to cause significant effects were conflated into a spreadsheet and visually compared to our results.

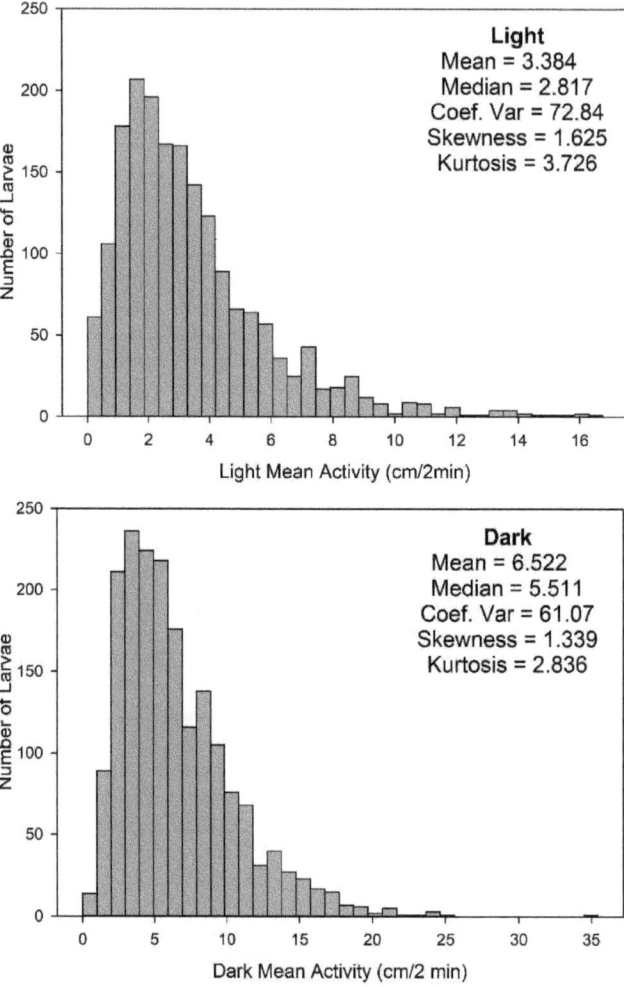

Figure 4. Histograms of the Distribution of the Control Activity for the Light and Dark Periods of Testing. The sum of the activity of the control animals (n = 1851) was plotted as a histogram to visualize the non-normal distribution of the data. Note that the activity intervals are different for the Light and Dark periods. Plots and data calculations were performed using SigmaPlot.

3. Results

Sixty-one chemicals were tested for both developmental toxicity and behavioral disruption. To determine whether developmental toxicity occurred, animals were assessed for death, non-hatching, or morphological abnormalities, including uninflated swim bladders. Normal looking embryos, such as the one depicted on Day 6 of our Experimental Design (Figure 1), have no obvious malformations, are of normal size and have an inflated swim bladder. Developmental toxicity data are shown in the inset graph on each box plot in Figure 5, and also in the summary figure for each chemical (Supplemental Figure S2). These data show the percent of normal larvae for each concentration tested. The red dashed line marks the 75% behavioral data inclusion cutoff with values that fall below that line indicated by a red circle. The black triangle represents negative control data for that chemical.

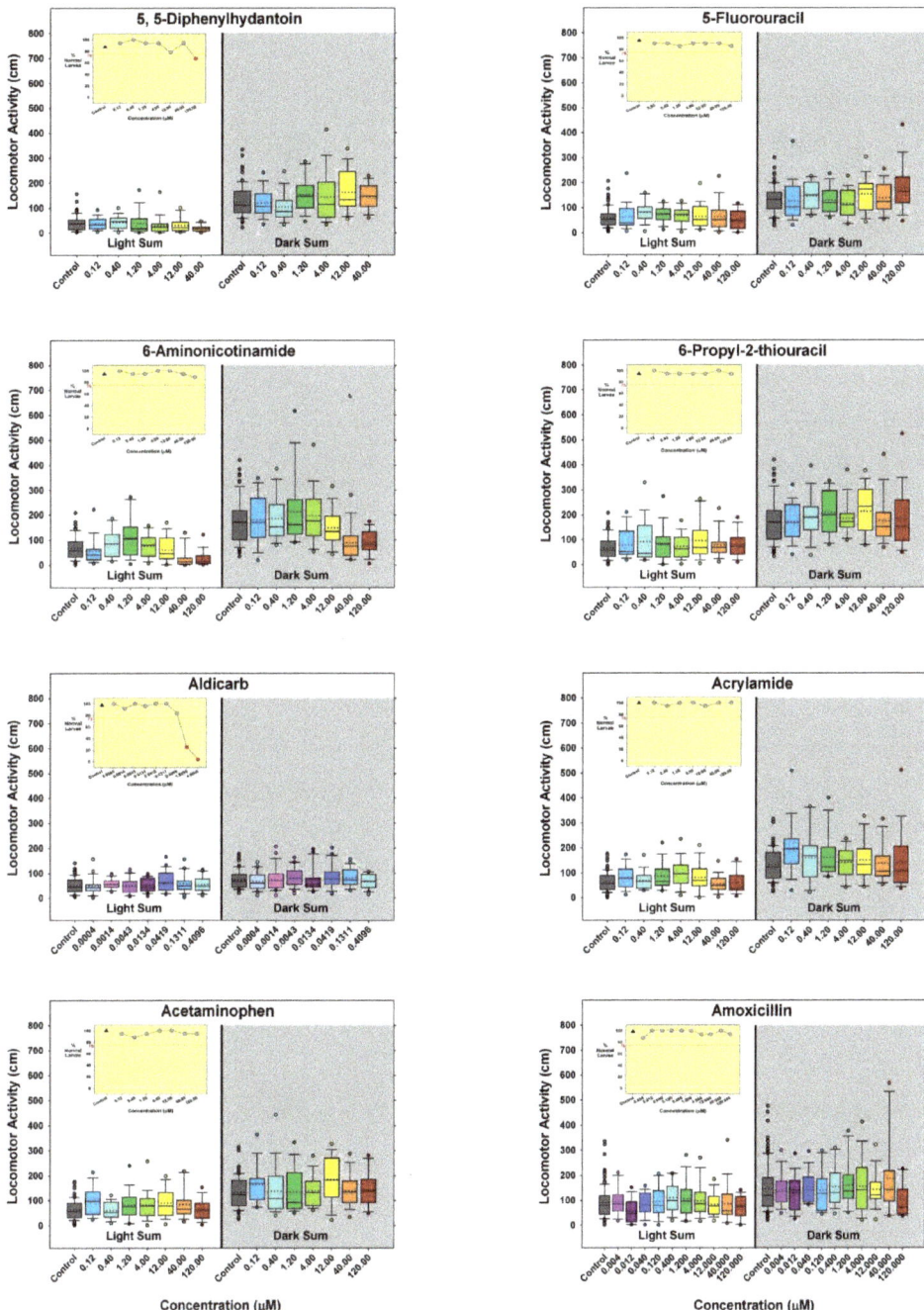

Figure 5. *Cont.*

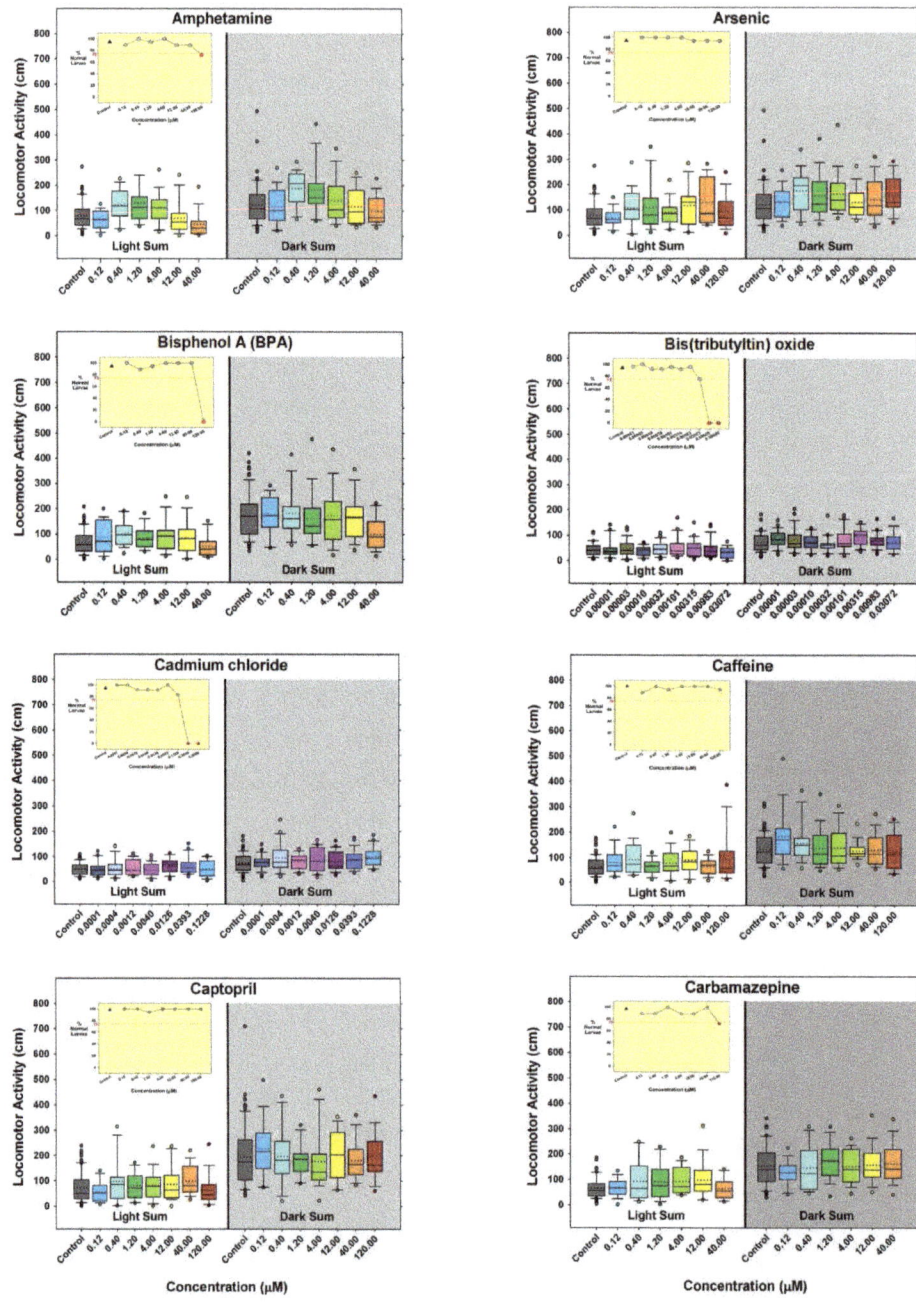

Figure 5. *Cont.*

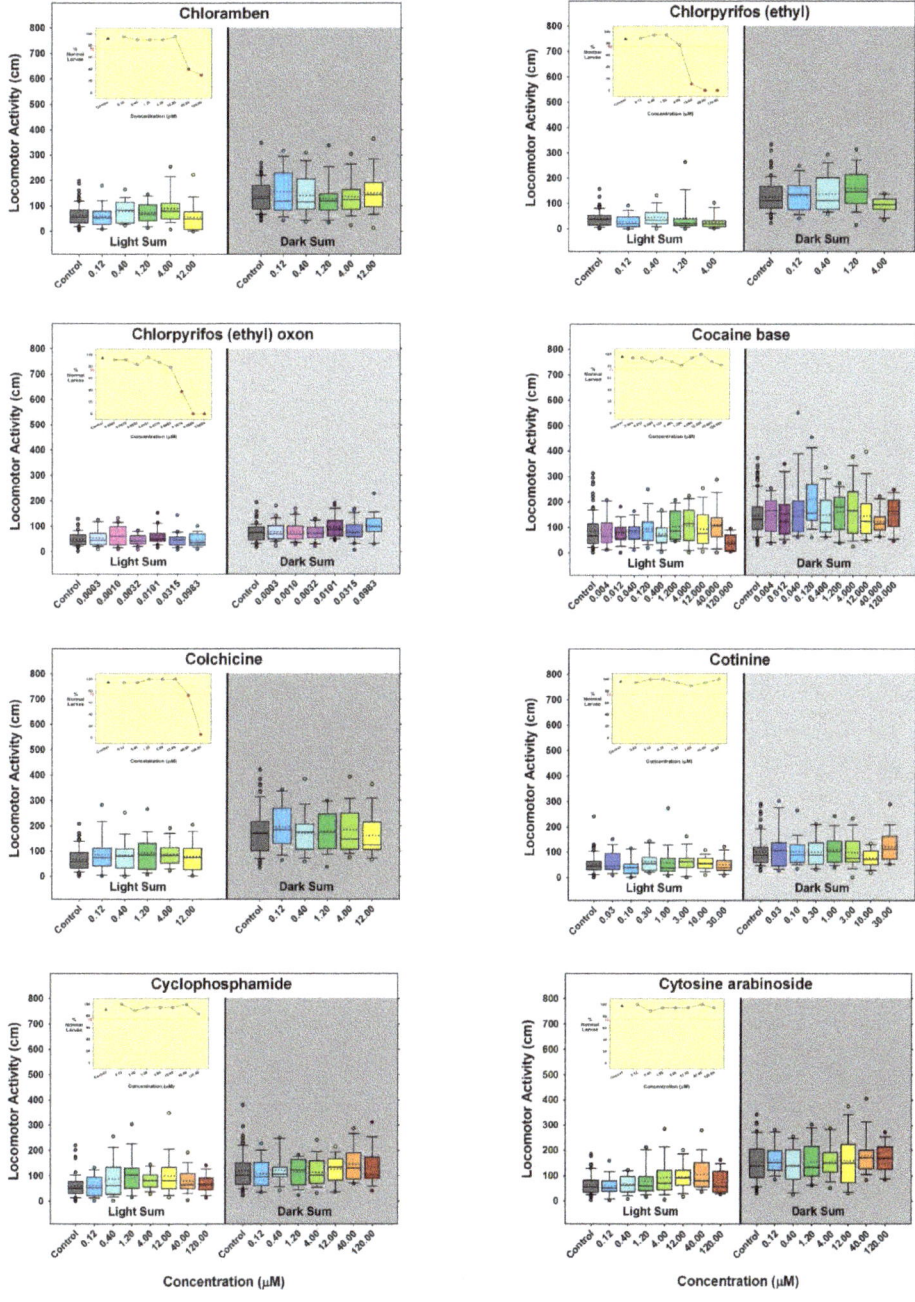

Figure 5. *Cont.*

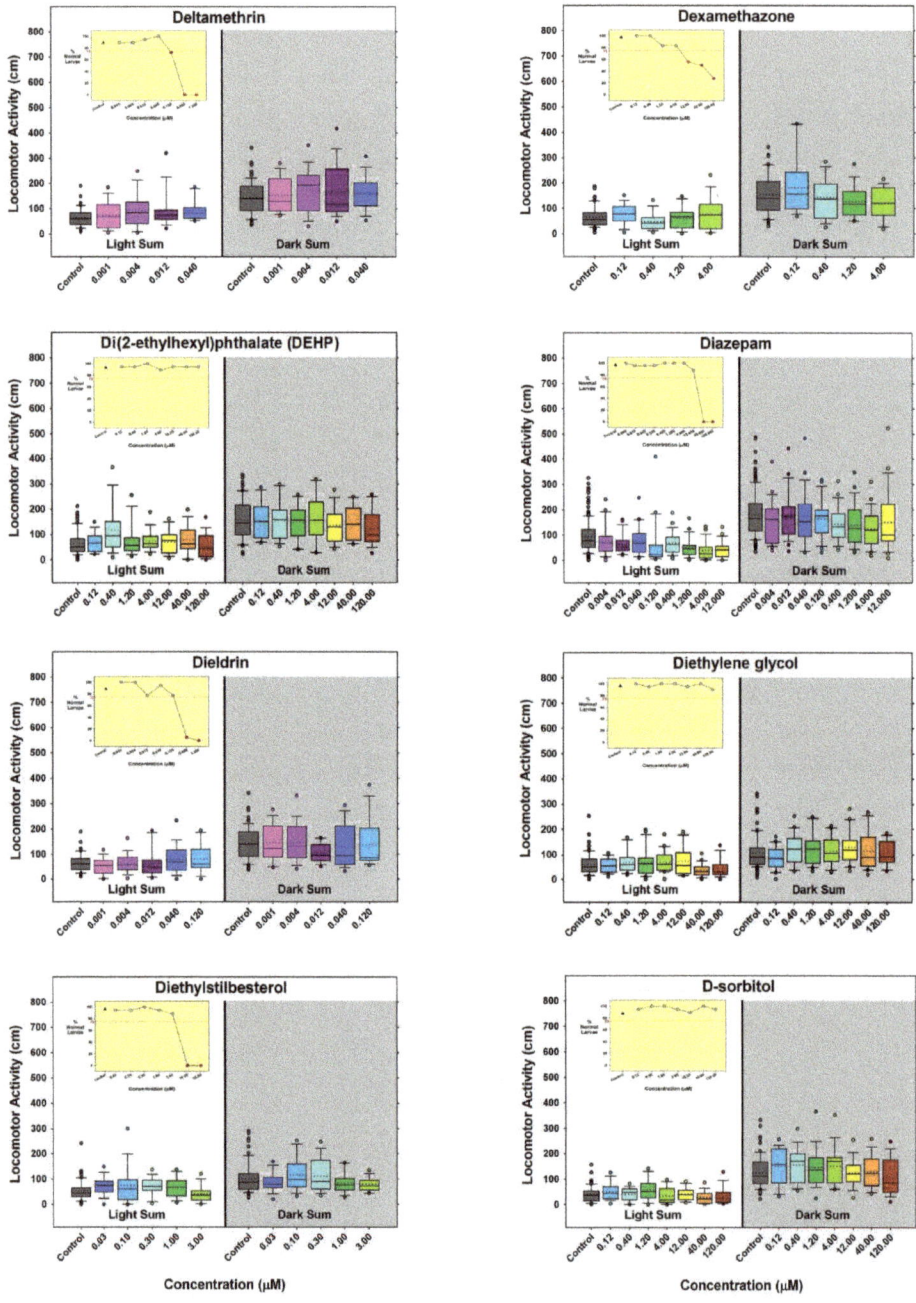

Figure 5. *Cont.*

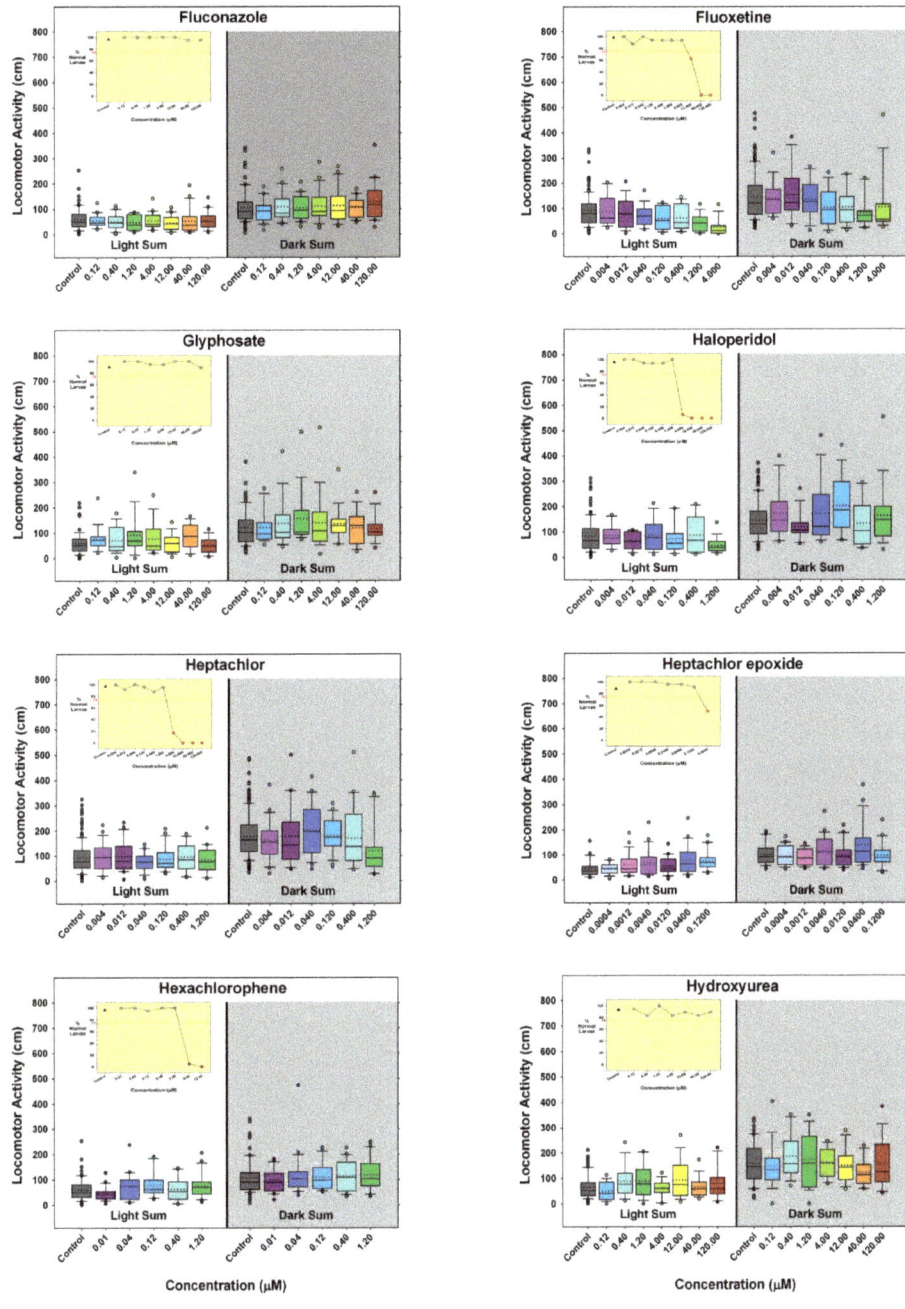

Figure 5. Cont.

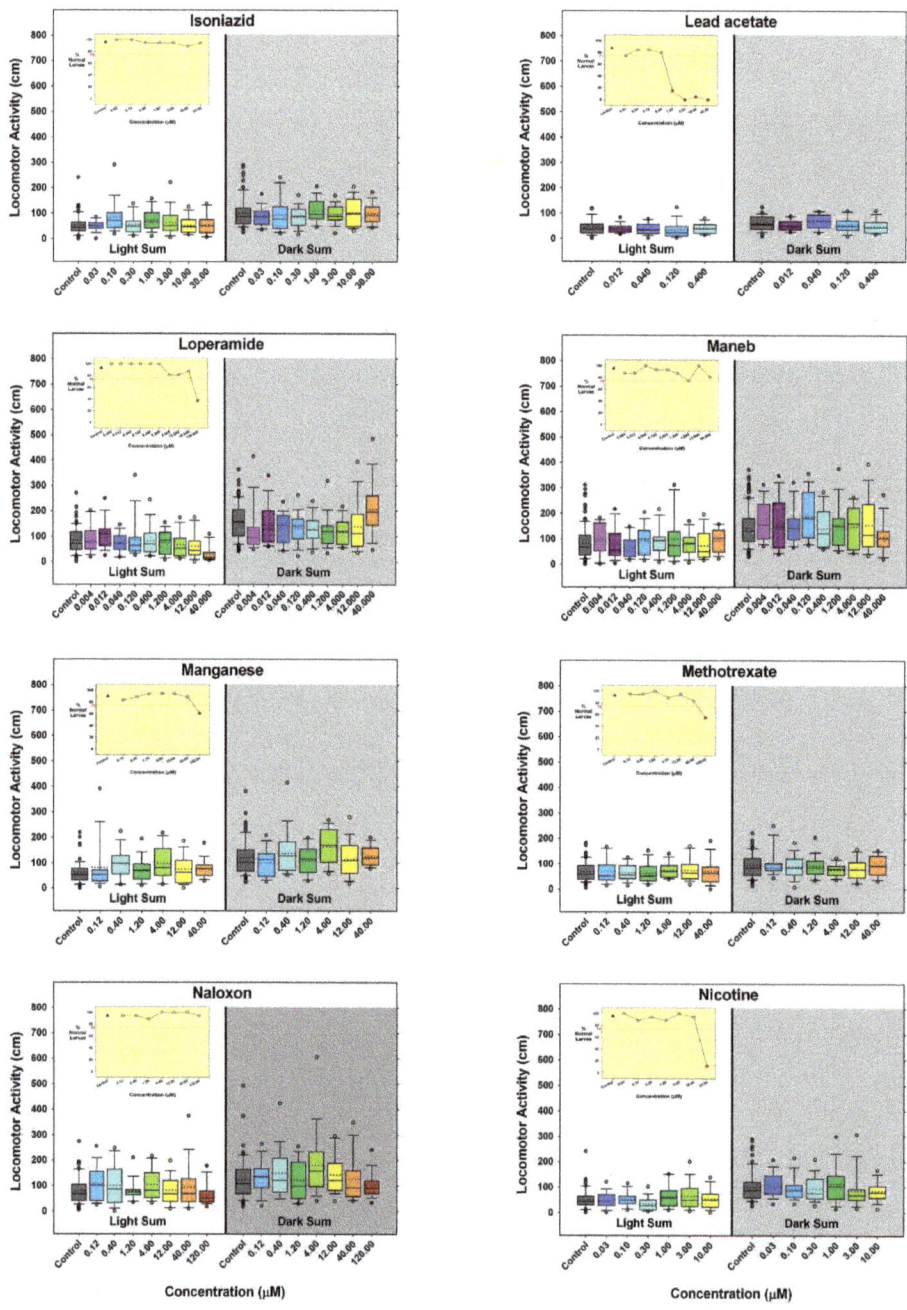

Figure 5. Cont.

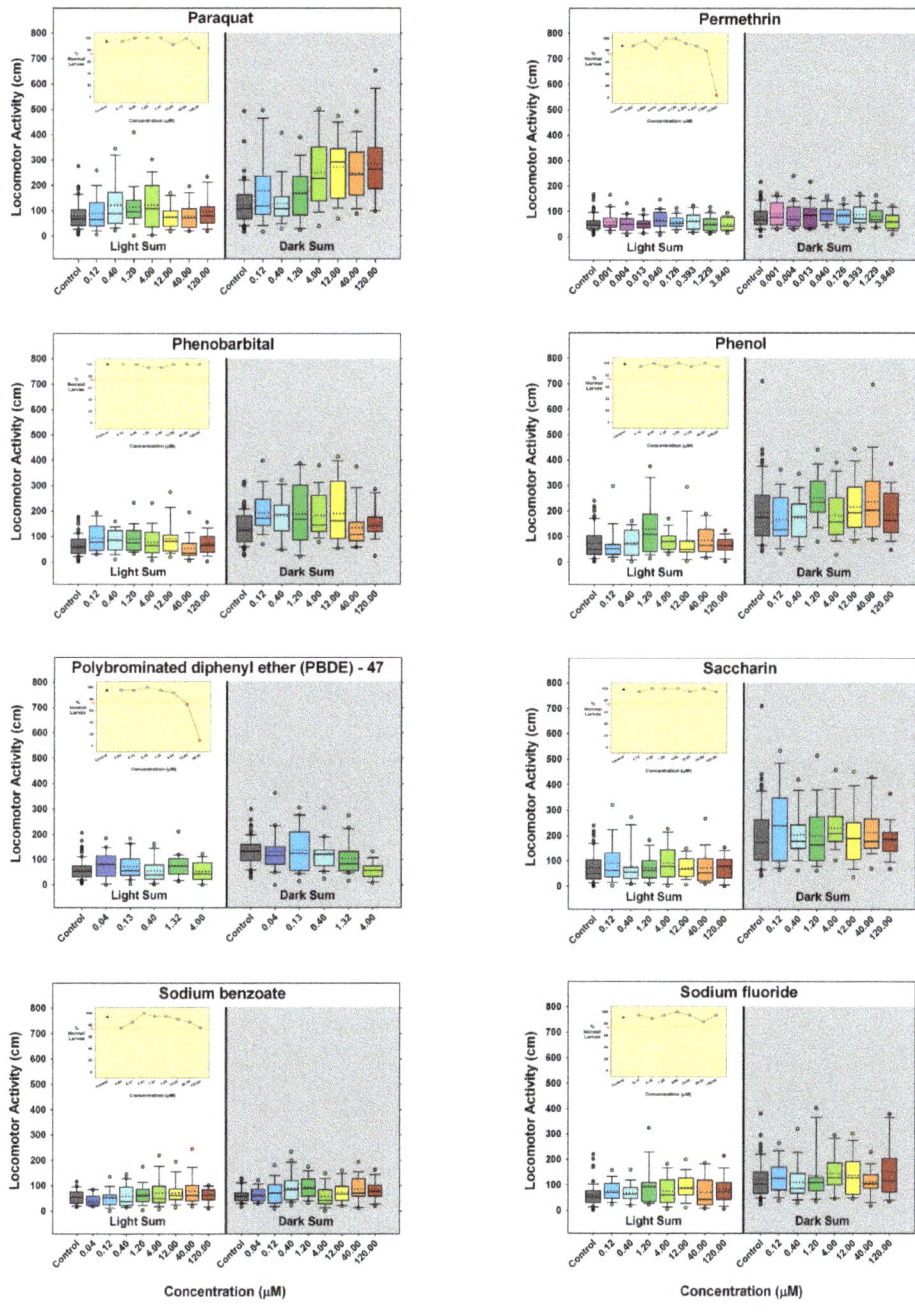

Figure 5. *Cont.*

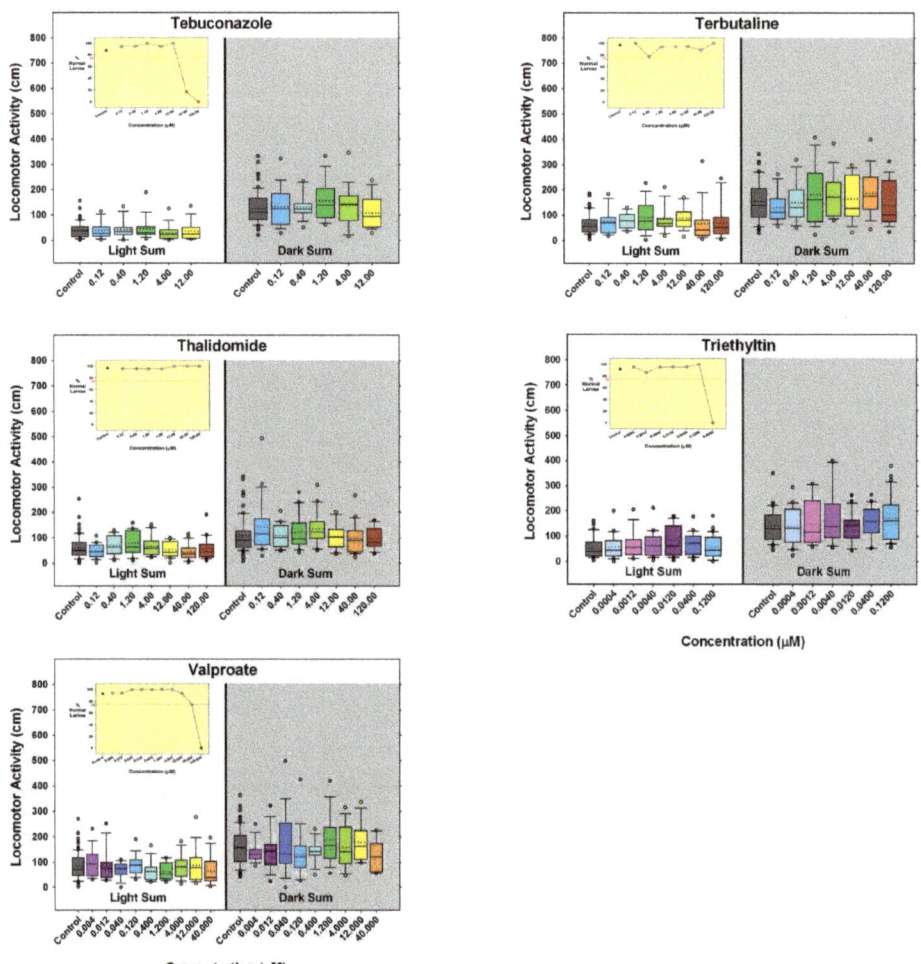

Figure 5. Behavioral Concentration Response for Each Chemical Presented in a Box Plot with Developmental Toxicity as a Line Graph in the Inset. Box plots show locomotor activity for both the Light and Dark (gray background) periods. The box represents the interquartile range (middle 50%), the top of the box to the top error bar is the upper quartile (75th percentile) while the bottom of the box to lower error bar is the lower quartile (25th percentile). The solid line in the middle of the box is the median and the dotted line in the middle of the box is the mean. The top whisker/error bar indicates the maximum and the bottom whisker/error bar indicates the minimum. The developmental toxicity inset shows the percent of normal larvae for the control and for each concentration. The dotted red line is the 75% line and concentration groups that fall below are considered developmentally toxic and not included in behavioral analyses. The triangle represents control data, and the gray circles indicate results at each concentration. All concentrations are in micromolar (μM).

The highest concentration tested was 120 μM; if there was considerable developmental toxicity, the tested concentrations were decreased until at least four concentrations showed no developmental toxicity (i.e., the number of dead, malformed and uninflated swim bladders exceeded 25%). Of the 61 chemicals tested, approximately half (n = 30) did not show any toxicity at the highest concentration tested (Figures 4 and 5). For the majority of the chemicals that did not cause toxicity, the highest concentration administered was

120 µM; however, for three chemicals (cotinine, isoniazid, maneb) the highest concentration was lower and ranged from 30–40 µM, due to solubility issues. Thirty-one chemicals did elicit developmental toxicity; twelve were toxic at the highest concentration only. For four of those twelve (heptachlor epoxide, nicotine, permethrin, triethyltin), the highest concentration tested was less than 120 µM, ranging from (0.4 to 30 µM).

Looking at the lower concentrations, a total of thirteen chemicals were developmentally toxic at concentrations of 10 µM or lower. Seven chemicals (bis(tributyltin)oxide, cadmium chloride, chlorpyrifos oxon, deltamethrin, dieldrin, heptachlor epoxide and triethyltin) showed toxicity at the lowest range (0.1 and 1 µM). Six other chemicals (aldicarb, diethylstilbesterol, haloperidol, heptachlor, hexachlorophene, and lead acetate) were developmentally toxic in the 1 to 10 µM range.

The developmental toxicity data on 6 dpf was used to determine which larvae would be included or removed from behavioral analyses. Concentrations with more than 25% dead or malformed larvae, and normal appearing larvae with uninflated swim bladders, were excluded from behavioral analyses. Furthermore, any individual larva that was not deemed normal was also removed from behavioral analysis, regardless of the concentration group.

The behavior data are also presented in Figure 5 as well as the supplementary summary figure (Supplementary Figure S2). Box plots showing the Light and Dark periods for each concentration were chosen to present the behavior data because of the amount of information they convey. Each box plot contains the minimum, maximum, median, and mean values, the interquartile range, the upper (75th percentile) and lower (25th percentile) quartiles, as well as outliers. Concentrations that were developmentally toxic (more than 25% of any concentration group was abnormal) appear on the inset graph, but not on the behavioral data box plot in this figure because that concentration was removed from behavior analysis.

The behavior data are also presented in Figure 5 as well as the supplementary summary figure (Supplementary Figure S2). Box plots showing the Light and Dark periods for each concentration were chosen to present the behavior data because of the amount of information they convey. Each box plot contains the minimum, maximum, median, and mean values, the interquartile range, the upper (75th percentile) and lower (25th percentile) quartiles, as well as outliers. Concentrations that were developmentally toxic (more than 25% of the test group was abnormal) appear on the inset graph, but not on the behavioral data box plot in this figure because that concentration was removed from behavior analysis. In addition to the box plots, Supplementary Figure S2 also presents the behavior data as the mean of each 2 min epoch ± SEM. For normal behaving embryos, the 2-min data behavior pattern shows a gradual increase, then activity leveling off in the Light, followed by a characteristic sharp increase in behavior when changing from Light to Dark, which is then followed by a gradual decrease and leveling off.

Eleven chemicals showed behavioral effects at concentrations that did not produce any developmental toxicity. For seven of them (amphetamine, diazepam, diethylstilbesterol, fluoxetine, heptachlor, loperamide, polybrominated diphenyl ether (PBDE-47)), developmental toxicity was observed at the highest concentration administered, so those concentrations were removed from behavioral analyses, and behavioral disruption in the otherwise normal looking embryos was observed at lower concentrations for those toxicants. In four chemicals (6-aminonicotinamide, cyclophosphamide, paraquat, phenobarbital) where no developmental toxicity (i.e., the number of dead, malformed and uninflated swim bladders exceeded 25%) was found at the tested concentrations, behavioral disruption was revealed.

Behavioral results showed differences for five chemicals in both the Light and Dark periods, while three (cyclophosphamide, diazepam, diethylstilbesterol) only produced effects in the Light, and three others (heptachlor, paraquat, PBDE-47) only produced behavioral effects in the Dark. Commonly, though not always, lower concentrations resulted in an increase in locomotion (hyperactivity) while higher concentrations decreased locomotion (hypoactivity). Four chemicals produced hyperactivity only, while six resulted in hypoactivity. One chemical (amphetamine) resulted in hyperactivity during both the

Dark and Light periods at lower concentrations, and hypoactivity during the Light period at the highest concentration.

Results from the Kruskal-Wallis nonparametric test are listed in Figure 3. A comparison summary of the nonparametric results and percent change values are presented in Figure 6. In this Figure, the degree of change from control is identified in 50% increments, using different colors. We introduced the percent change summary as another way of looking at the data, and potentially identifying effects overlooked by traditional statistics. Overall, comparing percent change calculations to nonparametric statistical results showed that the two techniques were mostly in agreement.

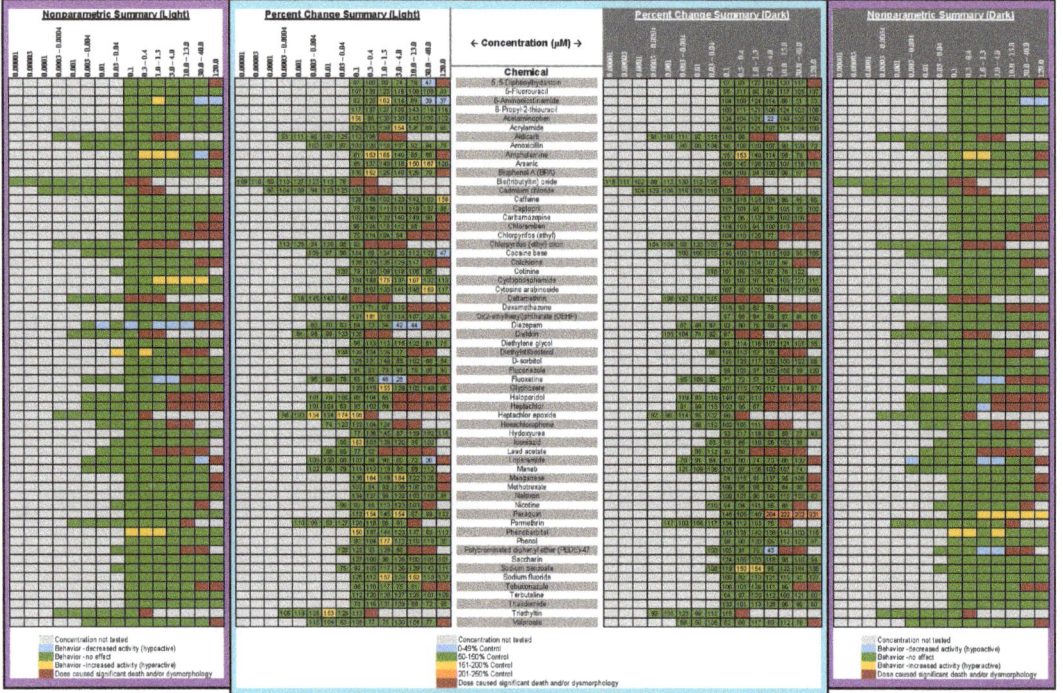

Figure 6. Comparison of Nonparametric Statistical Results and Percent Change Calculations. Comparison of the nonparametric statistical results (from Figure 3) and percent change calculations showing the degree of change in each concentration group compared to the controls. The middle column lists the chemical name, the outside columns show the nonparametric results for the Light and Dark periods, with the percent change columns next to them. Chemical concentrations are listed at the top of each column. Colored shading represents the following: light gray = concentration not tested; blue = decrease in activity; green = no effect; yellow = increase in activity; red = developmental toxicity. The percent change value is indicated in each cell and the data are color coded by 50% increments.

The comparison of our results with the results from other laboratories performing similar behavioral assays with larval zebrafish treated with the same chemicals during development is summarized in Figure 7. For 24 out of the total 61 chemicals, we were unable to find any published papers investigating the behavioral toxicity of those chemicals in larval zebrafish. We were, however, able to report information for 37 of the chemicals, and in many cases (29/37), found multiple papers that investigated the same chemical.

Legend:
- Noted as causing an effect, but the direction of effect was not clear.
- Noted as causing a **decrease** in activity.
- Noted as causing a **decrease and increase** in activity.
- Noted as causing an **increase** in activity.
- Reported results unclear, but it was reported that this concentration was tested
- Noted as causing developmental toxicity (i.e., lethality, non-hatching, malformations, uninflated swim bladder) (present study only).

The results for the current study are on the first line for each chemical, with bolded text and separated from comparison data by a double line
Superscript identifies the reference number for the entry.
All numerical entries are chemical concentrations in micromolar (µM).

5,5-Diphenylhydantoin	0.12	0.4	1.2	4	12	40	120					

5-Fluorouracil	0.12	0.4	1.2	4		12	40	120				
5-Fluorouracil[ ]		0.3	1	3		10	30					
5-Fluorouracil[ ]			1	2	4.5	9	18	34	67			

6-Aminonicotinamide	0.12	0.4	1.2	4	12	40	120					

6-Propyl-2-thiouracil	0.12	0.4	1.2		4	12	40	120				
6-Propyl-2-thiouracil[ ]		0.3	1	3		10	30					
6-Propyl-2-thiouracil[ ]			1	2	4.5	9	18	34	67			
6-Propyl-2-thiouracil[ ]						11.3	22.5	45		90	180	

Acetaminophen		0.12	0.4	1.2		4	12	40	120				
Acetaminophen[ ]				1	2	4.5	9	18	34	67			
Acetaminophen[ ]										100	210	410	830	1650
Acetaminophen[ ]							6.62			66.2		662		1323	2646	5292
Acetaminophen[ ]	0.03	0.17		0.81		4.13	20.67						
Acetaminophen[ ]	0.03		0.33			3.31							

Acrylamide	0.12	0.4	1.2		4	12	40	120				
Acrylamide[ ]		0.3	1	3		10	30					
Acrylamide[ ]			1	2	4.5	9	18	34	67			
Acrylamide[ ]						13.8	27.5	55	110	220		

Aldicarb	0.0004	0.0010	0.0043	0.0134	0.0419	0.1311	0.4096	1.2	4				
Aldicarb[ ]							0.3	1					
Aldicarb[ ]								1	2	4.5	9	18	34	67

Amoxicillin	0.004	0.012	0.04	0.12	0.4	1.2		4	12		40		120
Amoxicillin[ ]					0.3	1	3		10		30		
Amoxicillin[ ]						1	2	4.5	9	18	34	67	

Amphetamine	0.12	0.4	1.2	4	12	40	120

Arsenic	0.12	0.4	1.2	4	12	40	120

Bisphenol A (BPA)			0.12	0.4	1.2		4	12	40		120	
Bisphenol A[ ]				0.3	1		3	10				
Bisphenol A[ ]					1	2	4.5	9	18	34	67	
Bisphenol A[ ]			0.68	1.37	2.74	5.48	10.95					
Bisphenol A[ ]	0.001	0.01	0.1		1			10				
Bisphenol A[ ]					1		5	15				
Bisphenol A[ ]					1			10		30		

Bis(tributyltin) oxide	0.00001	0.00003	0.0001	0.0003	0.0010	0.0031	0.0098		0.0307	0.096		0.3
Bis(tributyltin) oxide[ ]						0.003	0.010		0.030	0.100		0.300
Bis(tributyltin) oxide[ ]				0.002	0.004	0.009	0.019	0.042	0.090	0.194		

Cadmium chloride	0.0001	0.0004	0.0012	0.0040	0.0126	0.0393	0.1228	0.384	1.2			
Cadmium chloride[ ]									5.45	54.5	109	218	436

Caffeine				0.12	0.4	1.2	4	12	40	120		
Caffeine[ ]										500	1000	2000

Captopril	0.12	0.4	1.2	4	12	40	120

Carbamazepine			0.12	0.4	1.2	4	12	40	120		
Carbamazepine[ ]										350	400
Carbamazepine[ ]	0.0042	0.042		0.42							
Carbamazepine[ ]							105.8	211.6		423.2	

Chloramben	0.12	0.4	1.2	4	12	40	120

Chlorpyrifos (ethyl)			0.12	0.4		1.2		4	12	40		120
Chlorpyrifos (+)				0.3		1		3				
Chlorpyrifos[ ]				0.3		1		3	10	30		
Chlorpyrifos[ ]						1	2	4.5	9	18	34	67
Chlorpyrifos[ ]			0.31	0.56	0.98	1.76	3.14					
Chlorpyrifos[ ]			0.3									
Chlorpyrifos[ ]			0.285									
Chlorpyrifos[ ]			0.285									
Chlorpyrifos[ ]		0.03		0.3								
Chlorpyrifos[ ]	0.001	0.01	0.1									
Chlorpyrifos[ ]		0.0285	0.0856	0.2852		0.8557						

Chlorpyrifos (ethyl) oxon	0.0003	0.0010	0.0032	0.0101	0.0315	0.0983	0.3072	0.96	3		

Cocaine base	0.004	0.012	0.04	0.12	0.4	1.2	4	12	40	120

Colchicine	0.12	0.4	1.2		4	12	40	120
Colchicine[ ]		0.3	1	3		10	30	

Figure 7. *Cont.*

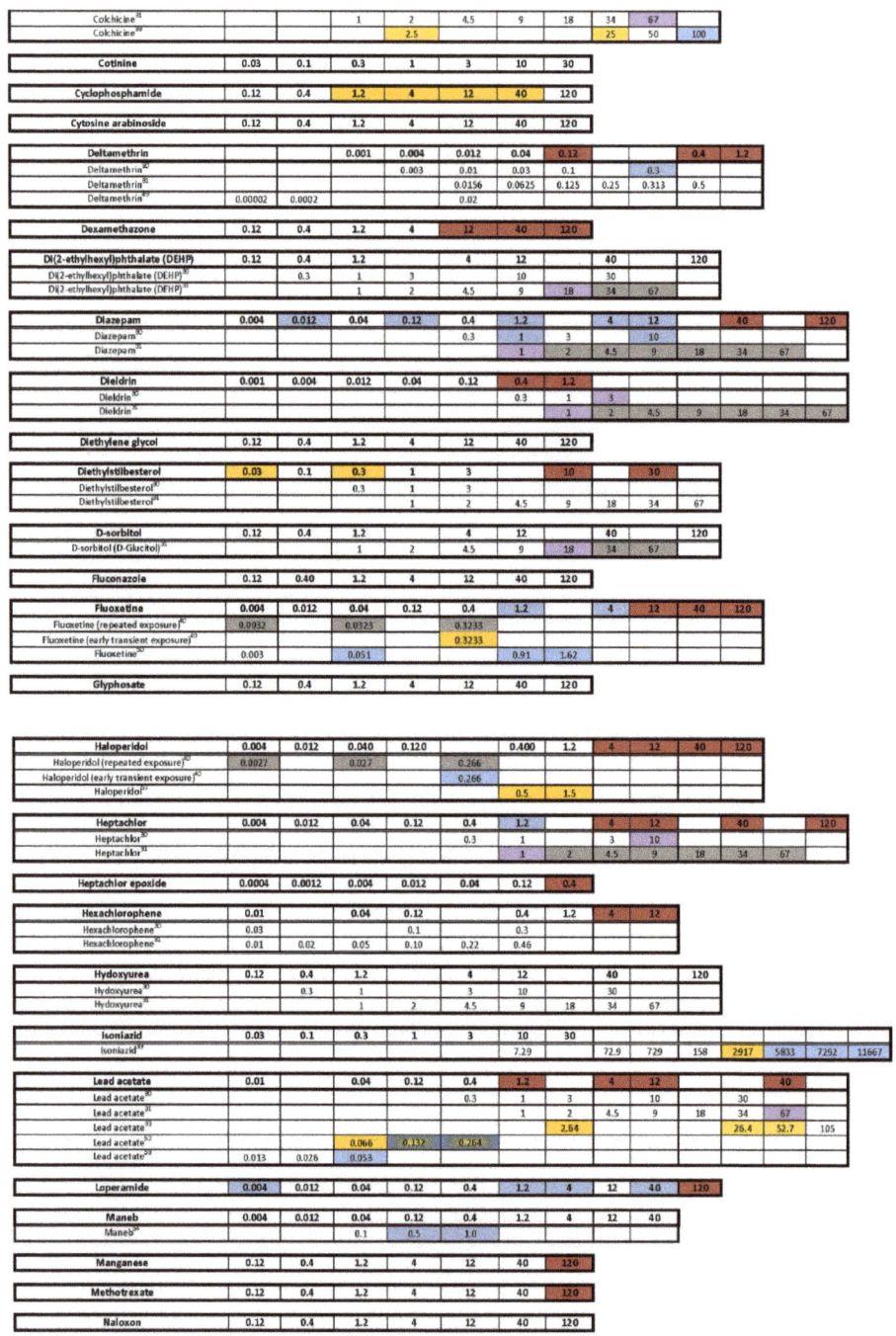

Figure 7. Cont.

Figure 7. **Comparison of the Present Behavioral Results with Previous Studies from the Literature that Included Similar Experimental Conditions Testing the Same Chemicals.** To be included, all studies met the following criteria: chemical exposure began during 0–3 dpf and lasted at least 24 h; behavior was tested 5–7 dpf, included an acclimation period prior to testing and at least one transition from Light to Dark during the testing protocol. This figure was populated based on information reported by other researchers; the results were not interpreted or inferred. Some studies only reported the lowest effect dose and did not report results for other concentrations that may also have had an effect. Effects may have occurred in the acclimation, Light or Dark periods. Colored shading represents the following: blue = decrease in activity; yellow = increase in activity; blue with yellow center = both decrease and increase in activity; purple = direction of the effect could not be determined; gray = chemical concentration was tested, but results were unclear and effect could not be determined; red = developmental toxicity for the current study only. Superscript refers to publication number in the reference section of this manuscript. The results for the current study are on the first line for each chemical, with bold text. Chlorpyrifos was both a test chemical and a positive control in our study; the results for the positive control are indicated by the (+) in this figure. All concentrations are in micromolar (μM).

4. Discussion

The current research evaluated a relatively large chemical library for gross developmental toxicity and behavioral effects (neurodevelopmental toxicity) following developmental exposure in embryonic/larval zebrafish. Then a subsequent comparison of our results to similar studies from other laboratories testing the same chemicals was made, and considerable variability among results was noted. We believe that part of this variability could be due to a general lack of comprehensive reporting of the experimental design and analyses. We have, therefore, endeavored to be detailed and measured in our experimental design and reporting.

In the present experiments, we took a rigorous approach to the experimental design and analysis of the data. Regarding the experimental design, we attempted to remove any chemical from the solution the larva was reared in by replacing 100% of the solution twice before behavioral testing. If the chemical is still present during behavioral testing, it is difficult to determine whether the behavioral effects are due to the chemical's action on brain development or are due to neuropharmacological actions. We suspect that this removal of the chemical before testing does separate developmental from pharmacological effects because when studying flame retardant chemicals [64] we obtained very different behavioral profiles depending on whether the chemicals were given acutely at the larval stage versus given during development and washed out before testing. Our approach to data analysis could also be regarded as conservative. If any concentration group presented with more than 25% abnormal animals, that entire concentration group was not included in the behavioral analysis; we believe those concentrations should be labeled as developmentally toxic. Within the concentration groups where there were $\geq$75% normal larvae, only larvae that presented as completely normal were included in any of the behavioral analyses. Moreover, our definition of normal appears to be stricter than some other laboratories: not only did the larva need to present without malformations, but the swim bladder had to be inflated. If the animal appeared normal with an uninflated swim bladder, that animal was not included in the behavioral analysis, as it is known that a zebrafish larva with an uninflated swim bladder does not behave normally in some assays [69,70]. In fact, if they do not inflate their swim bladder by 9 dpf, there is a high likelihood the larva will die [81]. Our approach to data analysis could also be regarded as conservative: because the data for the Light and Dark periods are not normally distributed (Figure 4), and because the number of independent observations in the control group was often more than the treated groups (Figure 3), nonparametric statistics were used. As the behavioral data in the Light and Dark periods are generated from the same animal, they are not independent observations and must be treated as repeated measures, so a Bonferroni correction was applied such that the α for the overall dose-response relationship for the Light or Dark was set to $\leq$0.025. Only if that overall dose response relationship was significant were step-down analyses conducted to determine which concentration groups were different from controls. In addition, because there have been admonishments in other publications to move beyond p values [82,83], we have included a table which shows the degree of change in each concentration group (Figure 6). This figure also includes a graphical representation of the results from the statistical analyses for comparison. The % change section of Figure 6 is color coded by 50% increments so that readers can judge for themselves about their degree of concern.

One other issue with zebrafish behavioral data analysis that has been discussed is the issue of endpoints. In the present analyses, only two endpoints (total locomotor activity in either the Light or Dark period) are used to assess the Light/Dark locomotor response data. As the full 100 min, light/dark behavioral profiles are quite complex (Supplementary Figure S2), there are many other behavioral endpoints to be captured and analyzed (e.g., [84–86]). Perhaps the larval zebrafish behavioral assessment community can capture those other behavioral characteristics in an organized and consistent manner so that "behavioral barcodes" linked to modes of action can be developed, much like how the acute effects of neuroactive chemicals have been indexed to unique behavioral patterns (e.g., [87,88]). In addition to a deeper analysis of the Light/Dark locomotor assay in larval zebrafish, perhaps we should augment the larval testing battery with other behavioral assays delving into other sectors of nervous system function. Both anxiety and pre-pulse inhibition are two behavioral assessments that have been developed for larval zebrafish and associated with neuropsychiatric disorders in humans (reviewed in: [89,90]).

Comparing our results to those previously published (Figure 7), we are prompted to ask some important questions:

(1) **Are there chemicals among multiple publications that consistently cause or do not cause behavioral effects? This would allow us to identify possible positive and negative controls.** There were five chemicals that appear to be candidates for positive controls: diazepam, fluoxetine, paraquat, PBDE-47, and chlorpyrifos. One publication reported decreased activity for diazepam in a similar concentration range as the present study, and the other paper reported behavioral changes, but whether it was an increase or decrease in activity was unclear as only a lowest effective dose was reported. As diazepam is known to be pharmacologically active at the gamma-aminobutyric acid receptor (reviewed in [91]), perhaps diazepam could be regarded as a positive control for GABAergic chemicals. For fluoxetine, one publication, as well as our own, reported decreased activity in larvae treated with fluoxetine during development, while another publication reported increased activity in animals treated with fluoxetine transiently during an early developmental window. Some of the effective concentration ranges aligned. As fluoxetine is a serotonin reuptake inhibitor, this chemical may serve as a positive control for the serotonergic disrupting class of chemicals. Although only one other publication tested paraquat in a developmental neurotoxicity test using zebrafish, the results were very similar to the present study, with both reporting markedly increased activity in the same dosage range. As paraquat has been reported to disrupt the development of the dopaminergic nervous system (reviewed in [92]), this chemical may serve as a positive control for the dopaminergic disrupting class of chemicals. The data for PBDE-47 as a possible positive control are a bit weaker mainly because only one other publication investigated the behavioral effects of developmental exposure to PBDE-47, and the effective concentration range did not overlap with our own data; however, both noted decreased activity. The fifth chemical that might serve as a positive control among testing publications is chlorpyrifos. There are multiple reports of developmental chlorpyrifos exposure producing behavioral alterations in larval zebrafish assays, but the range of effective concentrations spans four orders of magnitude. Because chlorpyrifos is an anticholinesterase, this chemical could serve as a positive control for the disruption of the cholinergic nervous system during development. In contrast, there are four chemicals that are candidates for negative controls, although the number of observations is smaller: aldicarb, amoxicillin, hexachlorophene and hydroxyurea. In all cases, there are two publications as well as the present study showing that developmental exposure to these chemicals in approximately the same concentration range did not produce behavioral alterations in the larval locomotor assay.

(2) **Are there chemicals that other publications have shown to produce behavioral changes after developmental exposure, but at concentrations that exceeded our concentration range or at concentrations that we deemed developmentally toxic?** Eight chemicals (aldicarb, cadmium chloride, caffeine, carbamazepine, deltamethrin, dieldrin, isoniazid, nicotine) would fall into that category. In fact, caffeine and isoniazid did not appear to produce behavioral effects unless tested in the millimolar range.

(3) **Are there unique chemicals that only our laboratory has tested that produced changes in larval locomotor activity after developmental exposure?** There were four chemicals that were tested in this publication that produced changes in locomotor activity after developmental exposure that other publications appear not to have tested: developmental exposure to 6-aminonicotinamide or loperamide produced decreased activity in the larvae, and developmental exposure to amphetamine produced an inverted "U" biphasic pattern of increased activity at lower concentrations and decreased activity at the higher concentrations. Cyclophosphamide also showed increased activity in the middle concentrations. In our laboratory embryos treated with diethylstilbesterol

during development showed increased activity at concentrations below those that tested negative in other publications.

(4) **Are there chemicals that have shown behavioral effects in other studies, but were not positive in our study?** There were three chemicals, valproate, chlorpyrifos, and lead (Pb) that fall into this category. Positive results were expected for chlorpyrifos and valproate because they have tested positive previous times in our laboratory [42,62–64]. Specifically, we have published two papers showing developmental valproate exposure elicits behavioral changes in larval zebrafish [62,63]. It appears that the developmental toxicity profile in the present study is similar to the previously published papers: 120 µM concentration caused malformations and death in a large portion of the larvae, and 40 µM was on the cusp of developmental toxicity. The behavioral toxicity, however, was not apparent in this present study as it had been in the previous studies. The other two publications i.e., [62,63] tested about twice as many animals per concentration, so perhaps this present result is an issue of statistical power. Statistical power may have also played a role in the disparate results for chlorpyrifos in the present study. In this study we tested chlorpyrifos in two different scenarios: one as a positive control throughout the study and the other as one of the chemicals under investigation. The results are summarized for both in Figure 7 with the positive control data listed as "Chlorpyrifos+", and the data for the test chemical listed as "Chlorpyrifos (ethyl)." As our positive control with many more observations (n = 115–132 per concentration), chlorpyrifos produced positive results in the same pattern that we often see: hypoactivity in both the Light and Dark periods, with the Dark period activity being less sensitive than the Light period activity (details in Figure 2). When testing chlorpyrifos as one of our test chemicals, however, with fewer observations (n = 14–16 per concentration), we hypothesize that there was less statistical power to detect the change. These negative results for chlorpyrifos or valproate indicate that we may need to increase the number of observations at each concentration in future developmental neurotoxicity screens. A power analysis was done when setting up our experimental design, but because the behavioral data are skewed, and require nonparametric analysis, it is difficult to perform an accurate power analysis for non-normally distributed endpoints. Lead (Pb) was another chemical where we expected a positive result given that four out of the five previous publications reported behavioral changes in larval zebrafish (Figure 7). Our results showed developmental toxicity ≥ 1.2 µM. Many of the larvae in the 1.2 and 4.0 µM concentrations showed a preponderance of uninflated swim bladders in the absence of other malformations, and therefore were not included in the behavioral assessment. If these animals had been tested in the behavioral protocol, there would have been markedly decreased activity in the Light period. One possibility to consider would be that swim bladder inflation may be a neurotoxic endpoint. Inflation of the swim bladder not only requires innervation [93,94], but it also requires a behavioral repertoire where the larva seeks out the air/water interface to take a gulp of air [95]. So perhaps swim bladder non-inflation belongs intercalated between a morphological and behavioral endpoint, and if an animal presents with an uninflated swim bladder, this could be logged as a potentially neurotoxic endpoint without behavioral confirmation.

(5) **Were there chemicals that showed considerable variation in the published results?** Four publications, including ours, tested 6-propyl-2-thiouracil with overlapping concentrations spanning about three orders of magnitude; only one publication out of the four reported changes in behavior. Six publications including our own tested acetaminophen, again with many testing in the same concentration ranges, and yet only three of the publications reporting changes in behavior. There was some overlap in the positive concentrations in two of the publications, but the third publication only found behavioral changes at millimolar concentrations. Only one out of four publications found that developmental carbamazepine produced behavioral alterations in larval zebrafish tested in the Light/Dark transition assay. For deltamethrin,

only one out of four publications found behavioral changes, whereas our laboratory reported developmental toxicity in the concentration range where the behavioral changes were reported. Three out of four publications did not find behavioral alterations after developmental saccharin exposure. Interestingly, saccharin is one of the few chemicals in this testing library that was classified as a "favorable" negative control chemical for developmental neurotoxicity screens [96], meaning that an expert panel's assessment of the chemical showed very little to no evidence that the chemical produces developmental neurotoxicity. Five publications studied the effects of tebuconazole on behavioral profiles in larval zebrafish with two publications reporting a positive result, and the other three publications testing in that same concentration range reported negative results. There were also contrasting results with thalidomide, where one out of three publications reported behavioral changes, but the other two publications reported a negative result in the same concentration range. These types of discrepancies indicate that the zebrafish larval Light/Dark locomotor assay will require more protocol and analysis standardization among laboratories.

Even though an effort was made to target similar assays for composing the summary in Figure 7, differences among the assay procedures and analyses could lead to the differing results. A lack of standardized reporting of specific experimental conditions created challenges in cataloging the results. Surprisingly, many experimental factors such as age, temperature, duration of chemical exposure, presence/absence of chemical during testing or presence/absence of the chorion were not specified in many publications. Rarely were the larval assessment criteria (i.e., morphological features that classified a larva as abnormal or not) clearly specified. Lack of standardization in reporting also makes it difficult to understand the specifics of the experimental design and subsequent analyses. Even with these omissions and differences, some chemicals have been identified that appear to be consistent positives or negatives across multiple laboratories.

In this publication we tested a relatively large group of chemicals for developmental neurotoxicity potential using a zebrafish behavioral assay and compared our results to publications using the same chemicals and employing a similar experimental design. There appears to be considerable variability within the literature regarding larval zebrafish behavioral alterations after developmental exposure to some of the chemicals. This comparison also allowed identification of some chemicals that are consistent positives and negatives across publications and prompts us to identify ways to improve the experimental design and interpretation of the assay that we conduct in our own laboratory. As a step toward data transparency and inter-laboratory collaboration, we have included all of our raw behavioral data to allow exploration of the data by other investigators and to encourage more zebrafish behavioral data sharing in the future.

Supplementary Materials: The following supporting information can be downloaded at: https://www.mdpi.com/article/10.3390/toxics10050256/s1. Figure S1, Effect of DMSO on Light/Dark Locomotor Activity; Figure S2, Time Course Behavioral Graph, Activity Box Plots and Developmental Toxicity for each Chemical; Table S1, Raw Data.

Author Contributions: Conceptualization, K.A.J. and S.P.; methodology, D.L.H., M.R.W. and S.P.; software, D.L.H., B.N.H. and M.R.W.; validation, D.L.H., B.N.H. and J.K.O.; formal analysis, K.A.J. and S.P.; investigation, B.N.H. and M.R.W.; data curation, K.A.J., D.L.H., B.N.H., J.K.O., M.R.W. and S.P.; writing—original draft preparation, K.A.J., B.N.H., K.N.B. and S.P.; writing—review and editing, K.A.J., D.L.H., B.N.H., J.K.O., K.N.B., M.R.W. and S.P.; visualization, K.A.J., J.K.O., K.N.B.; supervision, K.A.J. and S.P.; project administration, B.N.H. and M.R.W. All authors have read and agreed to the published version of the manuscript.

Funding: This research received no external funding.

Institutional Review Board Statement: The animal study protocol was approved by the EPA Office of Research and Development's Institutional Animal Care and Use Committee (IACUC) in RTP, NC (protocol #20-08-003, approved on 29 August 2017).

Informed Consent Statement: Not applicable.

Data Availability Statement: All raw data are included in the supplementary information and will also be uploaded to https://edg.epa.gov/metadata/catalog/main/home.page.

Acknowledgments: This manuscript has been subjected to review by the Center for Computational Toxicology and Exposure and approved for publication. Approval does not signify that the contents reflect the views of the Agency, nor does mention of trade names or commercial products constitute endorsement or recommendation for use. This project was supported, in part, by an appointment to the Research Participation Program at the Office of Research and Development administered by the Oak Ridge Institute for Science and Education through an interagency agreement with the U.S. Environmental Protection Agency. The authors wish to thank Drs. Tim Shafer, William Boyes, Aimen Farraj, and Kimberly Slentz-Kesler for critically reviewing earlier versions of this paper. Thanks also to Morgan Lowery for assembling the tables of raw data for publication, and Joan Hedge for expert management of the zebrafish facility.

Conflicts of Interest: The authors declare no conflict of interest.

References

1. Ghassabian, A.; Vandenberg, L.; Kannan, K.; Trasande, L. Endocrine-Disrupting Chemicals and Child Health. *Annu. Rev. Pharmacol. Toxicol.* **2022**, *62*, 573–594. [CrossRef]
2. Zablotsky, B.; Black, L.I.; Maenner, M.J.; Schieve, L.A.; Danielson, M.L.; Bitsko, R.H.; Blumberg, S.J.; Kogan, M.D.; Boyle, C.A. Prevalence and Trends of Developmental Disabilities among Children in the United States: 2009–2017. *Pediatrics* **2019**, *144*, e20190811. [CrossRef]
3. Bennett, D.; Bellinger, D.C.; Birnbaum, L.S.; Bradman, A.; Chen, A.; Cory-Slechta, D.A.; Engel, S.M.; Fallin, M.D.; Halladay, A.; Hauser, R.; et al. Project TENDR: Targeting Environmental Neuro-Developmental Risks The TENDR Consensus Statement. *Environ. Health Perspect.* **2016**, *124*, A118–A122. [CrossRef]
4. Balaguer-Trias, J.; Deepika, D.; Schuhmacher, M.; Kumar, V. Impact of Contaminants on Microbiota: Linking the Gut-Brain Axis with Neurotoxicity. *Int. J. Environ. Res. Public Health* **2022**, *19*, 1368. [CrossRef]
5. Dórea, J.G. Exposure to environmental neurotoxic substances and neurodevelopment in children from Latin America and the Caribbean. *Environ. Res.* **2021**, *192*, 110199. [CrossRef]
6. Preston, M.A.; Finseth, L.T.; Bourne, J.N.; Macklin, W.B. A novel myelin protein zero transgenic zebrafish designed for rapid readout of in vivo myelination. *Glia* **2019**, *67*, 650–667. [CrossRef]
7. Baier, H.; Wullimann, M.F. Anatomy and function of retinorecipient arborization fields in zebrafish. *J. Comp. Neurol.* **2021**, *529*, 3454–3476. [CrossRef]
8. Thyme, S.B.; Pieper, L.M.; Li, E.H.; Pandey, S.; Wang, Y.; Morris, N.S.; Sha, C.; Choi, J.W.; Herrera, K.J.; Soucy, E.R.; et al. Phenotypic Landscape of Schizophrenia-Associated Genes Defines Candidates and Their Shared Functions. *Cell* **2019**, *177*, 478–491.e420. [CrossRef]
9. Wood, J.D.; Bonath, F.; Kumar, S.; Ross, C.A.; Cunliffe, V.T. Disrupted-in-schizophrenia 1 and neuregulin 1 are required for the specification of oligodendrocytes and neurones in the zebrafish brain. *Hum. Mol. Genet.* **2009**, *18*, 391–404. [CrossRef]
10. Pilorge, M.; Fassier, C.; Le Corronc, H.; Potey, A.; Bai, J.; De Gois, S.; Delaby, E.; Assouline, B.; Guinchat, V.; Devillard, F.; et al. Genetic and functional analyses demonstrate a role for abnormal glycinergic signaling in autism. *Mol. Psychiatry* **2016**, *21*, 936–945. [CrossRef]
11. Goldshtein, H.; Muhire, A.; Petel Legare, V.; Pushett, A.; Rotkopf, R.; Shefner, J.M.; Peterson, R.T.; Armstrong, G.A.B.; Russek-Blum, N. Efficacy of Ciprofloxacin/Celecoxib combination in zebrafish models of amyotrophic lateral sclerosis. *Ann. Clin. Transl. Neurol.* **2020**, *7*, 1883–1897. [CrossRef]
12. Tilson, H.A. Neurobehavioral methods used in neurotoxicological research. *Toxicol. Lett.* **1993**, *68*, 231–240. [CrossRef]
13. Bushnell, P.J. Testing for cognitive function in animals in a regulatory context. *Neurotoxicol. Teratol.* **2015**, *52*, 68–77. [CrossRef]
14. Bownik, A.; Wlodkowic, D. Applications of advanced neuro-behavioral analysis strategies in aquatic ecotoxicology. *Sci. Total Environ.* **2021**, *772*, 145577. [CrossRef]
15. Kalueff, A.V.; Gebhardt, M.; Stewart, A.M.; Cachat, J.M.; Brimmer, M.; Chawla, J.S.; Craddock, C.; Kyzar, E.J.; Roth, A.; Landsman, S.; et al. Towards a comprehensive catalog of zebrafish behavior 1.0 and beyond. *Zebrafish* **2013**, *10*, 70–86. [CrossRef]
16. Fero, K.; Yokogawa, T.; Burgess, H.A. The behavioral repertoire of larval zebrafish. In *Zebrafish Models in Neurobehavioral Research*; Kalueff, A., Cachat, J., Eds.; Neuromethods; Humana Press: Totowa, NJ, USA, 2011; Volume 52.
17. Couderq, S.; Leemans, M.; Fini, J.B. Testing for thyroid hormone disruptors, a review of non-mammalian in vivo models. *Mol. Cell. Endocrinol.* **2020**, *508*, 110779. [CrossRef]
18. Porazzi, P.; Calebiro, D.; Benato, F.; Tiso, N.; Persani, L. Thyroid gland development and function in the zebrafish model. *Mol. Cell. Endocrinol.* **2009**, *312*, 14–23. [CrossRef]

19. Walter, K.M.; Miller, G.W.; Chen, X.; Yaghoobi, B.; Puschner, B.; Lein, P.J. Effects of thyroid hormone disruption on the ontogenetic expression of thyroid hormone signaling genes in developing zebrafish (*Danio rerio*). *Gen. Comp. Endocrinol.* **2019**, *272*, 20–32. [CrossRef]
20. Eachus, H.; Choi, M.K.; Ryu, S. The Effects of Early Life Stress on the Brain and Behaviour: Insights From Zebrafish Models. *Front. Cell Dev. Biol.* **2021**, *9*, 657591. [CrossRef]
21. Goldstone, J.V.; McArthur, A.G.; Kubota, A.; Zanette, J.; Parente, T.; Jonsson, M.E.; Nelson, D.R.; Stegeman, J.J. Identification and developmental expression of the full complement of Cytochrome P450 genes in Zebrafish. *BMC Genom.* **2010**, *11*, 643. [CrossRef]
22. Nawaji, T.; Yamashita, N.; Umeda, H.; Zhang, S.; Mizoguchi, N.; Seki, M.; Kitazawa, T.; Teraoka, H. Cytochrome P450 Expression and Chemical Metabolic Activity before Full Liver Development in Zebrafish. *Pharmaceuticals* **2020**, *13*, 456. [CrossRef]
23. de Esch, C.; Slieker, R.; Wolterbeek, A.; Woutersen, R.; de Groot, D. Zebrafish as potential model for developmental neurotoxicity testing: A mini review. *Neurotoxicol. Teratol.* **2012**, *34*, 545–553. [CrossRef]
24. Fitzgerald, J.A.; Konemann, S.; Krumpelmann, L.; Zupanic, A.; Vom Berg, C. Approaches to Test the Neurotoxicity of Environmental Contaminants in the Zebrafish Model: From Behavior to Molecular Mechanisms. *Environ. Toxicol. Chem.* **2021**, *40*, 989–1006. [CrossRef]
25. Nishimura, Y.; Murakami, S.; Ashikawa, Y.; Sasagawa, S.; Umemoto, N.; Shimada, Y.; Tanaka, T. Zebrafish as a systems toxicology model for developmental neurotoxicity testing. *Congenit. Anom. (Kyoto)* **2015**, *55*, 1–16. [CrossRef]
26. Mundy, W.R.; Padilla, S.; Breier, J.M.; Crofton, K.M.; Gilbert, M.E.; Herr, D.W.; Jensen, K.F.; Radio, N.M.; Raffaele, K.C.; Schumacher, K.; et al. Expanding the test set: Chemicals with potential to disrupt mammalian brain development. *Neurotoxicol. Teratol.* **2015**, *52*, 25–35. [CrossRef]
27. Harrill, J.A.; Freudenrich, T.; Wallace, K.; Ball, K.; Shafer, T.J.; Mundy, W.R. Testing for developmental neurotoxicity using a battery of in vitro assays for key cellular events in neurodevelopment. *Toxicol. Appl. Pharmacol.* **2018**, *354*, 24–39. [CrossRef]
28. Shafer, T.J.; Brown, J.P.; Lynch, B.; Davila-Montero, S.; Wallace, K.; Friedman, K.P. Evaluation of Chemical Effects on Network Formation in Cortical Neurons Grown on Microelectrode Arrays. *Toxicol. Sci.* **2019**, *169*, 436–455. [CrossRef]
29. Carstens, K.E.; Carpenter, A.F.; Martin, M.M.; Harrill, J.A.; Shafer, T.J.; Paul Friedman, K. Integrating data from in vitro New Approach Methodologies for Developmental Neurotoxicity. *Toxicol. Sci.* **2022**, *187*, 62–79. [CrossRef]
30. Dach, K.; Yaghoobi, B.; Schmuck, M.R.; Carty, D.R.; Morales, K.M.; Lein, P.J. Teratological and Behavioral Screening of the National Toxicology Program 91-Compound Library in Zebrafish (*Danio rerio*). *Toxicol. Sci.* **2019**, *167*, 77–91. [CrossRef]
31. Hagstrom, D.; Truong, L.; Zhang, S.; Tanguay, R.; Collins, E.S. Comparative Analysis of Zebrafish and Planarian Model Systems for Developmental Neurotoxicity Screens Using an 87-Compound Library. *Toxicol. Sci.* **2019**, *167*, 15–25. [CrossRef]
32. Selderslaghs, I.W.; Hooyberghs, J.; Blust, R.; Witters, H.E. Assessment of the developmental neurotoxicity of compounds by measuring locomotor activity in zebrafish embryos and larvae. *Neurotoxicol. Teratol.* **2013**, *37*, 44–56. [CrossRef]
33. Ali, S.; Champagne, D.L.; Richardson, M.K. Behavioral profiling of zebrafish embryos exposed to a panel of 60 water-soluble compounds. *Behav. Brain Res.* **2012**, *228*, 272–283. [CrossRef]
34. Nogueira, A.F.; Pinto, G.; Correia, B.; Nunes, B. Embryonic development, locomotor behavior, biochemical, and epigenetic effects of the pharmaceutical drugs paracetamol and ciprofloxacin in larvae and embryos of *Danio rerio* when exposed to environmental realistic levels of both drugs. *Environ. Toxicol.* **2019**, *34*, 1177–1190. [CrossRef]
35. Xia, L.; Zheng, L.; Zhou, J.L. Effects of ibuprofen, diclofenac and paracetamol on hatch and motor behavior in developing zebrafish (*Danio rerio*). *Chemosphere* **2017**, *182*, 416–425. [CrossRef]
36. Fraser, T.W.K.; Khezri, A.; Jusdado, J.G.H.; Lewandowska-Sabat, A.M.; Henry, T.; Ropstad, E. Toxicant induced behavioural aberrations in larval zebrafish are dependent on minor methodological alterations. *Toxicol. Lett.* **2017**, *276*, 62–68. [CrossRef]
37. Wang, X.; Dong, Q.; Chen, Y.; Jiang, H.; Xiao, Q.; Wang, Y.; Li, W.; Bai, C.; Huang, C.; Yang, D. Bisphenol A affects axonal growth, musculature and motor behavior in developing zebrafish. *Aquat. Toxicol.* **2013**, *142–143*, 104–113. [CrossRef]
38. Olsvik, P.A.; Whatmore, P.; Penglase, S.J.; Skjaerven, K.H.; Angles d'Auriac, M.; Ellingsen, S. Associations Between Behavioral Effects of Bisphenol A and DNA Methylation in Zebrafish Embryos. *Front. Genet.* **2019**, *10*, 184. [CrossRef]
39. Pruvot, B.; Quiroz, Y.; Voncken, A.; Jeanray, N.; Piot, A.; Martial, J.A.; Muller, M. A panel of biological tests reveals developmental effects of pharmaceutical pollutants on late stage zebrafish embryos. *Reprod. Toxicol.* **2012**, *34*, 568–583. [CrossRef]
40. Huang, I.J.; Sirotkin, H.I.; McElroy, A.E. Varying the exposure period and duration of neuroactive pharmaceuticals and their metabolites modulates effects on the visual motor response in zebrafish (*Danio rerio*) larvae. *Neurotoxicol. Teratol.* **2019**, *72*, 39–48. [CrossRef]
41. Pohl, J.; Ahrens, L.; Carlsson, G.; Golovko, O.; Norrgren, L.; Weiss, J.; Orn, S. Embryotoxicity of ozonated diclofenac, carbamazepine, and oxazepam in zebrafish (*Danio rerio*). *Chemosphere* **2019**, *225*, 191–199. [CrossRef]
42. Dishaw, L.V.; Hunter, D.L.; Padnos, B.; Padilla, S.; Stapleton, H.M. Developmental exposure to organophosphate flame retardants elicits overt toxicity and alters behavior in early life stage zebrafish (*Danio rerio*). *Toxicol. Sci.* **2014**, *142*, 445–454. [CrossRef] [PubMed]
43. Glazer, L.; Wells, C.N.; Drastal, M.; Odamah, K.A.; Galat, R.E.; Behl, M.; Levin, E.D. Developmental exposure to low concentrations of two brominated flame retardants, BDE-47 and BDE-99, causes life-long behavioral alterations in zebrafish. *Neurotoxicology* **2018**, *66*, 221–232. [CrossRef] [PubMed]
44. Li, R.; Zhang, L.; Shi, Q.; Guo, Y.; Zhang, W.; Zhou, B. A protective role of autophagy in TDCIPP-induced developmental neurotoxicity in zebrafish larvae. *Aquat. Toxicol.* **2018**, *199*, 46–54. [CrossRef] [PubMed]

45. Li, R.; Wang, H.; Mi, C.; Feng, C.; Zhang, L.; Yang, L.; Zhou, B. The adverse effect of TCIPP and TCEP on neurodevelopment of zebrafish embryos/larvae. *Chemosphere* **2019**, *220*, 811–817. [CrossRef] [PubMed]
46. Oliveri, A.N.; Bailey, J.M.; Levin, E.D. Developmental exposure to organophosphate flame retardants causes behavioral effects in larval and adult zebrafish. *Neurotoxicol. Teratol.* **2015**, *52*, 220–227. [CrossRef] [PubMed]
47. Richendrfer, H.; Pelkowski, S.D.; Colwill, R.M.; Creton, R. Developmental sub-chronic exposure to chlorpyrifos reduces anxiety-related behavior in zebrafish larvae. *Neurotoxicol. Teratol.* **2012**, *34*, 458–465. [CrossRef]
48. Sun, L.; Xu, W.; Peng, T.; Chen, H.; Ren, L.; Tan, H.; Xiao, D.; Qian, H.; Fu, Z. Developmental exposure of zebrafish larvae to organophosphate flame retardants causes neurotoxicity. *Neurotoxicol. Teratol.* **2016**, *55*, 16–22. [CrossRef]
49. Awoyemi, O.M.; Kumar, N.; Schmitt, C.; Subbiah, S.; Crago, J. Behavioral, molecular and physiological responses of embryo-larval zebrafish exposed to types I and II pyrethroids. *Chemosphere* **2019**, *219*, 526–537. [CrossRef]
50. de Farias, N.O.; Oliveira, R.; Sousa-Moura, D.; de Oliveira, R.C.S.; Rodrigues, M.A.C.; Andrade, T.S.; Domingues, I.; Camargo, N.S.; Muehlmann, L.A.; Grisolia, C.K. Exposure to low concentration of fluoxetine affects development, behaviour and acetylcholinesterase activity of zebrafish embryos. *Comp. Biochem. Physiol. Part C Toxicol. Pharmacol.* **2019**, *215*, 1–8. [CrossRef]
51. Oliveri, A.N.; Levin, E.D. Dopamine D1 and D2 receptor antagonism during development alters later behavior in zebrafish. *Behav. Brain Res.* **2019**, *356*, 250–256. [CrossRef]
52. Chen, J.; Chen, Y.; Liu, W.; Bai, C.; Liu, X.; Liu, K.; Li, R.; Zhu, J.H.; Huang, C. Developmental lead acetate exposure induces embryonic toxicity and memory deficit in adult zebrafish. *Neurotoxicol. Teratol.* **2012**, *34*, 581–586. [CrossRef] [PubMed]
53. Zhu, B.; Wang, Q.; Shi, X.; Guo, Y.; Xu, T.; Zhou, B. Effect of combined exposure to lead and decabromodiphenyl ether on neurodevelopment of zebrafish larvae. *Chemosphere* **2016**, *144*, 1646–1654. [CrossRef] [PubMed]
54. Cao, F.; Souders, C.L., 2nd; Li, P.; Pang, S.; Liang, X.; Qiu, L.; Martyniuk, C.J. Developmental neurotoxicity of maneb: Notochord defects, mitochondrial dysfunction and hypoactivity in zebrafish (*Danio rerio*) embryos and larvae. *Ecotoxicol. Environ. Saf.* **2019**, *170*, 227–237. [CrossRef] [PubMed]
55. Holden, L.L.; Truong, L.; Simonich, M.T.; Tanguay, R.L. Assessing the hazard of E-Cigarette flavor mixtures using zebrafish. *Food Chem. Toxicol.* **2020**, *136*, 110945. [CrossRef] [PubMed]
56. Crosby, E.B.; Bailey, J.M.; Oliveri, A.N.; Levin, E.D. Neurobehavioral impairments caused by developmental imidacloprid exposure in zebrafish. *Neurotoxicol. Teratol.* **2015**, *49*, 81–90. [CrossRef]
57. Wang, X.H.; Souders, C.L., 2nd; Zhao, Y.H.; Martyniuk, C.J. Paraquat affects mitochondrial bioenergetics, dopamine system expression, and locomotor activity in zebrafish (*Danio rerio*). *Chemosphere* **2018**, *191*, 106–117. [CrossRef]
58. Perez-Rodriguez, V.; Souders, C.L., 2nd; Tischuk, C.; Martyniuk, C.J. Tebuconazole reduces basal oxidative respiration and promotes anxiolytic responses and hypoactivity in early-staged zebrafish (*Danio rerio*). *Comp. Biochem. Physiol. Part C Toxicol. Pharmacol.* **2019**, *217*, 87–97. [CrossRef] [PubMed]
59. Kumar, N.; Awoyemi, O.; Willis, A.; Schmitt, C.; Ramalingam, L.; Moustaid-Moussa, N.; Crago, J. Comparative Lipid Peroxidation and Apoptosis in Embryo-Larval Zebrafish Exposed to 3 Azole Fungicides, Tebuconazole, Propiconazole, and Myclobutanil, at Environmentally Relevant Concentrations. *Environ. Toxicol. Chem.* **2019**, *38*, 1455–1466. [CrossRef]
60. Bailey, J.M.; Oliveri, A.N.; Karbhari, N.; Brooks, R.A.; De La Rocha, A.J.; Janardhan, S.; Levin, E.D. Persistent behavioral effects following early life exposure to retinoic acid or valproic acid in zebrafish. *Neurotoxicology* **2016**, *52*, 23–33. [CrossRef]
61. Chen, J.; Lei, L.; Tian, L.; Hou, F.; Roper, C.; Ge, X.; Zhao, Y.; Chen, Y.; Dong, Q.; Tanguay, R.L.; et al. Developmental and behavioral alterations in zebrafish embryonically exposed to valproic acid (VPA): An aquatic model for autism. *Neurotoxicol. Teratol.* **2018**, *66*, 8–16. [CrossRef]
62. Cowden, J.; Padnos, B.; Hunter, D.; MacPhail, R.; Jensen, K.; Padilla, S. Developmental exposure to valproate and ethanol alters locomotor activity and retino-tectal projection area in zebrafish embryos. *Reprod. Toxicol.* **2012**, *33*, 165–173. [CrossRef]
63. Zellner, D.; Padnos, B.; Hunter, D.L.; MacPhail, R.C.; Padilla, S. Rearing conditions differentially affect the locomotor behavior of larval zebrafish, but not their response to valproate-induced developmental neurotoxicity. *Neurotoxicol. Teratol.* **2011**, *33*, 674–679. [CrossRef]
64. Jarema, K.A.; Hunter, D.L.; Shaffer, R.M.; Behl, M.; Padilla, S. Acute and developmental behavioral effects of flame retardants and related chemicals in zebrafish. *Neurotoxicol. Teratol.* **2015**, *52*, 194–209. [CrossRef]
65. Padilla, S.; Hunter, D.L.; Padnos, B.; Frady, S.; MacPhail, R.C. Assessing motor activity in larval zebrafish: Influence of extrinsic and intrinsic variables. *Neurotoxicol. Teratol.* **2011**, *33*, 624–630. [CrossRef]
66. Chen, T.H.; Wang, Y.H.; Wu, Y.H. Developmental exposures to ethanol or dimethylsulfoxide at low concentrations alter locomotor activity in larval zebrafish: Implications for behavioral toxicity bioassays. *Aquat. Toxicol.* **2011**, *102*, 162–166. [CrossRef]
67. Christou, M.; Kavaliauskis, A.; Ropstad, E.; Fraser, T.W.K. DMSO effects larval zebrafish (*Danio rerio*) behavior, with additive and interaction effects when combined with positive controls. *Sci. Total Environ.* **2020**, *709*, 134490. [CrossRef]
68. Huang, Y.; Cartlidge, R.; Walpitagama, M.; Kaslin, J.; Campana, O.; Wlodkowic, D. Unsuitable use of DMSO for assessing behavioral endpoints in aquatic model species. *Sci. Total Environ.* **2018**, *615*, 107–114. [CrossRef]
69. Hill, B.N.; Coldsnow, K.D.; Hunter, D.L.; Hedge, J.M.; Korest, D.; Jarema, K.A.; Padilla, S. Assessment of Larval Zebrafish Locomotor Activity for Developmental Neurotoxicity Screening. In *Experimental Neurotoxicology Methods*; Llorens, J., Barenys, M., Eds.; Humana: New York, NY, USA, 2021; Volume 172, pp. 327–351. [CrossRef]
70. Burgess, H.A.; Granato, M. The neurogenetic frontier—lessons from misbehaving zebrafish. *Brief. Funct. Genom. Proteom.* **2008**, *7*, 474–482. [CrossRef]

71. Westerfield, M. *The Zebrafish Book: A Guide for the Laboratory Use of Zebrafish (Danio Rerio)*, 4th ed.; University of Oregon Press: Eugene, OR, USA, 2000.
72. Burgess, H.A.; Granato, M. Modulation of locomotor activity in larval zebrafish during light adaptation. *J. Exp. Biol.* **2007**, *210*, 2526–2539. [CrossRef]
73. Emran, F.; Rihel, J.; Adolph, A.R.; Wong, K.Y.; Kraves, S.; Dowling, J.E. OFF ganglion cells cannot drive the optokinetic reflex in zebrafish. *Proc. Natl. Acad. Sci. USA* **2007**, *104*, 19126–19131. [CrossRef]
74. Fernandes, A.M.; Fero, K.; Arrenberg, A.B.; Bergeron, S.A.; Driever, W.; Burgess, H.A. Deep brain photoreceptors control light-seeking behavior in zebrafish larvae. *Curr. Biol.* **2012**, *22*, 2042–2047. [CrossRef]
75. Prober, D.A.; Rihel, J.; Onah, A.A.; Sung, R.J.; Schier, A.F. Hypocretin/orexin overexpression induces an insomnia-like phenotype in zebrafish. *J. Neurosci.* **2006**, *26*, 13400–13410. [CrossRef]
76. Tufi, S.; Leonards, P.; Lamoree, M.; de Boer, J.; Legler, J.; Legradi, J. Changes in Neurotransmitter Profiles during Early Zebrafish (*Danio rerio*) Development and after Pesticide Exposure. *Environ. Sci. Technol.* **2016**, *50*, 3222–3230. [CrossRef]
77. MacPhail, R.C.; Brooks, J.; Hunter, D.L.; Padnos, B.; Irons, T.D.; Padilla, S. Locomotion in larval zebrafish: Influence of time of day, lighting and ethanol. *Neurotoxicology* **2009**, *30*, 52–58. [CrossRef]
78. Baker, N.; Knudsen, T.; Williams, A. Abstract Sifter: A comprehensive front-end system to PubMed. *F1000Research* **2017**, *6*, 2164. [CrossRef]
79. Hamm, J.T.; Ceger, P.; Allen, D.; Stout, M.; Maull, E.A.; Baker, G.; Zmarowski, A.; Padilla, S.; Perkins, E.; Planchart, A.; et al. Characterizing sources of variability in zebrafish embryo screening protocols. *ALTEX* **2019**, *36*, 103–120. [CrossRef]
80. Legradi, J.; el Abdellaoui, N.; van Pomeren, M.; Legler, J. Comparability of behavioural assays using zebrafish larvae to assess neurotoxicity. *Environ. Sci. Pollut. Res.* **2015**, *22*, 16277–16289. [CrossRef]
81. Goolish, E.M.; Okutake, K. Lack of gas bladder inflation by the larvae of zebrafish in the absence of an air-water interface. *J. Fish Biol.* **1999**, *55*, 1054–1063. [CrossRef]
82. Amrhein, V.; Greenland, S.; McShane, B. Scientists rise up against statistical significance. *Nature* **2019**, *567*, 305–307. [CrossRef]
83. Nuzzo, R. Scientific method: Statistical errors. *Nature* **2014**, *506*, 150–152. [CrossRef]
84. Haigis, A.C.; Ottermanns, R.; Schiwy, A.; Hollert, H.; Legradi, J. Getting more out of the zebrafish light dark transition test. *Chemosphere* **2022**, *295*, 133863. [CrossRef]
85. Hsieh, J.H.; Ryan, K.; Sedykh, A.; Lin, J.A.; Shapiro, A.J.; Parham, F.; Behl, M. Application of Benchmark Concentration (BMC) Analysis on Zebrafish Data: A New Perspective for Quantifying Toxicity in Alternative Animal Models. *Toxicol. Sci.* **2019**, *167*, 92–104. [CrossRef]
86. Liu, Y.; Ma, P.; Cassidy, P.A.; Carmer, R.; Zhang, G.; Venkatraman, P.; Brown, S.A.; Pang, C.P.; Zhong, W.; Zhang, M.; et al. Statistical Analysis of Zebrafish Locomotor Behaviour by Generalized Linear Mixed Models. *Sci. Rep.* **2017**, *7*, 2937. [CrossRef]
87. Ellis, L.D.; Seibert, J.; Soanes, K.H. Distinct models of induced hyperactivity in zebrafish larvae. *Brain Res.* **2012**, *1449*, 46–59. [CrossRef]
88. Irons, T.D.; MacPhail, R.C.; Hunter, D.L.; Padilla, S. Acute neuroactive drug exposures alter locomotor activity in larval zebrafish. *Neurotoxicol. Teratol.* **2010**, *32*, 84–90. [CrossRef]
89. Powell, S.B.; Weber, M.; Geyer, M.A. Genetic models of sensorimotor gating: Relevance to neuropsychiatric disorders. *Curr. Top. Behav. Neurosci.* **2012**, *12*, 251–318. [CrossRef]
90. Hanswijk, S.I.; Spoelder, M.; Shan, L.; Verheij, M.M.M.; Muilwijk, O.G.; Li, W.; Liu, C.; Kolk, S.M.; Homberg, J.R. Gestational Factors throughout Fetal Neurodevelopment: The Serotonin Link. *Int. J. Mol. Sci.* **2020**, *21*, 5850. [CrossRef]
91. Cerne, R.; Lippa, A.; Poe, M.M.; Smith, J.L.; Jin, X.; Ping, X.; Golani, L.K.; Cook, J.M.; Witkin, J.M. GABAkines-Advances in the discovery, development, and commercialization of positive allosteric modulators of GABAA receptors. *Pharmacol. Ther.* **2021**, *16*, 108035. [CrossRef]
92. Bastías-Candia, S.; Zolezzi, J.M.; Inestrosa, N.C. Revisiting the Paraquat-Induced Sporadic Parkinson's Disease-Like Model. *Mol. Neurobiol.* **2019**, *56*, 1044–1055. [CrossRef]
93. Finney, J.L.; Robertson, G.N.; McGee, C.A.; Smith, F.M.; Croll, R.P. Structure and autonomic innervation of the swim bladder in the zebrafish (*Danio rerio*). *J. Comp. Neurol.* **2006**, *495*, 587–606. [CrossRef]
94. Robertson, G.N.; McGee, C.A.; Dumbarton, T.C.; Croll, R.P.; Smith, F.M. Development of the swimbladder and its innervation in the zebrafish, *Danio rerio*. *J. Morphol.* **2007**, *268*, 967–985. [CrossRef]
95. Lindsey, B.W.; Smith, F.M.; Croll, R.P. From inflation to flotation: Contribution of the swimbladder to whole-body density and swimming depth during development of the zebrafish (*Danio rerio*). *Zebrafish* **2010**, *7*, 85–96. [CrossRef]
96. Martin, M.M.; Baker, N.C.; Boyes, W.K.; Carstens, K.E.; Culbreth, M.E.; Gilbert, M.E.; Harrill, J.A.; Nyffeler, J.; Padilla, S.; Paul Friedman, K.; et al. An expert-driven literature review of "negative" reference chemicals for developmental neurotoxicity (DNT) in vitro assay evaluation. *Neurotoxicol. Teratol.* **2021**, *submitted*.

Article

Chronic Perigestational Exposure to Chlorpyrifos Induces Perturbations in Gut Bacteria and Glucose and Lipid Markers in Female Rats and Their Offspring

Narimane Djekkoun [1,2], Flore Depeint [3], Marion Guibourdenche [1], Hiba El Khayat El Sabbouri [1], Aurélie Corona [1], Larbi Rhazi [3], Jerome Gay-Queheillard [1], Leila Rouabah [2], Farida Hamdad [4], Véronique Bach [1], Moncef Benkhalifa [1,4] and Hafida Khorsi-Cauet [1,*]

[1] PeriTox UMR_I 01 Laboratory, University Center for Health Research, CURS-UPJV, Picardy Jules Verne University, CEDEX 1, 80054 Amiens, France; djekkoun.narimane@gmail.com (N.D.); marion.guibourdenche@outlook.fr (M.G.); hiba.el-khayat-el-sabbouri@univ-cotedazur.fr (H.E.K.E.S.); aurelie.corona@u-picardie.fr (A.C.); jerome.gay@u-picardie.fr (J.G.-Q.); veronique.bach@u-picardie.fr (V.B.); benkhalifamoncef78@gmail.com (M.B.)
[2] Laboratory of Cellular and Molecular Biology, University of the Brothers Mentouri Constantine 1, Constantine 2500, Algeria; leilarouabah27@yahoo.fr
[3] Transformations & Agro-Ressources ULR7519, Institut Polytechnique UniLaSalle—Université d'Artois, 60026 Beauvais, France; flore.depeint@unilasalle.fr (F.D.); larbi.rhazi@unilasalle.fr (L.R.)
[4] Center for Human Biology, CHU Amiens-Picardie, 80000 Amiens, France; hamdadfarida2002@yahoo.fr
* Correspondence: hafida.khorsi@u-picardie.fr; Tel.: +33-322-827-896

Abstract: An increasing burden of evidence is pointing toward pesticides as risk factors for chronic disorders such as obesity and type 2 diabetes, leading to metabolic syndrome. Our objective was to assess the impact of chlorpyrifos (CPF) on metabolic and bacteriologic markers. Female rats were exposed before and during gestation and during lactation to CPF (1 mg/kg/day). Outcomes such as weight, glucose and lipid profiles, as well as disturbances in selected gut bacterial levels, were measured in both the dams (at the end of the lactation period) and in their female offspring at early adulthood (60 days of age). The results show that the weight of CPF dams were lower compared to the other groups, accompanied by an imbalance in blood glucose and lipid markers, and selected gut bacteria. Intra-uterine growth retardation, as well as metabolic disturbances and perturbation of selected gut bacteria, were also observed in their offspring, indicating both a direct effect on the dams and an indirect effect of CPF on the female offspring. Co-treatment with inulin (a prebiotic) prevented some of the outcomes of the pesticide. Further investigations could help better understand if those perturbations mimic or potentiate nutritional risk factors for metabolic syndrome through high fat diet.

Keywords: pesticides; prebiotic; intestinal dysbiosis; perigestational; dysmetabolism; risk factor

1. Introduction

Chemical pollution of the environment by insecticides has become a global phenomenon [1]. They are defined as chemicals used to prevent and control pests, including vectors of human or animal diseases. They are used to control unwanted plant or animal species that interfere with agricultural products [2–4]. Pesticides can include herbicides, insecticides, fungicides, disinfectants and rodenticides [3]. According to the most recent statistics on agriculture, forestry and fisheries for the European Union, the total quantity of pesticides sold in Europe amounted to almost 360,000 tons with a significant use of fungicides and bactericides (44%), and herbicides (32%) [5].

The use of pesticides in both developing and developed countries has increased dramatically in recent times. They are widely used in agriculture to increase the yields, quality and appearance of products and reduce the need for agricultural labor [6]. Thus, the potential contamination of the environment by pesticides raises concerns for the public and regulators. In the process of reducing phytochemical products, France has implemented National Health and Environment Plans (PNSE1: 2004–2008, PNSE2: 2009–2013, PNSE3: 2015–2019, PNSE 4: 2020–2024), initiated by the law of 9 August 2004 relating to public health policy. They aim to study the health consequences of exposure to various environmental pollution and include, among other things, estimating the population's exposure to pesticides, improving knowledge of pesticide exposure and the monitoring of occupational exposures. The Ecophyto plan was created in 2008 to reduce the use of phytosanitary products in France whilst maintaining economically efficient agriculture. Initially, the Ecophyto plan aimed to reduce the use of these substances by 50% by 2020. The target has been extended to 2025 in the face of implementation difficulties [7,8].

Some pesticides are considered endocrine disruptors [9]; therefore, daily exposure is likely to have serious and irreversible effects on the health of individuals [10]. Organophosphates (OP) represent the largest category used in the world and the most widespread due to their bioaccumulation in the environment [11–14], although their use in France is currently declining due to their consequences on animal and human health. Human exposure to these pesticides is mainly oral, by ingestion of pesticide residues in fruits and vegetables [15]. The extensive and indiscriminate use of OP pesticides in agriculture has been of major concern due to its potentially known or suspected harmful effects on humans. Among the most widely used OP is chlorpyrifos (CPF). Chlorpyrifos was first synthesized by German researchers in 1930 and first introduced in the United States in 1965 as a household insecticide by The Dow Chemical Company (Midland, MI, USA) [16,17]. According to the classification of the World Health Organization (WHO), chlorpyrifos is a class II pesticide of moderate toxicity [18]. CPF is a potent OP insecticide with low water solubility (80.9 mg/L) and a high adsorption coefficient in soil, has a longer half-life in soil (ranging from 65 to 360 days) [19], a wide spectrum of insecticidal action and relative safety compared to other organophosphates, which has led to its intensive use in agriculture. After the application of chlorpyrifos, less than 0.1% of the pesticides applied have a real impact on the intended target. The rest of the residue remains in the environment [16].

CPF use has been closely monitored in recent years, and while it is officially still authorized in 20 EU member states, renewal has been postponed pending further safety reports. The UK government adopted strict restrictions on the use of chlorpyrifos in 2016 due to new safety issues for human health. In the United States, the use of chlorpyrifos indoors has been banned since 2001. The ban process for agricultural use began in 2015 in California, and Hawaii banned the sale and use of chlorpyrifos at the end of 2020 [20]. However, the situation remains unclear and CPF can still be found in a number of settings.

CPF is recognized as a neurotoxic agent due to its inhibitory effect on various cholinesterase (ChE) enzymes at the central nervous (CNS) and systemic level [21]. High levels of pesticide residues have been found in several human cohorts [22–26]. According to the United States Environmental Protection Agency (EPA), the no-observed-adverse-effect level (NOAEL) of CPF for acute dietary exposure for the inhibition of red blood cell cholinesterase is 0.5 mg/kg/day [27]. Exposure to CPF has been linked to significant alterations in metabolites that interfere with cellular energy production and amino acid metabolism [28–30]; it also acts on hormonal signaling [31,32] and metabolism [33,34], and leads to the disruption of glucose and lipid metabolisms [35–38], leading to weight gain [39–42] and increasing the risk of developing chronic non-communicable diseases.

Considering that the oral route is the main cause of human exposure, the impact of a pesticide on the digestive tract is of particular interest. Since it is the first physiological barrier to come into contact with ingested food contaminants, many studies began to focus on the impact of CPF on the gut barrier and gut microbiota [35,43–46]. The gut microbiota refers to billions of microorganisms residing in the intestine and has a mutualistic relationship with its host [47]. It has several functional roles and impacts on human physiology. It modulates the host's nutrition through the production of some vitamins and the fermentation of non-digestible food components by the host, protects against pathogens [48] and drug metabolism, and influences intestinal epithelial homeostasis [49].

Exposure to CPF induced by disturbance is often characterized by a decrease in the number of beneficial microorganisms and a simultaneous increase in the number of potentially pathogenic microorganisms leading to dysbiosis [50–53]. In addition, CPF has been shown to increase intestinal permeability in rats [43,45] or in vitro [54,55], inducing a bacterial translocation which corresponds to the passage of viable bacteria of the gastrointestinal flora through the barrier of the intestinal mucosa (the lamina propria), to the mesenteric nodes, then to normally sterile internal organs such as the spleen and the liver [56].

Pregnancy is a very sensitive period of life where epigenetic marks can impact the long-term development of chronic disorders in the next generation. This is known as the concept of "developmental origin of health and disease" (DOHaD). Contamination of CPF during this period can lead to delayed maturation of the gut microbiota, affecting the bowel function [57]. This may contribute to the onset of obesity and type 2 diabetes (T2D) later in life [58].

Compelling evidence suggests that oral prebiotic supplementation improves these metabolic disorders [59,60]. Additionally, prebiotics are likely associated with increased bifidobacteria and lactobacilli, and the production of short chain fatty acids (SCFA), which are involved in modulating host metabolism [61]. They also strengthen the intestinal barrier, increase satiety by promoting intestinal hormones, improve glucose tolerance, and counteract fatty liver disease (lipogenesis) and insulin resistance [62].

The main objective of the study was to evaluate the effect of perigestional exposure to CPF on the metabolic regulations of dams and their female offspring in early adulthood. For this purpose, weight, lipid and glucose metabolism, levels of selected intestinal bacteria and bacterial translocation were assessed. The secondary objective was to study the protective effect of a prebiotic (inulin) on the same parameters.

2. Materials and Methods

2.1. Experimental Conditions

Chlorpyrifos (*O*,*O*-diethyl-*O*-(3,5,6-trichloro-2-pyridinyl)-phosphorothioate) with a purity of 99.8% was supplied by LGC Standards (Molsheim, France). Inulin was a kind gift from Cosucra (Belgium). Wistar rats were purchased from Janvier laboratories (Le Genest Saint Isle, France). Animal standard diet (Serlab3436, 3.1 kcal/g, Serlab, Montataire, France) was identical throughout the study.

The experiment was carried out according to the protocol approved by the Regional Directorate for Health, Animal Protection and the Environment (Amiens, France) and the Ministry of Research (reference number APAFIS # 8207-2016121322563594 v2 approved on 5 September 2017). All animals were treated in accordance with the EU Directive 2010/63.

Animals were housed in a NexGen Max cage system mounted on an EcoFlow rack system (Allentown Inc, Bussy Saint Georges, France) under constant conditions in a temperature- and air-controlled room (23 °C) with a 12-h light/dark cycle. The size of each group was set to four females for the reproduction protocol in order to reach a minimum of 6–10 female offspring for the passive impact analysis. After acclimation, sixteen 7-week old female Wistar rats (body weight (b.w.) 225 ± 4.9 g) were randomly assigned to four groups ($n = 4$/group) and housed two per cage.

Chlorpyrifos was dissolved in commercially available organic rapeseed oil at a concentration of 10 mg/mL. CPF solution or rapeseed oil only were administered daily (5 consecutive days followed by a 2-day break) by gavage at a dose of 1 mL/kg b.w. to the animals. This was equivalent to a final concentration of 1 mg/kg b.w. (the oral "no observed adverse effect level", or NOAEL, for inhibition of cerebral cholinesterase activity in rats [63]) for the CPF groups. Chicory inulin was dissolved in water at a concentration of 10 g/L. Animals were given access to the inulin-enriched or regular drinking water ad libitum. The treatments associated with each group is detailed in Figure 1.

	Dams				Female offspring			
	Control	CPF	Control inulin	CPF inulin	Control	CPF	Control inulin	CPF inulin
Standard diet	X	X	X	X	X	X	X	X
Gavage with rapeseed oil (1 mL/kg b.w.)	X		X					
Gavage CPF in rapeseed oil (1 mg/kg b.w.)		X		X				
Drinking water	X	X			X	X	X	X
Inulin in drinking water (10 g/L)			X	X				

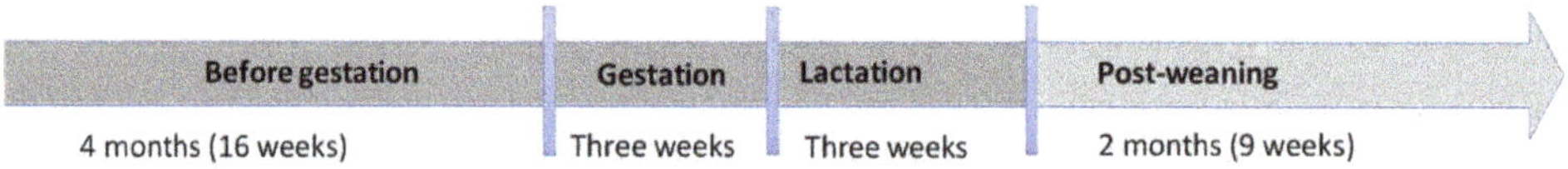

Before gestation	Gestation	Lactation	Post-weaning
4 months (16 weeks)	Three weeks	Three weeks	2 months (9 weeks)

Figure 1. Treatment groups. Dams were fed standard diet and various treatments per os before gestation until weaning at which time they were sacrificed. Their female offspring were fed a standard diet only until 60 days of age at which time they were sacrificed. CPF: Chlorpyrifos; b.w.: body weight.

Females were fed a standard diet and the corresponding gavage and drinking water for four consecutive months before mating them throughout gestation and lactation. Male rats were housed under the same conditions as females and received a standard diet. Gestation was assessed daily after mating (two females per male) by the presence of sperm in vaginal smears. The dams were then housed individually until birth and with their pups until weaning. The day of parturition was considered postnatal day 1 (PND1). The average litter size was 6 pups per dam. At the time of weaning (PND21), the female offspring were separated from their mother and housed with their littermates. Male offspring were included in a separate project [64,65]. They were housed according to density requirement in the EU legislation and fed a standard diet until the end of the experiment on PND60. Food and water consumption were measured. Dams were weighted daily throughout the experiment. To determine whether or not exposure during gestation and lactation induced growth retardation, pups were weighed at birth (PND3), at weaning (PND21) and at PND60.

At the end of the experiment (PND21 for dams, PND60 for offspring), the rats were euthanized by intraperitoneal administration of sodium pentobarbital, and plasma as well as various intestinal segments (ileum, colon and cecum) and internal organs (spleen, liver, mesenteric fat tissue and gonadal fat tissue) were removed under sterile conditions.

2.2. Metabolic Perturbations

Approximately 3 mL of intracardiac blood was collected from each animal into a test tube and centrifuged at $1500 \times g$ for 10 min at 4 °C. The harvested serum samples were transferred to clean test tubes before analyses.

Standard spectrophotometric methods based on an automation program from the University Hospital of Amiens (KT-6400 analyzer (Genrui Biotech Inc., Shenzhen, China)) were used to measure the following serum parameters: Cholesterol, High Density Lipoprotein (HDL), Low Density Lipoprotein (LDL), triglycerides (TG), blood sugar.

2.3. Disruption of Key Bacteria

2.3.1. Concentrations of Selected Intestinal Bacteria and Bacterial Translocation

Organs were weighed, placed in a sterile stomacher bag and homogenized in Ringer's saline solution (Bio-Rad, Marnes-la-Coquette, France) before serial 1/10 dilutions. Then, 1 mL of homogenate of intestinal segments and sterile organs was spread on different selective and non-selective media for qualitative and quantitative cultures of selected aerobic and anaerobic intestinal bacteria (Bactron Anaerobic, Sheldon Manufacturing, Cornelius, OR, USA) and incubated at 37 °C for 48 h to 4 days [51,52]. After incubation, bacteria were identified using standard microbiological techniques [45,50,66]. The isolated colonies grown on Petri dishes were counted using an automatic colony counter (Scan® 500, Interscience, St Nom la Bretèche, France) and expressed in log (CFU)/g of tissue.

2.3.2. Microbial Metabolites

Primary dilution samples for selected gut-bacteria analyses were used for short chain fatty acid (SCFA) measures. The supernatant was acidified to pH 2 using 25 µL of H_2SO_4 (2 M) and injected into a BP21 gas chromatography (GC) column (length: 30 m, inner diameter: 0.53 µm, film thickness: 1 µm) with an internal standard (4-hydroxy-4-methyl-2-pentanone). H_2 was used as a carrier gas at a flow rate of 1.5 mL/min. The initial oven temperature was 135 °C and was held there for 6 min, then raised by 25 °C/min to 180 °C and held for 1 min, then increased by 25 °C/min to 230 °C, and finally maintained at 230 °C for 1 min. Glass liner ultra-inert was used for the split injection. The temperatures of the flame ionization detector and the injection port were 280 °C and 240 °C, respectively. The flow rates of H_2 and air as makeup gas were 40 and 400 mL/min, respectively. The sample volume injected for the GC analysis (AutoSystem XL; PerkinElmer) was 1 µL and the run time for each analysis was approximately 10 min. The three main short-chain fatty acids (acetic acid, propionic acid, butyric acid) were identified as a function of the retention time of the different elution peaks. The quantification was obtained by comparison with a standard curve [67].

2.3.3. Serum Lipopolysaccharide (LPS)

Plasma LPS is a useful marker for the identification of increased intestinal permeability and thus of intestinal injury. Plasma assay was performed with the Rat Lipopolysaccharide ELISA kit according to the manufacturer instructions (#CSB E14247r, CliniSciences, Nanterre, France).

2.4. Statistical Analyses

Statistical analyses were performed with StatView software (version 5.0, Abacus Concepts Inc., Berkeley, CA, USA) and SPSS Statistics software (version 25.0). Data were analyzed using one-way and two-way ANOVA and KHI2 assay. The independence of endpoints measured in littermates was evaluated by one-way ANOVA for each parameter and treatment group to assess intra and intergroup effects. Except for body weight at PND3 and PND21 in control group, litter effect was not detected. Offspring were thus considered a valid experimental unit for the F1 impact analyses. In all analyses, the threshold of statistical significance was set at $p < 0.05$.

3. Results

3.1. Animal Weight and Weight Gain

To mimic chronic exposure, we exposed dams for 4 months before gestation, as well as during gestation and lactation (Table 1). No significant differences for weight gain were observed in dams. At PND3 (Figure 2A) the pups from dams exposed to CPF alone (7.5 g) were significantly smaller than the corresponding inulin group (9.9 g for CPFI, $p = 0.043$). This was still significant at weaning (45.3 g for CPF vs. 55.2 g for C and 56.0 g for CPFI, $p = 0.042$ and $p = 0.050$, respectively, Figure 2B). No significant variation remained at early adulthood (PND60, Figure 2C). To complete these observations, no variation was observed in drinking and eating patterns either.

Table 1. Body weight (g) before gestation (1st month and 4th month), during gestation (21st day of gestation) and during lactation (21st day of lactation) of dams exposed to CPF and inulin. The values are expressed as mean ± SEM ($n = 4$). Significantly different ($p < 0.05$) using analysis of variance (ANOVA) and Tukey's test. C: Control; CPF: Chlorpyrifos; CI: Control inulin; CPFI: Chlorpyrifos + inulin.

	Before Gestation (g)				Gestation (g)		Lactation (g)	
	1st Month		4th Month		21st Day of Pregnancy		21st Day of Lactation	
Control	268.7 ± 28.28		308.7 ± 31.75		396.6 ± 88.79		334.6 ± 17.47	
CPF	253.5 ± 23.86	F value = 0.980	286.7 ± 30.20	F value = 1.289	413.5 ± 105.35	F value = 0.347	311.0 ± 19.89	F value = 3.140
Control + Inulin	260.7 ± 33.62	p value = 0.44	297.2 ± 35.77	p value = 0.32	443.0 ± 68.06	p value = 0.793	392.0 ± 56.56	p value = 0.096
CPFI	283.0 ± 10.42		325.0 ± 11.63		447.0 ± 46.61		390.7 ± 42.46	

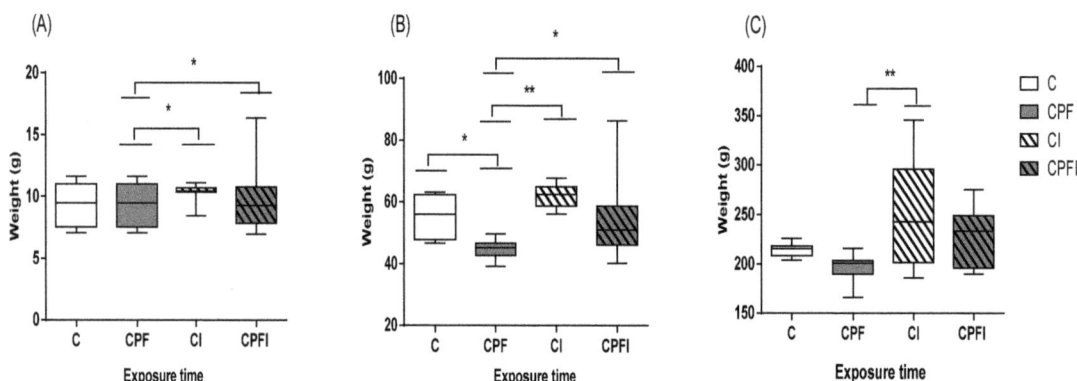

Figure 2. Effects of perigestational exposure to CPF and inulin on the body weight of the offspring of female rats 3 days after birth (PND3, (**A**)), as juveniles (PND21, (**B**)) and young adults (PND60, (**C**)). Values are expressed as mean ± SEM ($n = 7$–10) using analysis of variance (ANOVA) and Tukey's test. Significance * $p < 0.05$; ** $p < 0.01$; C: Control; CPF: Chlorpyrifos; CI: Control inulin; CPFI: Chlorpyrifos + inulin; PND: post-natal day.

3.2. Metabolic Perturbations

In dams and offspring to PND60, chronic exposure to CPF significantly altered the glycemic and lipid profile in rats (Table 2 and Figure 3).

Table 2. Effects of exposure to CPF and inulin on blood sugar (A) and lipid (B) levels in dams. Blood glucose, total cholesterol (TC), plasma triglycerides (TG), high density lipoproteins (HDL) or low density lipoproteins (LDL) were measured in plasma. The values are expressed as mean ± SEM ($n = 4$). Significantly different ($p < 0.05$) using analysis of variance (ANOVA) and Tukey's test. C: Control; CPF: Chlorpyrifos; CI: Control inulin; CPFI: Chlorpyrifos + inulin.

	Glucose (g/L)		Cholesterol (g/L)		Triglycerides (g/L)		HDL (g/L)		LDL (g/L)	
Control	7.8 ± 0.83	F value = 40.018	1.7 ± 0.25	F value = 6.480	1.4 ± 0.02	F value = 8.573	0.6 ± 0.02	F value = 3.817	0.8 ± 0.27	F value = 5.549
CPF	11.6 ± 0.08	p value = 0.0001	2.4 ± 0.24	p value = 0.02	1.8 ± 0.15	p value = 0.01	0.4 ± 0.08	p value = 0.066	1.6 ± 0.32	p value = 0.029
Control + inulin	7.4 ± 0.15		1.4 ± 0.17		1.3 ± 0.12		0.5 ± 0.10		0.5 ± 0.25	
CPFI	9.4 ± 0.40		2.0 ± 0.39		1.6 ± 0.12		0.6 ± 0.04		1.0 ± 0.36	

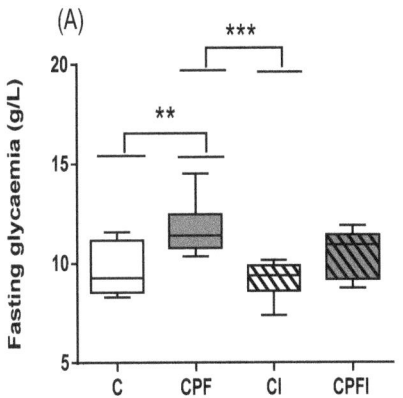
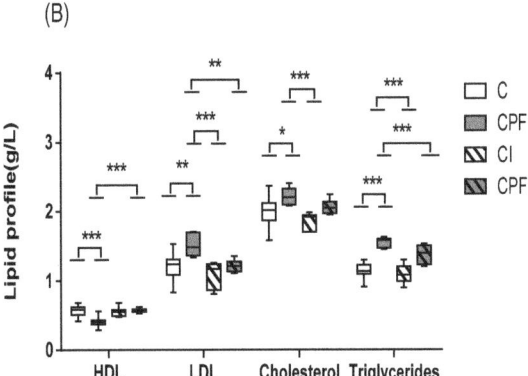

Figure 3. Effects of perigestational exposure to CPF and inulin on blood glucose (**A**) and lipid levels (**B**) in female offspring. Blood glucose, Total cholesterol (TC), plasma triglycerides (TG), high-density lipoproteins (HDL) or low-density lipoproteins (LDL) were measured in plasma. Values are expressed as mean ± SEM (n = 7–10) using analysis of variance (ANOVA) and Tukey's test. Significance * $p < 0.05$; ** $p < 0.01$, *** $p < 0.001$. C: Control; CPF: Chlorpyrifos; CI: Control inulin; CPFI: Chlorpyrifos + inulin.

Blood fasting glucose was significantly higher in dams exposed to CPF compared to the control ($p = 0.0001$) and the corresponding inulin-fed animals ($p = 0.004$), but only versus control group in offspring ($p = 0.002$, Figure 3A). Following direct exposure to CPF in dams or indirectly via lactation, lipid balance was significantly disturbed. In dams treated with CPF the total cholesterol increased by 20% ($p = 0.05$) and 28% for triglycerides ($p = 0.029$), while HDL-cholesterol levels decreased by 33% ($p = 0.048$). Moreover, in female offspring, as shown in Figure 3B, the CPF effect at PND60 remained strong with total cholesterol increased by 16% ($p = 0.012$), triglycerides by up to 15% ($p = 0.0001$ and $p = 0.009$ against C and CPFI, respectively), and LDL levels by up to 27% ($p = 0.001$ and $p = 0.002$ against C and CPFI, respectively). Finally, HDL levels decreased up to 30% in plasma of CPF offspring ($p = 0.0001$ and $p = 0.0001$ against C and CPFI, respectively).

3.3. Disturbances of the Selected Intestinal Bacteria

3.3.1. Concentrations of Selected Intestinal Bacteria

The pesticide altered the levels of selected gut bacteria in the dams. With regards to the total flora (total aerobic flora $p = 0.673$, total anaerobic flora $p = 0.673$), no significant difference between the groups was observed. When investigating specific bacterial populations (Tables 3 and 4), the concentration of *Lactobacillus* spp. decreases significantly by 1 log in the dams treated with CPF compared to the controls ($p = 0.05$). For *Bifidobacterium* spp., the results revealed no significant difference among the groups, but the level of *Bifidobacterium* decreased in the groups treated with CPF (0.3 log vs. control group). When considering potentially pathogenic flora, bacterial populations were more abundant for animals treated with CPF (0.4 log; $p = 0.009$ and 0.7 log; $p = 0.024$) against control group for *E. coli* and *Enterococcus*, respectively. Moreover, that the addition of the prebiotic decreased by 1 log the abundance of *Enterococcus* ($p = 0.003$). *Clostridium* population was increased in the CPF group (0.4 log vs. control) but the results showed no significant difference among the groups. In the offspring, no significant difference among the groups was observed for the total flora. The potentially beneficial flora, *Lactobacillus* spp. and *Bifidobacterium*, was significantly modulated in the different treatment groups (Figure 4A). Specifically, the level of *Lactobacillus* spp. decreased by 0.7 log with CPF against the control group ($p = 0.0001$) while inulin showed a protective 0.5 log increase in bacterial content for *Lactobacillus* ($p = 0.0001$). For potentially pathogenic flora (Figure 4B), *E. coli* was significantly more abundant (0.4 log) in rats treated with CPF ($p = 0.001$ vs. control group) and decreased by 0.5 log in the CPF group with inulin supplementation $p = 0.0001$ vs. CPF). Similarly, inulin reduced by 0.4 log the number of *Enterococcus* found in the CPF group ($p = 0.022$). The results of *Staphylococcus* and *Clostridium* showed no significant difference among the groups, but their abundance tended to increase with exposure to pesticides (0.2 log vs. control) and decrease with inulin supplementation (0.3 log vs. CPF only).

Table 3. Effects of exposure to CPF and inulin on beneficial flora in dams. The values are expressed as mean ± SEM ($n = 4$). Significantly different ($p < 0.05$) using analysis of variance (ANOVA) and Tukey's test. C: Control; CPF: Chlorpyrifos; CI: Control inulin; CPFI: Chlorpyrifos + inulin.

	Beneficial Flora (CFU/g)			
	Lactobacillus		*Bifidobacterium*	
Control	8.2 ± 0.09		7.2 ± 0.61	
CPF	7.2 ± 0.22	F value = 4.672	6.5 ± 0.32	F value = 2.192
Control + inulin	8.3 ± 0.66	p value = 0.04	7.6 ± 0.53	p value = 0.177
CPFI	7.9 ± 0.48		7.0 ± 0.60	

Table 4. Effects of exposure to CPF and inulin on potentially pathogenic flora in dams. The values are expressed as mean ± SEM ($n = 4$). Significantly different ($p < 0.05$) using analysis of variance (ANOVA) and Tukey's test. C: Control; CPF: Chlorpyrifos; CI: Control inulin; CPFI: Chlorpyrifos + inulin.

	Potentially Pathogenic Flora (CFU/g)							
	E.coli		*Enterococcus*		*Staphylococcus*		*Clostridium*	
Control	7.9 ± 0.26		7.6 ± 0.10		7.3 ± 0.28		6.7 ± 0.35	
CPF	8.3 ± 0.04	F value = 7.662	8.3 ± 0.14	F value = 13.012	8.2 ± 0.10	F value = 3.293	7.1 ± 0.20	F value = 0.935
Control + inulin	7.4 ± 0.31	p value = 0.013	7.4 ± 0.03	p value = 0.003	7.2 ± 0.28	p value = 0.088	7.0 ± 0.17	p value = 0.473
CPFI	7.9 ± 0.19		7.3 ± 0.33		7.5 ± 0.70		7.0 ± 0.36	

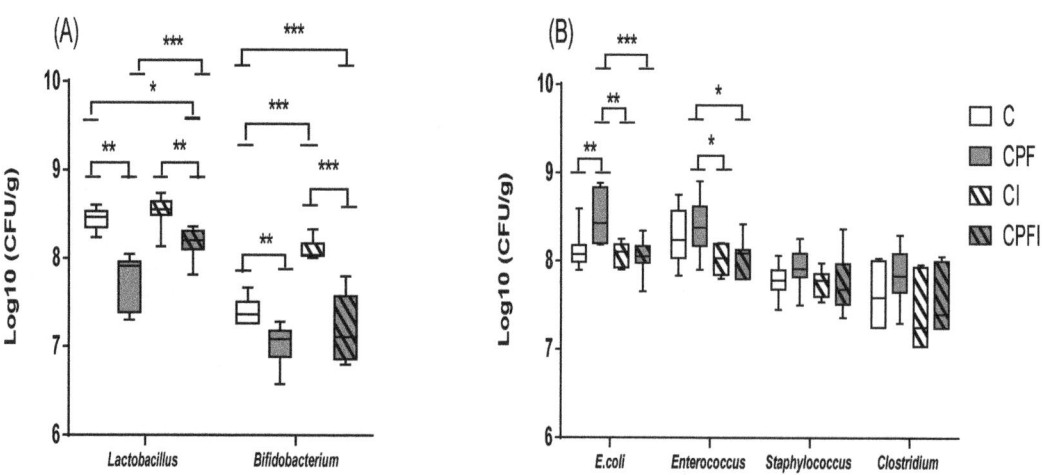

Figure 4. Effects of perigestational exposure to CPF and inulin on beneficial flora (**A**) and on potentially pathogenic flora (**B**) of female offspring. Values are expressed as mean ± SEM ($n = 7$–10) using analysis of variance (ANOVA) and Tukey's test. Significance * $p < 0.05$; ** $p < 0.01$, *** $p < 0.001$. C: Control; CPF: Chlorpyrifos; CI: Control inulin; CPFI: Chlorpyrifos + inulin.

3.3.2. Bacterial Metabolites

Another consequence of dysbiosis is the impaired production of short chain fatty acids (SCFA), which are dependent on various factors in the host microbiota. The profiles of individual SCFA differed depending on the treatment. The presence of CPF alone increased the acetic acid content (2%) in the dams, but not significantly. In the female offspring the CPF significantly increased the ratio of acetic acid and inulin prevented it ($p = 0.012$; CPF = 71.3% versus CPFI = 53.7%). The presence of inulin increased the levels of propionic and butyric acid in dams and offspring; variations were not significant for dams for either SCFA. In offspring, propionic acid was decreased with CPF and recovered with inulin (CPFI = 15.3% versus C = 15.4% versus CPF = 12.6%). Butyric acid inhibition by CPF and recovery by inulin, however, was significant in the offspring ($p = 0.008$; CPFI = 27.4% versus CPF = 16.0%).

3.3.3. Bacterial Translocation

Overall, a large aerobic and anaerobic translocation to the spleen was observed in the CPF groups (33% of dams and 25% of offspring), compared to the control and control inulin group (0% of dams and offspring) and CPF inulin (50% of dams and 0% of offspring). The CPF versus control differences were statistically significant only in the female offspring ($p = 0.046$). These results were corroborated by serum LPS concentrations in dams which were two-fold higher for the CPF group compared to the control and inulin, but no significant difference was observed. This was not observed in the offspring.

4. Discussion

Overall, whether it is through direct or indirect exposure to OP, there are changes in body weight, disruption of lipid metabolism and glucose metabolism, and changes in levels of selected gut bacteria. However, how CPF disrupts these metabolisms is not fully understood.

4.1. Animal Weight and Weight Gain

Weight gain for the dams as well as birth weight have been investigated extensively with regards to risk factors of chronic disorders (obesity, type 2 diabetes), but also as an indicator of the toxic capacity during gestation due to the development of the offspring. Several in vivo studies have suggested that CPF treatment can increase or decrease body weight and birth weight, depending on the dose tested.

The results of the present study revealed that the body weight of the CPF-treated groups was significantly lower than that of the other groups after chronic exposure to low levels of CPF. This is consistent with reports by several authors [68–72]. Researchers have suggested that a decrease in body weight may be due to increased oxidative stress and the degeneration of lipids and proteins [39,73,74]. Exposure to the 1 mg/kg dose did not significantly affect maternal weight gain compared to other groups. And the developmental assessment of female offspring indicated that all CPF-treated groups had low body mass gain compared to other PND60 groups, which is consistent with previous findings [38] and other studies [37,68,75,76].

4.2. Metabolic Perturbations

In this study, blood glucose concentration was markedly increased in both dams and female offspring. Consistent with these results, OP have been shown to induce hyperglycemia [37,38,77–80]. An increasing number of researchers have begun to focus on the mechanism of exposure to OP and hyperglycemia while other studies have reported that OP do not affect metabolic indices [35,81,82].

Changes in lipid profile due to CPF administration were observed in this study, other studies also showed an increase in TG in agreement with the present results [41,83] while others have observed lower or unchanged TG levels [37,84].

CPF also disrupts serum TC levels. In our study, cholesterol levels were higher, and several studies are in agreement with our results [28,37,41,83,84]. They suggested that serum TC levels may be increased in the groups treated with CPF due to blockage of the hepatic bile ducts, resulting in decreased or stopped TC secretion in the duodenum [39,85].

The result of the present study also demonstrated the dyslipidemic effect exerted by chronic administration of low levels of CPF. This was evident by the increased LDL levels and decreased HDL levels recorded in the serum of CPF-exposed rats. Some authors have suggested that CPF causes liver damage and lipoprotein synthesis [28,41,83,84].

In our experiment, chronic exposure to low doses of CPF was also responsible for developmental alterations in male pups, characterized by low birth weight, decreased plasma growth factor (IGF1) levels, leptin and a small increase in insulinemia. These effects were still present at weaning but disappeared in early adulthood when the animals were no longer exposed [64].

4.3. Disturbances of the Selected Intestinal Bacteria

To summarize, levels of key bacteria were unstable in dams and offspring with an increase in potentially pathogenic flora to the detriment of potentially beneficial flora in rats treated with CPF. CPF exposure was associated with significant microbial perturbation, showing the influence of CPF exposure on a number of bacteria, with reduced abundance of *Lactobacillus* spp. and *Bifidobacterium* spp., and a higher level of *Enterococcus* spp., *E. coli*, *Staphylococcus* spp. and *Clostridium* spp. in rats treated with CPF. Similar results were shown in previous work and other studies [50,84,86–89]. These results demonstrate that exposure to CPF induces disturbances in the gut microbiota and that early gut microbiota disruption could lead to long-term effects.

Another consequence of dysbiosis is the altered production of short-chain fatty acids (SCFA), which depend on various factors in the host microbiota. Butyrate, propionate and acetate represent 90 to 95% of SCFA present in the colon [90,91]. They are produced by fermentation of dietary fibers and are often associated with the prevention of several pathologies linked to inflammation or oxidative stress [92]. These results are in agreement with previous results on CPF [38] and other results on the different pesticides carbendazim [93], permethrin [94] and propamocarb [95], which suggests that the perturbation of the intestinal microbiota following pesticide exposure altered the proportions of SCFA, disturbed the energetic homeostasis and elicited multiple tissue inflammatory responses.

The bacterial translocation was predominantly higher in the CPF groups than in the other groups. It is recognized that microbial dysbiosis exerts a strong influence on the intestinal permeability [45,96,97]. The translocation phenomenon could be explained by the increase in permeability observed after exposure to CPF [45,50]. CPF alone inhibits the expression of tight junction and structural genes, while inulin and CPF/inulin co-treatment tended to increase expression [54]. Neonatal translocation is completely normal and even essential for the maturation of the immune system [98], but its persistence after weaning could become pathogenic. It is, therefore, likely that the pesticide disrupted the tight junctions, while the mixture enhanced the activity of the tight junctions [54]. Higher LPS concentrations in the serum of dams is a marker of permeability as well as a risk factor for low grade inflammation [99]. These are also consistent with C-reactive protein (CRP) serum concentrations in dams following CPF exposure (data not shown) as a sign of inflammation.

4.4. Nutritional Prevention

Prebiotic dietary fibers are likely to selectively modulate the gut microbiota and exert positive health effects [100]. The prevention approach used in this study was based on the use of inulin, acting as a prebiotic to counteract the effects of CPF on the gut microbiota [38,54], which in turn affects intestinal functions, such as metabolism and the integrity of the intestine [101]. Co-treatment with inulin prevents dysbiosis as well as early markers of dysmetabolism in rats chronically exposed to CPF. However, we could not find any other work in the literature studying this particular prevention strategy to counter pesticides except for an article on zebrafish [102].

Present results indicated that inulin consumption improved the lipid and glycemic profiles, consistent with other studies. The mechanism by which inulin acts on glucose and lipid metabolism remains unclear. A number of hypotheses have been proposed regarding the effect of prebiotics on improving dysmetabolism. One mechanism included decreased absorption of cholesterol via intestinal epithelial cells [103]. Inulin is a soluble, viscous compound that increases the thickness of the unstirred layer of the small intestine, thereby inhibiting absorption of cholesterol [104]. Inulin does not bind to bile acid in the upper digestive tract; however, it can help soluble bile acids bind to bacteria or insoluble compounds, such as calcium phosphate, by lowering the pH of the cecum [105]. As a result, fecal excretion of bile acids increases cholesterol utilization to rebuild bile acid in the liver and decrease the concentration of circulating and hepatic cholesterol [106]. Also, inulin treatment acts on glucose metabolism by improving serum GLP-1 levels to suppress IL-6 secretion and production, and hepatic gluconeogenesis and resulted in moderation of insulin tolerance. These results indicate that intestine-liver crosstalk is the primary mechanism for moderating insulin resistance by inulin [106–108].

Specific changes in the composition of the gut microbiota occurred in the prebiotic group; inulin increased the number of *Lactobacillus* and *Bifidobacterium* but also reduced the number of *Enterococcus* spp., *E. coli*, *Staphylococcus* spp. and *Clostridium* spp. The combination of CPF and inulin reduced the number of enterococci, while the numbers of *Bifidobacterium* and *Lactobacillus* increased. Another potential mechanistic pathway may be behind the change in the composition of the gut microbiota after inulin supplementation leading to an increase in SCFA and decrease in cecal pH. SCFA are important for health because they improve energy metabolism in the liver and muscles and immune function in the large intestine [109–112].

4.5. Perigestational Modulation of CPF

In the general population, new parents tend to look more closely into the food given to their infants, trying to aim for organic, pesticide-free options, even though their own diet may not be modified dramatically. It is not uncommon either that the diet of prospecting parents does not differ from that of the global population [113–115]. With this in mind, we set up this model to expose the dams to treatments before mating and continue it up to weaning but discontinue treatment for the offspring.

Differences in the toxicity of CPF between fetus and mother have already been reported due to the lipophilic power of CPF, which can cross the placental barrier [28,68,76,116]. CPF and its metabolite DETP (diethyl thiophosphonate) have been studied and published in the literature and residues have been detected in newborns [116,117]. 3,5,6-trichloro-2-pyridinol (TCPy), a specific metabolite of ethyl and methyl chlorpyrifos, was also recently found in the hair collected at birth in 311 new mothers [118]. This is, therefore, a critical window of exposure to toxic environment xenobiotics through aerosol or oral intake. During gestation and early childhood, the internal organs are still undergoing a process of maturation. Previous work showed that the intestine and microbiota maturation were delayed following CPF treatment [50]. The objective of the work presented herein was to better understand whether these perturbations would still have consequences if treatment was interrupted after weaning.

These data, together with the knowledge that a number of chronic disorders such as obesity or type 2 diabetes can be triggered by in utero imprinting, suggest that it is likely that CPF may lead to resilient effects event after treatment was discontinued. The results presented here for the offspring clearly support this hypothesis. It would be interesting to see if there were some epigenetic modifications in target tissues. Unfortunately sampling conditions did not allow for genomic analyses.

In addition, birth weight is thought to be predictive of chronic disorders. A low birth weight, often observed in experiments with OP, tends to be associated with adult risks of obesity and metabolic syndrome, whilst a larger birth weight would more likely be associated with increased risks of type 2 diabetes. Data show that both glucose and lipid profiles are modified by CPF, more targeted analyses or long-term experiments would be required to clearly differentiate between the two pathologies as they encounter a number of overlapping features.

5. Conclusions

The study allowed us to gather new evidence of resilient and indirect impact on offspring of CPF in metabolic disorders as summarized in Table 5. Modulations in specific bacterial populations as well as glucose and lipid profiles suggest mechanisms of dysmetabolism typical of obesity and diabetes [58]. The large number of endpoints measured would require statistical partitioning among variables, and this is a bias that may need to be highlighted when reaching overall conclusions. The strength of this work lies in the duration of the exposure of dams to the pesticide and the measure of indirect impact to the offspring to better mimic real-life settings. Further investigation in the epigenetic imprinting that could be associated with transgenerational effects would be complementary to these findings. In addition, while cultured microbiology may seem to be outdated for microbiota analyses compared to shotgun metagenomics, it is of great interest to follow up on bacterial translocation and perturbation of the intestinal permeability for a complete picture of the intestinal ecosystem. Finally, the nutritional prevention strategy can be further explored using other prebiotic molecules, probiotics or even testing different windows of exposure. A high-energy diet is also a risk factor leading to similar perturbations. It would be interesting to see whether mechanisms with CPF or high-fat diets are similar, and if a mix of dietary and environment factors would have synergistic effects. This hypothesis has been tested, samples are being analyzed and results are underway.

Table 5. Summary of direct and indirect impact of CPF and protective impact of inulin co-exposure.

		Direct Effect (Dams)		Indirect Effect (Offspring)	
		CPF	Inulin on CPF	CPF	Inulin on CPF
Metabolic	Weight	-	-	Loss (at PND21 only)	Recovery
	Glucose	Increased	Recovery	Increased	-
	Cholesterol	Increased (total) Increased (LDL) Decreased (HDL)	Recovery (total) - (LDL) Recovery (HDL)	Increased (total) Increased (LDL) Decreased (HDL)	- (total) Recovery (LDL) Recovery (HDL)
	Triglycerides	Increased	-	Increased	Recovery
Bacterial	Selected bacteria (+)	Decreased (*Lactobacillus*) - (*Bifidobacterium*)	- (*Lactobacillus*) - (*Bifidobacterium*)	Decreased (*Lactobacillus*) Decreased (*Bifidobacterium*)	Recovery (*Lactobacillus*) - (*Bifidobacterium*)
	Selected bacteria (−)	Increased (*E. coli*) Increased (*Enterococcus*)	- (*E. coli*) Recovery (*Enterococcus*)	Increased (*E. coli*) Increased (*Enterococcus*)	Recovery (*E. coli*) Recovery (*Enterococcus*)
	Metabolic ratio	-	-	Increased Acetate Decreased Butyrate	Recovery
	Translocation	-	-	Increased	-

Author Contributions: N.D.: Data curation; Investigation; Formal analysis; Visualization; Methodology; Writing—original draft; Writing—review & editing; F.D.: Conceptualization; Methodology; Validation; Writing—original draft; Writing—review & editing; M.G.: Investigation; Methodology; Writing—review & editing; H.E.K.E.S.: Investigation; Methodology; Writing—review & editing; A.C.: Investigation; Methodology; Writing—review & editing; L.R. (Larbi Rhazi): Investigation; Methodology; Writing—review & editing; J.G.-Q.: Methodology; Supervision; Funding acquisition; Writing—review & editing; L.R. (Leila Rouabah): Methodology; Writing—review & editing; F.H.: Investigation; Methodology; Writing—review & editing; V.B.: Supervision; Formal analysis; Writing—original draft; Writing—review & editing; M.B.: Investigation; Methodology; Writing—original draft; Writing—review & editing; H.K.-C.: Conceptualization; Visualization; Funding acquisition; Methodology; Project administration; Supervision; Validation; Writing—original draft; Writing—review & editing. All authors have read and agreed to the published version of the manuscript.

Funding: The animal experiment was supported by the Ministry of Higher Education, Research and Innovation and National Institutes of Health and Hauts-de-France region and was also funded by projet Fédératif Hospitalo-Universitaire "1000 days for Health" Proghomeo grant 2017.

Institutional Review Board Statement: The animal study protocol was approved by the Ethics Committee of the Regional Directorate for Health, Animal Protection and the Environment (Amiens, France) and the Ministry of Research (APAFIS # 8207-2016121322563594 v2 approved on 5 September 2017).

Informed Consent Statement: Not applicable.

Data Availability Statement: No open access to study data.

Acknowledgments: The authors thank Jérôme AUSEIL from the biochemistry laboratory, CHU Amiens for the metabolic assays.

Conflicts of Interest: The authors declare no conflict of interest.

Abbreviations

B.W.	body weight
C	Control
ChE	choline esterase
CI	control + inulin
CNS	central nervous system
CPF	chlorpyrifos
CPFI	CPF + inulin
CRP	C-reactive protein
DETP	diethyl thiophosphonate
DOHaD	Developmental origin of Health and Disease
EPA	United States Environmental Protection Agency
GC	gas chromatography
GLP1	glucose-like protein-1
HDL	high-density lipoprotein
IGF1	insulin growth factor-1
IL6	interleukin-6
LDL	low-density lipoprotein
LPS	lipopolysaccharide
NOAEL	no observed adverse effect level
OP	organophosphate
PND	post-natal day
PNSE	National Health and Environment Plans
SCFA	short chain fatty acid
SEM	standard error of the mean
T2D	type 2 diabetes
TCPy	3,5,6-trichloro-2-pyridinol
TG	triglyceride
WHO	World Health Organization

References

1. Wang, X.; Xing, H.; Li, X.; Xu, S.; Wang, X. Effects of atrazine and chlorpyrifos on the mRNA levels of IL-1 and IFN-γ2b in immune organs of common carp. *Fish. Shellfish Immunol.* **2011**, *31*, 126–133. [CrossRef] [PubMed]
2. OCSPP. *Opp Pesticides Industry Sales and Usage 2006 and 2007 Market Estimates*; OCSPP: Arlington, TX, USA, 2015.
3. World Health Organization; Safety International Programme on Chemical. *The WHO Recommended Classification of Pesticides by Hazard and Guidelines to Classification 2009*; World Health Organization: Geneva, Switzerland, 2010; ISBN 978-92-4-154796-3.
4. Solomon, K.R.; Williams, W.M.; Mackay, D.; Purdy, J.; Giddings, J.M.; Giesy, J.P. Properties and uses of chlorpyrifos in the United States. *Rev. Environ. Contam. Toxicol.* **2014**, *231*, 13–34. [CrossRef] [PubMed]
5. Sales of Pesticides in the EU. Available online: https://ec.europa.eu/eurostat/web/products-eurostat-news/-/ddn-20200603-1 (accessed on 23 September 2021).
6. Choudhary, S.; Yamini, N.R.; Yadav, S.K.; Kamboj, M.L.; Sharma, A. A review: Pesticide residue: Cause of many animal health problems. *J. Entomol. Zool. Stud.* **2018**, *6*, 330–333.
7. Guichard, L.; Dedieu, F.; Jeuffroy, M.-H.; Meynard, J.M.; Reau, R.; Savini, I. Le plan Ecophyto de réduction d'usage des pesticides en France: Décryptage d'un échec et raisons d'espérer. *Cah. Agric.* **2017**, *26*, 1–12. [CrossRef]
8. Renaudie, O. Les plans nationaux santé-environnement: Conciliation ou concurrence entre deux politiques publiques? *Rev. Droit Sanit. Soc.* **2019**, *HS*, 23–35.
9. Combarnous, Y. Endocrine Disruptor Compounds (EDCs) and agriculture: The case of pesticides. *Comptes Rendus Biol.* **2017**, *340*, 406–409. [CrossRef]
10. Jepson, P.C.; Murray, K.; Bach, O.; Bonilla, M.; Neumeister, L. *A Global Guideline for Pesticide Selection to Reduce Risks, and Establish a Minimum Pesticides List*; Social Science Research Network: Rochester, NY, USA, 2019.
11. Alamgir Zaman Chowdhury, M.; Fakhruddin, A.N.M.; Nazrul Islam, M.; Moniruzzaman, M.; Gan, S.H.; Khorshed Alam, M. Detection of the residues of nineteen pesticides in fresh vegetable samples using gas chromatography–mass spectrometry. *Food Control* **2013**, *34*, 457–465. [CrossRef]
12. Carr, R.L.; Ho, L.L.; Chambers, J.E. Selective toxicity of chlorpyrifos to several species of fish during an environmental exposure: Biochemical mechanisms. *Environ. Toxicol. Chem.* **1997**, *16*, 2369–2374. [CrossRef]
13. Casida, J.E.; Quistad, G.B. Organophosphate toxicology: Safety aspects of nonacetylcholinesterase secondary targets. *Chem. Res. Toxicol.* **2004**, *17*, 983–998. [CrossRef]
14. He, M.-J.; Lu, J.-F.; Wang, J.; Wei, S.-Q.; Hageman, K.J. Phthalate esters in biota, air and water in an agricultural area of western China, with emphasis on bioaccumulation and human exposure. *Sci. Total Environ.* **2020**, *698*, 134264. [CrossRef]
15. Grewal, A.S.; Grewal, A.S.; Singla, A.; Kamboj, P.; Dua, J.S.; Internationals, O. Pesticide Residues in Food Grains, Vegetables and Fruits: A Hazard to Human Health. *J. Med. Chem. Toxicol.* **2017**, *2*, 40–46. [CrossRef]
16. Chishti, Z.; Hussain, S.; Arshad, K.R.; Khalid, A.; Arshad, M. Microbial degradation of chlorpyrifos in liquid media and soil. *J. Environ. Manag.* **2013**, *114*, 372–380. [CrossRef]
17. Das, S.; Adhya, T.K. Degradation of chlorpyrifos in tropical rice soils. *J. Environ. Manag.* **2015**, *152*, 36–42. [CrossRef]
18. Kumar, A.; Correll, R.; Grocke, S.; Bajet, C. Toxicity of selected pesticides to freshwater shrimp, *Paratya australiensis* (Decapoda: Atyidae): Use of time series acute toxicity data to predict chronic lethality. *Ecotoxicol. Environ. Saf.* **2010**, *73*, 360–369. [CrossRef]
19. Li, D.; Huang, Q.; Lu, M.; Zhang, L.; Yang, Z.; Zong, M.; Tao, L. The organophosphate insecticide chlorpyrifos confers its genotoxic effects by inducing DNA damage and cell apoptosis. *Chemosphere* **2015**, *135*, 387–393. [CrossRef] [PubMed]
20. Hites, R.A. The Rise and Fall of Chlorpyrifos in the United States. *Environ. Sci. Technol.* **2021**, *55*, 1354–1358. [CrossRef] [PubMed]
21. Ur Rahman, S.; Xuebin, Q.; Yasin, G.; Cheng, H.; Mehmood, F.; Zain, M.; Shehzad, M.; Ahmad, M.I.; Riaz, L.; Rahim, A.; et al. Role of silicon on root morphological characters of wheat (*Triticum aestivum* L.) plants grown under Cd-contaminated nutrient solution. *Acta Physiol. Plant* **2021**, *43*, 60. [CrossRef]
22. Olisah, C.; Okoh, O.O.; Okoh, A.I. Occurrence of organochlorine pesticide residues in biological and environmental matrices in Africa: A two-decade review. *Heliyon* **2020**, *6*, e03518. [CrossRef]
23. Mulder, T.A.; van den Dries, M.A.; Korevaar, T.I.M.; Ferguson, K.K.; Peeters, R.P.; Tiemeier, H. Organophosphate pesticides exposure in pregnant women and maternal and cord blood thyroid hormone concentrations. *Environ. Int.* **2019**, *132*, 105124. [CrossRef]
24. Araki, A.; Miyashita, C.; Mitsui, T.; Goudarzi, H.; Mizutani, F.; Chisaki, Y.; Itoh, S.; Sasaki, S.; Cho, K.; Moriya, K.; et al. Prenatal organochlorine pesticide exposure and the disruption of steroids and reproductive hormones in cord blood: The Hokkaido study. *Environ. Int.* **2018**, *110*, 1–13. [CrossRef] [PubMed]
25. Lehmann, E.; Oltramare, C.; Nfon Dibié, J.-J.; Konaté, Y.; de Alencastro, L.F. Assessment of human exposure to pesticides by hair analysis: The case of vegetable-producing areas in Burkina Faso. *Environ. Int.* **2018**, *111*, 317–331. [CrossRef]
26. Iglesias-González, A.; Hardy, E.M.; Appenzeller, B.M.R. Cumulative exposure to organic pollutants of French children assessed by hair analysis. *Environ. Int.* **2020**, *134*, 105332. [CrossRef] [PubMed]
27. U.S. Environmental Protection Agency. Revised Human Health Risk Assessment on Chlorpyrifos. Available online: https://www.epa.gov/ingredients-used-pesticide-products/revised-human-health-risk-assessment-chlorpyrifos (accessed on 28 June 2021).

28. Perez-Fernandez, C.; Morales-Navas, M.; Aguilera-Sáez, L.M.; Abreu, A.C.; Guardia-Escote, L.; Fernández, I.; Garrido-Cárdenas, J.A.; Colomina, M.T.; Giménez, E.; Sánchez-Santed, F. Medium and long-term effects of low doses of Chlorpyrifos during the postnatal, preweaning developmental stage on sociability, dominance, gut microbiota and plasma metabolites. *Environ. Res.* **2020**, *184*, 109341. [CrossRef] [PubMed]
29. Wang, H.-P.; Liang, Y.-J.; Long, D.-X.; Chen, J.-X.; Hou, W.-Y.; Wu, Y.-J. Metabolic profiles of serum from rats after subchronic exposure to chlorpyrifos and carbaryl. *Chem. Res. Toxicol.* **2009**, *22*, 1026–1033. [CrossRef]
30. Xu, M.-Y.; Sun, Y.-J.; Wang, P.; Xu, H.-Y.; Chen, L.-P.; Zhu, L.; Wu, Y.-J. Metabolomics analysis and biomarker identification for brains of rats exposed subchronically to the mixtures of low-dose cadmium and chlorpyrifos. *Chem. Res. Toxicol.* **2015**, *28*, 1216–1223. [CrossRef] [PubMed]
31. Lassiter, T.L.; Brimijoin, S. Rats gain excess weight after developmental exposure to the organophosphorothionate pesticide, chlorpyrifos. *Neurotoxicol. Teratol.* **2008**, *30*, 125–130. [CrossRef] [PubMed]
32. Slotkin Theodore, A.; Brown Kathleen, K.; Seidler Frederic, J. Developmental Exposure of Rats to Chlorpyrifos Elicits Sex-Selective Hyperlipidemia and Hyperinsulinemia in Adulthood. *Environ. Health Perspect.* **2005**, *113*, 1291–1294. [CrossRef] [PubMed]
33. Lasram, M.M.; Dhouib, I.B.; Annabi, A.; El Fazaa, S.; Gharbi, N. A review on the molecular mechanisms involved in insulin resistance induced by organophosphorus pesticides. *Toxicology* **2014**, *322*, 1–13. [CrossRef] [PubMed]
34. Peris-Sampedro, F.; Salazar, J.G.; Cabré, M.; Reverte, I.; Domingo, J.L.; Sánchez-Santed, F.; Colomina, M.T. Impaired retention in AβPP Swedish mice six months after oral exposure to chlorpyrifos. *Food Chem. Toxicol.* **2014**, *72*, 289–294. [CrossRef]
35. Fang, B.; Li, J.W.; Zhang, M.; Ren, F.Z.; Pang, G.F. Chronic chlorpyrifos exposure elicits diet-specific effects on metabolism and the gut microbiome in rats. *Food Chem. Toxicol.* **2018**, *111*, 144–152. [CrossRef]
36. Li, J.; Ren, F.; Li, Y.; Luo, J.; Pang, G. Chlorpyrifos Induces Metabolic Disruption by Altering Levels of Reproductive Hormones. *J. Agric. Food Chem.* **2019**, *67*, 10553–10562. [CrossRef]
37. Peris-Sampedro, F.; Cabré, M.; Basaure, P.; Reverte, I.; Domingo, J.L.; Teresa Colomina, M. Adulthood dietary exposure to a common pesticide leads to an obese-like phenotype and a diabetic profile in apoE3 mice. *Environ. Res.* **2015**, *142*, 169–176. [CrossRef]
38. Reygner, J.; Lichtenberger, L.; Elmhiri, G.; Dou, S.; Bahi-Jaber, N.; Rhazi, L.; Depeint, F.; Bach, V.; Khorsi-Cauet, H.; Abdennebi-Najar, L. Inulin Supplementation Lowered the Metabolic Defects of Prolonged Exposure to Chlorpyrifos from Gestation to Young Adult Stage in Offspring Rats. *PLoS ONE* **2016**, *11*, e0164614. [CrossRef]
39. Goel, A.; Dani, V.; Dhawan, D.K. Protective effects of zinc on lipid peroxidation, antioxidant enzymes and hepatic histoarchitecture in chlorpyrifos-induced toxicity. *Chem. Biol. Interact.* **2005**, *156*, 131–140. [CrossRef]
40. Meggs, W.J.; Brewer, K.L. Weight gain associated with chronic exposure to chlorpyrifos in rats. *J. Med. Toxicol.* **2007**, *3*, 89–93. [CrossRef] [PubMed]
41. Uchendu, C.; Ambali, S.F.; Ayo, J.O.; Esievo, K.A.N. Body weight and hematological changes induced by chronic exposure to low levels of chlorpyrifos and deltamethrin combination in rats: The effect of alpha-lipoic acid. *Comp. Clin. Pathol.* **2018**, *27*, 1383–1388. [CrossRef]
42. Whyatt, R.M.; Rauh, V.; Barr, D.B.; Camann, D.E.; Andrews, H.F.; Garfinkel, R.; Hoepner, L.A.; Diaz, D.; Dietrich, J.; Reyes, A.; et al. Prenatal insecticide exposures and birth weight and length among an urban minority cohort. *Environ. Health Perspect.* **2004**, *112*, 1125–1132. [CrossRef] [PubMed]
43. Cook, T.J.; Shenoy, S.S. Intestinal permeability of chlorpyrifos using the single-pass intestinal perfusion method in the rat. *Toxicology* **2003**, *184*, 125–133. [CrossRef]
44. Gao, J.; Naughton, S.X.; Beck, W.D.; Hernandez, C.M.; Wu, G.; Wei, Z.; Yang, X.; Bartlett, M.G.; Terry, A.V. Chlorpyrifos and chlorpyrifos oxon impair the transport of membrane bound organelles in rat cortical axons. *NeuroToxicology* **2017**, *62*, 111–123. [CrossRef]
45. Joly Condette, C.; Khorsi-Cauet, H.; Morlière, P.; Zabijak, L.; Reygner, J.; Bach, V.; Gay-Quéheillard, J. Increased Gut Permeability and Bacterial Translocation after Chronic Chlorpyrifos Exposure in Rats. *PLoS ONE* **2014**, *9*, e102217. [CrossRef]
46. Velmurugan, G.; Ramprasath, T.; Gilles, M.; Swaminathan, K.; Ramasamy, S. Gut Microbiota, Endocrine-Disrupting Chemicals, and the Diabetes Epidemic. *Trends Endocrinol. Metab.* **2017**, *28*, 612–625. [CrossRef] [PubMed]
47. Bäckhed, F.; Ley, R.E.; Sonnenburg, J.L.; Peterson, D.A.; Gordon, J.I. Host-Bacterial Mutualism in the Human Intestine. *Science* **2005**, *307*, 1915–1920. [CrossRef]
48. Cani, P.D.; Everard, A.; Duparc, T. Gut microbiota, enteroendocrine functions and metabolism. *Curr. Opin. Pharmacol.* **2013**, *13*, 935–940. [CrossRef] [PubMed]
49. Tomas, J.; Reygner, J.; Mayeur, C.; Ducroc, R.; Bouet, S.; Bridonneau, C.; Cavin, J.-B.; Thomas, M.; Langella, P.; Cherbuy, C. Early colonizing Escherichia coli elicits remodeling of rat colonic epithelium shifting toward a new homeostatic state. *ISME J.* **2015**, *9*, 46–58. [CrossRef]
50. Condette, C.J.; Bach, V.; Mayeur, C.; Gay-Quéheillard, J.; Khorsi-Cauet, H. Chlorpyrifos Exposure During Perinatal Period Affects Intestinal Microbiota Associated With Delay of Maturation of Digestive Tract in Rats. *J. Pediatr. Gastroenterol. Nutr.* **2015**, *61*, 30–40. [CrossRef]
51. Condette, C.J.; Gay-Quéheillard, J.; Léké, A.; Chardon, K.; Delanaud, S.; Bach, V.; Khorsi-Cauet, H. Impact of chronic exposure to low doses of chlorpyrifos on the intestinal microbiota in the Simulator of the Human Intestinal Microbial Ecosystem (SHIME) and in the rat. *Environ. Sci. Pollut. Res. Int.* **2013**, *20*, 2726–2734. [CrossRef]

52. Reygner, J.; Joly Condette, C.; Bruneau, A.; Delanaud, S.; Rhazi, L.; Depeint, F.; Abdennebi-Najar, L.; Bach, V.; Mayeur, C.; Khorsi-Cauet, H. Changes in Composition and Function of Human Intestinal Microbiota Exposed to Chlorpyrifos in Oil as Assessed by the SHIME® Model. *Int. J. Environ. Res. Public Health* **2016**, *13*, 1088. [CrossRef] [PubMed]
53. Xia, J.; Jin, C.; Pan, Z.; Sun, L.; Fu, Z.; Jin, Y. Chronic exposure to low concentrations of lead induces metabolic disorder and dysbiosis of the gut microbiota in mice. *Sci. Total Environ.* **2018**, *631–632*, 439–448. [CrossRef]
54. Réquilé, M.; Gonzàlez Alvarez, D.O.; Delanaud, S.; Rhazi, L.; Bach, V.; Depeint, F.; Khorsi-Cauet, H. Use of a combination of in vitro models to investigate the impact of chlorpyrifos and inulin on the intestinal microbiota and the permeability of the intestinal mucosa. *Environ. Sci. Pollut. Res.* **2018**, *25*, 22529–22540. [CrossRef]
55. Tirelli, V.; Catone, T.; Turco, L.; Di Consiglio, E.; Testai, E.; De Angelis, I. Effects of the pesticide clorpyrifos on an in vitro model of intestinal barrier. *Toxicol. Vitr.* **2007**, *21*, 308–313. [CrossRef]
56. Steffen, E.K.; Berg, R.D. Relationship between cecal population levels of indigenous bacteria and translocation to the mesenteric lymph nodes. *Infect. Immun.* **1983**, *39*, 1252–1259. [CrossRef]
57. Sarron, E.; Pérot, M.; Barbezier, N.; Delayre-Orthez, C.; Gay-Quéheillard, J.; Anton, P.M. Early exposure to food contaminants reshapes maturation of the human brain-gut-microbiota axis. *World J. Gastroenterol.* **2020**, *26*, 3145. [CrossRef] [PubMed]
58. Djekkoun, N.; Lalau, J.-D.; Bach, V.; Depeint, F.; Khorsi-Cauet, H. Chronic oral exposure to pesticides and their consequences on metabolic regulation: Role of the microbiota. *Eur. J. Nutr.* **2021**, *60*, 4131–4149. [CrossRef]
59. Cani, P.D.; Joly, E.; Horsmans, Y.; Delzenne, N.M. Oligofructose promotes satiety in healthy human: A pilot study. *Eur. J. Clin. Nutr.* **2006**, *60*, 567–572. [CrossRef]
60. Everard, A.; Lazarevic, V.; Derrien, M.; Girard, M.; Muccioli, G.G.; Neyrinck, A.M.; Possemiers, S.; Holle, A.V.; François, P.; de Vos Willem, M.; et al. Responses of Gut Microbiota and Glucose and Lipid Metabolism to Prebiotics in Genetic Obese and Diet-Induced Leptin-Resistant Mice. *Diabetes* **2011**, *60*, 2775–2786. [CrossRef]
61. Byrne, C.S.; Chambers, E.S.; Morrison, D.J.; Frost, G. The role of short chain fatty acids in appetite regulation and energy homeostasis. *Int. J. Obes.* **2015**, *39*, 1331–1338. [CrossRef]
62. Geurts, L.; Neyrinck, A.M.; Delzenne, N.M.; Knauf, C.; Cani, P.D. Gut microbiota controls adipose tissue expansion, gut barrier and glucose metabolism: Novel insights into molecular targets and interventions using prebiotics. *Benef. Microbes.* **2014**, *5*, 3–17. [CrossRef] [PubMed]
63. Cochran, R.C.; Kishiyama, J.; Aldous, C.; Carr, W.C.; Pfeifer, K.F. Chlorpyrifos: Hazard assessment based on a review of the effects of short-term and long-term exposure in animals and humans. *Food Chem. Toxicol.* **1995**, *33*, 165–172. [CrossRef]
64. Guibourdenche, M.; El Khayat El Sabbouri, H.; Bonnet, F.; Djekkoun, N.; Khorsi-Cauet, H.; Corona, A.; Guibourdenche, J.; Bach, V.; Anton, P.M.; Gay-Quéheillard, J. Perinatal exposure to chlorpyrifos and/or a high-fat diet is associated with liver damage in male rat offspring. *Cells Dev.* **2021**, *166*, 203678. [CrossRef] [PubMed]
65. El Khayat El Sabbouri, H.; Gay-Quéheillard, J.; Joumaa, W.H.; Delanaud, S.; Guibourdenche, M.; Darwiche, W.; Djekkoun, N.; Bach, V.; Ramadan, W. Does the perigestational exposure to chlorpyrifos and/or high-fat diet affect respiratory parameters and diaphragmatic muscle contractility in young rats? *Food Chem. Toxicol.* **2020**, *140*, 111322. [CrossRef]
66. American Society of Microbiology. *Manual of Clinical Microbiology*, 10th ed.; American Society of Microbiology: Washington, DC, USA, 2011; ISBN 978-1-55581-672-8.
67. Lecerf, J.-M.; Dépeint, F.; Clerc, E.; Dugenet, Y.; Niamba, C.N.; Rhazi, L.; Cayzeele, A.; Abdelnour, G.; Jaruga, A.; Younes, H.; et al. Xylo-oligosaccharide (XOS) in combination with inulin modulates both the intestinal environment and immune status in healthy subjects, while XOS alone only shows prebiotic properties. *Br. J. Nutr.* **2012**, *108*, 1847–1858. [CrossRef]
68. De Felice, A.; Greco, A.; Calamandrei, G.; Minghetti, L. Prenatal exposure to the organophosphate insecticide chlorpyrifos enhances brain oxidative stress and prostaglandin E2 synthesis in a mouse model of idiopathic autism. *J. Neuroinflamm.* **2016**, *13*, 149. [CrossRef] [PubMed]
69. Li, J.-W.; Fang, B.; Pang, G.-F.; Zhang, M.; Ren, F.-Z. Age- and diet-specific effects of chronic exposure to chlorpyrifos on hormones, inflammation and gut microbiota in rats. *Pestic. Biochem. Physiol.* **2019**, *159*, 68–79. [CrossRef]
70. Xu, M.-Y.; Sun, Y.-J.; Wang, P.; Yang, L.; Wu, Y.-J. Metabolomic biomarkers in urine of rats following long-term low-dose exposure of cadmium and/or chlorpyrifos. *Ecotoxicol. Environ. Saf.* **2020**, *195*, 110467. [CrossRef] [PubMed]
71. Xu, M.-Y.; Wang, P.; Sun, Y.-J.; Yang, L.; Wu, Y.-J. Joint toxicity of chlorpyrifos and cadmium on the oxidative stress and mitochondrial damage in neuronal cells. *Food Chem. Toxicol.* **2017**, *103*, 246–252. [CrossRef] [PubMed]
72. Yang, F.; Li, J.; Pang, G.; Ren, F.; Fang, B. Effects of Diethyl Phosphate, a Non-Specific Metabolite of Organophosphorus Pesticides, on Serum Lipid, Hormones, Inflammation, and Gut Microbiota. *Molecules* **2019**, *24*, 2003. [CrossRef] [PubMed]
73. Goel, A.; Dani, V.; Dhawan, D.K. Zinc mediates normalization of hepatic drug metabolizing enzymes in chlorpyrifos-induced toxicity. *Toxicol. Lett.* **2007**, *169*, 26–33. [CrossRef]
74. Zhang, Q.; Zheng, S.; Wang, S.; Wang, W.; Xing, H.; Xu, S. Chlorpyrifos induced oxidative stress to promote apoptosis and autophagy through the regulation of miR-19a-AMPK axis in common carp. *Fish. Shellfish Immunol.* **2019**, *93*, 1093–1099. [CrossRef]
75. Ramirez-Vargas, M.A.; Flores-Alfaro, E.; Uriostegui-Acosta, M.; Alvarez-Fitz, P.; Parra-Rojas, I.; Moreno-Godinez, M.E. Effects of Exposure to Malathion on Blood Glucose Concentration: A Meta-Analysis. *Environ. Sci. Pollut. Res.* **2018**, *25*, 3233–3242. [CrossRef]

76. Silva, J.G.; Boareto, A.C.; Schreiber, A.K.; Redivo, D.D.B.; Gambeta, E.; Vergara, F.; Morais, H.; Zanoveli, J.M.; Dalsenter, P.R. Chlorpyrifos induces anxiety-like behavior in offspring rats exposed during pregnancy. *Neurosci. Lett.* **2017**, *641*, 94–100. [CrossRef]
77. Acker, C.I.; Nogueira, C.W. Chlorpyrifos acute exposure induces hyperglycemia and hyperlipidemia in rats. *Chemosphere* **2012**, *89*, 602–608. [CrossRef] [PubMed]
78. Joshi, A.K.R.; Rajini, P.S. Hyperglycemic and stressogenic effects of monocrotophos in rats: Evidence for the involvement of acetylcholinesterase inhibition. *Exp. Toxicol. Pathol.* **2012**, *64*, 115–120. [CrossRef] [PubMed]
79. Lasram, M.M.; Annabi, A.B.; Rezg, R.; Elj, N.; Slimen, S.; Kamoun, A.; El-Fazaa, S.; Gharbi, N. Effect of short-time malathion administration on glucose homeostasis in Wistar rat. *Pestic. Biochem. Physiol.* **2008**, *92*, 114–119. [CrossRef]
80. Sullivan, J.F.; Jetton, M.M.; Hahn, H.K.; Burch, R.E. Enhanced lipid peroxidation in liver microsomes of zinc-deficient rats. *Am. J. Clin. Nutr.* **1980**, *33*, 51–56. [CrossRef]
81. Akhtar, N.; Srivastava, M.K.; Raizada, R.B. Assessment of chlorpyrifos toxicity on certain organs in rat, Rattus norvegicus. *J. Environ. Biol.* **2009**, *30*, 1047–1053. [PubMed]
82. Ambali, S.F.; Ayo, J.O.; Esievo, K.A.N.; Ojo, S.A. Hemotoxicity Induced by Chronic Chlorpyrifos Exposure in Wistar Rats: Mitigating Effect of Vitamin C. *Vet. Med. Int.* **2011**, *2011*, e945439. [CrossRef]
83. Uchendu, C.; Ambali, S.F.; Ayo, J.O.; Esievo, K.A.N. Chronic co-exposure to chlorpyrifos and deltamethrin pesticides induces alterations in serum lipids and oxidative stress in Wistar rats: Mitigating role of alpha-lipoic acid. *Environ. Sci. Pollut. Res.* **2018**, *25*, 19605–19611. [CrossRef]
84. Lukowicz, C.; Ellero-Simatos, S.; Régnier, M.; Polizzi, A.; Lasserre, F.; Montagner, A.; Yannick, L.; Jamin Emilien, L.; Martin, J.-F.; Naylies, C.; et al. Metabolic Effects of a Chronic Dietary Exposure to a Low-Dose Pesticide Cocktail in Mice: Sexual Dimorphism and Role of the Constitutive Androstane Receptor. *Environ. Health Perspect.* **2018**, *126*, 067007. [CrossRef]
85. Newairy, A.A.; Abdou, H.M. Effect of propolis consumption on hepatotoxicity and brain damage in male rats exposed to chlorpyrifos. *Afr. J. Biotechnol.* **2013**, *12*, 5232–5243. [CrossRef]
86. Dong, T.; Guan, Q.; Hu, W.; Zhang, M.; Zhang, Y.; Chen, M.; Wang, X.; Xia, Y. Prenatal exposure to glufosinate ammonium disturbs gut microbiome and induces behavioral abnormalities in mice. *J. Hazard. Mater.* **2020**, *389*, 122152. [CrossRef]
87. Gao, B.; Mahbub, R.; Lu, K. Sex-Specific Effects of Organophosphate Diazinon on the Gut Microbiome and Its Metabolic Functions. *Environ. Health Perspect.* **2017**, *125*, 198–206. [CrossRef] [PubMed]
88. Perez-Fernandez, C.; Morales-Navas, M.; Guardia-Escote, L.; Garrido-Cárdenas, J.A.; Colomina, M.T.; Giménez, E.; Sánchez-Santed, F. Long-term effects of low doses of Chlorpyrifos exposure at the preweaning developmental stage: A locomotor, pharmacological, brain gene expression and gut microbiome analysis. *Food Chem. Toxicol.* **2020**, *135*, 110865. [CrossRef]
89. Zhao, Y.; Zhang, Y.; Wang, G.; Han, R.; Xie, X. Effects of chlorpyrifos on the gut microbiome and urine metabolome in mouse (*Mus musculus*). *Chemosphere* **2016**, *153*, 287–293. [CrossRef]
90. Blaut, M. Gut microbiota and energy balance: Role in obesity. *Proc. Nutr. Soc.* **2015**, *74*, 227–234. [CrossRef] [PubMed]
91. Morrison, D.J.; Preston, T. Formation of short chain fatty acids by the gut microbiota and their impact on human metabolism. *Gut Microbes.* **2016**, *7*, 189–200. [CrossRef] [PubMed]
92. Chen, T.; Kim, C.Y.; Kaur, A.; Lamothe, L.; Shaikh, M.; Keshavarzian, A.; Hamaker, B.R. Dietary fibre-based SCFA mixtures promote both protection and repair of intestinal epithelial barrier function in a Caco-2 cell model. *Food Funct.* **2017**, *8*, 1166–1173. [CrossRef]
93. Jin, C.; Xia, J.; Wu, S.; Tu, W.; Pan, Z.; Fu, Z.; Wang, Y.; Jin, Y. Insights Into a Possible Influence on Gut Microbiota and Intestinal Barrier Function During Chronic Exposure of Mice to Imazalil. *Toxicol. Sci.* **2018**, *162*, 113–123. [CrossRef] [PubMed]
94. Nasuti, C.; Coman, M.M.; Olek, R.A.; Fiorini, D.; Verdenelli, M.C.; Cecchini, C.; Silvi, S.; Fedeli, D.; Gabbianelli, R. Changes on fecal microbiota in rats exposed to permethrin during postnatal development. *Environ. Sci. Pollut. Res. Int.* **2016**, *23*, 10930–10937. [CrossRef] [PubMed]
95. Wu, S.; Jin, C.; Wang, Y.; Fu, Z.; Jin, Y. Exposure to the fungicide propamocarb causes gut microbiota dysbiosis and metabolic disorder in mice. *Environ. Pollut.* **2018**, *237*, 775–783. [CrossRef]
96. Carroll, I.M.; Maharshak, N. Enteric bacterial proteases in inflammatory bowel disease- pathophysiology and clinical implications. *World J. Gastroenterol.* **2013**, *19*, 7531–7543. [CrossRef]
97. Martinez-Medina, M.; Denizot, J.; Dreux, N.; Robin, F.; Billard, E.; Bonnet, R.; Darfeuille-Michaud, A.; Barnich, N. Western diet induces dysbiosis with increased E coli in CEABAC10 mice, alters host barrier function favouring AIEC colonisation. *Gut* **2014**, *63*, 116–124. [CrossRef] [PubMed]
98. Yajima, M.; Nakayama, M.; Hatano, S.; Yamazaki, K.; Aoyama, Y.; Yajima, T.; Kuwata, T. Bacterial translocation in neonatal rats: The relation between intestinal flora, translocated bacteria, and influence of milk. *J. Pediatr. Gastroenterol. Nutr.* **2001**, *33*, 592–601. [CrossRef] [PubMed]
99. Cani, P.D.; Delzenne, N.M. The Role of the Gut Microbiota in Energy Metabolism and Metabolic Disease. *Curr. Pharm. Des.* **2009**, *15*, 1546–1558. [CrossRef] [PubMed]
100. Roberfroid, M.; Gibson, G.R.; Hoyles, L.; McCartney, A.L.; Rastall, R.; Rowland, I.; Wolvers, D.; Watzl, B.; Szajewska, H.; Stahl, B.; et al. Prebiotic effects: Metabolic and health benefits. *Br. J. Nutr.* **2010**, *104* (Suppl. S2), S1–S63. [CrossRef]
101. Davani-Davari, D.; Negahdaripour, M.; Karimzadeh, I.; Seifan, M.; Mohkam, M.; Masoumi, S.J.; Berenjian, A.; Ghasemi, Y. Prebiotics: Definition, Types, Sources, Mechanisms, and Clinical Applications. *Foods* **2019**, *8*, 92. [CrossRef]

102. Yousefi, S.; Hoseinifar, S.H. Protective effects of prebiotic in zebrafish, Danio rerio, under experimental exposure to Chlorpyrifos. *Int. J. Aquat. Biol.* **2018**, *6*, 49–54. [CrossRef]
103. Ooi, L.-G.; Liong, M.-T. Cholesterol-Lowering Effects of Probiotics and Prebiotics: A Review of in Vivo and in Vitro Findings. *Int. J. Mol. Sci.* **2010**, *11*, 2499–2522. [CrossRef] [PubMed]
104. Dikeman, C.L.; Murphy, M.R.; Fahey, G.C., Jr. Dietary Fibers Affect Viscosity of Solutions and Simulated Human Gastric and Small Intestinal Digesta. *J. Nutr.* **2006**, *136*, 913–919. [CrossRef]
105. Levrat, M.-A.; Rémésy, C.; Demigné, C. High Propionic Acid Fermentations and Mineral Accumulation in the Cecum of Rats Adapted to Different Levels of Inulin. *J. Nutr.* **1991**, *121*, 1730–1737. [CrossRef]
106. Liu, F.; Prabhakar, M.; Ju, J.; Long, H.; Zhou, H.-W. Effect of inulin-type fructans on blood lipid profile and glucose level: A systematic review and meta-analysis of randomized controlled trials. *Eur. J. Clin. Nutr.* **2017**, *71*, 9–20. [CrossRef]
107. Song, X.; Zhong, L.; Lyu, N.; Liu, F.; Li, B.; Hao, Y.; Xue, Y.; Li, J.; Feng, Y.; Ma, Y.; et al. Inulin Can Alleviate Metabolism Disorders in ob/ob Mice by Partially Restoring Leptin-related Pathways Mediated by Gut Microbiota. *Genom. Proteom. Bioinform.* **2019**, *17*, 64–75. [CrossRef] [PubMed]
108. Zhang, Q.; Yu, H.; Xiao, X.; Hu, L.; Xin, F.; Yu, X. Inulin-type fructan improves diabetic phenotype and gut microbiota profiles in rats. *PeerJ* **2018**, *6*, e4446. [CrossRef] [PubMed]
109. Cummings, J.H.; Pomare, E.W.; Branch, W.J.; Naylor, C.P.; Macfarlane, G.T. Short chain fatty acids in human large intestine, portal, hepatic and venous blood. *Gut* **1987**, *28*, 1221–1227. [CrossRef]
110. Kimura, I.; Ozawa, K.; Inoue, D.; Imamura, T.; Kimura, K.; Maeda, T.; Terasawa, K.; Kashihara, D.; Hirano, K.; Tani, T.; et al. The gut microbiota suppresses insulin-mediated fat accumulation via the short-chain fatty acid receptor GPR43. *Nat. Commun.* **2013**, *4*, 1829. [CrossRef] [PubMed]
111. Okada, T.; Fukuda, S.; Hase, K.; Nishiumi, S.; Izumi, Y.; Yoshida, M.; Hagiwara, T.; Kawashima, R.; Yamazaki, M.; Oshio, T.; et al. Microbiota-derived lactate accelerates colon epithelial cell turnover in starvation-refed mice. *Nat. Commun.* **2013**, *4*, 1654. [CrossRef]
112. Sasaki, H.; Miyakawa, H.; Watanabe, A.; Nakayama, Y.; Lyu, Y.; Hama, K.; Shibata, S. Mice Microbiota Composition Changes by Inulin Feeding with a Long Fasting Period under a Two-Meals-Per-Day Schedule. *Nutrients* **2019**, *11*, E2802. [CrossRef] [PubMed]
113. Liu, Y.; Sam, A.G. *The Organic Premium of Baby Food Estimated with National Level Scanner Data*; Social Science Research Network: Rochester, NY, USA, 2020.
114. Maguire, K.B.; Owens, N.N.; Simon, N.B. Focus on Babies: A Note on Parental Attitudes and Preferences for Organic Babyfood. *J. Agribus.* **2006**, *24*, 187–195.
115. Maguire, K.B.; Owens, N.; Simon, N.B. *Focus on Babies: Evidence on Parental Attitudes towards Pesticide Risks*; AgEcon Search: Minneapolis, MN, USA, 2004.
116. Berton, T.; Mayhoub, F.; Chardon, K.; Duca, R.-C.; Lestremau, F.; Bach, V.; Tack, K. Development of an analytical strategy based on LC–MS/MS for the measurement of different classes of pesticides and theirs metabolites in meconium: Application and characterisation of foetal exposure in France. *Environ. Res.* **2014**, *132*, 311–320. [CrossRef]
117. Haraux, E.; Tourneux, P.; Kouakam, C.; Stephan-Blanchard, E.; Boudailliez, B.; Leke, A.; Klein, C.; Chardon, K. Isolated hypospadias: The impact of prenatal exposure to pesticides, as determined by meconium analysis. *Environ. Int.* **2018**, *119*, 20–25. [CrossRef] [PubMed]
118. Béranger, R.; Hardy, E.M.; Binter, A.-C.; Charles, M.-A.; Zaros, C.; Appenzeller, B.M.R.; Chevrier, C. Multiple pesticides in mothers' hair samples and children's measurements at birth: Results from the French national birth cohort (ELFE). *Int. J. Hyg. Environ. Health* **2020**, *223*, 22–33. [CrossRef]

Article

Triiodothyronine or Antioxidants Block the Inhibitory Effects of BDE-47 and BDE-49 on Axonal Growth in Rat Hippocampal Neuron-Glia Co-Cultures

Hao Chen [†], Rhianna K. Carty [†], Adrienne C. Bautista, Keri A. Hayakawa and Pamela J. Lein *

Department of Molecular Biosciences, University of California, Davis, CA 95616, USA; hachen@ionisph.com (H.C.); rhianna.k.carty@gmail.com (R.K.C.); abcashion@ucdavis.edu (A.C.B.); kahayakawa@ucdavis.edu (K.A.H.)
* Correspondence: pjlein@ucdavis.edu; Tel.: +1-530-752-1970
† These authors contributed equally to this work.

Abstract: We previously demonstrated that polybrominated diphenyl ethers (PBDEs) inhibit the growth of axons in primary rat hippocampal neurons. Here, we test the hypothesis that PBDE effects on axonal morphogenesis are mediated by thyroid hormone and/or reactive oxygen species (ROS)-dependent mechanisms. Axonal growth and ROS were quantified in primary neuronal-glial co-cultures dissociated from neonatal rat hippocampi exposed to nM concentrations of BDE-47 or BDE-49 in the absence or presence of triiodothyronine (T3; 3–30 nM), N-acetyl-cysteine (NAC; 100 μM), or α-tocopherol (100 μM). Co-exposure to T3 or either antioxidant prevented inhibition of axonal growth in hippocampal cultures exposed to BDE-47 or BDE-49. T3 supplementation in cultures not exposed to PBDEs did not alter axonal growth. T3 did, however, prevent PBDE-induced ROS generation and alterations in mitochondrial metabolism. Collectively, our data indicate that PBDEs inhibit axonal growth via ROS-dependent mechanisms, and that T3 protects axonal growth by inhibiting PBDE-induced ROS. These observations suggest that co-exposure to endocrine disruptors that decrease TH signaling in the brain may increase vulnerability to the adverse effects of developmental PBDE exposure on axonal morphogenesis.

Keywords: axonal growth; developmental neurotoxicity; neuronal morphogenesis; PBDE; reactive oxygen species; thyroid hormone

1. Introduction

The brominated flame retardants, polybrominated diphenyl ethers (PBDEs), are considered to be likely environmental risk factors for neurodevelopmental disorders [1–4]. Epidemiologic studies have identified a negative association between developmental exposure to PBDEs and executive function, motor behavior, and attention in infants and children [5–12]. These findings are of significant public health concern given the documented widespread human exposure to PBDEs with significantly higher body burdens in infants and toddlers relative to adults [13,14]. However, there remains significant uncertainty regarding the underlying mechanism(s) by which PBDEs interfere with neurodevelopment.

It has been hypothesized that PBDE developmental neurotoxicity reflects altered patterns of neuronal connectivity [12,15,16]. A critical determinant of the patterns of connections formed between neurons during development is axonal morphology. Interference with temporal and/or spatial aspects of axonal morphogenesis has been shown to cause functional deficits in experimental models [17–19]. Moreover, altered patterns of axonal growth are implicated in the pathogenesis of various neurodevelopmental disorders [20,21]. Recently, we demonstrated that BDE-47, a PBDE congener that is highly abundant in human tissues, and BDE-49, an understudied PBDE congener with levels comparable to BDE-47 in gestational tissues of women living in southeast Michigan [22], inhibited axonal growth in primary hippocampal neuron-glia co-cultures, in part by delaying neuronal polarization [23].

BDE-47 and BDE-49 effects on axonal growth in primary hippocampal neurons were prevented by pharmacological blockade of ryanodine receptors (RyR) or siRNA knockdown of RyR, implicating RyR-dependent mechanisms in PBDE developmental neurotoxicity [23]. However, an unexpected finding from our previous studies was that the axon inhibitory effects of BDE-47 and BDE-49 exhibited comparable concentration-effect relationships despite significant differences in their potency at the RyR [24]. This observation raised the possibility that the RyR is not the primary molecular target but rather a downstream effector in the adverse outcome pathway (AOP) linking PBDEs to axonal growth inhibition. PBDEs have been shown to interfere with thyroid hormone (TH) signaling and to cause oxidative stress via increased levels of intracellular reactive oxygen species (ROS) [25,26], and both TH and ROS are reported to modulate RyR activity [27] and to influence axonal growth [28,29]. Therefore, in this study, we leveraged a primary rat hippocampal neuron-glia co-culture model to assess the relative contributions of TH and ROS-dependent mechanisms in mediating the axon inhibitory activity of BDE-49 and BDE-47. Our findings support the hypothesis that PBDEs inhibit axonal growth via ROS-dependent mechanisms, and that the TH, triiodothyronine (T3), protects against the effects of PBDEs on axonal growth by blocking PBDE-induced ROS.

2. Materials and Methods

2.1. Materials

Neat certified BDE-47 (2,2′,4,4′-tetrabromodiphenyl ether, >99% pure) and BDE-49 (2,2′,4,5′-tetrabromodiphenyl ether, >99% pure) were purchased from AccuStandard Inc. (New Haven, CT, USA), and verified for purity and composition by GC/MS by the UC Davis Superfund Research Program Analytical Core. Stock solutions of each BDE were made in dry dimethyl sulfoxide (DMSO, Sigma-Aldrich, St. Louis, MO, USA). 3,3′,5-Triiodo-L-thyronine (T3), N-acetyl-L-cysteine (NAC) and DL-α-tocopherol acetate were purchased from Sigma-Aldrich.

2.2. Animals

All procedures involving animals were approved by the University of California Davis Animal Care and Use Committee and conformed to the NIH Guide for the Care and Use of Laboratory Animals, and the ARRIVE guidelines [30]. Timed-pregnant Sprague Dawley rats were purchased from Charles River Laboratory (Hollister, CA, USA) and individually housed in clear plastic cages with corn cob bedding at 22 ± 2 °C under a 12 h dark–light cycle. Food and water were provided ad libitum.

2.3. Cell Culture

Primary neuron-glia co-cultures were prepared from hippocampi harvested from postnatal day (P) 0–1 male and female rat pups as previously described [31]. Briefly, rat pups were separated from the dam and anesthetized by placing them on a gauze pad on ice. Once pups ceased moving, they were euthanized by decapitation using sterile scissors. Hippocampi were harvested from the pup's head by sterile dissection and then dissociated using trypsin (1 mg/mL) and DNAse (0.3 mg/mL). Dissociated hippocampal cells were plated on poly-L-lysine (0.5 mg/ML, Sigma Aldrich) coated glass coverslips (BellCo, Vineland, NJ, USA) and maintained at 37 °C in NeuralQ Basal Medium supplemented with 2% (v/v) GS21 (MTI-GlobalStem, Gaithersburg, MD, USA) and GlutaMAX (ThermoScientific, Waltham, MA, USA). The concentration of T3 in the complete medium used to maintain cultures was ~2.6 nM [32,33]. For studies of axonal growth, neurons were plated at 27,000 cells/cm^2; for qPCR and Western blot experiments, neurons were plated at 105,000 cells/cm^2. Cultures were exposed to varying concentrations of BDE-47 or BDE-49 diluted in culture medium from 1000× stocks; vehicle control cultures were exposed to

DMSO (1:1000 dilution). A subset of cultures was co-exposed to T3, NAC, or α-tocopherol diluted 1:1000 directly into cell cultures from 1000× stocks in sterile distilled water.

2.4. Quantification of Axonal Outgrowth

Cultures were exposed to BDE-47, BDE-49, or vehicle (1:1000 DMSO) for 48 h beginning 3 h post-plating, and then fixed with 4% (w/v) paraformaldehyde (Sigma Aldrich) in 0.2 M phosphate buffer. To visualize axons, hippocampal cultures were immunostained with antibody specific for tau-1 (1:1000, Millipore, Billerica, MA, USA, RRID AB_94855). Our previous studies [23] demonstrated that exposure to BDE-47 or BDE-49 did not alter the expression of tau, as determined by Western blotting. Axonal lengths of tau-1 immunopositive neurons were manually quantified by an individual blinded to experimental condition using ImageJ software with the NeuronJ plugin [34]. As previously defined [35], in any given neuron, the axon was identified as the neurite whose length was >2.5× the cell body diameter and exceeded that of the other minor processes of the same neuron. Only non-overlapping neurons were quantified as proximity to other neurons can affect neuronal morphology.

2.5. Quantitative Polymerase Chain Reaction (qPCR)

Total RNA was isolated from cell cultures using TRIzol Reagent (ThermoScientific) per the manufacturer's instructions, and cDNA was synthesized using the SuperScriptTMViloTM MasterMix containing SuperScriptTM III Reverse Transcriptase (Invitrogen, Carlsbad, CA, USA). Samples were mixed with Power SYBR Green MasterMix and forward and reverse primers (see Supplemental Table S1 for primer sequences and amplification efficiencies) and then loaded into a MicroAmp 384 Reaction Plate (ThermoScientific). qPCR plates were run on a 7900HT System by the Real-Time PCR Research and Diagnostics Core Facility at UC Davis. qPCR primers and probes were ordered from Integrated DNA Technologies (Coralville, IA, USA) using PrimeTime® Predesigned qPCR Assays. Transcript levels were normalized to the average of the reference genes Ppia and Hprt1 and expression ratios were calculated by Pfaffl method [36] using REST 2009 software (Qiagen, Valencia, CA, USA).

2.6. ROS Measurements

Rat hippocampal neurons cultures were exposed to BDE-47, BDE-49, or vehicle (1:1000 DMSO) in the absence or presence of T3, NAC, or α-tocopherol 3 h post-plating. Global ROS production was measured 1 h following exposures using ROS-Glo assay (Promega, Madison, WI, USA) according to manufacturer's protocol, which specified using H_2O_2 as a positive technical control. Luminescence was recording using an H1 hybrid microplate reader (BioTek Instruments, Winooski, VT, USA).

2.7. Mitochondrial Metabolism Kinetics

Primary rat hippocampal neuron cultures were plated in 96-well plates at 27,000 cells/cm^2 for 48 h. Cells were then exposed to BDE-47, BDE49, or vehicle (1:1000) in the absence or presence of T3 in combination with a mitochondrial substrate library, MitoPlate-S (Biolog, Inc., Hayward, CA, USA). Mitochondrial substrate metabolism was characterized according to the manufacturer's protocol. Kinetics was recorded on the H1 hybrid microplate reader at a wavelength of 590 nm.

2.8. Statistics

All data are presented as mean ± SE unless otherwise indicated. Graphs were created in GraphPad Prism 8.3.0. Statistical analyses were performed with GraphPad Prism using one-way ANOVA with post hoc Tukey's or Dunnett's or post hoc Kruskal–Wallis with Dunn's as appropriate for the normality of the data as measured by Shapiro–Wilk. qPCR data were analyzed using SDS 2.4 (ThermoScientific) and REST 2009 software (Qiagen, Valencia, CA, USA) with statistical analyses performed using REST 2009 pairwise reallocation randomization test. Significant differences between single and co-exposures or positive controls and vehicle were determined using Student's *t*-test. Statistical significance was defined as $p < 0.05$.

3. Results

3.1. T3 Blocked the Axon Inhibitory Effects of BDE-47 and BDE-49

We previously demonstrated that exposure to either BDE-47 or BDE-49 at concentrations ranging from 200 pM to 2 µM inhibited axonal growth in primary rat hippocampal neurons [23]. To address the question of whether these PBDE congeners modulated axonal growth via effects on TH signaling, we first tested whether the axon inhibitory activity of PBDEs could be blocked by supplementation of the culture medium with T3. Axon lengths were quantified on day in vitro (DIV) 2 after a 48 h exposure to BDE-47 or BDE-49 at 2 or 200 nM in the absence or presence of exogenous T3 at 3 or 30 nM. Consistent with our previous findings, BDE-47 or BDE-49 did not alter the number of axons extended by an individual neuron, but these PBDEs did significantly reduce axonal length relative to vehicle controls (Figure 1A,B). Addition of exogenous T3 at 3 or 30 nM, which raised T3 concentrations in the culture medium to ~5.6 and 32.6 nM, respectively, prevented the inhibition of axonal growth by BDE-47 or BDE-49, as indicated by the fact that axon lengths of neurons exposed to PBDEs in culture medium supplemented with T3 were not significantly different from those of vehicle controls (Figure 1A,B).

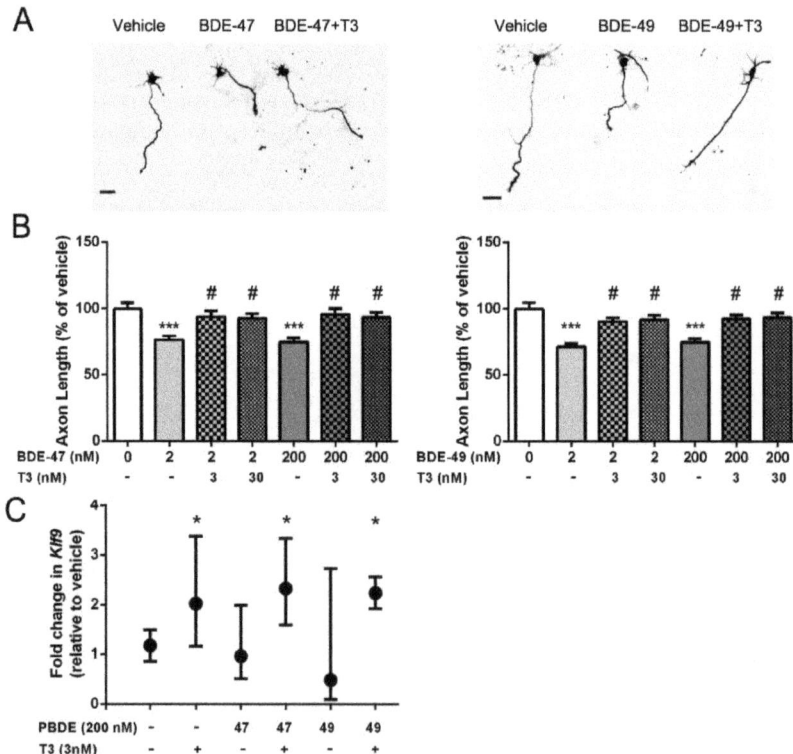

Figure 1. T3 supplementation prevented BDE-47 and BDE-49 inhibition of axonal growth in primary hippocampal neurons. Primary neuron-glia co-cultures dissociated from the hippocampi of P0-1 rats were exposed to vehicle (DMSO diluted 1:1000) or varying concentrations of BDE-47 or BDE-49 in the absence or presence of T3 beginning 3 h after plating. After 48 h exposure, cultures were fixed and immunostained for the axon-selective cytoskeletal protein tau-1. (**A**) Representative photomicrographs

of DIV 2 hippocampal neurons exposed to vehicle, BDE 47 at 2 nM ± exogenous T3 at 3 nM. Scale bar = 25 µm. (**B**) Quantification of axon length in tau-1 immunopositive neurons. Data presented as the mean ± SE (n = 70–90 neurons from three independent dissections). *** Significantly different from vehicle at $p < 0.001$; # significantly different from the corresponding BDE treatment in the absence of T3 at $p < 0.05$ as determined by one-way ANOVA followed by Tukey's post hoc test. (**C**) Fold-change in transcript levels of *Klf9* (as a % of vehicle control). Data are presented as the mean ± SE of *Klf9* expression normalized to the average of the reference genes *Ppia* and *Hprt1*. * Significantly different from vehicle at $p < 0.05$ as determined by REST 2009 pairwise randomization test.

T3 is a component of many neuronal cell culture medias [34,35], and the medium used in these studies contained T3 at ~2.6 nM [34,35]. Thus, our observation that T3 supplementation protected against PBDE inhibition of axonal growth raised the possibility that PBDEs inhibited axonal growth by interfering with TH signaling. As one test of this possibility, we determined whether PBDEs interfered with TH-mediated gene expression. The gene Kruppel-like factor 9 (*Klf9*), previously known as Basic transcription element-binding protein (*Bteb*), has been shown to be a sensitive TH-responsive gene in the developing brain [36,37]. Analyses of *Klf9* transcripts in 2 DIV hippocampal cell cultures confirmed that *Klf9* expression is significantly upregulated in cultures exposed to exogenous T3 at 3 nM for 48 h (Figure 1C). In contrast, exposure to BDE-47 or BDE-49 at 200 nM for 48 h had no significant effect on *Klf9* transcript levels relative to vehicle control cultures and did not inhibit the upregulation of *Klf9* by T3 (Figure 1C).

To determine whether the protective effect of T3 on PBDE inhibition of axonal growth was mediated via direct effects of T3 on axonal growth, we quantified the effect of supplementing the culture medium with T3 on axonal growth in cultures not exposed to PBDEs. As seen in representative photomicrographs (Figure 2A), supplementation with T3 at either 3 or 30 nM had no obvious effect on axonal morphology in terms of the number, length, or branching of axons in DIV 2 hippocampal neurons. Quantification of axon length confirmed that 48 h exposure to medium supplemented with T3 did not significantly alter axon length relative to that observed in vehicle control cultures (Figure 2B).

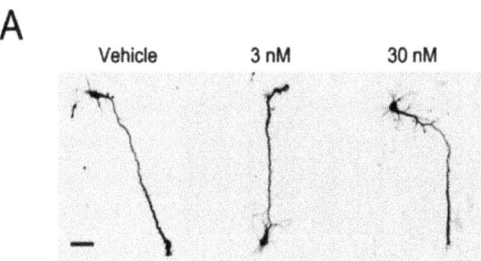

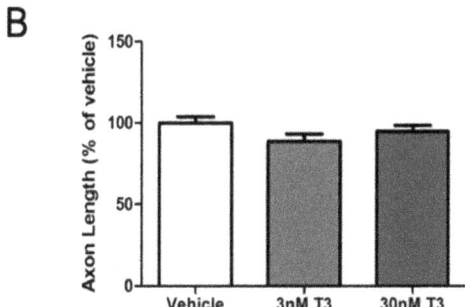

Figure 2. T3 did not influence axonal growth. Primary neuron-glia co-cultures dissociated from the hippocampi of P0-1 rat hippocampi were exposed to vehicle (DMSO diluted 1:1000) or T3 and/or

BDE-47 or BDE-49 beginning 3 h after plating. After 48 h exposure, cultures were fixed and immunostained for tau-1. Representative photomicrographs (**A**) and quantification of axon length (**B**) in tau-1 immunopositive neurons at DIV 2. Data are presented as the mean ± SE (n = 30–40 neurons per group from one dissection; results repeated in 3 independent dissections). There were no significant differences between neurons exposed to vehicle vs. T3 as determined by one-way ANOVA ($p < 0.05$). Scale bar = 25 µm.

3.2. Antioxidants Blocked PBDE Inhibition of Axonal Growth

Previous reports have demonstrated that PBDEs increase levels of ROS in cultured neurons [38–40] and that PBDE-induced ROS can be blocked by mechanistically diverse antioxidants, specifically the NADPH oxidase inhibitor, NAC, or the ROS scavenger, α-tocopherol [41,42]. To evaluate a role for ROS in the axon inhibitory effects of PBDEs, we thus determined whether co-exposure to NAC or α-tocopherol blocked the inhibition of axonal growth by BDE-47 or BDE-49. No significant changes in axon length were observed with antioxidant treatment alone (Supplemental material Figure S1). As shown in representative photomicrographs (Figure 3A) and confirmed by quantitative morphometric analyses of axons (Figure 3B), axon lengths of hippocampal neurons exposed to BDE-47 or BDE-49 at 200 nM in the presence of 100 µM NAC or 100 µM α-tocopherol were not significantly different from those of vehicle control neurons. Cultures co-exposed to PBDEs and antioxidants were significantly longer than axon lengths of hippocampal neurons exposed to the corresponding BDE alone.

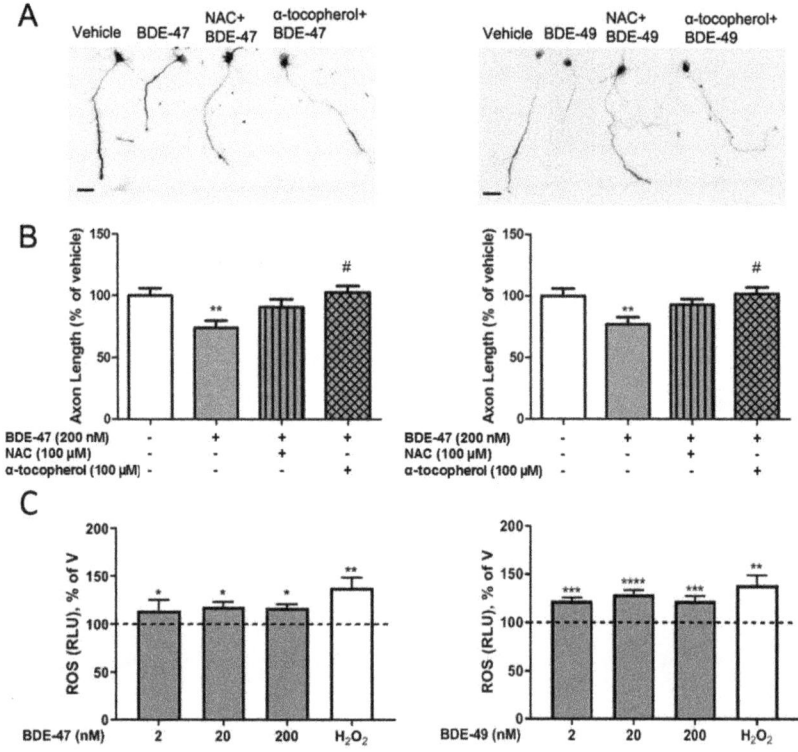

Figure 3. *Cont.*

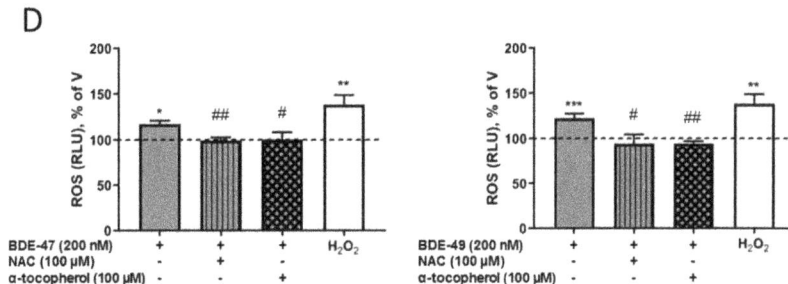

Figure 3. Antioxidants prevented BDE-47 and BDE-49 inhibition of axonal growth and production of ROS. Primary neuron-glia co-cultures dissociated from the hippocampi of P0-1 rat pups were exposed to vehicle, BDE-47 or BDE-49 in the absence or presence of N-acetyl cysteine (NAC) or α-tocopherol. After 48 h exposure, cultures were fixed and immunostained for tau-1. (**A**) Representative photomicrographs of DIV 2 hippocampal neurons from different experimental groups. Scale bar = 25 µm. (**B**) Quantification of axon length in tau-1 immunopositive cells (n = 70–90 neurons from three independent dissections). Quantification of ROS levels following exposure to vehicle, BDE-47 or BDE-49 alone (**C**) or in the presence of an antioxidant (**D**) (n = three independent dissections). H_2O_2 was included as a positive technical control for the ROS-Glo assay per the manufacturer's instructions. Data presented as the mean ± SE. * Significantly different from vehicle at * $p < 0.05$, ** $p < 0.01$, *** $p < 0.001$, **** $p < 0.0001$; # significantly different from PBDE treatment alone at # $p < 0.05$, ## $p < 0.01$, as determined by one-way ANOVA followed by Tukey's post hoc test; †significantly different from individual PBDE treatment at $p < 0.05$ as determined by Student's t-test.

To determine whether nM concentrations of BDE-47 or BDE-49 that inhibit axonal growth increased intracellular ROS, ROS were measured in cultures acutely exposed to BDE-47 or BDE-49. Both BDE-47 and BDE-49-exposed cultures had higher amounts of ROS compared to vehicle control cultures (Figure 3C). We next evaluated whether antioxidants blocked the inhibitory effects of PBDEs on axonal growth by providing protection against ROS generation (Figure 3D). In the presence of NAC or α-tocopherol, PBDEs did not produce significant amounts of ROS compared to vehicle. However, ROS production was substantially reduced compared to BDE-47 or BDE-49 alone.

It is posited that ROS generation largely originates from mitochondrial damage [43]. BDE-47 can disrupt the mitochondrial membrane potential [44], while both BDE-47 [45,46] and BDE-49 [47] can decrease mitochondrial bioenergetics. Thus, we next sought to determine whether acute exposure to nM concentrations of BDE-47 or BDE-49 altered mitochondrial metabolism. Compared to vehicle control cultures, mitochondrial metabolism was significantly impacted in cultures acutely exposed to either BDE-47 or BDE-49 at 200 nM (Figure 4B).

3.3. T3 Blocked PBDE Axon Inhibition by Blocking PBDE-Induced ROS

To determine whether T3 conferred protection against the axon inhibitory effects of PBDEs via upregulation of endogenous antioxidant molecules, we quantified the effects of T3, BDE-47, and BDE-49, alone and in combination, on the production of ROS (Figure 4A). In contrast to cultures exposed to PBDEs in the absence of T3, in cultures co-exposed for 1 h to one of these PBDEs and T3 exhibited no significant change in ROS levels relative to vehicle controls. Moreover, ROS levels were significantly reduced in cultures co-exposed to PBDEs and T3 relative to cultures exposed to PBDEs in the absence of T3. We then explored whether T3 protected against disrupted mitochondrial bioenergetics (Figure 4B). Following acute exposure to either BDE-47 or BDE-49 in combination with T3, there were no marked alterations in mitochondrial substrate metabolism relative to vehicle. In addition, any effects observed with individual PBDE exposure were eliminated in cultures co-exposed to PBDEs and T3.

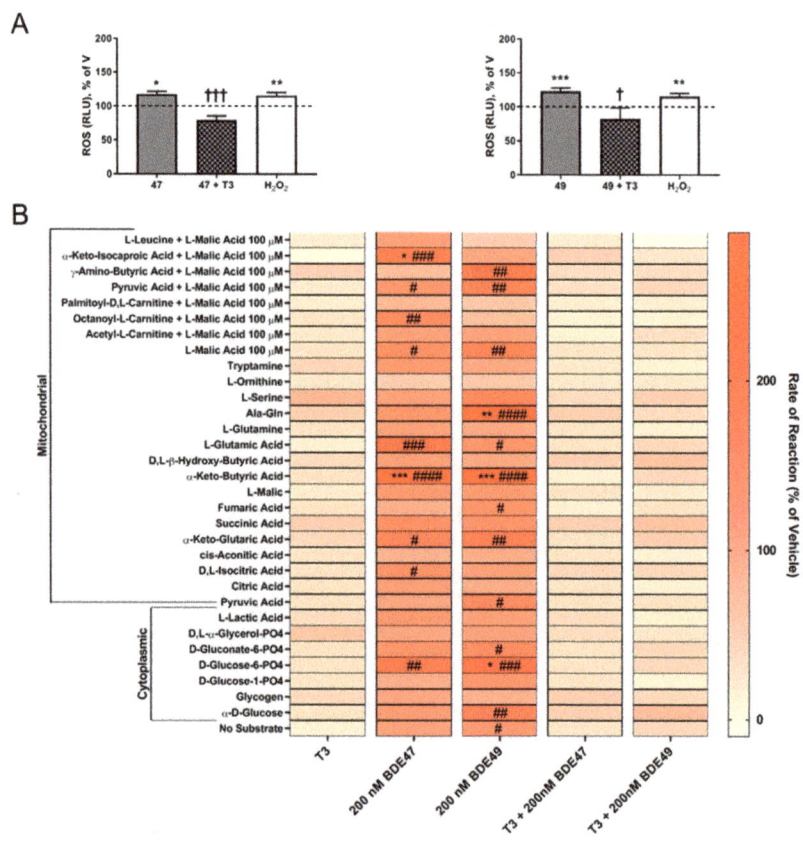

Figure 4. T3 normalized ROS levels and mitochondrial substrate metabolism in cultures exposed to BDE-47 or BDE-49. Hippocampal neuron-glia co-cultures were exposed to vehicle, T3, BDE-47 and/or BDE-49 for 1 h on DIV 2. (**A**) Quantification of ROS production following co-exposure to T3 and PBDEs. (**B**) Mitochondrial substrate metabolism kinetics immediately following PBDE exposure alone and in the presence of T3. Data presented as the mean ± SE (n = three independent dissections). * Significantly different from vehicle at * $p < 0.05$, ** $p < 0.01$, *** $p < 0.001$; # significantly different from T3 at # $p < 0.05$, ## $p < 0.01$, ### $p < 0.001$, #### $p < 0.0001$ as determined by one-way ANOVA followed by Dunnett's post hoc test; †significantly different from individual PBDE treatment at † $p < 0.05$, ††† $p < 0.001$ as determined by Student's t-test.

4. Discussion

The findings from this study extend our previous report that BDE-47 and BDE-49 inhibited axonal growth in primary rat hippocampal neurons [23] by demonstrating that the axon inhibitory activity of these PBDE congeners is mediated by increased levels of intracellular ROS. The evidence in support of this conclusion includes: (1) BDE-47 and BDE-49 increased ROS in primary rat hippocampal neurons at nM concentrations that also inhibited axonal growth; and (2) co-exposure to either the NADPH oxidase inhibitor, NAC, or the ROS scavenger, α-tocopherol, blocked the axon inhibitory effects of BDE-47 and BDE-49. Additionally, we observed that supplementation of the culture medium with exogenous T3 blocked the inhibition of axonal growth in PBDE-exposed neuronal cultures, coincident with mitigation of PBDE effects on intracellular ROS and metabolic substrate production from the mitochondria. These findings suggest a role for T3 in maintaining intracellular redox homeostasis in response to pro-oxidants, which if true, represents a novel mechanism by which thyroid hormone disruption contributes to adverse neurodevelopmental outcomes.

Our observations are consistent with previous reports that PBDEs upregulated biomarkers of oxidative stress in the brain of adult and developing rodent models [48,49] and increased intracellular ROS levels in cultured neural cells [38,39,50,51]. This earlier work demonstrated that μM concentrations of PBDEs increased intracellular ROS in cultured neurons to levels that triggered apoptosis [24,52,53]. Here, we found that exposure of primary hippocampal neuron-glia co-cultures to BDE-47 or BDE-49 at nM concentrations also increased intracellular ROS, but this was associated with inhibited axonal growth. Our data extend reports in the literature indicating that physiologic levels of ROS regulate axonal specification and axonal growth in primary hippocampal neurons, and modulation of ROS synthesis in axonal growth cones cause cytoskeletal rearrangements that alter axonal morphogenesis [54]. Collectively, these observations suggest a model in which nM PBDE concentrations increase ROS locally in the axonal growth cone to modulate signaling pathways that regulate axonal growth [55,56], whereas μM PBDE concentrations increase intracellular ROS globally to trigger cell death. Confirmation of this model will require the adaptation of sensitive technologies to detect localized changes in ROS in subcellular domains of neurons [57] exposed to PBDEs at concentrations that inhibit axonal growth.

PBDEs can interfere with thyroid hormone signaling and thyroid hormone disruption is widely posited to contribute to the developmental neurotoxicity of these environmental contaminants [25,26]. PBDEs have been shown to suppress dendritic growth in Purkinje cells by disrupting TH receptor-mediated transcription [58], and we observed that co-exposure to T3 blocked inhibition of axonal growth by BDE-47 or BDE-49. However, several lines of evidence argue against the hypothesis that PBDEs inhibit axonal growth in hippocampal neurons via direct interference with TH signaling. First, in hippocampal cultures not exposed to PBDEs, T3 supplementation of the culture medium did not promote axonal growth. Second, exposure of hippocampal cultures to BDE-47 or BDE-49 did not alter expression of *Klf9*, a gene known to be highly sensitive to upregulation by TH in the developing brain [59]. Nor did BDE-47 or BDE-49 significantly block T3-induced *Klf9* expression. These findings are in agreement with previous studies [58] in which qPCR analyses detected no significant changes in transcript levels of TH-responsive genes, including TRα1 or TRβ, in primary rat Purkinje cells exposed to PBDEs. Moreover, since the affinity of T3 to the thyroid hormone receptor (THR) is approximately 0.1 nM, the observation that T3 present in the medium without addition of extra T3 is not sufficient to prevent the axon inhibitory effects of PBDEs suggests that the neuroprotective effect of exogenous T3 in this model is mediated by THR-independent mechanisms. NH-3, a pharmacological THR modulator with mixed agonist/antagonistic activity [60], may be useful for addressing this question, but given experimental evidence that the concentration–response relationship for antagonistic vs. agonist effects of NH-3 vary across models, its effectiveness in mechanistic studies of the axon inhibitory activity of PBDEs will require identification of a concentration that antagonizes THR in this model system [60,61].

Our data suggest that T3 supplementation prevented PBDE inhibition of axonal growth by mitigating PBDE-induced ROS. Specifically, we observed that T3 supplementation ameliorated PBDE-induced ROS generation. A key question is how. TH has been reported to upregulate expression of endogenous antioxidant molecules [62,63]. However, preliminary qPCR analyses failed to detect significant upregulation of several endogenous antioxidants in primary hippocampal neuron-glia co-cultures exposed to BDE-47 or BDE-49 in the presence of T3 (Supplemental Table S2). This observation does not rule out the possibility that T3 upregulated expression of cellular antioxidants other than those we assessed and/or that T3 increased the activity of enzymatic antioxidants. In addition, *Klf9* upregulation by 5 or 10 nM T3 supplementation has previously been shown to protect the axons of primary cortical murine neurons from hypoxic injury [64]. Whether *Klf9* or other T3-regulated targets are directly involved in mitigating PBDE axon inhibition remains to be investigated. However, it is now clear that TH can also signal via non-transcriptional mechanisms [65–67], including direct influence on mitochondrial respiration [68]. Consistent with this literature, our data support a model in which T3 protects mitochondrial metabolism against PBDE-mediated disruption of mitochondrial bioenergetics.

The observation that T3 prevented PBDE-induced changes in mitochondrial substrate utilization at concentrations that also blocked PBDE inhibition of axon growth, yet had no effect on basal axonogenesis, suggested that PBDEs increased ROS as a consequence of altered mitochondrial metabolism. In support of this proposed mechanism, at concentrations that increased intracellular ROS levels, BDE-47 and BDE-49 increased utilization of metabolic substrates (α-keto-isocaproic acid, α-keto-butyric acid, Ala-Gln, D-glucose-6-PO4) used to produce NADH, and disruption of NADH production has been linked to increased ROS generation [69]. However, the mechanism(s) by which PBDEs interfere with mitochondrial metabolism remain to be elucidated.

Findings from our previous studies suggested RyR was a downstream effector in PBDE-induced axon growth inhibition [23]. Given the redox-sensitive nature of RyR [70,71] and spatial relationship with mitochondria [72], a potential indirect mechanism presents itself wherein mitochondrial ROS production alters RyR gating and, consequently, calcium signaling to interfere with axon growth. This model is supported by experimental evidence demonstrating that disruption of mitochondrial function affects calcium homeostasis, which in turn delays polarization of developing neurons and inhibits axonal growth [73]. As we previously reported [23], PBDE inhibition of axonal growth is due in part to delayed neuronal polarization. The role of RyR as a downstream key event rather than the molecular initiating event in PBDE developmental neurotoxicity may explain the differential response of dendrites vs. axons to non-dioxin-like polychlorinated biphenyls (PCBs) vs. PBDEs. Specifically, in primary rat hippocampal and cortical neuron-glia co-cultures, non-dioxin-like PCBs were observed to promote dendritic growth, but have no effect on axonal growth, and the dendrite promoting activity was mediated by RyR sensitization [74,75]. In contrast, PBDEs were observed to inhibit axonal growth but have no effect on dendritic growth in the same neuronal cell culture model [23].

In summary, our study provides novel insight into the interplay between ROS, TH, and axonal growth in PBDE developmental neurotoxicity. Whether PBDE interference with axonal growth contributes to adverse neurodevelopmental outcomes in vivo is still to be determined; however, clinical [20,21] and experimental evidence [17–19] demonstrate that altered spatiotemporal patterns of axonal growth during brain development can cause functional deficits. Susceptibility to this neurotoxic activity of PBDEs may be enhanced in populations with heritable mutations that alter mitochondrial and redox signaling, which are themselves associated with increased risk of neurodevelopmental disorders [76,77]. The finding that T3 protects against axon growth inhibition by BDE-47 and BDE-49 in vitro suggests that PBDE-mediated TH dysregulation [2,78] also has the potential to amplify PBDE effects on axonal growth in vivo. Further studies into gene × environment interactions associated with these mechanisms may lead to a better understanding of populations with increased vulnerability to PBDE developmental neurotoxicity.

Supplementary Materials: The following supporting information can be downloaded at: https://www.mdpi.com/article/10.3390/toxics10020092/s1, Figure S1: The antioxidants NAC and α-tocopherol do not alter axonal outgrowth relative to vehicle control cultures. Primary neuron-glia co-cultures dissociated from the hippocampi of P0-1 rat pups were exposed to vehicle, N-acetyl cysteine (NAC) or α-tocopherol. After a 48 h exposure, cultures were fixed and immunostained for tau-1. Axon length was quantified in tau-1 immunopositive cells ($n = 70$–90 neurons from three independent dissections). Data presented as the mean ± SE. No significant differences between groups was detected using one-way ANOVA ($p < 0.05$); Table S1: Primer sequences and amplification efficiencies; Table S2: Average fold changes relative to vehicle of levels of transcripts encoding cellular antioxidants.

Author Contributions: Conceptualization, H.C., R.K.C. and P.J.L.; data curation, H.C., R.K.C. and A.C.B.; formal analysis, H.C. and R.K.C.; funding acquisition, P.J.L.; investigation, H.C., R.K.C., A.C.B. and K.A.H.; methodology, H.C., R.K.C. and K.A.H.; project administration, P.J.L.; supervision, P.J.L.; visualization, H.C. and R.K.C.; writing—original draft, H.C. and R.K.C.; writing—review and editing,

H.C., R.K.C., A.C.B., K.A.H. and P.J.L. All authors have read and agreed to the published version of the manuscript.

Funding: This research was supported by the National Institute of Environmental Health Sciences, grant numbers R01 ES014901, P30 ES023513, P42 ES04699, and T32 ES007059, and the National Institute of Child Health and Development, grant number P50 HD103526. The contents of this work do not necessarily represent the official views of the NIEHS or NICHD, and these institutes do not endorse the purchase of any commercial products or services mentioned in the publication.

Institutional Review Board Statement: The animal study protocol was approved by the Institutional Animal Care and Use Committee of the University of California, Davis (protocol #18813 and #18853, approved 18 April 2013).

Informed Consent Statement: Not applicable.

Data Availability Statement: Data will be made available upon reasonable request.

Conflicts of Interest: The authors declare no conflict of interest. The funders had no role in the design of the study; in the collection, analyses, or interpretation of data; in the writing of the manuscript, or in the decision to publish the results.

References

1. Lam, J.; Lanphear, B.P.; Bellinger, D.; Axelrad, D.A.; McPartland, J.; Sutton, P.; Davidson, L.; Daniels, N.; Sen, S.; Woodruff, T.J. Developmental PBDE Exposure and IQ/ADHD in Childhood: A Systematic Review and Meta-analysis. *Environ. Health Perspect.* **2017**, *125*, 086001. [CrossRef]
2. Vuong, A.M.; Braun, J.M.; Webster, G.M.; Thomas Zoeller, R.; Hoofnagle, A.N.; Sjödin, A.; Yolton, K.; Lanphear, B.P.; Chen, A. Polybrominated Diphenyl Ether (PBDE) Exposures and Thyroid Hormones in Children at Age 3 Years. *Environ. Int.* **2018**, *117*, 339–347. [CrossRef]
3. Gibson, E.A.; Siegel, E.L.; Eniola, F.; Herbstman, J.B.; Factor-Litvak, P. Effects of Polybrominated Diphenyl Ethers on Child Cognitive, Behavioral, and Motor Development. *Int. J. Environ. Res. Public Health* **2018**, *15*, 1636. [CrossRef] [PubMed]
4. Dorman, D.C.; Chiu, W.; Hales, B.F.; Hauser, R.; Johnson, K.J.; Mantus, E.; Martel, S.; Robinson, K.A.; Rooney, A.A.; Rudel, R.; et al. Polybrominated Diphenyl Ether (PBDE) Neurotoxicity: A Systematic Review and Meta-Analysis of Animal Evidence. *J. Toxicol. Environ. Health B Crit. Rev.* **2018**, *21*, 269–289. [CrossRef] [PubMed]
5. Chao, H.R.; Tsou, T.C.; Huang, H.L.; Chang-Chien, G.P. Levels of Breast Milk Pbdes from Southern Taiwan and Their Potential Impact on Neurodevelopment. *Pediatr. Res.* **2011**, *70*, 596–600. [CrossRef] [PubMed]
6. Eskenazi, B.; Chevrier, J.; Rauch, S.A.; Kogut, K.; Harley, K.G.; Johnson, C.; Trujillo, C.; Sjodin, A.; Bradman, A. In Utero and Childhood Polybrominated Diphenyl Ether (PBDE) Exposures and Neurodevelopment in the CHAMACOS Study. *Environ. Health Perspect.* **2013**, *121*, 257–262. [CrossRef]
7. Berghuis, S.A.; Bos, A.F.; Sauer, P.J.; Roze, E. Developmental Neurotoxicity of Persistent Organic Pollutants: An Update on Childhood Outcome. *Arch. Toxicol.* **2015**, *89*, 687–709. [CrossRef]
8. Cowell, W.J.; Lederman, S.A.; Sjodin, A.; Jones, R.; Wang, S.; Perera, F.P.; Wang, R.; Rauh, V.A.; Herbstman, J.B. Prenatal Exposure to Polybrominated Diphenyl Ethers and Child Attention Problems at 3–7 Years. *Neurotoxicol. Teratol.* **2015**, *52*, 143–150. [CrossRef]
9. Linares, V.; Belles, M.; Domingo, J.L. Human Exposure to PBDE and Critical Evaluation of Health Hazards. *Arch. Toxicol.* **2015**, *89*, 335–356. [CrossRef]
10. Tsai, M.H.; Chao, H.R.; Hsu, W.L.; Tsai, C.C.; Lin, C.W.; Chen, C.H. Analysis of Polybrominated Diphenyl Ethers and Lipid Composition in Human Breast Milk and Their Correlation with Infant Neurodevelopment. *Int. J. Environ. Res. Public Health* **2021**, *18*, 11501. [CrossRef]
11. Azar, N.; Booij, L.; Muckle, G.; Arbuckle, T.E.; Seguin, J.R.; Asztalos, E.; Fraser, W.D.; Lanphear, B.P.; Bouchard, M.F. Prenatal Exposure to Polybrominated Diphenyl Ethers (PBDEs) and Cognitive Ability in Early Childhood. *Environ. Int.* **2021**, *146*, 106296. [CrossRef] [PubMed]
12. de Water, E.; Curtin, P.; Zilverstand, A.; Sjodin, A.; Bonilla, A.; Herbstman, J.B.; Ramirez, J.; Margolis, A.E.; Bansal, R.; Whyatt, R.M.; et al. A Preliminary Study on Prenatal Polybrominated Diphenyl Ether Serum Concentrations and Intrinsic Functional Network Organization and Executive Functioning in Childhood. *J. Child Psychol. Psychiatry* **2019**, *60*, 1010–1020. [CrossRef] [PubMed]
13. Schecter, A.; Papke, O.; Tung, K.C.; Joseph, J.; Harris, T.R.; Dahlgren, J. Polybrominated Diphenyl Ether Flame Retardants in the U.S. Population: Current Levels, Temporal Trends, and Comparison with Dioxins, Dibenzofurans, and Polychlorinated Biphenyls. *J. Occup. Environ. Med.* **2005**, *47*, 199–211. [CrossRef] [PubMed]
14. Frederiksen, M.; Vorkamp, K.; Thomsen, M.; Knudsen, L.E. Human Internal and External Exposure to Pbdes—A Review of Levels and Sources. *Int. J. Hyg. Environ. Health* **2009**, *212*, 109–134. [CrossRef] [PubMed]
15. Kodavanti, P.R.; Curras-Collazo, M.C. Neuroendocrine Actions of Organohalogens: Thyroid Hormones, Arginine Vasopressin, and Neuroplasticity. *Front. Neuroendocrinol.* **2010**, *31*, 479–496. [CrossRef] [PubMed]

16. Stamou, M.; Streifel, K.M.; Goines, P.E.; Lein, P.J. Neuronal Connectivity as A Convergent Target of Gene X Environment Interactions that Confer Risk for Autism Spectrum Disorders. *Neurotoxicol. Teratol.* **2013**, *36*, 3–16. [CrossRef]
17. Berger-Sweeney, J.; Hohmann, C.F. Behavioral Consequences of Abnormal Cortical Development: Insights into Developmental Disabilities. *Behav. Brain Res.* **1997**, *86*, 121–142. [CrossRef]
18. Cremer, H.; Chazal, G.; Carleton, A.; Goridis, C.; Vincent, J.D.; Lledo, P.M. Long-Term but not Short-Term Plasticity at Mossy Fiber Synapses is Impaired in Neural Cell Adhesion Molecule-Deficient Mice. *Proc. Natl. Acad. Sci. USA* **1998**, *95*, 13242–13247. [CrossRef]
19. Barone, S., Jr.; Das, K.P.; Lassiter, T.L.; White, L.D. Vulnerable Processes of Nervous System Development: A Review of Markers and Methods. *Neurotoxicology* **2000**, *21*, 15–36.
20. Geschwind, D.H.; Levitt, P. Autism Spectrum Disorders: Developmental Disconnection Syndromes. *Curr. Opin. Neurobiol.* **2007**, *17*, 103–111. [CrossRef]
21. Engle, E.C. Human Genetic Disorders of Axon Guidance. *Cold Spring Harb. Protoc.* **2010**, *2*, a001784. [CrossRef] [PubMed]
22. Miller, M.F.; Chernyak, S.M.; Batterman, S.; Loch-Caruso, R. Polybrominated Diphenyl Ethers in Human Gestational Membranes from Women in Southeast Michigan. *Environ. Sci. Technol.* **2009**, *43*, 3042–3046. [CrossRef] [PubMed]
23. Chen, H.; Streifel, K.M.; Singh, V.; Yang, D.; Mangini, L.; Wulff, H.; Lein, P.J. From the Cover: BDE-47 and BDE-49 Inhibit Axonal Growth in Primary Rat Hippocampal Neuron-Glia Co-Cultures via Ryanodine Receptor-Dependent Mechanisms. *Toxicol. Sci.* **2017**, *156*, 375–386. [CrossRef]
24. Kim, K.H.; Bose, D.D.; Ghogha, A.; Riehl, J.; Zhang, R.; Barnhart, C.D.; Lein, P.J.; Pessah, I.N. Para- and Ortho-Substitutions are Key Determinants of Polybrominated Diphenyl Ether Activity toward Ryanodine Receptors and Neurotoxicity. *Environ. Health Perspect.* **2011**, *119*, 519–526. [CrossRef] [PubMed]
25. Costa, L.G.; de Laat, R.; Tagliaferri, S.; Pellacani, C. A Mechanistic View of Polybrominated Diphenyl Ether (PBDE) Developmental Neurotoxicity. *Toxicol. Lett.* **2014**, *230*, 282–294. [CrossRef]
26. Hendriks, H.S.; Westerink, R.H. Neurotoxicity and Risk Assessment of Brominated and Alternative Flame Retardants. *Neurotoxicol. Teratol.* **2015**, *52*, 248–269. [CrossRef]
27. Pessah, I.N.; Cherednichenko, G.; Lein, P.J. Minding the calcium store: Ryanodine Receptor Activation as a Convergent Mechanism of PCB Toxicity. *Pharmacol. Ther.* **2010**, *125*, 260–285. [CrossRef]
28. Batistuzzo, A.; Ribeiro, M.O. Clinical and Subclinical Maternal Hypothyroidism and their Effects on Neurodevelopment, Behavior and Cognition. *Arch. Endocrinol. Metab.* **2020**, *64*, 89–95. [CrossRef]
29. Terzi, A.; Suter, D.M. The Role of NADPH Oxidases in Neuronal Development. *Free Radic. Biol. Med.* **2020**, *154*, 33–47. [CrossRef]
30. Kilkenny, C.; Browne, W.J.; Cuthill, I.C.; Emerson, M.; Altman, D.G. Improving Bioscience Research Reporting: The ARRIVE Guidelines for Reporting Animal Research. *PLoS Biol.* **2010**, *8*, e1000412. [CrossRef]
31. Brewer, G.J. Isolation and Culture of Adult Rat Hippocampal Neurons. *J. Neurosci. Methods* **1997**, *71*, 143–155. [CrossRef]
32. Brewer, G.J.; Torricelli, J.R.; Evege, E.K.; Price, P.J. Optimized Survival of Hippocampal Neurons in B27-Supplemented Neurobasal, a New Serum-Free Medium Combination. *J. Neurosci. Res.* **1993**, *35*, 567–576. [CrossRef] [PubMed]
33. Chen, Y.; Stevens, B.; Chang, J.; Milbrandt, J.; Barres, B.A.; Hell, J.W. NS21: Re-Defined and Modified Supplement B27 for Neuronal Cultures. *J. Neurosci. Methods* **2008**, *171*, 239–247. [CrossRef]
34. Meijering, E.; Jacob, M.; Sarria, J.C.; Steiner, P.; Hirling, H.; Unser, M. Design and Validation of a Tool for Neurite Tracing and Analysis in Fluorescence Microscopy Images. *Cytometry A* **2004**, *58*, 167–176. [CrossRef] [PubMed]
35. Dotti, C.G.; Sullivan, C.A.; Banker, G.A. The Establishment of Polarity by Hippocampal Neurons in Culture. *J. Neurosci.* **1988**, *8*, 1454–1468. [CrossRef]
36. Denver, R.J.; Ouellet, L.; Furling, D.; Kobayashi, A.; Fujii-Kuriyama, Y.; Puymirat, J. Basic Transcription Element-Binding Protein (BTEB) is a Thyroid Hormone-Regulated Gene in the Developing Central Nervous System. Evidence for a role in neurite outgrowth. *Int. J. Biol. Chem.* **1999**, *274*, 23128–23134. [CrossRef]
37. Gilbert, M.E.; Sanchez-Huerta, K.; Wood, C. Mild Thyroid Hormone Insufficiency during Development Compromises Activity-Dependent Neuroplasticity in the Hippocampus of Adult Male Rats. *Endocrinology* **2016**, *157*, 774–787. [CrossRef]
38. Chen, J.; Liufu, C.; Sun, W.; Sun, X.; Chen, D. Assessment of the Neurotoxic Mechanisms of Decabrominated Diphenyl Ether (PBDE-209) in Primary Cultured Neonatal Rat Hippocampal Neurons Includes Alterations in Second Messenger Signaling and Oxidative Stress. *Toxicol. Lett.* **2010**, *192*, 431–439. [CrossRef]
39. Tagliaferri, S.; Caglieri, A.; Goldoni, M.; Pinelli, S.; Alinovi, R.; Poli, D.; Pellacani, C.; Giordano, G.; Mutti, A.; Costa, L.G. Low Concentrations of the Brominated Flame Retardants BDE-47 and BDE-99 Induce Synergistic Oxidative Stress-Mediated Neurotoxicity in Human Neuroblastoma Cells. *Toxicol. In Vitro* **2010**, *24*, 116–122. [CrossRef]
40. Costa, L.G.; Tagliaferri, S.; Roque, P.J.; Pellacani, C. Role of Glutamate Receptors in Tetrabrominated Diphenyl Ether (BDE-47) Neurotoxicity in Mouse Cerebellar Granule Neurons. *Toxicol. Lett.* **2016**, *241*, 159–166. [CrossRef]
41. An, J.; Yin, L.; Shang, Y.; Zhong, Y.; Zhang, X.; Wu, M.; Yu, Z.; Sheng, G.; Fu, J.; Huang, Y. The Combined Effects of BDE47 and Bap on Oxidatively Generated DNA Damage in L02 Cells and the Possible Molecular Mechanism. *Mutat. Res.* **2011**, *721*, 192–198. [CrossRef] [PubMed]
42. Park, H.R.; Kamau, P.W.; Loch-Caruso, R. Involvement of Reactive Oxygen Species in Brominated Diphenyl Ether-47-Induced Inflammatory Cytokine Release from Human Extravillous Trophoblasts in vitro. *Toxicol. Appl. Pharmacol.* **2014**, *274*, 283–292. [CrossRef] [PubMed]

43. Rigoulet, M.; Yoboue, E.D.; Devin, A. Mitochondrial ROS Generation and its Regulation: Mechanisms involved in H_2O_2 Signaling. *Antioxid. Redox Signal.* **2011**, *14*, 459–468. [CrossRef] [PubMed]
44. Yan, C.; Huang, D.; Zhang, Y. The Involvement of ROS Overproduction and Mitochondrial Dysfunction in PBDE-47-Induced Apoptosis on Jurkat Cells. *Exp. Toxicol. Pathol.* **2011**, *63*, 413–417. [CrossRef] [PubMed]
45. Pazin, M.; Pereira, L.C.; Dorta, D.J. Toxicity of Brominated Flame Retardants, BDE-47 and BDE-99 Stems from Impaired Mitochondrial Bioenergetics. *Toxicol. Mech. Methods* **2015**, *25*, 34–41. [CrossRef]
46. Lefevre, P.L.; Wade, M.; Goodyer, C.; Hales, B.F.; Robaire, B. A Mixture Reflecting Polybrominated Diphenyl Ether (PBDE) Profiles Detected in Human Follicular Fluid Significantly Affects Steroidogenesis and Induces Oxidative Stress in a Female Human Granulosa Cell Line. *Endocrinology* **2016**, *157*, 2698–2711. [CrossRef]
47. Napoli, E.; Hung, C.; Wong, S.; Giulivi, C. Toxicity of the Flame-Retardant BDE-49 on Brain Mitochondria and Neuronal Progenitor Striatal Cells Enhanced by a PTEN-Deficient Background. *Toxicol. Sci.* **2013**, *132*, 196–210. [CrossRef]
48. Belles, M.; Alonso, V.; Linares, V.; Albina, M.L.; Sirvent, J.J.; Domingo, J.L.; Sanchez, D.J. Behavioral Effects and Oxidative Status in Brain Regions of Adult Rats Exposed to BDE-99. *Toxicol. Lett.* **2010**, *194*, 1–7. [CrossRef]
49. Costa, L.G.; Pellacani, C.; Dao, K.; Kavanagh, T.J.; Roque, P.J. The Brominated Flame Retardant BDE-47 Causes Oxidative Stress and Apoptotic Cell Death in vitro and in vivo in Mice. *Neurotoxicology* **2015**, *48*, 68–76. [CrossRef]
50. Giordano, G.; Kavanagh, T.J.; Costa, L.G. Neurotoxicity of A Polybrominated Diphenyl Ether Mixture (DE-71) in Mouse Neurons and Astrocytes is Modulated by Intracellular Glutathione Levels. *Toxicol. Appl. Pharmacol.* **2008**, *232*, 161–168. [CrossRef]
51. He, P.; He, W.; Wang, A.; Xia, T.; Xu, B.; Zhang, M.; Chen, X. PBDE-47-Induced Oxidative Stress, DNA Damage and Apoptosis in Primary Cultured Rat Hippocampal Neurons. *Neurotoxicology* **2008**, *29*, 124–129. [CrossRef] [PubMed]
52. Huang, S.C.; Giordano, G.; Costa, L.G. Comparative Cytotoxicity and Intracellular Accumulation of Five Polybrominated Diphenyl Ether Congeners in Mouse Cerebellar Granule Neurons. *Toxicol. Sci.* **2010**, *114*, 124–132. [CrossRef] [PubMed]
53. Zhang, S.; Kuang, G.; Zhao, G.; Wu, X.; Zhang, C.; Lei, R.; Xia, T.; Chen, J.; Wang, Z.; Ma, R.; et al. Involvement of the Mitochondrial P53 Pathway in PBDE-47-Induced SH-SY5Y Cells Apoptosis and its Underlying Activation Mechanism. *Food Chem. Toxicol.* **2013**, *62*, 699–706. [CrossRef] [PubMed]
54. Wilson, C.; Gonzalez-Billault, C. Regulation of Cytoskeletal Dynamics by Redox Signaling and Oxidative Stress: Implications for Neuronal Development and Trafficking. *Front. Cell. Neurosci.* **2015**, *9*, 381. [CrossRef] [PubMed]
55. Olguin-Albuerne, M.; Moran, J. ROS Produced by NOX2 Control in vitro Development of Cerebellar Granule Neurons Development. *ASN Neuro* **2015**, *7*. [CrossRef]
56. Wilson, C.; Munoz-Palma, E.; Henriquez, D.R.; Palmisano, I.; Nunez, M.T.; Di Giovanni, S.; Gonzalez-Billault, C. A Feed-Forward Mechanism Involving the NOX Complex and RyR-Mediated Ca2+ Release During Axonal Specification. *Neurosci. Res.* **2016**, *36*, 11107–11119. [CrossRef]
57. Belousov, V.V.; Fradkov, A.F.; Lukyanov, K.A.; Staroverov, D.B.; Shakhbazov, K.S.; Terskikh, A.V.; Lukyanov, S. Genetically Encoded Fluorescent Indicator for Intracellular Hydrogen Peroxide. *Nat. Methods* **2006**, *3*, 281–286. [CrossRef]
58. Ibhazehiebo, K.; Iwasaki, T.; Kimura-Kuroda, J.; Miyazaki, W.; Shimokawa, N.; Koibuchi, N. Disruption of Thyroid Hormone Receptor-Mediated Transcription and Thyroid Hormone-Induced Purkinje Cell Dendrite Arborization by Polybrominated Diphenyl Ethers. *Environ. Health Perspect.* **2011**, *119*, 168–175. [CrossRef]
59. Denver, R.J.; Williamson, K.E. Identification of A Thyroid Hormone Response Element in the Mouse Kruppel-Like Factor 9 Gene to Explain its Postnatal Expression in the Brain. *Endocrinology* **2009**, *150*, 3935–3943. [CrossRef]
60. Walter, K.M.; Singh, L.; Singh, V.; Lein, P.J. Investigation of NH3 as A Selective Thyroid Hormone Receptor Modulator in Larval Zebrafish (*Danio rerio*). *Neurotoxicology* **2021**, *84*, 96–104. [CrossRef]
61. Sethi, S.; Morgan, R.K.; Feng, W.; Lin, Y.; Li, X.; Luna, C.; Koch, M.; Bansal, R.; Duffel, M.W.; Puschner, B.; et al. Comparative Analyses of the 12 Most Abundant PCB Congeners Detected in Human Maternal Serum for Activity at the Thyroid Hormone Receptor and Ryanodine Receptor. *Environ. Sci. Technol.* **2019**, *53*, 3948–3958. [CrossRef] [PubMed]
62. Das, K.; Chainy, G.B. Thyroid Hormone Influences Antioxidant Defense System in Adult Rat Brain. *Neurochem. Res.* **2004**, *29*, 1755–1766. [CrossRef] [PubMed]
63. Fernandez, V.; Tapia, G.; Varela, P.; Romanque, P.; Cartier-Ugarte, D.; Videla, L.A. Thyroid Hormone-Induced Oxidative Stress in Rodents and Humans: A Comparative View and Relation to Redox Regulation of Gene Expression. *Comp. Biochem. Physiol. C. Toxicol. Pharmacol.* **2006**, *142*, 231–239. [CrossRef] [PubMed]
64. Li, J.; Abe, K.; Milanesi, A.; Liu, Y.-Y.; Brent, G.A. Thyroid Hormone Protects Primary Cortical Neurons Exposed to Hypoxia by Reducing DNA Methylation and Apoptosis. *Endocrinology* **2019**, *160*, 2243–2256. [CrossRef] [PubMed]
65. Cheng, S.Y.; Leonard, J.L.; Davis, P.J. Molecular Aspects of Thyroid Hormone Actions. *Endocr. Rev.* **2010**, *31*, 139–170. [CrossRef]
66. Davis, P.J.; Lin, H.Y.; Mousa, S.A.; Luidens, M.K.; Hercbergs, A.A.; Wehling, M.; Davis, F.B. Overlapping Nongenomic and Genomic Actions of Thyroid Hormone and Steroids. *Steroids* **2011**, *76*, 829–833. [CrossRef] [PubMed]
67. Flamant, F. Futures Challenges in Thyroid Hormone Signaling Research. *Front. Endocrinol.* **2016**, *7*, 58. [CrossRef] [PubMed]
68. Wrutniak-Cabello, C.; Casas, F.; Cabello, G. Thyroid Hormone Action in Mitochondria. *J. Mol. Endocrinol.* **2001**, *26*, 67–77. [CrossRef]
69. Murphy, M.P. How Mitochondria Produce Reactive Oxygen Species. *Biochem. J.* **2009**, *417*, 1–13. [CrossRef]
70. Feng, W.; Liu, G.; Allen, P.D.; Pessah, I.N. Transmembrane Redox Sensor of Ryanodine Receptor Complex. *Int. J. Biol. Chem.* **2000**, *275*, 35902–35907. [CrossRef]

71. Donoso, P.; Sanchez, G.; Bull, R.; Hidalgo, C. Modulation of Cardiac Ryanodine Receptor Activity by ROS and RNS. *Front. Biosci.* **2011**, *16*, 553–567. [CrossRef] [PubMed]
72. Eisner, V.; Csordás, G.; Hajnóczky, G. Interactions between Sarco-Endoplasmic Reticulum and Mitochondria in Cardiac and Skeletal Muscle—Pivotal Roles in Ca2+ and Reactive Oxygen Species Signaling. *J. Cell Sci.* **2013**, *126*, 2965. [CrossRef] [PubMed]
73. Mattson, M.P.; Partin, J. Evidence for Mitochondrial Control of Neuronal Polarity. *J. Neurosci. Res.* **1999**, *56*, 8–20. [CrossRef]
74. Wayman, G.A.; Yang, D.; Bose, D.D.; Lesiak, A.; Ledoux, V.; Bruun, D.; Pessah, I.N.; Lein, P.J. PCB-95 Promotes Dendritic Growth via Ryanodine Receptor-Dependent Mechanisms. *Environ. Health Perspect.* **2012**, *120*, 997–1002. [CrossRef]
75. Yang, D.; Kania-Korwel, I.; Ghogha, A.; Chen, H.; Stamou, M.; Bose, D.D.; Pessah, I.N.; Lehmler, H.J.; Lein, P.J. PCB 136 Atropselectively Alters Morphometric and Functional Parameters of Neuronal Connectivity in Cultured Rat Hippocampal Neurons via Ryanodine Receptor-Dependent Mechanisms. *Toxicol. Sci.* **2014**, *138*, 379–392. [CrossRef]
76. Yui, K.; Sato, A.; Imataka, G. Mitochondrial Dysfunction and Its Relationship with mTOR Signaling and Oxidative Damage in Autism Spectrum Disorders. *Mini-Rev. Med. Chem.* **2015**, *15*, 373–389. [CrossRef]
77. Song, G.; Napoli, E.; Wong, S.; Hagerman, R.; Liu, S.; Tassone, F.; Giulivi, C. Altered Redox Mitochondrial Biology in the Neurodegenerative Disorder Fragile X-Tremor/Ataxia Syndrome: Use of Antioxidants in Precision Medicine. *Mol. Med.* **2016**, *22*, 548–559. [CrossRef]
78. Chevrier, J.; Harley, K.G.; Bradman, A.; Gharbi, M.; Sjodin, A.; Eskenazi, B. Polybrominated Diphenyl Ether (PBDE) Flame Retardants and Thyroid Hormone during Pregnancy. *Environ. Health Perspect.* **2010**, *118*, 1444–1449. [CrossRef]

Review

Male Lower Urinary Tract Dysfunction: An Underrepresented Endpoint in Toxicology Research

Nelson T. Peterson [1] and Chad M. Vezina [1,2,*]

[1] Molecular and Environmental Toxicology Graduate Program, University of Wisconsin-Madison, Madison, WI 53706, USA; ntpeterson3@wisc.edu
[2] Department of Comparative Biosciences, University of Wisconsin-Madison, Madison, WI 53706, USA
* Correspondence: chad.vezina@wisc.edu

Abstract: Lower urinary tract dysfunction (LUTD) is nearly ubiquitous in men of advancing age and exerts substantial physical, mental, social, and financial costs to society. While a large body of research is focused on the molecular, genetic, and epigenetic underpinnings of the disease, little research has been dedicated to the influence of environmental chemicals on disease initiation, progression, or severity. Despite a few recent studies indicating a potential developmental origin of male LUTD linked to chemical exposures in the womb, it remains a grossly understudied endpoint in toxicology research. Therefore, we direct this review to toxicologists who are considering male LUTD as a new aspect of chemical toxicity studies. We focus on the LUTD disease process in men, as well as in the male mouse as a leading research model. To introduce the disease process, we describe the physiology of the male lower urinary tract and the cellular composition of lower urinary tract tissues. We discuss known and suspected mechanisms of male LUTD and examples of environmental chemicals acting through these mechanisms to contribute to LUTD. We also describe mouse models of LUTD and endpoints to diagnose, characterize, and quantify LUTD in men and mice.

Keywords: lower urinary tract dysfunction; lower urinary tract symptoms; BPH; prostate

---

Citation: Peterson, N.T.; Vezina, C.M. Male Lower Urinary Tract Dysfunction: An Underrepresented Endpoint in Toxicology Research. *Toxics* **2022**, *10*, 89. https://doi.org/10.3390/toxics10020089

Academic Editor: Soisungwan Satarug

Received: 4 January 2022
Accepted: 11 February 2022
Published: 16 February 2022

Publisher's Note: MDPI stays neutral with regard to jurisdictional claims in published maps and institutional affiliations.

Copyright: © 2022 by the authors. Licensee MDPI, Basel, Switzerland. This article is an open access article distributed under the terms and conditions of the Creative Commons Attribution (CC BY) license (https://creativecommons.org/licenses/by/4.0/).

1. Introduction

LUTD is a deviation from normal urinary voiding. While LUTD occurs in males and females, disease mechanisms differ between sexes. The prostate plays a considerable role in male LUTD, the focus of this review. For such a pervasive disease, male LUTD has suffered from a surprising lack of research attention. Part of the problem is the disease's complexity, driven by a constellation of underlying factors across multiple organs that are incompletely understood. Another problem is that the historical research record for LUTD is muddled by vast and inconsistent nomenclature used to describe the disease, decentralizing the resource of primary peer-reviewed literature. Several vocabulary terms are used to describe histological, anatomical, physiological, and clinical pathologies in the lower urinary tract. The following terms are sometimes conflated or interchanged with LUTD, and often used inappropriately: benign prostatic hyperplasia (BPH), benign prostatic enlargement (BPE), bladder outlet obstruction (BOO), partial bladder outlet obstruction (pBOO), lower urinary tract symptoms (LUTS), and others. These terms are defined in Table 1.

Male LUTD can be confirmed by specialized urodynamic studies at the urology clinic (diagnostic and experimental approaches used to identify LUTD mechanisms in mice and humans are described in Table 2). However, male LUTD is most often identified in the primary care clinic based on patient reported symptoms. LUTS can include but are not limited to weak stream, incomplete bladder emptying and more frequent voiding, especially at night. Male LUTD frequently begins in the fifth decade of life or later and is a progressive disease that can result in a loss of bladder function, bladder and kidney stones, acute urinary retention, and renal injury/failure [1–7]. LUTD disrupts sleep and has been linked to depression, decreased workplace productivity, and a reduced quality of life [8–12]. If not successfully managed, LUTD can be fatal.

Table 1. Definitions of terms used to describe anatomical and physiological disorders of the male lower urinary tract.

Acronym	Term	Definition
BPE	Benign Prostatic Enlargement	Non-malignant enlargement of the prostate, defined by imaging or digital rectal exam, and usually caused by BPH.
BPH	Benign Prostatic Hyperplasia	Histologically defined benign growth within the prostate. In humans, the growth pattern is nodular and can be primarily epithelial, stromal or mixed patterns of hyperplasia. BPH is often responsible for BPE.
BPO	Benign Prostatic Obstruction	BOO secondary to BPE.
BOO	Bladder Outlet Obstruction	Blockage of urine passage from an obstruction at the base of the bladder or bladder neck.
	Clinical Prostatitis	A spectrum of conditions characterized by differing degrees of inflammation, bacterial and abacterial, of the prostate, genitourinary tract or pelvis and may not include the prostate.
DO	Detrusor Overactivity	A urodynamic observation characterized by involuntary detrusor contractions during the filling phase that may be spontaneous or provoked.
DSD	Detrsor Spincter Dyssynergia	A disorder where the detrusor muscle contracts while the urethral and/or periurethral sphincter is involuntarily contracted and closed, resulting in bladder outlet obstruction.
	Histological Prostatitis	Prostate inflammation detected in a biopsy specimen.
LUTD	Lower Urinary Tract Dysfunction	A detrimental deviation from normal voiding function. Examples include decreased flow rate, increased voiding frequency, increased or decreased sensation associated with filling, an inability to completely void urine, and an inability to store urine until voluntary release.
LUTS	Lower Urinary Tract Symptoms	Patient described symptoms, scored using the international prostate symptom score, the American Urological Association Symptom index, or other indices that may (or may not) include bother.
OAB	Overactive Bladder	Urgency to urinate with or without urge incontinence, and usually associated with increased voiding frequency.
OVD	Obstruction Voiding Disorder	Lower urinary tract dysfunction deriving from an obstruction in the lower urinary tract.
pBOO	Partial Bladder Outlet Obstruction	Partial blockage of urine passage from an obstruction at the base of the bladder or bladder neck.
	Prostatitism	Male LUTD deriving from a prostatic mechanism
	Prostatomegaly	Prostate enlargement from malignant or non-malignant mechanisms.

Table 2. Strengths and limitations of methods to evaluate lower urinary tract dysfunction in men and male mice.

Method in Men	Method in Male Mice	Method Description	Strengths and Limitations
Cell and tissue-based calcium flux assays	Cell and tissue-based calcium flux assays	Calcium indicator dyes or genetically encoded calcium sensors are used to measure intracellular calcium concentrations in response to pharmacological agents and electrical field stimuli.	This method has been applied in vitro with human and mouse tissues and cells, and in vivo with mice, penetration can be limited for calcium indicator dyes and genetically encoded sensors are generally limited to mouse tissues.
Cystometry	Cystometry	A catheter is placed in the bladder and the bladder is filled with water or saline while measuring pressures associated with bladder filling and emptying. The catheter can also be used to collect post-void residual urine in the bladder.	Effective at measuring bladder pressure, but catheter is placed retropublicly in mice and transurethrally in humans which can contribute to intraspecies variability. Baseline pattern can vary by strain in mice.
Cystoscopy	Not available	A cystoscope is inserted into the urethra to visualize the lower urinary tract.	Effective in identifying prostatic enlargement, urethral and bladder inflammation, and some urological cancers, but this method is not available for mice.
Histology and immunohistochemistry	Histology and immunohistochemistry	Tissues sections are evaluated for BPH, inflammation and collagen accumulation (definitive diagnosis of BPH, histological prostatitis, fibrosis) and can be used to assess LUTD mechanisms.	Effective for assessing anatomical and cellular changes in lower urinary tract tissues and definitive diagnosis for some urological diseases but is invasive and therefore control tissues are difficult to obtain for healthy men for experimental comparisons; definitive identification of cell types requires complex multiplex protocols.
Isometric contractility	Isometric Contractility	Bladder, prostate, or urethral tissue is mounted in saline bath, pharmacological agents or electrical field stimuli are applied and force displacement is measured.	Quantitative and can reveal specific receptor mediated mechanisms of muscle function but is invasive and destructive to tissue (cannot be easily multiplexed with other methods.
Magnetic resonance imaging	Magnetic resonance imaging	Quantifies bladder wall thickness, detrusor and bladder volume, bladder neck angle, urethral length and diameter and prostate volume.	Can identify mechanisms of LUTD (bladder decompensation, BPE), but time consuming and expensive.
Symptom score	Not applicable	Standardized surveys such as the American Urological Association Symptom Index, the International Prostate Symptom Score, LURN, the National Institutes of Health-chronic prostatitis symptom index (NIH-CPSI) and others are used to quantify urinary symptoms and quality of life	Rapid, inexpensive and can be given repeatedly to monitor disease progression or responsiveness to therapy; limited to humans and not applicable for mice.
Ultrasound	Ultrasound	Quantifies bladder volume and wall thickness, urethral lumen diameter and in mice, velocity of urine as it passes through the urethra.	Fast but high-resolution imaging (for mice) requires expensive equipment.

Table 2. Cont.

Method in Men	Method in Male Mice	Method Description	Strengths and Limitations
Uroflowmetry	Uroflowmetry	Performed by measuring voided urine flow and volume.	Non-invasive, but requires specialized equipment, and operator experience and cannot distinguish between anatomical (bladder, prostate, or urethra) mechanisms of LUTD.
Voiding diary	Void spot assay	Men use a journal to record urinary void frequency, timing, and use a capture container to record volume; for mice, a filter paper is placed at the bottom of the cage and later illuminated to quantify void spot number, size, and pattern.	Inexpensive, noninvasive, but can vary by day and individual and cannot distinguish mechanism (bladder, urethra, prostate) of voiding dysfunction.

LUTD is extremely common. A 2008 study estimated that 1.9 billion people, representing 45% of the population, are affected by LUTD [9]. The economic burden of LUTD is staggering. The disease affects more than half of men over 50 years of age in the Western world, resulting in $4 billion for the pharmacological treatment and $2 billion for the surgical treatment of LUTD and associated prostatic problems [13–15]. The most common therapies for male LUTD are directed to block alpha adrenoreceptor function (alpha blockers) and dihydrotestosterone synthesis (steroid 5 alpha reductase inhibitors), factors which contribute to prostatic smooth muscle contraction and prostatic enlargement, respectively. Unfortunately, these therapies are incompletely effective. Their magnitude of effect is marginal, not all patients respond, and existing therapies are only moderately protective against disease progression [16–18]. It is becoming clear that male LUTD derives from many different mechanisms, not all of which are addressed by current therapies. Factors responsible for severe drug-refractory disease are not understood. Recent studies reveal potential roles for environmental chemical exposures, during the fetal period when the lower urinary tract is developing [19–21] and during other stages, in driving LUTD susceptibility and progression, opening an entirely new line of toxicology research towards understanding environmental factors that contribute to LUTD processes.

This review is intended as a resource for toxicologists and other discipline specialists who are considering entry into the urologic disease research space and wishing to examine LUTD as a toxicology research endpoint. We describe the anatomy, cellular composition, and physiology of male lower urinary tract organs including the bladder, urethra, and prostate. We describe known and emerging disease mechanisms. We also highlight the limited examples of how environmental chemicals influence male LUTD and list opportunities for future research.

2. Overview of Male Lower Urinary Tract Anatomy and Physiology

Several benign diseases of the lower urinary tract are accompanied by a change in distribution, type, or state of cells that comprise lower urinary tract tissues [22–24]. Therefore, we describe the cellular anatomy of the male lower urinary tract to give toxicologists an appreciation of the normal cellular organization and changes which occur in response to chemical insults and disease The male lower urinary tract consists of the bladder, prostate, and urethra (Figure 1). Urine flows from the kidney to the bladder via the ureter and passes through the prostatic urethra and prostate before continuing through the penile urethra and exiting the body as voided urine (Figure 1).

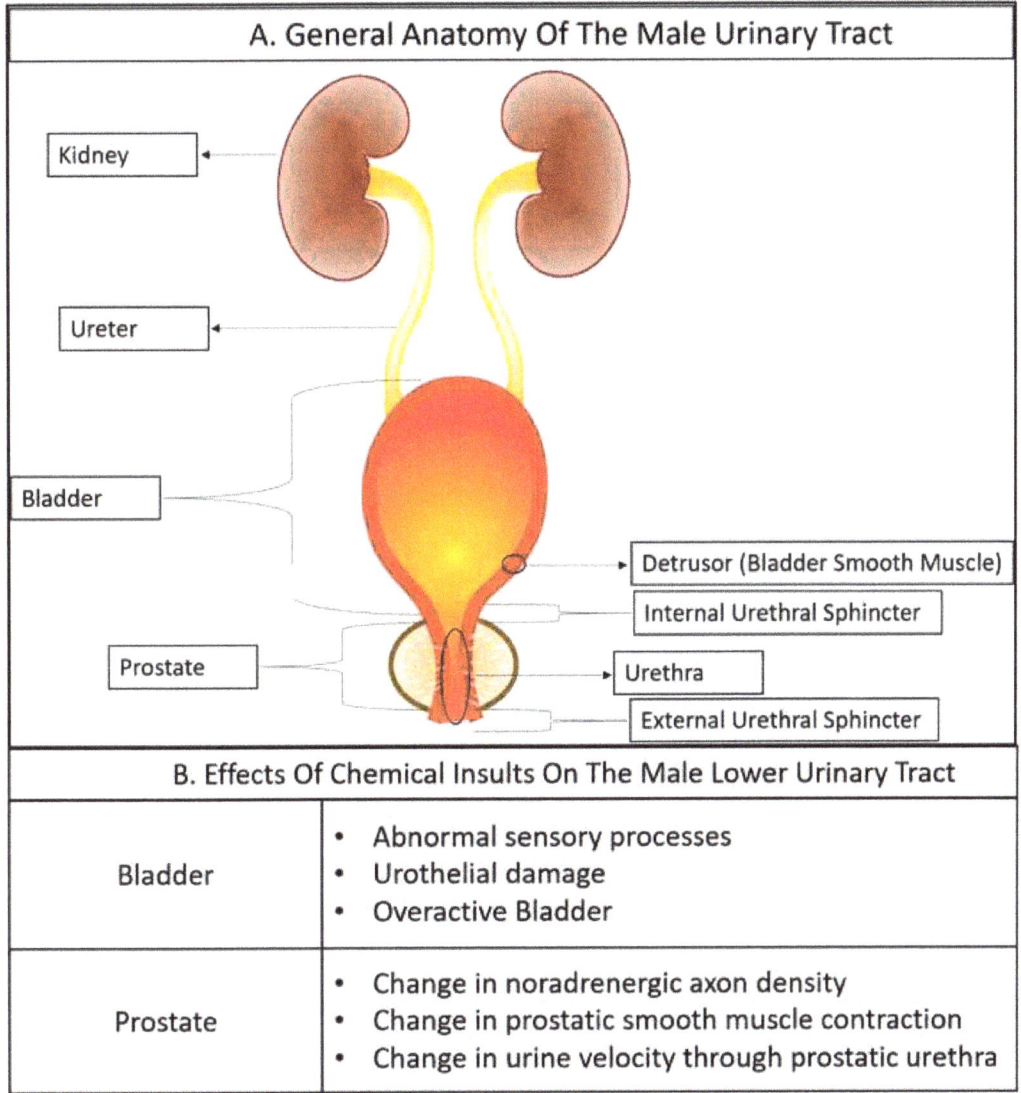

Figure 1. General anatomy of the male urinary tract and effects of chemical insults on the male lower urinary tract. (**A**) A general depiction of the male lower urinary tract. (**B**) Known effects of environmental chemicals on the lower urinary tract of either the man or male mouse.

2.1. The Bladder

The bladder's primary functions are to store and expel urine. The bladder wall consists of three tissue layers: a specialized epithelium known as the urothelium, the lamina propria, and the bladder smooth muscle (detrusor) [25,26].

The mature urothelium is comprised of basal, intermediate, and superficial cells [27]. Bladder epithelial cell differentiation begins early in fetal development (weeks 7–8 in humans), and the trajectory of urothelial cell differentiation during development and regeneration is susceptible to epigenetic modification [28] revealing a potential mechanism of toxicity for epigenetic modifying chemicals. The mature urothelium must achieve three unique functions. The first is to maintain distensibility to accommodate bladder filling and emptying. Bladder volume increases significantly during the storage cycle, a process which would normally challenge the integrity of an epithelial lining [29]. Bladder distensibility is achieved by urothelial cell junction rearrangements and cell sliding during bladder filling [29].

The second role of the urothelium is to protect sub-urothelial tissue from toxins, microorganisms, and urine solutes [27,30]. Barrier function is facilitated by secreted uroplakins [31]. Uroplakins are transmembrane proteins which assemble into a crystalline structure and are interrupted by hinge regions to allow bladder distension [32]. Uroplakins assemble to form uroplaques, rigid bio-membrane structures which cover 90% of bladder lumen [32]. Uroplaques are integral to the integrity, flexibility, and solubility of the urothelium [33]. The control of urothelial cell division is integral to maintaining functional uroplaques and restoring them after bladder damage. Although the urothelial cell turnover is normally slow with a labeling index of 1% in mice, the urothelium is reconstituted quickly after injury through the progenitor activities of basal and intermediate cells [32–36]. The epithelium of the urothelium can be completely repaired in 4 weeks in guinea pigs and 6 weeks in men [36]. Some mechanisms by which the bladder restores barrier function are surprising. For example, we found that under certain circumstances when widespread urothelial cell death depletes the bladder of its own progenitors, it can recruit non-resident, non-bladder (Wolffian duct) epithelial progenitors, drive their differentiation into uroplakin secreting superficial cells and restore barrier function [37]. Barrier function is crucial because sub-epithelial bladder cells are severely compromised by urine exposure. The experimental use of cyclophosphamide, an antineoplastic used therapeutically for Hodgkin's lymphoma, multiple myeloma, and other cancers, has widened the understanding of barrier function and consequences of barrier function loss. Cyclophosphamide is bio-transformed into acrolein, which accumulates in the urine and drives urothelial cell death, resulting in hemorrhagic cystitis and changes in physiology [38–40]. Environmental chemicals with urothelial cell toxicity are expected to drive bladder inflammation and dysfunction like that of cyclophosphamide.

The third role of the urothelium is that of a sensor. In combination with nerve terminals within the bladder, the urothelium detects and responds to mechanical and chemical stimuli to alter detrusor contractility and moderate bladder afferent nerve activity [41,42]. Factors released by urothelial cells include acetylcholine, adenosine triphosphate (ATP), nerve growth factor, nitric oxide (NO), prostaglandins, and others [41,43].

The lamina propria contains a fibroelastic connective tissue with intervening afferent and efferent nerve fibers, a vast vascular network and dispersed fibroblasts, a loose smooth muscle layer (the muscularis mucosa), and myofibroblasts [26,44]. The elastic fibers within the lamina propria allow the bladder to recover its original shape after voiding [45].

The detrusor is the major smooth muscle component of the bladder [46]. The detrusor is organized as a circular muscle inner layer sandwiched between longitudinal muscle outer layers [46]. Muscle bundles are surrounded by collagen [46–48]. Detrusor contraction is predominantly controlled by cholinergic neurons [49,50], but can also be induced by purinergic neurons and relatively rare sympathetic neurons [49,50].

The normal voiding cycle is divided into filling and voiding phases [51]. Urine expands the bladder during the filling phase, while bladder pressure remains lower than urethral pressure [50,51]. There is still uncertainty about how the bladder relays the perception of fullness to the brain. One possibility is that mechanoreceptors and mechanosensitive ion channels within the bladder transmit information about fullness to afferent neurons [52–56]. There is also evidence that urothelial cells, stretched during bladder filling, release ATP to activate purinergic receptors on bladder afferents and relay bladder fullness to the brain [57–59]. Another possibility is that the perception of fullness is not driven by a slow increase in bladder pressure (intravesicular pressure), but rather by an increasing rate of spontaneous transient contractions, also called micromotions, which exist throughout the filling phase. Micromotions drive the major portion of afferent outflow to the central nervous system during bladder filling, acting in part through a mechanism involving calcium-activated potassium (SK type) channels [60].

In 1925, F.J.F. Barrington identified a brain stem region which controls micturition, including sensation of bladder fullness and the contractions leading to urination [61]. Studies using retrograde and anterograde neuronal labeling pinpointed the location of this micturition center in the pontine tegmentum [62–67]. This site of micturition control is referred to as Barrington's nucleus, the pontine micturition center, and the M-region [62]. Afferent and efferent urinary voiding pathways are integrated in Barrington's nucleus. During the storage phase, glutamatergic neurons in the periaqueductal gray and hypothalamus relay information about bladder fullness and bladder volume threshold for voiding to Barrington's nucleus [68,69]. During the voiding phase, corticotropin releasing hormone-positive and estrogen receptor 1-positive neurons within Barrington's nucleus activate efferent pathways to drive detrusor contraction [62,70,71]. Additional neurons in Barrington's nucleus send inhibitory signals to the external urethral sphincter, driving its relaxation and allowing urine to flow unimpeded from the bladder into the urethra [71–73]. Though there is widespread evidence that environmental contaminants can disrupt connectivity, complexity, arborization, and signaling of neurons within the peripheral and central nervous system, whether environmental chemicals impact bladder ascending and descending neural pathways is rarely examined [74–80].

There is limited evidence that environmental chemical exposures can disrupt bladder neural circuitry as it is established during the fetal and neonatal periods, raising concerns about a developmental basis of bladder health and disease. A recent study tested the impact of exposure to a polychlorinated biphenyl (PCB) mixture on bladder structure and function [19]. The PCB mixture used in this study mimics the most encountered congeners in women who are at risk for having a child with a neurodevelopmental disorder [81,82]. PCBs were delivered orally to nulliparous female mice (75% C57BL/6J/25% SVJ129) starting two weeks before mating, through pregnancy and lactation, and continuing in offspring before their bladders were analyzed at postnatal days 28–31. The PCB mixture increased densities of sub-urothelial beta-3 tubulin (general neural fiber marker) fibers and calcitonin gene-related peptide positive (peptidergic fiber marker) fibers in male mice but not female mice, and these changes were accompanied by an increase in male bladder volume [19], suggesting they were sufficient to drive a change in bladder function.

2.2. The Urethra

The human male urethra is divided into two parts, consisting of five segments: the anterior urethra (fossa, penile, and bulbar segments) and the posterior urethra (membranous and prostatic segments) [83]. The rodent male urethra is divided into two parts—penile and pelvic [84]. The human and rodent urethra are populated by epithelial cells, smooth and striated muscle cells, blood vessels, and sensory and motor neurons [85]. While the cellular components of the anterior/penile urethra have not been extensively characterized, single cell ribonucleic acid (RNA) sequencing approaches have been used to determine the cellular components of the prostatic urethra [24,86,87]. Urethral epithelium consists of club cells, hillock cells, basal epithelial cells, and neuroendocrine cells [86]. Urethral club and hillock cells were recently identified, but their functional characterization is incomplete and represents a future research opportunity. Lung club cells, which are transcriptionally like those in the urethra, act as progenitors and mediate anti-inflammatory and antioxidant processes [88–90]. Lung hillock cells, which are transcriptionally like those in the urethra, serve as progenitors, and participate in barrier function and immunomodulation [91,92].

2.3. The Urethral Sphincter

The urethral sphincter serves as a valve to regulate urine flow between the bladder and urethra [93]. During the bladder storage phase, urethral pressure exceeds bladder pressure to maintain continence [50]. During the voiding phase, the urethral sphincter falls open, urethral pressure decreases while bladder pressure increases, the urethra distends and urine flows through the prostatic urethra and penile urethra to become voided urine [50]. The urethral sphincter is divided into two parts: the external sphincter and the internal urethral sphincter [93,94]. The external sphincter consists of striated muscle circumscribing the urethra and is under voluntary control [93,95]. The internal urethral sphincter is indistinct from the rest of the lower urinary tract smooth muscle (bladder smooth muscle is continuous with urethral and prostatic smooth muscle), but is physiologically defined by its autonomic regulation, connected via a reflex arc to the bladder [95,96]. Urethral smooth muscle is organized as a thin longitudinal superficial layer, a dense circular layer, and a thin longitudinal deep layer [94].

2.4. The Prostate

The prostate synthesizes a portion of the ejaculate [97]. Prostatic smooth muscle contracts during ejaculation to propel prostatic fluid into the urethra [98]. The prostatic urethra also distends to accommodate urine during voiding. Benign prostatic disease changes the prostate's histology and cellular composition and can prevent prostatic urethral distention during voiding, causing BOO, a common etiology for LUTD (defined in Table 1).

The human prostate is a spherical gland encapsulated by a fibromuscular sheath known as the prostatic capsule [24,96,99]. The base of the prostate is adjacent to the bladder and the prostatic urethra courses through its center [100]. The prostatic ductal network is like that of a branched tree: the main ducts drain directly into the urethra and divide into primary, secondary, and tertiary branches as they extend towards acini concentrated in the gland's periphery [101]. The human prostate is organized into zones, differing in cellular composition and responsiveness to disease, and includes the transition zone (most susceptible to histological BPH, defined in Table 1) [24,102], the central zone and the peripheral zone (most susceptible to prostate cancer) [100,102]. The rodent prostate, often used as a disease model for humans, is anatomically distinct from the human prostate in that it is not spherical, but instead divided into four bilaterally symmetrical lobes: the anterior, dorsal, lateral, and ventral prostate [102]. While spontaneous cancer is not observed in the mouse prostate, a variety of genetically engineered mouse models are susceptible to prostate cancer and disease incidence differs by lobe [103]. The mouse prostate gland develops BPH spontaneously with age, but lesions are diffuse, like those that contribute to clinical disease in the dog, and unlike nodular BPH in the humans [104,105]. The rodent prostate ductal network is organized as a branched tree, like that of the human prostate, but ducts are surrounded by a looser stroma than in human prostate and the rodent gland is encapsulated in a thin adventitia instead of the thick capsule that surrounds the human prostate.

Human prostatic epithelium is made up of luminal, basal cells; neuroendocrine, club and hillock cells are also present, but are rare in prostate compared to urethral epithelium [22,86,106]. Human prostate stroma consists of three smooth muscle cell types (peri-prostatic, vascular smooth muscle and pericyte), two fibroblast cell types (peri-epithelial and interstitial), leukocytes, endothelial cells, and sensory and autonomic nerve fibers [86]. Mouse prostate stroma contains three fibroblast cell subtypes distributed in distinct proximal–distal and lobe-specific patterns and smooth muscle [24,106]. The transcriptomes of mouse prostatic and urethral fibroblasts are like human interstitial fibroblasts [24]. However, mouse urethral and ductal fibroblasts evoke Wingless related-integration site (Wnt) and Transforming growth factor beta (TGFβ) signaling pathways that are less abundant in human prostate fibroblasts [24]. Human peri-epithelial fibroblasts instead express Wnt inhibitors that could buffer Wnt ligands produced by other stromal or epithelial cells [24]. Human prostatic fibroblasts are organized in layers that center around epithelial structures, while mouse prostatic fibroblasts are not layered and differ by lobe [24]. Human and mouse prostate fibroblasts are most abundant in the proximal regions of prostatic ducts and least abundant in acini in the distal regions [24,107].

The recent observation, derived from single cell RNA-sequencing data, that human and mouse prostate cellular landscapes are similar, is also supported by previous microarray data [108]. Similarly, mouse prostate organogenesis is like that of the human prostate [109]. These data support the use of mice as a relevant model species for studying cellular and molecular mechanisms of benign prostatic disease. The key to understanding the differences in prostate architecture and benign prostate hyperplasia manifestation between these species may lie in the function of the specialized prostate epithelial and stromal cells of these species [109].

Prostate disease can be detected by changes in the spatial distribution and frequency of prostate cells [24]. Prostate cell immunophenotyping has proven difficult, as disease processes frequently lead to changes in cell state and cell type that cannot be easily distinguished by simple immunohistochemical staining protocols. New and validated RNA-Sequencing approaches, as well as cell sorting protocols deriving from them, have recently been described [86] and will be essential for elucidating prostate cell functions in future studies.

3. LUTD Mechanisms

3.1. Benign Prostatic Diseases

A variety of benign prostatic conditions contribute to male LUTD, many of which are believed to cause LUTD by driving BOO (defined in Table 1). The impacts of BOO extend beyond the prostate and into the bladder. A prolonged intravesicular pressure increase and bladder contraction against resistance reprograms the bladder in a process known as bladder compensation: the detrusor becomes thicker [110], it undergoes functional changes in ion channel physiology [111] and efferent signaling is reprogrammed [112]. If BOO is not effectively addressed, the bladder decompensates, much like a heart undergoing hypertrophic cardiomyopathy: the detrusor thins, is replaced by fibrotic tissue, and becomes incapable of mounting an effective contraction to fully evacuate urine from the bladder. There is evidence in rabbits that bladder decompensation is at least partially reversed by relief of bladder outlet obstruction [113]. Recovery from BOO likely depends on the severity of bladder decompensation at the time of surgery [113–115]. Thus, BOO must be effectively addressed before it permanently impairs bladder function.

BPH is a leading cause of LUTD in men of advancing age. Human BPH is defined by prostate histology, specifically the presence of stromal, epithelial, or mixed nodules in the central and transition zones (Table 1) [22,116–120]. Small hyperplastic nodules can form as early as the 3rd decade of life and increase in frequency and volume with advancing age [121]. BPH mechanisms are not fully understood, but it has been hypothesized that BPH arises from a reawakening of embryonic signaling pathways [121] or disrupted homeostatic regulation of cell growth and death programs [116–120].

Aging-related changes in circulating testosterone and 17-beta-estradiol concentrations are another mechanism linked to male LUTD. Serum and prostate tissue concentrations of testosterone and 17-beta-estradiol change with age in men [122,123] and the changes are associated temporally and mechanistically with male LUTD [124–126]. Pharmacological alterations in testosterone and 17-beta-estradiol are a proven cause of LUTD in non-human male primates, canines, rats, and mice [124,127–133]. In mice, slow-release implants of testosterone and estradiol drive an increase in voiding frequency, a reduction in voided volume, an increase in collagen deposition, and a change in velocity of urine flow through the prostatic urethra [124]. The mechanism by which changes in circulating testosterone and 17-beta-estradiol drive voiding dysfunction are not clear but may include direct actions on the bladder [134,135], changes in prostatic desmin and smooth muscle actin content or function [136–139].

The fact that LUTD arises from natural changes in circulating sex hormone concentration raises questions about impacts of endocrine disrupting chemicals on male voiding function, and this area of toxicology research is in its infancy. For example, subcutaneous implants of the estrogenic chemical bisphenol A (BPA, 25 mg), combined with testosterone (2.5 mg) and given to C57BL/6N adult (6–8 weeks old) male mice, increase bladder mass and volume, increase voiding frequency, and reduce the volume of voided urine, suggestive of BOO [140]. BPA may act more broadly in the lower urinary tract, affecting the bladder as well as the prostate. Delivery of BPA (0.05–0.5 mg/kg/day) to Pietrain × Duroc mixed-breed juvenile female pigs increases the number and thickness of vasoactive intestinal polypeptide (VIP) expressing neurons in the bladder wall [141], raising questions about the influence of BPA, and the larger class of environmental estrogens to which it belongs, on detrusor recovery after contraction.

Prostate inflammation, also called prostatitis (defined in Table 1), is extremely common and has been closely associated with LUTD. Approximately 50% of prostate biopsy, surgical or autopsy specimens harbor evidence of histological inflammation, most typically characterized as chronic (lymphocytic) inflammation [142]. The incidence of prostate histological inflammation is even higher (75%) in men with LUTD [143]. The presence of prostate inflammation in a biopsy specimen correlates with risk of symptomatic progression, urinary retention, and need for surgery [142,144–146]. A significant proportion of men with histologically defined prostate inflammation will develop urinary dysfunction [147]. Two placebo-controlled drug trials, Reduction by Dutasteride of Prostate Cancer Events (REDUCE) and Medical Therapy of Prostatic Symptoms (MTOPS), correlate histological prostate inflammation in human male prostate with increased prostate volume [144]. MTOPS study outcomes reveal that men with histological inflammation are more likely to progress to advanced LUTD, including acute urinary retention [144]. A separate study found that men with prostatitis were 2.4 times more likely to develop BPH and the presence of histological prostate inflammation in baseline biopsies was associated with 70% increased odds of requiring later treatment for LUTD [146]. Despite clear evidence that some environmental chemicals can drive inflammation and modulate autoimmunity, there is little information about environmental impacts on prostate inflammation and this represents a future opportunity that can be examined using immunohistochemical and physiological methods in Table 2.

There is a distinction between histological and clinical prostatitis: histological prostatitis is identified in histological tissue sections, while clinical prostatitis is diagnosed by physical examination, urinalysis, imaging, cystoscopy, or patient questionnaire (for example, The National Institute of Health Chronic Prostatitis Symptom Index (NIH-CPSI)) [148]. Clinical prostatitis accounts for a significant proportion of outpatient visits [149]. Clinical prostatitis includes acute and chronic bacterial prostatitis, nonbacterial prostatitis, and asymptomatic prostatitis [148].

Prostate fibrosis is a recently identified mechanism of male LUTD. Fibrosis is an abnormal, detrimental version of the wound-healing process and is characterized by collagen deposition and tissue stiffening [150]. Macoska et al. [151] were the first to report fibrosis in the human prostate and link collagen accumulation to tissue stiffness and LUTS severity. Subsequent reports linked prostate fibrosis to histological inflammation, LUTS, and resistance to a combination therapy of alpha blockers and 5 alpha reductase inhibitors [150,152]. Prostatic fibrosis is an evolutionarily conserved LUTD process, supported by the fact that collagens also accumulate within the prostates of aging intact dogs and mice [104,153]. Though triggers for prostate fibrosis are not fully known, and whether environmental contaminants drive prostate fibrosis has not been studied, prostatic fibrosis results from prostate inflammation secondary to *E. coli* infection or obesity in mice [154,155].

Prostatic smooth muscle dysfunction is the target of the most prescribed drug class for male LUTD, the alpha blockers, and can be studied experimentally using calcium flux assays and isometric contractility assays described in Table 2. A study by Baumgarten et al. [156] was the first to identify noradrenergic axons in the human prostate, a surprising discovery considering that autonomic outflow to the bladder is mediated instead by cholinergic axons. Receptor binding studies and isometric contractility assays showed that noradrenergic receptors in prostatic smooth muscle mediate prostate tissue contractility [157,158]. The outcomes of these studies ushered the hypothesis that prostatic smooth muscle hyperactivity impairs urine flow through the prostatic urethra to cause BOO in some men. While this hypothesis was the basis for developing alpha blockers for male LUTD, little research has been directed at identifying mechanisms of prostatic smooth muscle dysfunction, most notably dysfunction mediated by environmental chemicals. This area remains ripe for scientific exploration. Prostatic smooth muscle contraction is controlled by autonomic neurons and aging is one factor that may contribute to changes in prostatic innervation [124,159]. There is emerging evidence that environmental chemicals can also change prostatic innervation to cause prostatic smooth muscle dysfunction, specifically by acting during the fetal and neonatal periods when prostate autonomic innervation is established. For example, we recently showed in C57BL/6J mice that gestational exposure to the widespread environmental contaminant TCDD (a single 1 µg/kg oral maternal dose on the 13th day of gestation) increases noradrenergic fiber density (nerve terminals) in the prostate of male mouse fetuses without changing the density of cholinergic or peptidergic fibers [160]. TCDD-induced prostatic noradrenergic hyperinnervation persists into adulthood and is coupled to hyperactivity of prostatic smooth muscle and abnormal urinary function in mice, including increased urinary frequency [160]. These findings are important because they support the concept that prostate neuroanatomical development is malleable, at least in mice, and that intrauterine chemical exposures can permanently reprogram prostate neuromuscular function to cause male LUTD in adulthood. In contrast, exposure to TCDD and other aryl hydrocarbon receptor agonists during adulthood appear to protect against BPH in men [161,162]. Differing consequences of aryl hydrocarbon receptor activation in the fetal period, versus adulthood, highlight the need to control for age in studies that examine potential impacts of environmental chemicals on urinary function and LUTD.

3.2. Bladder Mechanisms of Male LUTD

A variety of bladder conditions can lead to urinary dysfunction. This section describes the most common causes of male LUTD.

Overactive bladder is characterized by involuntary detrusor contraction. Consistent changes in animal models of overactive bladder include patchy denervation of the bladder, enlarged sensory neurons, hypertrophic dorsal root ganglia, and an enhanced spinal micturition reflex [163]. Overactive bladder is often characterized by sensory dysfunction [163]. There is a role for muscarinic M2 receptors in the severity of urinary urgency [163]. Some individuals with overactive bladder have a thicker bladder wall, suggesting overactive bladder may derive from BOO in some men [163,164].

The etiology of overactive bladder is multifactorial, deriving from three major mechanisms: myogenic factors, urotheliogenic factors, and neurogenic factors [41]. Myogenic factors contributing to overactive bladder include spontaneous detrusor contractions in response to bladder distension, ischemia, and changes in smooth muscle properties over time [41]. Neurogenic factors may include abnormal sensory processes, abnormal afferent excitability, or in some cases, damage, or abnormalities in central processing [41]. Dimethylaminopropionitrile, used in the manufacture of polyurethane, is an inhalation hazard that acts through a neurogenic mechanism to cause overactive bladder [165]. Methyl mercury also causes overactive bladder through what appears to be a neurogenic mechanism [166,167]. Damage to the urothelium can also cause overactive bladder, as rupture of urothelial cells releases factors that can drive detrusor contractility and micturition [41]. Biphenyl, used as a resin, a heat transfer medium, and an anti-fungal, is an example of an environmental chemical that causes urothelial cell damage and death [168].

Underactive bladder, also known as detrusor underactivity, is defined by detrusor contraction of inadequate strength, and results in prolonged or incomplete bladder emptying [169]. Patients with underactive bladder have a diminished sense of bladder fullness and are unable to mount forceful bladder contractions [170]. Underactive bladder can occur after episodic overactive bladder, reminiscent of bladder decompensation after BOO. In fact, there is a documented relationship between LUTD, underactive bladder, and fibrosis of the bladder [171]. The interstitial cells of Cajal, a specialized cell population with smooth muscle pacemaking activity, have been implicated in underactive bladder. The frequency of interstitial cells of Cajal is reduced in mice with underactive bladder and is associated with reduced frequency and amplitude of detrusor contraction [172]. Rats driven by bladder outlet obstruction to develop underactive bladder are deficient in stem cell factor, a ligand for the receptor C-kit which controls proliferation and function of interstitial cells of Cajal, and an increase in stem cell factor restores detrusor contractility [172].

3.3. Urethral Mechanisms of Male LUTD

Detrusor sphincter dyssynergia is characterized by simultaneous contraction of the detrusor and urinary sphincter, thereby impairing urine outflow from the bladder [173]. Detrusor sphincter dyssynergia manifests in three distinct phenotypes: (Type 1) increased sphincter activity during detrusor contraction which then ceases, resulting in delayed urination, (Type 2) intermittent clonic contractions during voiding, resulting in intermittent stream, (Type 3) continuous sphincter activity during detrusor contraction, resulting in impaired voiding [174]. Detrusor sphincter dyssynergia is common in men with spinal cord injuries or multiple sclerosis and has the capability to drive bladder decompensation, elevate pressure in the ureter and pelvis, and cause hydronephrosis, renal scarring and terminal kidney failure [173,174].

Neurological disease commonly manifests in bladder dysfunction [175]. Autonomic nervous system lesions (stroke, tumor, traumatic spinal cord injury, myelopathies due to cervico-arthrosis spina bifida), disseminated lesions (Parkinson's disease, brain trauma, multiple sclerosis, meningo-encephalitis,) and peripheral neuropathies (diabetes mellitius) have all been identified as mechanisms of bladder dysfunction [175] and act in part by disrupting coordination between the detrusor, urinary sphincter, and central nervous system [173,174]. While there are many examples of environmental chemicals causing neuropathies, the consequences on lower urinary tract function are rarely examined.

3.4. The Relationship between LUTD and Comorbidities

Recent studies connect LUTD to other diseases. For example, people with cardiovascular disease, diabetes and obstructive sleep apnea are at increased risk of developing LUTD [176–179]. A common thread linking these diseases is a change in hemodynamics connected to ischemic injury [180–182], a factor that independently drives LUTD in mice [174]. Environmental chemical exposures have been linked to cardiovascular disease and diabetes [183–186], and this is another mechanism by which they may drive LUTD.

4. Mouse Research Models of Male LUTD

Here we describe animal models used to study various etiologies of LUTD. While it is important to realize that results from animal models are not always transferable to humans, it is also crucial to highlight that animal models are used in preclinical trails to test the safety and efficacy of drugs and are an invaluable tool to use in toxicological studies.

4.1. Benign Prostatic Hyperplasia

A variety of genetically engineered mouse models have been used to drive expression of growth factors or mitogenic hormones in the prostate. Androgen responsive promoter sequences, androgen-induced cre recombinase or viral promoters are used to target genetic modifications to mouse prostate tissue and overexpress fibroblast growth factor 2 or fibroblast growth factor 3 to drive epithelial BPH [187–189], overexpress prolactin [190–193] or interleukin 1 alpha [194] to drive epithelial and stromal BPH and prostate inflammation, delete serine/threonine kinase 11 to promote stromal BPH in the periurethral region [195], or genetically modify other sequences. Expression of an activated form of P110 alpha, the catalytic subunit of PI3K, in mouse prostate epithelium also drives epithelial BPH in mice but accompanied with a stark fibrotic response in prostatic stroma [196]. Many of these genetically engineered mouse models were created before contemporary methods were optimized for mouse urinary physiology phenotyping. The historical goal was to use genetically engineered mice to identify molecular mediators of BPH and test efficacy of drugs and dietary substances for relieving BPH in preclinical model species. While genetically engineered mouse models are useful for understanding homeostatic mechanisms of prostate cell proliferation, it is becoming clear that BPH is not always linked to LUTD in men [197], and it remains important to characterize urinary physiology in these mice as a more relevant endpoint for male LUTD.

4.2. Mouse Models of Prostate Inflammation

Histological inflammation of the human prostate is extremely common: in one study, it was detected in nearly 80% of prostate biopsy specimens from 60+ year old men and was strongly associated with urinary voiding symptoms [198]. Prostate infection by ascending microbes is one potential mechanism of prostate inflammation and supported by the frequent encounter of bacteria in human prostate tissue specimens [199–201]. One strategy for driving prostate inflammation in mice involves urethral catheterization and delivery of uropathogenic *E. coli*. A variety of isolates have been used (*E. coli* UTI89, 4017, 1677 and CP-1), ranging from those collected as clinical urine isolates from women with bladder infections, to others collected from men with pelvic pain [202,203]. The pattern of inflammation (acute vs. chronic) depends in part on mouse strain used [204] and method of *E. coli* delivery (single vs. multiple inoculations, catheter size, instillation volume and bacterial load). It is essential to control these variables carefully when considering experimental design, and mice instilled with sterile saline (sham operated mice) are an essential component of experimental design because urethral catheterization can itself induce trauma, urethritis, and changes in urinary voiding physiology [144]. Prostatic *E. coli* infection is linked to prostate fibrosis and changes in voiding patterns in mice, but voiding patterns differ between *E. coli* strains and methods of infection and can include high volume, low frequency voiding [144] or low volume, high frequency voiding [107,205,206].

Many men with histological prostatitis present with a pattern of prostatic infiltrate consistent with prostate autoimmunity [207–209], an observation co-opted for the design of mouse models. The prostate ovalbumin expressing transgenic-3 mouse expresses ovalbumin under the control of the androgen-responsive probasin promoter [210,211]. Autologous splenocytes are activated in vitro and transplanted to drive T-cell mediated prostate autoimmunity and inflammation [207]. While the pattern of inflammation and mechanisms of cell proliferation have been carefully studied in this mouse, the urinary physiology phenotype is not well characterized. The experimental autoimmune prostatitis mouse model involves repeated intradermal injections of rat prostate homogenate into mice to drive a T-cell based autoimmune reaction that has been used to examine mechanisms of male LUTD and chronic pelvic pain [212,213].

A non-bacterial mouse model of prostate inflammation was created based on observations that IL-1 beta abundance increases after intraprostatic injection of noxious agents or uropathogenic *E. coli* infection [214–217] and that Prostatic IL-1 beta abundance is elevated in humans with histological BPH and correlates with LUTS and chronic pelvic pain [218–222]. This mouse model utilizes the Tet-On system which induces expression of a gene in the presence of doxycycline and is tunable, with stronger transgene expression with doxycycline dose [223]. A double transgene of Hoxb13-rTA transgene and a TetO-IL1 beta responder is used to drive IL-1 beta in prostatic epithelial cells [223,224]. The urinary metabolomic proteomic signatures of this mouse have been described, but the urinary physiology phenotype remains to be determined [224,225].

A recent mouse model of prostate inflammation is based off observations that prostate secretory proteins are leaked into prostate stroma of some men with LUTD and accompanied with patchy loss of the adherens junction protein e-cadherin, suggesting a loss of prostate barrier function [226]. Genetic depletion of e-cadherin in mouse prostate epithelium increases prostate mass and cell proliferation, thickens prostate stroma, and increases voiding frequency while reducing voided urine volume, and increases spontaneous bladder contractions [227].

4.3. Mouse Models of Partial Bladder Outlet Obstruction (pBOO)

Surgical approaches were first used to model pBOO in male mice. One approach involves a retropubic incision to apply and cinch a suture or metal ring around the bladder neck or pelvic urethra to drive bladder compensation and overactive bladder, and later bladder decompensation, detrusor underactivity, fibrosis, and loss of muscle mass [228–231]. The mouse model has been essential for recognizing new druggable pathways for restoring function to the decompensated bladder [232].

Treatment with exogenous androgens combined with estrogens is a non-surgical method to drive BOO in mice. Mice are given slow-release implants of androgen (testosterone or dihydrotestosterone) in combination with slow-release implants of estrogen (17beta-estradiol or diethylstilbesterol). The combination of androgen plus estrogen is necessary for prostate gland maintenance, as estrogens delivered to male mice in the absence of androgens disrupts hypothalamic/pituitary/gonadal signaling and cause prostate gland atrophy [128]. Genetically engineered mice that overexpress aromatase are also used to recapitulate the endocrine environment of advancing age [129,233]. Male mice treated with androgen and estrogen develop progressive LUTD, with evidence of disease processes (increased bladder weight as evidence of hypertrophy/compensation for BOO) occurring as early as two weeks after treatment [124]. Sustained exposure to exogenous androgens and estrogens elicits a variety of changes to the male lower urinary tract of multiple species, including prostatic hypertrophy and inflammation, urethral narrowing and abnormal urethral muscle tone, urinary dysfunction with progressive onset, bladder overactivity and eventual decompensation [124,127–131,133,234–238]. Estrogen receptor activation is a key driver of urinary dysfunction, as exogenous estradiol given to male mice drives urinary retention in the absence of exogenous testosterone [239] and estrogen receptor 1 is required for urinary retention and voiding dysfunction from exogenous testosterone and 17beta-estradiol in mice [237]. An important consideration when exogenous androgen and estrogen are used to drive male LUTD, especially when incorporating genetic changes to identify mechanisms, is that hormone responsiveness, disease onset, progression and severity are influenced by genetic background and mouse strain [234]. The delivery system of exogenous androgens and estrogens should be considered if using hormones to drive LUTD for a toxicology study. Compressed pellets of androgens and estrogens can be crushed when animals are restrained for chemical exposure (injection) [21]. Silastic capsule preparations of androgens and estrogens are more durable [240].

4.4. Mouse Models of Overactive Bladder (OAB)

OAB can be induced by ischemic injury [24,93,241]. A balloon catheter is passed through the iliac artery and inflated, then withdrawn to cause endothelial damage [241]. This injury is combined with a cholesterol enriched diet to cause bladder arterial occlusions and chronic bladder ischemia [206]. This model results in increased voiding frequency but decreased voided volume, and more frequent non-voiding bladder contractions in rats [241].

OAB can also be induced by the introduction of noxious stimuli (acetic acid, hydrochloric acid, and others) into the bladder [242,243]. Chemical induced OAB decreases the inter-voiding interval of anesthetized mice, reduces bladder capacity, and sensitizes afferent nerves [243–246].

4.5. Mouse Models of Detrusor Sphincter Dyssynergia (DSD)

The most common approach to evoke detrusor sphincter dyssynergia is to induce spinal cord injury under anesthesia [247]. Urine must be manually expressed at least three times per day until the micturition reflexes recover (10–14 days), then once per day [247]. Cystometry profiling of injured mice reveals increased activity of the external urethral sphincter coupled with increased urethral pressure and voiding pressure, increased frequency and magnitude of non-voiding contractions, and increased bladder capacity [247,248].

5. Conclusions

Lower urinary physiology is extremely complex, shaped by contributions from the urethra, prostate, bladder, ascending and descending neural pathways, and the brain. Despite an extremely high prevalence of male LUTD and devastating impacts on society, LUTD mechanisms and factors that influence LUTD severity are poorly understood. Environmental contributions to LUTD remain almost completely unexamined. We provided this overview of male lower urinary tract anatomy, physiology, and cell biology, described known disease mechanisms, and highlighted knowledge gaps that require additional research to direct new attention from toxicologists and environmental health specialists to this widespread disease. We detailed examples of environmental chemicals that perturb urinary tract function and described mouse models of LUTD with the intention that public health specialists, epidemiologists and toxicologists will consider LUTD research in toxicity assessments. Future risk mitigation strategies will likely be critical to reducing the burden and severity of LUTD in aging adults.

Author Contributions: Conceptualization, N.T.P. and C.M.V. writing—original draft preparation, N.T.P. and C.M.V.; writing—review and editing, N.T.P. and C.M.V. All authors have read and agreed to the published version of the manuscript.

Funding: This research was funded by National Institutes of Health Grants R01 ES001332 and U54 DK104310, T32 ES007015, and the UW-Madison Advanced Opportunity Fellowship. The content is solely the responsibility of the authors and does not necessarily represent the official views of the National Institutes of Health.

Institutional Review Board Statement: Not Applicable.

Informed Consent Statement: Not Applicable.

Data Availability Statement: This paper does not report any data that has not been published elsewhere.

Conflicts of Interest: The authors declare no conflict of interest.

References

1. Jacobsen, S.J.; Jacobson, D.J.; Girman, C.J.; Roberts, R.O.; Rhodes, T.; Guess, H.A.; Lieber, M.M. Natural history of prostatism: Risk factors for acute urinary retention. *J. Urol.* **1997**, *158*, 481–487. [CrossRef]
2. Roehrborn, C.G.; Malice, M.-P.; Cook, T.J.; Girman, C.J. Clinical predictors of spontaneous acute urinary retention in men with luts and clinical bph: A comprehensive analysis of the pooled placebo groups of several large clinical trials. *Urology* **2001**, *58*, 210–216. [CrossRef]
3. Winters, J.C.; Dmochowski, R.R.; Goldman, H.B.; Herndon, C.D.; Kobashi, K.C.; Kraus, S.R.; Lemack, G.E.; Nitti, V.W.; Rovner, E.S.; Wein, A.J.; et al. Urodynamic studies in adults: AUA/SUFU guideline. *J. Urol.* **2012**, *188*, 2464–2472. [CrossRef]
4. Agarwal, A.; Eryuzlu, L.N.; Cartwright, R.; Thorlund, K.; Tammela, T.L.; Guyatt, G.H.; Auvinen, A.; Tikkinen, K.A. What is the most bothersome lower urinary tract symptom? Individual- and population-level perspectives for both men and women. *Eur. Urol.* **2014**, *65*, 1211–1217. [CrossRef]
5. Shah, H.N. Benign prostate hyperplasia and bladder stones: An update. *Curr. Bladder Dysfunct. Rep.* **2013**, *8*, 250–260. [CrossRef]
6. Wei, J.T.; Calhoun, E.; Jacobsen, S.J. Urologic diseases in america project: Benign prostatic hyperplasia. *J. Urol.* **2005**, *173*, 1256–1261. [CrossRef]
7. Launer, B.M.; McVary, K.T.; Ricke, W.A.; Lloyd, G.L. The rising worldwide impact of benign prostatic hyperplasia. *BJU Int.* **2021**, *127*, 722–728. [CrossRef]
8. Kobelt, G.; Borgstrom, F.; Mattiasson, A. Productivity, vitality and utility in a group of healthy professionally active individuals with nocturia. *BJU Int.* **2003**, *91*, 190–195. [CrossRef]
9. Cakir, O.O.; McVary, K.T. LUTS and sleep disorders: Emerging risk factor. *Curr. Urol. Rep.* **2012**, *13*, 407–412. [CrossRef]
10. Ancoli-Israel, S.; Bliwise, D.L.; Norgaard, J.P. The effect of nocturia on sleep. *Sleep Med. Rev.* **2011**, *15*, 91–97. [CrossRef]
11. Johnson, T.V.; Abbasi, A.; Ehrlich, S.S.; Kleris, R.S.; Owen-Smith, A.; Raison, C.L.; Master, V.A. IPSS quality of life question: A possible indicator of depression among patients with lower urinary tract symptoms. *Can. J. Urol.* **2012**, *19*, 6100–6104.
12. Rom, M.; Schatzl, G.; Swietek, N.; Rucklinger, E.; Kratzik, C. Lower urinary tract symptoms and depression. *BJU Int.* **2012**, *110*, E918–E921. [CrossRef]
13. Jepsen, J.V.; Bruskewitz, R.C. Recent developments in the surgical management of benign prostatic hyperplasia. *Urology* **1998**, *51*, 23–31. [CrossRef]
14. Chute, C.G.; Panser, L.A.; Girman, C.J.; Oesterling, J.E.; Guess, H.A.; Jacobsen, S.J.; Lieber, M.M. The prevalence of prostatism: A population-based survey of urinary symptoms. *J. Urol.* **1993**, *150*, 85–89. [CrossRef]

15. Berry, S.J.; Coffey, D.S.; Walsh, P.C.; Ewing, L.L. The development of human benign prostatic hyperplasia with age. *J. Urol.* **1984**, *132*, 474–479. [CrossRef]
16. Hutchison, A.; Farmer, R.; Verhamme, K.; Berges, R.; Navarrete, R.V. The efficacy of drugs for the treatment of LUTS/BPH, a study in 6 European countries. *Eur. Urol.* **2007**, *51*, 207–216. [CrossRef]
17. Kirby, R.S.; Roehrborn, C.; Boyle, P.; Bartsch, G.; Jardin, A.; Cary, M.M.; Sweeney, M.; Grossman, E.B. Efficacy and tolerability of doxazosin and finasteride, alone or in combination, in treatment of symptomatic benign prostatic hyperplasia: The Prospective European doxazosin and combination therapy (PREDICT) trial. *Urology* **2003**, *61*, 119–126. [CrossRef]
18. Roehrborn, C.G. Efficacy of alpha-adrenergic receptor blockers in the treatment of male lower urinary tract symptoms. *Rev. Urol.* **2009**, *11*, S1–S8.
19. Keil Stietz, K.P.; Kennedy, C.L.; Sethi, S.; Valenzuela, A.; Nunez, A.; Wang, K.; Wang, Z.; Wang, P.; Spiegelhoff, A.; Puschner, B.; et al. In utero and lactational pcb exposure drives anatomic changes in the juvenile mouse bladder. *Curr. Res. Toxicol.* **2021**, *2*, 1–18. [CrossRef]
20. Turco, A.E.; Oakes, S.R.; Stietz, K.P.K.; Dunham, C.L.; Joseph, D.B.; Chathurvedula, T.S.; Girardi, N.M.; Schneider, A.J.; Gawdzik, J.; Sheftel, C.M.; et al. A neuroanatomical mechanism linking perinatal TCDD exposure to lower urinary tract dysfunction in adulthood. *Dis. Model Mech.* **2021**, *14*, 1–10. [CrossRef]
21. Turco, A.E.; Thomas, S.; Crawford, L.K.; Tang, W.; Peterson, R.E.; Li, L.; Ricke, W.A.; Vezina, C.M. In utero and lactational 2,3,7,8-tetrachlorodibenzo-p-dioxin (TCDD) exposure exacerbates urinary dysfunction in hormone-treated c57bl/6j mice through a non-malignant mechanism involving proteomic changes in the prostate that differ from those elicited by testosterone and estradiol. *Am. J. Clin. Exp. Urol.* **2020**, *8*, 59–72. [PubMed]
22. Middleton, L.W.; Shen, Z.; Varma, S.; Pollack, A.S.; Gong, X.; Zhu, S.; Zhu, C.; Foley, J.W.; Vennam, S.; Sweeney, R.T.; et al. Genomic analysis of benign prostatic hyperplasia implicates cellular re-landscaping in disease pathogenesis. *JCI Insight* **2019**, *5*, e129749. [CrossRef] [PubMed]
23. Joseph, D.B.; Henry, G.H.; Malewska, A.; Reese, J.C.; Mauck, R.J.; Gahan, J.C.; Hutchinson, R.C.; Mohler, J.L.; Roehrborn, C.G.; Strand, D.W. 5-alpha reductase inhibitors induce a prostate luminal to club cell transition in human benign prostatic hyperplasia. *J. Pathol.* **2021**, 1–15. [CrossRef] [PubMed]
24. Joseph, D.B.; Henry, G.H.; Malewska, A.; Iqbal, N.S.; Ruetten, H.M.; Turco, A.E.; Abler, L.L.; Sandhu, S.K.; Cadena, M.T.; Malladi, V.S.; et al. Urethral luminal epithelia are castration-insensitive cells of the proximal prostate. *Prostate* **2020**, *80*, 872–884. [CrossRef] [PubMed]
25. Andersson, K.-E.; McCloskey, K.D. Lamina propria: The functional center of the bladder? *Neurourol. Urodyn.* **2014**, *33*, 9–16. [CrossRef] [PubMed]
26. Apodaca, G. The uroepithelium: Not just a passive barrier. *Traffic* **2004**, *5*, 117–128. [CrossRef] [PubMed]
27. Graham, E.; Chai, T.C. Dysfunction of bladder urothelium and bladder urothelial cells in interstitial cystitis. *Curr. Urol. Rep.* **2006**, *7*, 440–446. [CrossRef]
28. Guo, C.; Balsara, Z.R.; Hill, W.G.; Li, X. Stage- and subunit-specific functions of polycomb repressive complex 2 in bladder urothelial formation and regeneration. *Development* **2017**, *144*, 400–408. [CrossRef]
29. Khandelwal, P.; Abraham, S.N.; Apodaca, G. Cell biology and physiology of the uroepithelium. *Am. J. Physiol. Ren. Physiol.* **2009**, *297*, F1477–F1501. [CrossRef]
30. Hurst, R.E. Structure, function, and pathology of proteoglycans and glycosaminoglycans in the urinary tract. *World J. Urol.* **1994**, *12*, 3–10. [CrossRef]
31. Hu, P.; Deng, F.-M.; Liang, F.-X.; Hu, C.-M.; Auerbach, A.B.; Shapiro, E.; Wu, X.-R.; Kachar, B.; Sun, T.-T. Ablation of uroplakin III gene results in small urothelial plaques, urothelial leakage, and vesicoureteral reflux. *J. Cell Biol.* **2000**, *151*, 961–972. [CrossRef]
32. Wu, X.R.; Kong, X.P.; Pellicer, A.; Kreibich, G.; Sun, T.T. Uroplakins in urothelial biology, function, and disease. *Kidney Int.* **2009**, *75*, 1153–1165. [CrossRef]
33. Kątnik-Prastowska, I.; Lis, J.; Matejuk, A. Glycosylation of uroplakins. Implications for bladder physiopathology. *Glycoconj. J.* **2014**, *31*, 623–636. [CrossRef]
34. Gandhi, D.; Molotkov, A.; Batourina, E.; Schneider, K.; Dan, H.; Reiley, M.; Laufer, E.; Metzger, D.; Liang, F.; Liao, Y.; et al. Retinoid signaling in progenitors controls specification and regeneration of the urothelium. *Develop. Cell* **2013**, *26*, 469–482. [CrossRef]
35. Sutherland Ronald, S.; Baskin Laurence, S.; Hayward Simon, W.; Cunha Gerald, R. Regeneration of bladder urothelium, smooth muscle, blood vessels and nerves into an acellular tissue matrix. *J. Urol.* **1996**, *156*, 571–577. [CrossRef]
36. Hicks, R.M. The mammalian urinary bladderan accommodating organ. *Biol. Rev.* **1975**, *50*, 215–246. [CrossRef]
37. Joseph, D.B.; Chandrashekar, A.S.; Abler, L.L.; Chu, L.F.; Thomson, J.A.; Mendelsohn, C.; Vezina, C.M. In vivo replacement of damaged bladder urothelium by Wolffian duct epithelial cells. *Proc. Natl. Acad. Sci. USA* **2018**, *115*, 8394–8399. [CrossRef]
38. Cox, P.J. Cyclophosphamide cystitis—Identification of acrolein as the causative agent. *Biochem. Pharmacol.* **1979**, *28*, 2045–2049. [CrossRef]
39. Brock, N.; Stekar, J.; Pohl, J.; Niemeyer, U.; Scheffler, G. Acrolein, the causative factor of urotoxic side-effects of cyclophosphamide, ifosfamide, trofosfamide and sufosfamide. *Arzneimittelforschung* **1979**, *29*, 659–661.
40. Al-Rawithi, S.; El-Yazigi, A.; Ernst, P.; Al-Fiar, F.; Nicholls, P.J. Urinary excretion and pharmacokinetics of acrolein and its parent drug cyclophosphamide in bone marrow transplant patients. *Bone Marrow Transplant.* **1998**, *22*, 485–490. [CrossRef]

41. Andersson, K.E.; Nomiya, M.; Yamaguchi, O. Chronic pelvic ischemia: Contribution to the pathogenesis of lower urinary tract symptoms (LUTS): A new target for pharmacological treatment? *Low Urin. Tract Symptoms* **2015**, *7*, 1–8. [CrossRef]
42. Birder, L.A.; Kanai, A.J.; de Groat, W.C.; Kiss, S.; Nealen, M.L.; Burke, N.E.; Dineley, K.E.; Watkins, S.; Reynolds, I.J.; Caterina, M.J. Vanilloid receptor expression suggests a sensory role for urinary bladder epithelial cells. *Proc. Natl. Acad. Sci. USA* **2001**, *98*, 13396–13401. [CrossRef]
43. Ferguson, D.R.; Kennedy, I.; Burton, T.J. ATP is released from rabbit urinary bladder epithelial cells by hydrostatic pressure changes—Possible sensory mechanism? *J. Physiol.* **1997**, *505*, 503–511. [CrossRef]
44. Wiseman, O.J.; Fowler, C.J.; Landon, D.N. The role of the human bladder lamina propria myofibroblast. *BJU Int.* **2003**, *91*, 89–93. [CrossRef]
45. Aitken, K.J.; Bägli, D.J. The bladder extracellular matrix. Part I: Architecture, development and disease. *Nat. Rev. Urol.* **2009**, *6*, 596–611. [CrossRef]
46. Andersson, K.-E.; Arner, A. Urinary bladder contraction and relaxation: Physiology and pathophysiology. *Physiol. Rev.* **2004**, *84*, 935–986. [CrossRef]
47. Sjuve, R.; Haase, H.; Ekblad, E.; Malmqvist, U.; Morano, I.; Arner, A. Increased expression of non-muscle myosin heavy chain-B in connective tissue cells of hypertrophic rat urinary bladder. *Cell Tissue Res.* **2001**, *304*, 271–278. [CrossRef]
48. Seki, T.; Naito, I.; Oohashi, T.; Sado, Y.; Ninomiya, Y. Differential expression of type IV collagen isoforms, alpha5(IV) and alpha6(IV) chains, in basement membranes surrounding smooth muscle cells. *Histochem. Cell Biol.* **1998**, *110*, 359–366. [CrossRef]
49. Gilpin, S.A.; Gilpin, C.J.; Dixon, J.S.; Gosling, J.A.; Kirby, R.S. The effect of age on the autonomic innervation of the urinary bladder. *Br. J. Urol.* **1986**, *58*, 378–381. [CrossRef]
50. Erdem, N.; Chu, F.M. Management of overactive bladder and urge urinary incontinence in the elderly patient. *Am. J. Med.* **2006**, *119*, 29–36. [CrossRef]
51. Drake, M.J.; Mills, I.W.; Gillespie, J.I. Model of peripheral autonomous modules and a myovesical plexus in normal and overactive bladder function. *Lancet* **2001**, *358*, 401–403. [CrossRef]
52. Häbler, H.J.; Jänig, W.; Koltzenburg, M. Activation of unmyelinated afferent fibres by mechanical stimuli and inflammation of the urinary bladder in the cat. *J. Physiol.* **1990**, *425*, 545–562. [CrossRef]
53. Morrison, J. The activation of bladder wall afferent nerves. *Exp. Physiol.* **1999**, *84*, 131–136. [CrossRef]
54. Tennyson, L.E.; Tai, C.; Chermansky, C.J. Using the native afferent nervous system to sense bladder fullness: State of the art. *Curr. Bladder Dysfunct. Rep.* **2016**, *11*, 346–349. [CrossRef]
55. Marshall, K.L.; Saade, D.; Ghitani, N.; Coombs, A.M.; Szczot, M.; Keller, J.; Ogata, T.; Daou, I.; Stowers, L.T.; Bönnemann, C.G.; et al. PIEZO2 in sensory neurons and urothelial cells coordinates urination. *Nature* **2020**, *588*, 290–295. [CrossRef]
56. Dalghi, M.G.; Ruiz, W.G.; Clayton, D.R.; Montalbetti, N.; Daugherty, S.L.; Beckel, J.M.; Carattino, M.D.; Apodaca, G. Functional roles for PIEZO1 and PIEZO2 in urothelial mechanotransduction and lower urinary tract interoception. *JCI Insight* **2021**, *6*, e152984. [CrossRef]
57. Ford, A.P.; Cockayne, D.A. ATP and P2X purinoceptors in urinary tract disorders. In *Handbook of Experimental Pharmacology*; Springer: Berlin/Heidelberg, Germany, 2011; pp. 485–526.
58. Vlaskovska, M.; Kasakov, L.; Rong, W.; Bodin, P.; Bardini, M.; Cockayne, D.A.; Ford, A.P.; Burnstock, G. P2X3 knock-out mice reveal a major sensory role for urothelially released. *ATP J. Neurosci.* **2001**, *21*, 5670–5677. [CrossRef]
59. Ferguson, A.C.; Sutton, B.W.; Boone, T.B.; Ford, A.P.; Munoz, A. Inhibition of urothelial P2X3 receptors prevents desensitization of purinergic detrusor contractions in the rat bladder. *BJU Int.* **2015**, *116*, 293–301. [CrossRef]
60. Heppner, T.J.; Tykocki, N.R.; Hill-Eubanks, D.; Nelson, M.T. Transient contractions of urinary bladder smooth muscle are drivers of afferent nerve activity during filling. *J. Gen. Physiol.* **2016**, *147*, 323–335. [CrossRef]
61. Barrington, F.J.F. The effect of lesions of the hind- and mid-brain on micturition in the cat. *Q. J. Exp. Physiol.* **1925**, *15*, 81–102. [CrossRef]
62. Sasaki, M. Role of Barrington's nucleus in micturition. *J. Comp. Neurol.* **2005**, *493*, 21–26. [CrossRef]
63. Westlund, K.N.; Coulter, J.D. Descending projections of the locus coeruleus and subcoeruleus/medial parabrachial nuclei in monkey: Axonal transport studies and dopamine-β-hydroxylase immunocytochemistry. *Brain Res. Rev.* **1980**, *2*, 235–264. [CrossRef]
64. Martin, G.F.; Humbertson, A.O., Jr.; Laxson, L.C.; Panneton, W.M.; Tschismadia, I. Spinal projections from the mesencephalic and pontine reticular formation in the north American opossum: A study using axonal transport techniques. *J. Comp. Neurol.* **1979**, *187*, 373–399. [CrossRef]
65. Loewy, A.D.; Saper, C.B.; Baker, R.P. Descending projections from the pontine micturition center. *Brain Res.* **1979**, *172*, 533–538. [CrossRef]
66. Holstege, G.; Kuypers, H.G.J.M.; Boer, R.C. Anatomical evidence for direct brain stem projections to the somatic motoneuronal cell groups and autonomic preganglionic cell groups in cat spinal cord. *Brain Res.* **1979**, *171*, 329–333. [CrossRef]
67. Blok, B.F.M.; Holstege, G. Direct projections from the periaqueductal gray to the pontine micturition center (M-region). An anterograde and retrograde tracing study in the cat. *Neurosci. Lett.* **1994**, *166*, 93–96. [CrossRef]
68. Verstegen, A.M.J.; Vanderhorst, V.; Gray, P.A.; Zeidel, M.L.; Geerling, J.C. Barrington's nucleus: Neuroanatomic landscape of the mouse "pontine micturition center". *J. Comp. Neurol.* **2017**, *525*, 2287–2309. [CrossRef]

69. Verstegen, A.M.J.; Klymko, N.; Zhu, L.; Mathai, J.C.; Kobayashi, R.; Venner, A.; Ross, R.A.; VanderHorst, V.G.; Arrigoni, E.; Geerling, J.C.; et al. Non-crh glutamatergic neurons in barrington's nucleus control micturition via glutamatergic afferents from the midbrain and hypothalamus. *Curr. Biol.* **2019**, *29*, 2775–2789.e2777. [CrossRef]
70. Keller, J.A.; Chen, J.; Simpson, S.; Wang, E.H.-J.; Lilascharoen, V.; George, O.; Lim, B.K.; Stowers, L. Voluntary urination control by brainstem neurons that relax the urethral sphincter. *Nat. Neurosci.* **2018**, *21*, 1229–1238. [CrossRef]
71. Hou, X.H.; Hyun, M.; Taranda, J.; Huang, K.W.; Todd, E.; Feng, D.; Atwater, E.; Croney, D.; Zeidel, M.L.; Osten, P. Central control circuit for context-dependent micturition. *Cell* **2016**, *167*, 73–86. [CrossRef]
72. Griffiths, D.; Holstege, G.; Dalm, E.; Wall, H.D. Control and coordination of bladder and urethral function in the brainstem of the cat. *Neurourol. Urodyn.* **1990**, *9*, 63–82. [CrossRef]
73. Van Batavia, J.P.; Butler, S.; Lewis, E.; Fesi, J.; Canning, D.A.; Vicini, S.; Valentino, R.J.; Zderic, S.A. Corticotropin-releasing hormone from the pontine micturition center plays an inhibitory role in micturition. *J. Neurosci.* **2021**, *41*, 7314–7325. [CrossRef]
74. Siblerud, R.; Mutter, J.; Moore, E.; Naumann, J.; Walach, H. A hypothesis and evidence that mercury may be an etiological factor in Alzheimer's disease. *Int. J. Environ. Res. Public Health* **2019**, *16*, 5152. [CrossRef]
75. Tchounwou, P.B.; Ayensu, W.K.; Ninashvili, N.; Sutton, D. Review: Environmental exposure to mercury and its toxicopathologic implications for public health. *Environ. Toxicol.* **2003**, *18*, 149–175. [CrossRef]
76. Sindhu, K.K.; Sutherling, W.W. Role of lead in the central nervous system: Effect on electroencephlography, evoked potentials, electroretinography, and nerve conduction. *Neurodiagn. J.* **2015**, *55*, 107–121. [CrossRef]
77. Mason, L.H.; Harp, J.P.; Han, D.Y. Pb neurotoxicity: Neuropsychological effects of lead toxicity. *Biomed. Res. Int.* **2014**, *2014*, 840547. [CrossRef]
78. Naughton, S.X.; Terry, A.V. Neurotoxicity in acute and repeated organophosphate exposure. *Toxicology* **2018**, *408*, 101–112. [CrossRef]
79. Mochizuki, H. Arsenic neurotoxicity in humans. *Int. J. Mol. Sci.* **2019**, *20*, 3418. [CrossRef]
80. Richardson, J.R.; Fitsanakis, V.; Westerink, R.H.S.; Kanthasamy, A.G. Neurotoxicity of pesticides. *Acta Neuropathol.* **2019**, *138*, 343–362. [CrossRef]
81. Sethi, S.; Morgan, R.K.; Feng, W.; Lin, Y.; Li, X.; Luna, C.; Koch, M.; Bansal, R.; Duffel, M.W.; Puschner, B.; et al. Comparative analyses of the 12 most abundant PCB congeners detected in human maternal serum for activity at the thyroid hormone receptor and ryanodine receptor. *Environ. Sci. Technol.* **2019**, *53*, 3948–3958. [CrossRef]
82. Rude, K.M.; Keogh, C.E.; Gareau, M.G. The role of the gut microbiome in mediating neurotoxic outcomes to PCB exposure. *Neurotoxicology* **2019**, *75*, 30–40. [CrossRef] [PubMed]
83. Galgano, S.J.; Sivils, C.; Selph, J.P.; Sanyal, R.; Lockhart, M.E.; Zarzour, J.G. The male urethra: Imaging and surgical approach for common pathologies. *Curr. Probl. Diagn. Radiol.* **2021**, *50*, 410–418. [CrossRef] [PubMed]
84. Georgas, K.M.; Armstrong, J.; Keast, J.R.; Larkins, C.E.; McHugh, K.M.; Southard-Smith, E.M.; Cohn, M.J.; Batourina, E.; Dan, H.; Schneider, K.; et al. An illustrated anatomical ontology of the developing mouse lower urogenital tract. *Development* **2015**, *142*, 1893–1908. [CrossRef] [PubMed]
85. Morales, O.; Romanus, R. Urethrography in the male: The boundaries of the different urethral parts and detail studies of the urethral mucous membrane and its motility. *J. Urol.* **1955**, *73*, 162–171. [CrossRef]
86. Henry, G.H.; Malewska, A.; Joseph, D.B.; Malladi, V.S.; Lee, J.; Torrealba, J.; Mauck, R.J.; Gahan, J.C.; Raj, G.V.; Roehrborn, C.G.; et al. A cellular anatomy of the normal adult human prostate and prostatic urethra. *Cell Rep.* **2018**, *25*, 3530–3542.e3535. [CrossRef]
87. Crowley, L.; Cambuli, F.; Aparicio, L.; Shibata, M.; Robinson, B.D.; Xuan, S.; Li, W.; Hibshoosh, H.; Loda, M.; Rabadan, R. A single-cell atlas of the mouse and human prostate reveals heterogeneity and conservation of epithelial progenitors. *Elife* **2020**, *9*, e59465. [CrossRef]
88. Zhai, J.; Insel, M.; Addison, K.J.; Stern, D.A.; Pederson, W.; Dy, A.; Rojas-Quintero, J.; Owen, C.A.; Sherrill, D.L.; Morgan, W.; et al. Club cell secretory protein deficiency leads to altered lung function. *Am. J. Respir. Crit. Care Med.* **2019**, *199*, 302–312. [CrossRef]
89. Rawlins, E.L.; Okubo, T.; Xue, Y.; Brass, D.M.; Auten, R.L.; Hasegawa, H.; Wang, F.; Hogan, B.L. The role of Scgb1a1+ Clara cells in the long-term maintenance and repair of lung airway, but not alveolar, epithelium. *Cell Stem Cell* **2009**, *4*, 525–534. [CrossRef]
90. Hong, K.U.; Reynolds, S.D.; Giangreco, A.; Hurley, C.M.; Stripp, B.R. Clara cell secretory protein-expressing cells of the airway neuroepithelial body microenvironment include a label-retaining subset and are critical for epithelial renewal after progenitor cell depletion. *Am. J. Respir. Cell Mol. Biol.* **2001**, *24*, 671–681. [CrossRef]
91. Montoro, D.T.; Haber, A.L.; Biton, M.; Vinarsky, V.; Lin, B.; Birket, S.E.; Yuan, F.; Chen, S.; Leung, H.M.; Villoria, J.; et al. A revised airway epithelial hierarchy includes CFTR-expressing ionocytes. *Nature* **2018**, *560*, 319–324. [CrossRef]
92. Deprez, M.; Zaragosi, L.E.; Truchi, M.; Becavin, C.; Ruiz García, S.; Arguel, M.J.; Plaisant, M.; Magnone, V.; Lebrigand, K.; Abelanet, S.; et al. A single-cell atlas of the human healthy airways. *Am. J. Respir. Crit. Care Med.* **2020**, *202*, 1636–1645. [CrossRef] [PubMed]
93. Jung, J.; Ahn, H.K.; Huh, Y. Clinical and functional anatomy of the urethral sphincter. *Int. Neurourol. J.* **2012**, *16*, 102–106. [CrossRef] [PubMed]
94. Oelrich, T.M. The urethral sphincter muscle in the male. *Am. J. Anat.* **1980**, *158*, 229–246. [CrossRef]
95. Nyangoh Timoh, K.; Moszkowicz, D.; Creze, M.; Zaitouna, M.; Felber, M.; Lebacle, C.; Diallo, D.; Martinovic, J.; Tewari, A.; Lavoué, V.; et al. The male external urethral sphincter is autonomically innervated. *Clin. Anat.* **2021**, *34*, 263–271. [CrossRef] [PubMed]

96. McNeal, J.E. The zonal anatomy of the prostate. *Prostate* **1981**, *2*, 35–49. [CrossRef]
97. Verze, P.; Cai, T.; Lorenzetti, S. The role of the prostate in male fertility, health and disease. *Nat. Rev. Urol.* **2016**, *13*, 379–386. [CrossRef] [PubMed]
98. Medved, M.; Sammet, S.; Yousuf, A.; Oto, A. MR imaging of the prostate and adjacent anatomic structures before, during, and after ejaculation: Qualitative and quantitative evaluation. *Radiology* **2014**, *271*, 452–460. [CrossRef]
99. Joseph, D.B.; Henry, G.H.; Malewska, A.; Reese, J.C.; Mauck, R.J.; Gahan, J.C.; Hutchinson, R.C.; Malladi, V.S.; Roehrborn, C.G.; Vezina, C.M.; et al. Single-cell analysis of mouse and human prostate reveals novel fibroblasts with specialized distribution and microenvironment interactions. *J. Pathol.* **2021**, *255*, 141–154. [CrossRef]
100. McNeal, J.; Kindrachuk, R.; Freiha, F.; Bostwick, D.; Redwine, E.; Stamey, T. Patterns of progression in prostate cancer. *Lancet* **1986**, *327*, 60–63. [CrossRef]
101. Frick, J.; Aulitzky, W. Physiology of the prostate. *Infection* **1991**, *19*, S115–S118. [CrossRef]
102. Lee, C.H.; Akin-Olugbade, O.; Kirschenbaum, A. Overview of prostate anatomy, histology, and pathology. *Endocrinol. Metab. Clin. N. Am.* **2011**, *40*, 565–575. [CrossRef] [PubMed]
103. Yu, C.; Hu, K.; Nguyen, D.; Wang, Z.A. From genomics to functions: Preclinical mouse models for understanding oncogenic pathways in prostate cancer. *Am. J. Cancer Res.* **2019**, *9*, 2079–2102.
104. Liu, T.T.; Thomas, S.; McLean, D.T.; Roldan-Alzate, A.; Hernando, D.; Ricke, E.A.; Ricke, W.A. Prostate enlargement and altered urinary function are part of the aging process. *Aging* **2019**, *11*, 2653–2669. [CrossRef]
105. Huggins, C. The etiology of benign prostatic hypertrophy. *Bull. N. Y. Acad. Med.* **1947**, *23*, 696–704. [PubMed]
106. Ittmann, M. Anatomy and histology of the human and murine prostate. *Cold Spring Harb. Perspect. Med.* **2018**, *8*, a030346. [CrossRef] [PubMed]
107. Pattabiraman, G.; Bell-Cohn, A.J.; Murphy, S.F.; Mazur, D.J.; Schaeffer, A.J.; Thumbikat, P. Mast cell function in prostate inflammation, fibrosis, and smooth muscle cell dysfunction. *Am. J. Physiol. Ren. Physiol.* **2021**, *321*, F466–F479. [CrossRef] [PubMed]
108. Thielen, J.L.; Volzing, K.G.; Collier, L.S.; Green, L.E.; Largaespada, D.A.; Marker, P.C. Markers of prostate region-specific epithelial identity define anatomical locations in the mouse prostate that are molecularly similar to human prostate cancers. *Differentiation* **2007**, *75*, 49–61. [CrossRef]
109. Cunha, G.R.; Vezina, C.M.; Isaacson, D.; Ricke, W.A.; Timms, B.G.; Cao, M.; Franco, O.; Baskin, L.S. Development of the human prostate. *Differentiation* **2018**, *103*, 24–45. [CrossRef]
110. Tubaro, A.; Mariani, S.; De Nunzio, C.; Miano, R. Bladder weight and detrusor thickness as parameters of progression of benign prostatic hyperplasia. *Curr. Opin. Urol.* **2010**, *20*, 37–42. [CrossRef]
111. Kita, M.; Yunoki, T.; Takimoto, K.; Miyazato, M.; Kita, K.; de Groat, W.C.; Kakizaki, H.; Yoshimura, N. Effects of bladder outlet obstruction on properties of Ca2+-activated K+ channels in rat bladder. *Am. J. Physiol. Regul. Integr. Comp. Physiol.* **2010**, *298*, R1310–R1319. [CrossRef]
112. Sugaya, K.; de Goat, W.C. Excitatory and inhibitory influence of pathways in the pelvic nerve on bladder activity in rats with bladder outlet obstruction. *LUTS Low. Urin. Tract Symptoms* **2009**, *1*, 51–55. [CrossRef] [PubMed]
113. Gosling, J.A.; Kung, L.S.; Dixon, J.S.; Horan, P.; Whitbeck, C.; Levin, R.M. Correlation between the structure and function of the rabbit urinary bladder following partial outlet obstruction. *J. Urol.* **2000**, *163*, 1349–1356. [CrossRef]
114. Kojima, M.; Inui, E.; Ochiai, A.; Naya, Y.; Kamoi, K.; Ukimura, O.; Watanabe, H. Reversible change of bladder hypertrophy due to benign prostatic hyperplasia after surgical relief of obstruction. *J. Urol.* **1997**, *158*, 89–93. [CrossRef] [PubMed]
115. Tubaro, A.; Carter, S.; Hind, A.; Vicentini, C.; Miano, L. A prospective study of the safety and efficacy of suprapubic transvesical prostatectomy in patients with benign prostatic hyperplasia. *J. Urol.* **2001**, *166*, 172–176. [CrossRef]
116. Fibbi, B.; Penna, G.; Morelli, A.; Adorini, L.; Maggi, M. Chronic inflammation in the pathogenesis of benign prostatic hyperplasia. *Int. J. Androl.* **2010**, *33*, 475–488. [CrossRef]
117. Bierhoff, E.; Vogel, J.; Benz, M.; Giefer, T.; Wernert, N.; Pfeifer, U. Stromal nodules in benign prostatic hyperplasia. *Eur. Urol.* **1996**, *29*, 345–354. [CrossRef]
118. Meigs, J.B.; Mohr, B.; Barry, M.J.; Collins, M.M.; McKinlay, J.B. Risk factors for clinical benign prostatic hyperplasia in a community-based population of healthy aging men. *J. Clin. Epidemiol.* **2001**, *54*, 935–944. [CrossRef]
119. Michel, M.C.; Mehlburger, L.; Schumacher, H.; Bressel, H.U.; Goepel, M. Effect of diabetes on lower urinary tract symptoms in patients with benign prostatic hyperplasia. *J. Urol.* **2000**, *163*, 1725–1729. [CrossRef]
120. Verhamme, K.M.; Dieleman, J.P.; Bleumink, G.S.; van der Lei, J.; Sturkenboom, M.C.; Artibani, W.; Begaud, B.; Berges, R.; Borkowski, A.; Chappel, C.R.; et al. Incidence and prevalence of lower urinary tract symptoms suggestive of benign prostatic hyperplasia in primary care—The Triumph project. *Eur. Urol.* **2002**, *42*, 323–328. [CrossRef]
121. McNeal, J. Pathology of benign prostatic hyperplasia. Insight into etiology. *Urol. Clin. N. Am.* **1990**, *17*, 477–486. [CrossRef]
122. Vermeulen, A. Decreased androgen levels and obesity in men. *Ann. Med.* **1996**, *28*, 13–15. [CrossRef] [PubMed]
123. Krieg, M.; Nass, R.; Tunn, S. Effect of aging on endogenous level of 5 alpha-dihydrotestosterone, testosterone, estradiol, and estrone in epithelium and stroma of normal and hyperplastic human prostate. *J. Clin. Endocrinol. Metab.* **1993**, *77*, 375–381. [PubMed]

124. Nicholson, T.M.; Ricke, E.A.; Marker, P.C.; Miano, J.M.; Mayer, R.D.; Timms, B.G.; vom Saal, F.S.; Wood, R.W.; Ricke, W.A. Testosterone and 17β-estradiol induce glandular prostatic growth, bladder outlet obstruction, and voiding dysfunction in male mice. *Endocrinology* **2012**, *153*, 5556–5565. [CrossRef] [PubMed]
125. Belanger, A.; Candas, B.; Dupont, A.; Cusan, L.; Diamond, P.; Gomez, J.L.; Labrie, F. Changes in serum concentrations of conjugated and unconjugated steroids in 40- to 80-year-old men. *J. Clin. Endocrinol. Metab.* **1994**, *79*, 1086–1090. [PubMed]
126. Schatzl, G.; Brossner, C.; Schmid, S.; Kugler, W.; Roehrich, M.; Treu, T.; Szalay, A.; Djavan, B.; Schmidbauer, C.P.; Soregi, S.; et al. Endocrine status in elderly men with lower urinary tract symptoms: Correlation of age, hormonal status, and lower urinary tract function. The prostate study group of the Austrian society of urology. *Urology* **2000**, *55*, 397–402. [CrossRef]
127. Byron, J.K.; Taylor, K.H.; Phillips, G.S.; Stahl, M.S. Urethral sphincter mechanism incompetence in 163 neutered female dogs: Diagnosis, treatment, and relationship of weight and age at neuter to development of disease. *J. Vet. Intern. Med.* **2017**, *31*, 442–448. [CrossRef]
128. Bernoulli, J.; Yatkin, E.; Konkol, Y.; Talvitie, E.M.; Santti, R.; Streng, T. Prostatic inflammation and obstructive voiding in the adult Noble rat: Impact of the testosterone to estradiol ratio in serum. *Prostate* **2008**, *68*, 1296–1306. [CrossRef]
129. Streng, T.; Lehtoranta, M.; Poutanen, M.; Talo, A.; Lammintausta, R.; Santti, R. Developmental, estrogen induced infravesical obstruction is reversible in adult male rodents. *J. Urol.* **2002**, *168*, 2263–2268. [CrossRef]
130. Streng, T.K.; Talo, A.; Andersson, K.E.; Santti, R. A dose-dependent dual effect of oestrogen on voiding in the male mouse? *BJU Int.* **2005**, *96*, 1126–1130. [CrossRef]
131. Walsh, P.C.; Wilson, J.D. The induction of prostatic hypertrophy in the dog with androstanediol. *J. Clin. Investig.* **1976**, *57*, 1093–1097. [CrossRef]
132. Jeyaraj, D.A.; Udayakumar, T.S.; Rajalakshmi, M.; Pal, P.C.; Sharma, R.S. Effects of long-term administration of androgens and estrogen on rhesus monkey prostate: Possible induction of benign prostatic hyperplasia. *J. Androl.* **2000**, *21*, 833–841. [PubMed]
133. Tam, N.N.; Zhang, X.; Xiao, H.; Song, D.; Levin, L.; Meller, J.; Ho, S.M. Increased susceptibility of estrogen-induced bladder outlet obstruction in a novel mouse model. *Lab. Investig.* **2015**, *95*, 546–560. [CrossRef] [PubMed]
134. Shenfeld, O.Z.; McCammon, K.A.; Blackmore, P.F.; Ratz, P.H. Rapid effects of estrogen and progesterone on tone and spontaneous rhythmic contractions of the rabbit bladder. *Urol. Res.* **1999**, *27*, 386–392. [CrossRef] [PubMed]
135. Yasay, G.D.; Kau, S.T.; Li, J.H. Mechanoinhibitory effect of estradiol in guinea pig urinary bladder smooth muscles. *Pharmacology* **1995**, *51*, 273–280. [CrossRef]
136. Zhang, J.; Hess, M.W.; Thurnher, M.; Hobisch, A.; Radmayr, C.; Cronauer, M.V.; Hittmair, A.; Culig, Z.; Bartsch, G.; Klocker, H. Human prostatic smooth muscle cells in culture: Estradiol enhances expression of smooth muscle cell-specific markers. *Prostate* **1997**, *30*, 117–129. [CrossRef]
137. Scarano, W.R.; Cordeiro, R.S.; Goes, R.M.; Carvalho, H.F.; Taboga, S.R. Tissue remodeling in Guinea pig lateral prostate at different ages after estradiol treatment. *Cell Biol. Int.* **2005**, *29*, 778–784. [CrossRef]
138. Holterhus, P.M.; Zhao, G.Q.; Aumuller, G. Effects of androgen deprivation and estrogen treatment on the structure and protein expression of the rat coagulating gland. *Anat. Rec.* **1993**, *235*, 223–232. [CrossRef]
139. Tam, C.C.; Wong, Y.C. Ultrastructural study of the effects of 17 beta-oestradiol on the lateral prostate and seminal vesicle of the castrated guinea pig. *Acta Anat.* **1991**, *141*, 51–62.
140. Nicholson, T.M.; Nguyen, J.L.; Leverson, G.E.; Taylor, J.A.; Vom Saal, F.S.; Wood, R.W.; Ricke, W.A. Endocrine disruptor bisphenol A is implicated in urinary voiding dysfunction in male mice. *Am. J. Physiol. Ren. Physiol.* **2018**, *315*, F1208–F1216. [CrossRef]
141. Makowska, K.; Lech, P.; Majewski, M.; Rychlik, A.; Gonkowski, S. Bisphenol A affects vipergic nervous structures in the porcine urinary bladder trigone. *Sci. Rep.* **2021**, *11*, 12147. [CrossRef]
142. Bushman, W.A.; Jerde, T.J. The role of prostate inflammation and fibrosis in lower urinary tract symptoms. *Am. J. Physiol. Ren. Physiol.* **2016**, *311*, F817–F821. [CrossRef] [PubMed]
143. Lee, S.; Yang, G.; Bushman, W. Prostatic inflammation induces urinary frequency in adult mice. *PLoS ONE* **2015**, *10*, e0116827. [CrossRef] [PubMed]
144. Lloyd, G.L.; Ricke, W.A.; McVary, K.T. Inflammation, voiding and benign prostatic hyperplasia progression. *J. Urol.* **2019**, *201*, 868–870. [CrossRef] [PubMed]
145. Kaplan, S.A.; Roehrborn, C.G.; McConnell, J.D.; Meehan, A.G.; Surynawanshi, S.; Lee, J.Y.; Rotonda, J.; Kusek, J.W.; Nyberg, L.M., Jr. Long-term treatment with finasteride results in a clinically significant reduction in total prostate volume compared to placebo over the full range of baseline prostate sizes in men enrolled in the MTOPS trial. *J. Urol.* **2008**, *180*, 1030–1032. [CrossRef] [PubMed]
146. St Sauver, J.L.; Jacobson, D.J.; McGree, M.E.; Girman, C.J.; Lieber, M.M.; Jacobsen, S.J. Longitudinal association between prostatitis and development of benign prostatic hyperplasia. *Urology* **2008**, *71*, 475–479. [CrossRef]
147. Penna, G.; Fibbi, B.; Amuchastegui, S.; Cossetti, C.; Aquilano, F.; Laverny, G.; Gacci, M.; Crescioli, C.; Maggi, M.; Adorini, L. Human benign prostatic hyperplasia stromal cells as inducers and targets of chronic immuno-mediated inflammation. *J. Immunol.* **2009**, *182*, 4056–4064. [CrossRef]
148. Doiron, R.C.; Shoskes, D.A.; Nickel, J.C. Male cp/cpps: Where do we stand? *World J. Urol.* **2019**, *37*, 1015–1022. [CrossRef]
149. Roberts, R.O.; Lieber, M.M.; Bostwick, D.G.; Jacobsen, S.J. A review of clinical and pathological prostatitis syndromes. *Urology* **1997**, *49*, 809–821. [CrossRef]

150. Cantiello, F.; Cicione, A.; Salonia, A.; Autorino, R.; Tucci, L.; Madeo, I.; Damiano, R. Periurethral fibrosis secondary to prostatic inflammation causing lower urinary tract symptoms: A prospective cohort study. *Urology* **2013**, *81*, 1018–1024. [CrossRef]
151. Macoska, J. Prostatic fibrosis is associated with lower urinary tract symptoms. *J. Urol.* **2012**, *188*, 1375–1381. [CrossRef]
152. Macoska, J.A.; Uchtmann, K.S.; Leverson, G.E.; McVary, K.T.; Ricke, W.A. Prostate transition zone fibrosis is associated with clinical progression in the MTOPS study. *J. Urol.* **2019**, *202*, 1240–1247. [CrossRef] [PubMed]
153. Ruetten, H.; Wegner, K.A.; Romero, M.F.; Wood, M.W.; Marker, P.C.; Strand, D.; Colopy, S.A.; Vezina, C.M. Prostatic collagen architecture in neutered and intact canines. *Prostate* **2018**, *78*, 839–848. [CrossRef] [PubMed]
154. Lee, S.; Yang, G.; Mulligan, W.; Gipp, J.; Bushman, W. Ventral prostate fibrosis in the Akita mouse is associated with macrophage and fibrocyte infiltration. *J. Diabetes Res.* **2014**, *2014*, 939053. [CrossRef] [PubMed]
155. Gharaee-Kermani, M.; Rodriguez-Nieves, J.A.; Mehra, R.; Vezina, C.A.; Sarma, A.V.; Macoska, J.A. Obesity-induced diabetes and lower urinary tract fibrosis promote urinary voiding dysfunction in a mouse model. *Prostate* **2013**, *73*, 1123–1133. [CrossRef]
156. Baumgarten, H.G.; Falck, B.; Holstein, A.F.; Owman, C.; Owman, T. Adrenergic innervation of the human testis, epididymis, ductus deferens and prostate: A fluorescence microscopic and fluorimetric study. *Z. Zellforsch. Und Mikrosk. Anat.* **1968**, *90*, 81–95. [CrossRef]
157. Caine, M.; Raz, S.; Zeigler, M. Adrenergic and cholinergic receptors in the human prostate, prostatic capsule and bladder neck. *Br. J. Urol.* **1975**, *47*, 193–202. [CrossRef]
158. Raz, S.; Zeigler, M.; Caine, M. Pharmacological receptors in the prostate. *Br. J. Urol.* **1973**, *45*, 663–667. [CrossRef]
159. Ricke, W.A.; Lee, C.W.; Clapper, T.R.; Schneider, A.J.; Moore, R.W.; Keil, K.P.; Abler, L.L.; Wynder, J.L.; Lopez Alvarado, A.; Beaubrun, I.; et al. In Utero and lactational TCDD exposure increases susceptibility to lower urinary tract dysfunction in adulthood. *Toxicol. Sci.* **2016**, *150*, 429–440. [CrossRef]
160. Turco, A.E.; Oakes, S.R.; Keil Stietz, K.P.; Dunham, C.L.; Joseph, D.B.; Chathurvedula, T.S.; Girardi, N.M.; Schneider, A.J.; Gawdzik, J.; Sheftel, C.M.; et al. A mechanism linking perinatal 2,3,7,8 tetrachlorodibenzo-p-dioxin exposure to lower urinary tract dysfunction in adulthood. *Dis. Models Mech.* **2021**, *14*, dmm049068. [CrossRef]
161. Gupta, A.; Ketchum, N.; Roehrborn, C.G.; Schecter, A.; Aragaki, C.C.; Michalek, J.E. Serum dioxin, testosterone, and subsequent risk of benign prostatic hyperplasia: A prospective cohort study of Air Force veterans. *Environ. Health Perspect.* **2006**, *114*, 1649–1654. [CrossRef]
162. Gupta, A.; Gupta, S.; Pavuk, M.; Roehrborn, C.G. Anthropometric and metabolic factors and risk of benign prostatic hyperplasia: A prospective cohort study of Air Force veterans. *Urology* **2006**, *68*, 1198–1205. [CrossRef]
163. Steers, W.D. Pathophysiology of overactive bladder and urge urinary incontinence. *Rev. Urol.* **2002**, *4*, S7. [PubMed]
164. De Nunzio, C.; Presicce, F.; Lombardo, R.; Carter, S.; Vicentini, C.; Tubaro, A. Detrusor overactivity increases bladder wall thickness in male patients: A urodynamic multicenter cohort study. *Neurourol. Urodyn.* **2017**, *36*, 1616–1621. [CrossRef] [PubMed]
165. Mumtaz, M.M.; Farooqui, M.Y.; Ghanayem, B.I.; Ahmed, A.E. The urotoxic effects of N,N′-dimethylaminopropionitrile. In vivo and in vitro metabolism. *Toxicol. Appl. Pharmacol.* **1991**, *110*, 61–69. [CrossRef]
166. Hara, N.; Saito, H.; Takahashi, K.; Takeda, M. Lower urinary tract symptoms in patients with Niigata Minamata disease: A case-control study 50 years after methyl mercury pollution. *Int. J. Urol.* **2013**, *20*, 610–615. [CrossRef]
167. Eto, K. Pathology of Minamata disease. *Toxicol. Pathol.* **1997**, *25*, 614–623. [CrossRef]
168. Smith, R.A.; Christenson, W.R.; Bartels, M.J.; Arnold, L.L.; St John, M.K.; Cano, M.; Garland, E.M.; Lake, S.G.; Wahle, B.S.; McNett, D.A.; et al. Urinary physiologic and chemical metabolic effects on the urothelial cytotoxicity and potential DNA adducts of o-phenylphenol in male rats. *Toxicol. Appl. Pharmacol.* **1998**, *150*, 402–413. [CrossRef]
169. Abrams, P.; Cardozo, L.; Fall, M.; Griffiths, D.; Rosier, P.; Ulmsten, U.; Van Kerrebroeck, P.; Victor, A.; Wein, A. The standardisation of terminology of lower urinary tract function: Report from the Standardisation Sub-committee of the International Continence Society. *Am. J. Obstet. Gynecol.* **2002**, *187*, 116–126. [CrossRef]
170. Miyazato, M.; Yoshimura, N.; Chancellor, M.B. The other bladder syndrome: Underactive bladder. *Rev. Urol.* **2013**, *15*, 11–22.
171. Kim, S.J.; Kim, J.; Na, Y.G.; Kim, K.H. Irreversible bladder remodeling induced by fibrosis. *Int. Neurourol. J.* **2021**, *25*, S3–S7. [CrossRef]
172. Feng, J.; Gao, J.; Zhou, S.; Liu, Y.; Zhong, Y.; Shu, Y.; Meng, M.S.; Yan, J.; Sun, D.; Fang, Q.; et al. Role of stem cell factor in the regulation of ICC proliferation and detrusor contraction in rats with an underactive bladder. *Mol. Med. Rep.* **2017**, *16*, 1516–1522. [CrossRef] [PubMed]
173. Castro-Diaz, D.; Taracena Lafuente, J.M. Detrusor-sphincter dyssynergia. *Int. J. Clin. Pract. Suppl.* **2006**, *151*, 17–21. [CrossRef]
174. Stoffel, J.T. Detrusor sphincter dyssynergia: A review of physiology, diagnosis, and treatment strategies. *Transl. Androl. Urol.* **2016**, *5*, 127–135. [PubMed]
175. Amarenco, G.; Sheikh Ismaël, S.; Chesnel, C.; Charlanes, A.; Le Breton, F. Diagnosis and clinical evaluation of neurogenic bladder. *Eur. J. Phys. Rehabil. Med.* **2017**, *53*, 975–980. [CrossRef] [PubMed]
176. Ito, H.; Yoshiyasu, T.; Yamaguchi, O.; Yokoyama, O. Male lower urinary tract symptoms: Hypertension as a risk factor for storage symptoms, but not voiding symptoms. *Lower Urin. Tract Symptoms* **2012**, *4*, 68–72. [CrossRef]
177. Wang, X.; Su, Y.; Yang, C.; Hu, Y.; Dong, J.-Y. Benign prostatic hyperplasia and cardiovascular risk: A prospective study among chinese men. *World J. Urol.* **2021**, *40*, 177–183. [CrossRef]
178. Hammarsten, J.; Högstedt, B. Clinical, anthropometric, metabolic and insulin profile of men with fast annual growth rates of benign prostatic hyperplasia. *Blood Press.* **1999**, *8*, 29–36.

179. Abler, L.L.; Vezina, C.M. Links between lower urinary tract symptoms, intermittent hypoxia and diabetes: Causes or cures? *Respir. Physiol. Neurobiol.* **2018**, *256*, 87–96. [CrossRef]
180. Lejay, A.; Fang, F.; John, R.; Van, J.A.D.; Barr, M.; Thaveau, F.; Chakfe, N.; Geny, B.; Scholey, J.W. Ischemia reperfusion injury, ischemic conditioning and diabetes mellitus. *J. Mol. Cell. Cardiol.* **2016**, *91*, 11–22. [CrossRef]
181. Vinik, A.I.; Maser, R.E.; Mitchell, B.D.; Freeman, R. Diabetic autonomic neuropathy. *Diabetes Care* **2003**, *26*, 1553–1579. [CrossRef]
182. Morgan, B.J. Vascular consequences of intermittent hypoxia. In *Hypoxia and the Circulation*; Springer: Berlin/Heidelberg, Germany, 2007; pp. 69–84.
183. Andersson, K.E.; Boedtkjer, D.B.; Forman, A. The link between vascular dysfunction, bladder ischemia, and aging bladder dysfunction. *Ther. Adv. Urol.* **2017**, *9*, 11–27. [CrossRef] [PubMed]
184. Lind, L.; Lind, P.M. Can persistent organic pollutants and plastic-associated chemicals cause cardiovascular disease? *J. Int. Med.* **2012**, *271*, 537–553. [CrossRef]
185. Meneguzzi, A.; Fava, C.; Castelli, M.; Minuz, P. Exposure to perfluoroalkyl chemicals and cardiovascular disease: Experimental and epidemiological evidence. *Front. Endocrinol.* **2021**, *12*, 850. [CrossRef] [PubMed]
186. Velmurugan, G.; Ramprasath, T.; Gilles, M.; Swaminathan, K.; Ramasamy, S. Gut microbiota, endocrine-disrupting chemicals, and the diabetes epidemic. *Trends Endocrinol. Metab.* **2017**, *28*, 612–625. [CrossRef] [PubMed]
187. Muller, W.J.; Lee, F.S.; Dickson, C.; Peters, G.; Pattengale, P.; Leder, P. The int-2 gene product acts as an epithelial growth factor in transgenic mice. *EMBO J.* **1990**, *9*, 907–913. [CrossRef]
188. Konno-Takahashi, N.; Takeuchi, T.; Nishimatsu, H.; Kamijo, T.; Tomita, K.; Schalken, J.A.; Teshima, S.; Kitamura, T. Engineered FGF-2 expression induces glandular epithelial hyperplasia in the murine prostatic dorsal lobe. *Eur. Urol.* **2004**, *46*, 126–132. [CrossRef]
189. Tutrone, R.F., Jr.; Ball, R.A.; Ornitz, D.M.; Leder, P.; Richie, J.P. Benign prostatic hyperplasia in a transgenic mouse: A new hormonally sensitive investigatory model. *J. Urol.* **1993**, *149*, 633–639. [CrossRef]
190. Wennbo, H.; Kindblom, J.; Isaksson, O.G.; Törnell, J. Transgenic mice overexpressing the prolactin gene develop dramatic enlargement of the prostate gland. *Endocrinology* **1997**, *138*, 4410–4415. [CrossRef]
191. Pigat, N.; Reyes-Gomez, E.; Boutillon, F.; Palea, S.; Barry Delongchamps, N.; Koch, E.; Goffin, V. Combined sabal and urtica extracts (WS((R)) 1541) exert anti-proliferative and anti-inflammatory effects in a mouse model of benign prostate hyperplasia. *Front. Pharmacol.* **2019**, *10*, 311. [CrossRef]
192. Bernichtein, S.; Pigat, N.; Camparo, P.; Latil, A.; Viltard, M.; Friedlander, G.; Goffin, V. Anti-inflammatory properties of lipidosterolic extract of serenoa repens (Permixon(R)) in a mouse model of prostate hyperplasia. *Prostate* **2015**, *75*, 706–722. [CrossRef]
193. Dillner, K.; Kindblom, J.; Flores-Morales, A.; Pang, S.T.; Tornell, J.; Wennbo, H.; Norstedt, G. Molecular characterization of prostate hyperplasia in prolactin-transgenic mice by using cDNA representational difference analysis. *Prostate* **2002**, *52*, 139–149. [CrossRef] [PubMed]
194. Vital, P.; Castro, P.; Tsang, S.; Ittmann, M. The senescence-associated secretory phenotype promotes benign prostatic hyperplasia. *Am. J. Pathol.* **2014**, *184*, 721–731. [CrossRef] [PubMed]
195. George, J.W.; Patterson, A.L.; Tanwar, P.S.; Kajdacsy-Balla, A.; Prins, G.S.; Teixeira, J.M. Specific deletion of LKB1/Stk11 in the Mullerian duct mesenchyme drives hyperplasia of the periurethral stroma and tumorigenesis in male mice. *Proc. Natl. Acad. Sci. USA* **2017**, *114*, 3445–3450. [CrossRef] [PubMed]
196. Wegner, K.A.; Mueller, B.R.; Unterberger, C.J.; Avila, E.J.; Ruetten, H.; Turco, A.E.; Oakes, S.R.; Girardi, N.M.; Halberg, R.B.; Swanson, S.M.; et al. Prostate epithelial-specific expression of activated PI3K drives stromal collagen production and accumulation. *J. Pathol.* **2020**, *250*, 231–242. [CrossRef]
197. Turkbey, B.; Huang, R.; Vourganti, S.; Trivedi, H.; Bernardo, M.; Yan, P.; Benjamin, C.; Pinto, P.A.; Choyke, P.L. Age-related changes in prostate zonal volumes as measured by high-resolution magnetic resonance imaging (MRI): A cross-sectional study in over 500 patients. *BJU Int.* **2012**, *110*, 1642–1647. [CrossRef]
198. Nickel, J.C.; Roehrborn, C.G.; O'Leary, M.P.; Bostwick, D.G.; Somerville, M.C.; Rittmaster, R.S. The relationship between prostate inflammation and lower urinary tract symptoms: Examination of baseline data from the REDUCE trial. *Eur. Urol.* **2008**, *54*, 1379–1384. [CrossRef]
199. Nickel, J.C. Prostatitis. *Can. Urol. Assoc. J.* **2011**, *5*, 306–315. [CrossRef]
200. Aiello, S.E.; Moses, M.A.; Allen, D.G. *The Merck Veterinary Manual*, 11th ed.; Merck & Co. Inc.: Kenilworth, NJ, USA, 2016; p. 3325.
201. Krieger, J.N.; Riley, D.E. Bacteria in the chronic prostatitis-chronic pelvic pain syndrome: Molecular approaches to critical research questions. *J. Urol.* **2002**, *167*, 2574–2583. [CrossRef]
202. Elkahwaji, J.E.; Ott, C.J.; Janda, L.M.; Hopkins, W.J. Mouse model for acute bacterial prostatitis in genetically distinct inbred strains. *Urology* **2005**, *66*, 883–887. [CrossRef]
203. Rudick, C.N.; Berry, R.E.; Johnson, J.R.; Johnston, B.; Klumpp, D.J.; Schaeffer, A.J.; Thumbikat, P. Uropathogenic *Escherichia coli* induces chronic pelvic pain. *Infect. Immun.* **2011**, *79*, 628–635. [CrossRef]
204. Nowicki, B.; Singhal, J.; Fang, L.; Nowicki, S.; Yallampalli, C. Inverse relationship between severity of experimental pyelonephritis and nitric oxide production in C3H/HeJ mice. *Infect. Immun.* **1999**, *67*, 2421–2427. [CrossRef] [PubMed]

205. Ruetten, H.; Sandhu, J.; Mueller, B.; Wang, P.; Zhang, H.L.; Wegner, K.A.; Cadena, M.; Sandhu, S.; Abler, L.L.; Zhu, J.; et al. A uropathogenic *E. coli* UTI89 model of prostatic inflammation and collagen accumulation for use in studying aberrant collagen production in the prostate. *Am. J. Physiol. Ren. Physiol.* **2021**, *320*, F31–F46. [CrossRef] [PubMed]
206. Bell-Cohn, A.; Mazur, D.J.; Hall, C.C.; Schaeffer, A.J.; Thumbikat, P. Uropathogenic *Escherichia coli*-induced fibrosis, leading to lower urinary tract symptoms, is associated with type-2 cytokine signaling. *Am. J. Physiol. Ren. Physiol.* **2019**, *316*, F682–F692. [CrossRef]
207. Wang, H.H.; Wang, L.; Jerde, T.J.; Chan, B.D.; Savran, C.A.; Burcham, G.N.; Crist, S.; Ratliff, T.L. Characterization of autoimmune inflammation induced prostate stem cell expansion. *Prostate* **2015**, *75*, 1620–1631. [CrossRef] [PubMed]
208. Motrich, R.D.; Maccioni, M.; Molina, R.; Tissera, A.; Olmedo, J.; Riera, C.M.; Rivero, V.E. Presence of INFgamma-secreting lymphocytes specific to prostate antigens in a group of chronic prostatitis patients. *Clin. Immunol.* **2005**, *116*, 149–157. [CrossRef]
209. Habermacher, G.M.; Chason, J.T.; Schaeffer, A.J. Prostatitis/chronic pelvic pain syndrome. *Annu. Rev. Med.* **2006**, *57*, 195–206. [CrossRef]
210. Lees, J.R.; Charbonneau, B.; Hayball, J.D.; Diener, K.; Brown, M.; Matusik, R.; Cohen, M.B.; Ratliff, T.L. T-cell recognition of a prostate specific antigen is not sufficient to induce prostate tissue destruction. *Prostate* **2006**, *66*, 578–590. [CrossRef]
211. Lees, J.R.; Charbonneau, B.; Swanson, A.K.; Jensen, R.; Zhang, J.; Matusik, R.; Ratliff, T.L. Deletion is neither sufficient nor necessary for the induction of peripheral tolerance in mature CD8+ T cells. *Immunology* **2006**, *117*, 248–261. [CrossRef]
212. Done, J.D.; Rudick, C.N.; Quick, M.L.; Schaeffer, A.J.; Thumbikat, P. Role of mast cells in male chronic pelvic pain. *J. Urol.* **2012**, *187*, 1473–1482. [CrossRef]
213. Rudick, C.N.; Schaeffer, A.J.; Thumbikat, P. Experimental autoimmune prostatitis induces chronic pelvic pain. *Am. J. Physiol. Regul. Integr. Comp. Physiol.* **2008**, *294*, R1268–R1275. [CrossRef]
214. Lang, M.D.; Nickel, J.C.; Olson, M.E.; Howard, S.R.; Ceri, H. Rat model of experimentally induced abacterial prostatitis. *Prostate* **2000**, *45*, 201–206. [CrossRef]
215. Funahashi, Y.; O'Malley, K.J.; Kawamorita, N.; Tyagi, P.; DeFranco, D.B.; Takahashi, R.; Gotoh, M.; Wang, Z.; Yoshimura, N. Upregulation of androgen-responsive genes and transforming growth factor-β1 cascade genes in a rat model of non-bacterial prostatic inflammation. *Prostate* **2014**, *74*, 337–345. [CrossRef] [PubMed]
216. Mizoguchi, S.; Mori, K.; Wang, Z.; Liu, T.; Funahashi, Y.; Sato, F.; DeFranco, D.B.; Yoshimura, N.; Mimata, H. Effects of estrogen receptor β stimulation in a rat model of non-bacterial prostatic inflammation. *Prostate* **2017**, *77*, 803–811. [CrossRef] [PubMed]
217. Funahashi, Y.; Wang, Z.; O'Malley, K.J.; Tyagi, P.; DeFranco, D.B.; Gingrich, J.R.; Takahashi, R.; Majima, T.; Gotoh, M.; Yoshimura, N. Influence of *E. coli*-induced prostatic inflammation on expression of androgen-responsive genes and transforming growth factor beta 1 cascade genes in rats. *Prostate* **2015**, *75*, 381–389. [CrossRef]
218. Torkko, K.C.; Wilson, R.S.; Smith, E.E.; Kusek, J.W.; van Bokhoven, A.; Lucia, M.S. Prostate biopsy markers of inflammation are associated with risk of clinical progression of benign prostatic hyperplasia: Finding for the MTOPS study. *J. Urol.* **2015**, *194*, 454–461. [CrossRef]
219. Pontari, M.A.; Ruggieri, M.R. Mechanisms in prostatitis/chronic pelvic pain syndrome. *J. Urol.* **2004**, *172*, 839–845. [CrossRef]
220. Nadler, R.B.; Koch, A.E.; Calhoun, E.A.; Campbell, P.L.; Pruden, D.L.; Bennett, C.L.; Yarnold, P.R.; Schaeffer, A.J. IL-1beta and TNF-alpha in prostatic secretions are indicators in the evaluation of men with chronic prostatitis. *J. Urol.* **2000**, *164*, 214–218. [CrossRef]
221. Huang, T.R.; Li, W.; Peng, B. Correlation of inflammatory mediators in prostatic secretion with chronic prostatitis and chronic pelvic pain syndrome. *Andrologia* **2018**, *50*, e12860. [CrossRef]
222. Ricote, M.; García-Tuñón, I.; Bethencourt, F.R.; Fraile, B.; Paniagua, R.; Royuela, M. Interleukin-1 (IL-1alpha and IL-1beta) and its receptors (IL-1RI, IL-1RII, and IL-1Ra) in prostate carcinoma. *Cancer* **2004**, *100*, 1388–1396. [CrossRef]
223. Ashok, A.; Keener, R.; Rubenstein, M.; Stookey, S.; Bajpai, S.; Hicks, J.; Alme, A.K.; Drake, C.G.; Zheng, Q.; Trabzonlu, L.; et al. Consequences of interleukin 1β-triggered chronic inflammation in the mouse prostate gland: Altered architecture associated with prolonged CD4(+) infiltration mimics human proliferative inflammatory atrophy. *Prostate* **2019**, *79*, 732–745. [CrossRef]
224. Hao, L.; Thomas, S.; Greer, T.; Vezina, C.M.; Bajpai, S.; Ashok, A.; De Marzo, A.M.; Bieberich, C.J.; Li, L.; Ricke, W.A. Quantitative proteomic analysis of a genetically induced prostate inflammation mouse model via custom 4-plex DiLeu isobaric labeling. *Am. J. Physiol. Ren. Physiol.* **2019**, *316*, F1236–F1243. [CrossRef]
225. Hao, L.; Shi, Y.; Thomas, S.; Vezina, C.M.; Bajpai, S.; Ashok, A.; Bieberich, C.J.; Ricke, W.A.; Li, L. Comprehensive urinary metabolomic characterization of a genetically induced mouse model of prostatic inflammation. *Int. J. Mass Spectrom.* **2018**, *434*, 185–192. [CrossRef] [PubMed]
226. O'Malley, K.J.; Eisermann, K.; Pascal, L.E.; Parwani, A.V.; Majima, T.; Graham, L.; Hrebinko, K.; Acquafondata, M.; Stewart, N.A.; Nelson, J.B.; et al. Proteomic analysis of patient tissue reveals PSA protein in the stroma of benign prostatic hyperplasia. *Prostate* **2014**, *74*, 892–900. [CrossRef] [PubMed]
227. Pascal, L.E.; Mizoguchi, S.; Chen, W.; Rigatti, L.H.; Igarashi, T.; Dhir, R.; Tyagi, P.; Wu, Z.; Yang, Z.; de Groat, W.C.; et al. Prostate-specific deletion of cdh1 induces murine prostatic inflammation and bladder overactivity. *Endocrinology* **2021**, *162*, bqaa212. [CrossRef]
228. Eljamal, K.; Kajioka, S.; Maki, T.; Ushijima, M.; Kawagoe, K.; Lee, K.; Sasaguri, T. New mouse model of underactive bladder developed by placement of a metal ring around the bladder neck. *Lower Urin. Tract Symptoms* **2021**, *13*, 299–307. [CrossRef] [PubMed]

229. Austin, J.C.; Chacko, S.K.; DiSanto, M.; Canning, D.A.; Zderic, S.A. A male murine model of partial bladder outlet obstruction reveals changes in detrusor morphology, contractility and Myosin isoform expression. *J. Urol.* **2004**, *172*, 1524–1528. [CrossRef]
230. Taylor, J.A.; Zhu, Q.; Irwin, B.; Maghaydah, Y.; Tsimikas, J.; Pilbeam, C.; Leng, L.; Bucala, R.; Kuchel, G.A. Null mutation in macrophage migration inhibitory factor prevents muscle cell loss and fibrosis in partial bladder outlet obstruction. *Am. J. Physiol. Ren. Physiol.* **2006**, *291*, F1343–F1353. [CrossRef]
231. Chen, J.; Drzewiecki, B.A.; Merryman, W.D.; Pope, J.C. Murine bladder wall biomechanics following partial bladder obstruction. *J. Biomechem.* **2013**, *46*, 2752–2755. [CrossRef]
232. Vasquez, E.; Cristofaro, V.; Lukianov, S.; Burkhard, F.C.; Gheinani, A.H.; Monastyrskaya, K.; Bielenberg, D.R.; Sullivan, M.P.; Adam, R.M. Deletion of neuropilin 2 enhances detrusor contractility following bladder outlet obstruction. *JCI Insight* **2017**, *2*, e90617. [CrossRef]
233. Streng, T.; Li, X.; Lehtoranta, M.; Makela, S.; Poutanen, M.; Talo, A.; Tekmal, R.R.; Santti, R. Infravesical obstruction in aromatase over expressing transgenic male mice with increased ratio of serum estrogen-to-androgen concentration. *J. Urol.* **2002**, *168*, 298–302. [CrossRef]
234. Wegner, K.A.; Ruetten, H.; Girardi, N.M.; O'Driscoll, C.A.; Sandhu, J.K.; Turco, A.E.; Abler, L.L.; Wang, P.; Wang, Z.; Bjorling, D.E.; et al. Genetic background but not prostatic epithelial beta-catenin influences susceptibility of male mice to testosterone and estradiol-induced urinary dysfunction. *Am. J. Clin. Exp. Urol.* **2021**, *9*, 121–131.
235. Konkol, Y.; Vuorikoski, H.; Streng, T.; Tuomela, J.; Bernoulli, J. Characterization a model of prostatic diseases and obstructive voiding induced by sex hormone imbalance in the Wistar and Noble rats. *Transl. Androl. Urol.* **2019**, *8*, S45–S57. [CrossRef] [PubMed]
236. Gopal, M.; Sammel, M.D.; Arya, L.A.; Freeman, E.W.; Lin, H.; Gracia, C. Association of change in estradiol to lower urinary tract symptoms during the menopausal transition. *Obstet. Gynecol.* **2008**, *112*, 1045–1052. [CrossRef]
237. Nicholson, T.M.; Moses, M.A.; Uchtmann, K.S.; Keil, K.P.; Bjorling, D.E.; Vezina, C.M.; Wood, R.W.; Ricke, W.A. Estrogen receptor-alpha is a key mediator and therapeutic target for bladder complications of benign prostatic hyperplasia. *J. Urol.* **2015**, *193*, 722–729. [CrossRef] [PubMed]
238. Buhl, A.E.; Yuan, Y.D.; Cornette, J.C.; Frielink, R.D.; Knight, K.A.; Ruppel, P.L.; Kimball, F.A. Steroid-induced urogenital tract changes and urine retention in laboratory rodents. *J. Urol.* **1985**, *134*, 1262–1267. [CrossRef]
239. Collins, D.E.; Mulka, K.R.; Hoenerhoff, M.J.; Taichman, R.S.; Villano, J.S. Clinical assessment of urinary tract damage during sustained-release estrogen supplementation in mice. *Comp. Med.* **2017**, *67*, 11–21. [PubMed]
240. Van Steenbrugge, G.J.; Groen, M.; de Jong, F.H.; Schroeder, F.H. The use of steroid-containing silastic implants in male nude mice: Plasma hormone levels and the effect of implantation on the weights of the ventral prostate and seminal vesicles. *Prostate* **1984**, *5*, 639–647. [CrossRef]
241. Nomiya, M.; Yamaguchi, O.; Andersson, K.-E.; Sagawa, K.; Aikawa, K.; Shishido, K.; Yanagida, T.; Kushida, N.; Yazaki, J.; Takahashi, N. The effect of atherosclerosis-induced chronic bladder ischemia on bladder function in the rat. *Neurourol. Urodyn.* **2012**, *31*, 195–200. [CrossRef]
242. Yoshida, A.; Kageyama, A.; Fujino, T.; Nozawa, Y.; Yamada, S. Loss of muscarinic and purinergic receptors in urinary bladder of rats with hydrochloric acid-induced cystitis. *Urology* **2010**, *76*, 1017-e7. [CrossRef]
243. Kakizaki, H.; de Groat William, C. Role of spinal nitric oxide in the facilitation of the micturition reflex by bladder irritation. *J. Urol.* **1996**, *155*, 355–360. [CrossRef]
244. Mitsui, T.; Kakizaki, H.; Matsuura, S.; Ameda, K.; Yoshioka, M.; Koyanagi, T. Afferent fibers of the hypogastric nerves are involved in the facilitating effects of chemical bladder irritation in rats. *J. Neurophysiol.* **2001**, *86*, 2276–2284. [CrossRef] [PubMed]
245. Tai, C.; Shen, B.; Chen, M.; Wang, J.; Liu, H.; Roppolo, J.R.; de Groat, W.C. Suppression of bladder overactivity by activation of somatic afferent nerves in the foot. *BJU Int.* **2011**, *107*, 303. [CrossRef] [PubMed]
246. Choudhary, M.; van Asselt, E.; van Mastrigt, R.; Clavica, F. Neurophysiological modeling of bladder afferent activity in the rat overactive bladder model. *J. Physiol. Sci.* **2015**, *65*, 329–338. [CrossRef] [PubMed]
247. Wang, Z.; Liao, L. Improvement in detrusor-sphincter dyssynergia by bladder-wall injection of replication-defective herpes simplex virus vector-mediated gene delivery of kynurenine aminotransferase II in spinal cord injury rats. *Spinal Cord* **2017**, *55*, 155–161. [CrossRef]
248. Saito, T.; Gotoh, D.; Wada, N.; Tyagi, P.; Minagawa, T.; Ogawa, T.; Ishizuka, O.; Yoshimura, N. Time-dependent progression of neurogenic lower urinary tract dysfunction after spinal cord injury in the mouse model. *Am. J. Physiol. Ren. Physiol.* **2021**, *321*, F26–F32. [CrossRef]

Article

Developmental Phenotypic and Transcriptomic Effects of Exposure to Nanomolar Levels of 4-Nonylphenol, Triclosan, and Triclocarban in Zebrafish (*Danio rerio*)

Jessica Phillips [1,2], Alex S. Haimbaugh [1,2], Camille Akemann [1,2], Jeremiah N. Shields [1], Chia-Chen Wu [1,3], Danielle N. Meyer [1,2,3], Bridget B. Baker [1,4], Zoha Siddiqua [5], David K. Pitts [5] and Tracie R. Baker [1,2,3,*]

1. Institute of Environmental Health Sciences, Wayne State University, Detroit, MI 48202, USA; jessica.montrief@wayne.edu (J.P.); alexhaim@wayne.edu (A.S.H.); cakemann14@gmail.com (C.A.); jshields@wayne.edu (J.N.S.); chiachenwu@ufl.edu (C.-C.W.); danielle.meyer@ufl.edu (D.N.M.); bridgetbaker@ufl.edu (B.B.B.)
2. Department of Pharmacology, Wayne State University, Detroit, MI 28201, USA
3. Department of Environmental and Global Health, University of Florida, Gainesville, FL 32610, USA
4. Department of Wildlife Ecology and Conservation, University of Florida, Gainesville, FL 32610, USA
5. Department of Pharmaceutical Sciences, College of Pharmacy and Health Sciences, Wayne State University, Detroit, MI 48202, USA; zoha.siddiqua@wayne.edu (Z.S.); pitts@wayne.edu (D.K.P.)
*   Correspondence: tracie.baker@ufl.edu

Abstract: Triclosan, triclocarban and 4-nonylphenol are all chemicals of emerging concern found in a wide variety of consumer products that have exhibited a wide range of endocrine-disrupting effects and are present in increasing amounts in groundwater worldwide. Results of the present study indicate that exposure to these chemicals at critical developmental periods, whether long-term or short-term in duration, leads to significant mortality, morphologic, behavioral and transcriptomic effects in zebrafish (*Danio rerio*). These effects range from total mortality with either long- or short-term exposure at 100 and 1000 nM of triclosan, to abnormalities in uninflated swim bladder seen with long-term exposure to triclocarban and short-term exposure to 4-nonylphenol, and cardiac edema seen with short-term 4-nonylphenol exposure. Additionally, a significant number of genes involved in neurological and cardiovascular development were differentially expressed after the exposures, as well as lipid metabolism genes and metabolic pathways after exposure to each chemical. Such changes in behavior, gene expression, and pathway abnormalities caused by these three known endocrine disruptors have the potential to impact not only the local ecosystem, but human health as well.

Keywords: triclosan; triclocarbon; detergents; 4-nonylphenol; *Danio rerio*; zebrafish; environmental toxicity; development; aquatic environment; ground water chemicals

1. Introduction

Since the Industrial Revolution, humans have created and used chemicals as a part of technological advancements necessary to meet the demands of the world's exponentially growing population. Unfortunately, most chemicals are put into use before the full extent of their health effects on humans and wildlife is known. In the past, chemicals have been released into the environment via dumping in lakes, rivers, or into the ground before regulations like the Resource Conservation and Recovery Act (RCRA) were put in place for chemical waste disposal. Despite regulations like these, chemicals are still used and released into the environment, and many are resistant to degradation with the potential to cause adverse effects on local ecosystems that extend to wildlife and humans. For instance, the U.S. Food and Drug Administration released a Final Rule in 2016 regarding the antimicrobial triclosan, that concluded human and ecosystem health is not sufficiently protected from the adverse impacts of antimicrobial and antiseptic chemicals by existing regulatory practices [1].

As research increases our knowledge of these environmental contaminants and their properties, we are discovering that many are endocrine disrupting chemicals (EDCs). According to the National Institute of Environmental Health Sciences (NIEHS), EDCs are "chemicals that may interfere with the body's endocrine system and produce adverse developmental, reproductive, neurological, and immune effects in both humans and wildlife" [1]. Sources of EDCs are wide ranging and include industrial processes, personal care products, pharmaceuticals, pesticides, and more. Triclosan and triclocarban are antimicrobial agents now banned in soaps in the United States due to evidence that they do not prevent disease or improve health but are toxic and carcinogenic, mainly via endocrine disruption [2]. Nonetheless, they are still found in personal care products such as lotions, deodorants, and toothpaste, such as Colgate, which contains 10 mM of triclosan, then enter the environment mainly through wastewater effluent [3,4] where they accumulate due to their resistance to biodegradation, with an approximate average of 200 ng/L found for both in U.S. surface waters [5]. Triclosan and triclocarban also have the ability to cross the placental barrier [6]. Pregnant rats exposed to triclosan had dramatically decreased serum levels of estradiol, progesterone, and prolactin, as well as differential expression of genes responsible for hormone biosynthesis targeting the placenta [7]. Studies have shown that triclosan exposure may be linked to decreased oocyte implantation in women struggling with fertility [8] and may have placental endocrine effects [9]. In fact, human fecundity decreases when triclosan concentrations exceed 75 ng/mL in urine [10]. Human studies also showed disruption of thyroid hormones, specifically triclosan has a positive association with circulating levels of triiodothyronine [11]. Similarly, triclocarban exposure disrupts thyroid hormones in human cell lines and frogs [12], increases androgenic activity in human cell lines [13], reduces female plasma vitellogenin and estradiol in female fathead minnows, resulting in approximately half the cumulative egg production compared to unexposed fish, and decreases testosterone while increasing estradiol in male fathead minnows at environmentally relevant levels [14].

4-nonylphenol is another known EDC that can be found in a wide range of products including fungicides, food packaging, toys, clothes, jewelry, and cosmetics [15]. 4-nonylphenol causes health effects such as liver toxicity and steatosis in male rats [16] as well as induces hormone disruption by inhibiting progesterone and androstenedione [17]. Additionally, exposure to 4-nonylphenol during critical developmental periods such as puberty leads to decreased testosterone production and spermatogenesis, as well as increased morphological sperm abnormalities in rats [18]. Humans are not only exposed to these chemicals directly in consumer products, but also via the environment where they are found ubiquitously in surface water, drinking water, wastewater effluent, soil, and wildlife [19–24].

Although these EDCs have been studied in many different organisms, little is known about the health consequences of exposure to environmentally relevant levels at critical developmental windows, and the acute and later life effects of these early life exposures. The purpose of this study is to evaluate the health risks of early life exposure to EDCs commonly found at low but persistent levels in the environment. Thus, we exposed zebrafish (*Danio rerio*) larvae to triclosan, triclocarban, or 4-nonylphenol during early development, and examined its effects on morphology, behavior, and gene expression. Zebrafish are an NIH-approved human model because their genome shares 70% homology with the human genome. Organogenesis takes place within the first 42 h post-fertilization, whereas hatching occurs around 48 h post-fertilization, at which point the larval stage begins [25]. Zebrafish can also produce hundreds of eggs per week, making them an excellent model for high-throughput screening for multiple chemicals and endpoints. Additionally, since zebrafish are swimming freely by 3 days post-fertilization (dpf), we can measure neurobehavioral endpoints in early life. We found that exposure to triclocarban, triclosan, and 4-nonylphenol during these two critical development periods, embryogenesis and the early larval stage, can have detrimental morphological, behavioral, and transcriptomic effects, providing insight into timing of exposure, targets, and mechanisms of EDC toxicity.

2. Material and Methods

2.1. Animal Husbandry

The adult zebrafish (wild-type AB strain) used to spawn the larval fish for the experiments were maintained on a 14:10 light/dark cycle in reverse osmosis (RO) water buffered with salt (Instant Ocean© Spectrum Brands, Blacksburg, VA, USA) with temperature maintained at 27 °C–30 °C in a recirculating system (Aquaneering, San Diego, CA, USA). Adult fish were monitored and fed fish flakes twice daily (Aquatox Fish Diet, Zeigler Bros Inc., Gardners, PA, USA), with feeding supplemented by brine shrimp. Adult zebrafish were bred in spawning tanks with a sex ratio of 1 male to 2 females, and embryos were collected 4 h post-fertilization (hpf). The embryos were cleaned with bleach 58 ppm for 5 min (Clorox Company, Oakland, CA, USA), rinsed with RO water and then egg water (600 mg/L salt in RO water), sorted into exposure groups for their respective chemicals, and incubated at 28 °C. Zebrafish use protocols were approved by the Institutional Animal Care and Use Committee at Wayne State University, according to the National Institutes of Health Guide to the Care and Use of Laboratory Animals (Protocol 16-03-054; approved 4 August 2016).

2.2. Chemical Exposures

4-nonylphenol (CAS# 104-40-5, Sigma Aldrich, USA), triclocarban (CAS# 101-20-2, U.S. Pharmacopeia, Rockville, MD, USA), and triclosan (CAS# 3380-34-5, U.S. Pharmacopeia, Rockville, MD, USA) were used to prepare stock solutions. The 4-nonylphenol solutions were prepared in fish water (60 mg/L salt in RO water) at concentrations of 0.1, 1, 10, 100 and 1000 nM. The triclosan solutions were prepared in acetone at concentrations of 0.1, 1, 10, 100 and 1000 nM. The triclocarbon solutions were prepared in acetone at concentrations of 0.01, 0.1, 1, 10 and 100 nM. Control fish for the triclosan and triclocarban exposures were placed in vehicle (0.01% acetone (v/v) in RO water). The chemical solutions for the exposures were prepared daily from aliquots of the stock solutions. Larval chemical exposures were performed in 6-well plates, with embryos at a density of 30 per well and per exposure concentration, at one of two different time periods: either 120 h from 4 hpf to 5 days post-fertilization (dpf; "long-term exposure"); or 24 h at 4–5 dpf ("short-term exposure"). This was replicated 5 times, for a total of 150 larval fish exposed for each chemical, concentration, and duration of exposure. Approximately 90% of the chemical solution was removed from each well daily and replenished with freshly prepared chemical solution. After the exposure period, larval fish were rinsed 3 times with egg water to end the chemical exposure.

2.3. Abnormality Screening

Zebrafish embryos were screened at 24, 48, 72, 96, and 120 hpf for mortality and morphological abnormalities under stereomicroscope (M165C, Leica Microsystem, Wetzlar, Germany). The mortality endpoints assessed were coagulation of embryo and lack of heartbeat. The abnormality endpoints assessed were number of unhatched embryos compared to hatched embryos, skeletal deformities, improperly inflated swim bladder, yolk sac edema, cardiac edema, and total abnormalities. Embryos were screened using 6.7× magnification, with detailed evaluation occurring at a magnification of 20×. Results were analyzed for each chemical using a Chi-Square test with significance set at $p < 0.05$, with pairwise comparison with Bonferroni corrections.

2.4. Behavioral Analysis

At 5 dpf, 24 larval zebrafish from each exposure group and control groups with inflated swim bladders and without any morphological abnormalities were tested with 1 larva per well in a 24-well plate with 2 mL of fish water per plate. Each plate was allowed to acclimate for at least an hour at 27 °C before being placed in to the DanioVision Chamber

(Noldus Information Technology, Wageningen, The Netherlands) to undergo the behavioral assay. The behavioral assay consisted of 3 min light and dark alternating periods, with a total of four light-dark cycles (24 min in total) and took place between 14:00 and 22:00. The integration time was set to 6 s and raw data files were processed using custom R scripts [26]. The behavioral endpoints assessed were as follows: Behavioral testing was performed between 1400 and 2200 h using fish that had acclimated in visible light. The raw data was exported from EthoVisionXT14 into a spreadsheet to perform quality control. The data series were not normally distributed, as normality was tested via a Shapiro–Wilks test. Interquartile range (IQR) method was used to remove outliers from light cycles. Data series were excluded from the overall light series if two serial data points were larger than 75th percentile plus 1.5 of IQR after the 1:00 min mark. In the dark cycle, data series were excluded if two serial data points were smaller than the median of the light data series. Finally, data series with a mean ratio of light:dark series equal or larger than 0.9 were removed. The behavioral data were then analyzed using ANOVA and Tukey's HSD tests. Significance was considered at p-value smaller than 0.05. The quality control and statistics were conducted using R (http://www.r-project.org accessed on: 13 July 2021).

2.5. Transcriptomics

At 5 dpf, five larval fish per chemical concentration were pooled per tube in RNALater™ (Thermo Fisher Scientific, Waltham, MA, USA) at 4 °C. RNALater™ was removed after 24 h and samples were kept at −80 °C until RNA extraction. RNA isolation was performed using the RNeasy Lipid Tissue Mini Kit (QIAGEN, Hilden, Germany) according to manufacturer recommendations. RNA purity was measured with Qubit® 3.0 Fluorometer (Invitrogen, Waltham, MA, USA) and RNA was stored at −80 °C until Quantseq library preparation. Quantseq 3′ mRNA-seq libraries were prepared from isolated RNA using QuantSeq 3′ mRNA-Seq Library Prep Kit FWD for Illumina (Lexogen, Vienna, Austria). Samples were normalized to 40 ng/μL (total input of 200 ng in 5 μL) and amplified at 17 cycles. Libraries were quantified using a Qubit® 2.0 Fluorometer and Qubit® dsDNA Broad Range Assay Kit (Invitrogen, Carlsbad, CA, USA), and run on an Agilent TapeStation 2200 (Agilent Technologies, Santa Clara, CA, USA) for quality control. The samples were sequenced on a HiSeq 2500 (Illumina, San Diego, CA, USA) in rapid mode (single-end 50 bp reads). Reads were aligned to D. rerio (Build danRer10) using the BlueBee Genomics Platform (BlueBee, Rijswijk, The Netherlands). Differential gene expression between the control and exposure lineage zebrafish was evaluated using DEseq2 (available through GenePattern; Broad Institute, Cambridge, Massachusetts). Genes with significant changes in expression, as defined by absolute log2-fold change value ≥ 0.75 and adjusted p-value < 0.1 were uploaded into Ingenuity Pathway Analysis software (IPA; QIAGEN Bioinformatics, Redwood City, CA, USA) for analysis using RefSeq IDs as identifiers.

3. Results

3.1. Triclosan

3.1.1. Larval Abnormalities and Mortality

No significant larval abnormalities were found at any concentration of triclosan following either the 24 or 120 h exposure (Figure 1A). The two highest triclosan concentrations (100 and 1000 nM) resulted in significant mortality following both the 24 and 120 h exposures, with all larval fish dying by 5 dpf in the 120 h exposure ($p < 0.001$). Because of the high mortality rate at these concentrations, abnormalities and behavior could not be evaluated. The percentage of unhatched eggs was significantly decreased in the 120 h exposure in the 10 nM concentration exposure group compared to control ($p < 0.01$).

A.

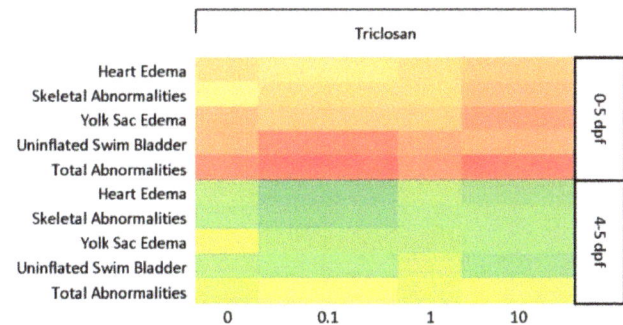

B.

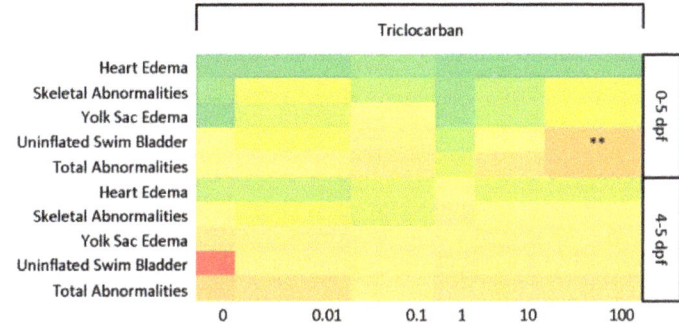

C.

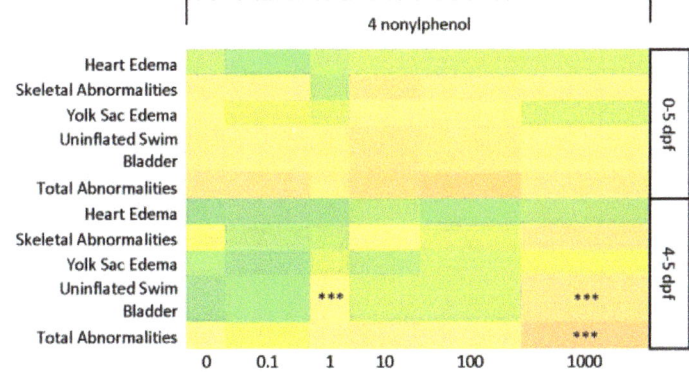

Figure 1. Heat map showing abnormality rate of zebrafish exposed to triclosan (**A**), triclocarban (**B**), or 4-nonylphenol (**C**) starting from 4 h post-fertilization to 5 days post-fertilization (0–5 dpf) or 4–5 dpf. ** indicates significant difference from control ($p < 0.10$), *** ($p < 0.001$).

3.1.2. Behavior

The 120 h triclosan exposure resulted in significant behavioral changes in each concentration compared to controls with significant decreases in distance moved during the dark cycle following 0.1 and 10 nM exposures ($p < 0.001$), as well as 1 nM exposure ($p < 0.05$; Figure 2A). No significant difference in distance moved was observed during the light cycle for the 120 h exposure groups. In the 24 h exposure, however, movement was increased in both the dark cycle for the 1 nM exposure group ($p < 0.001$) and in the light cycle for the 10 nM exposure group ($p < 0.05$; Figure 2A).

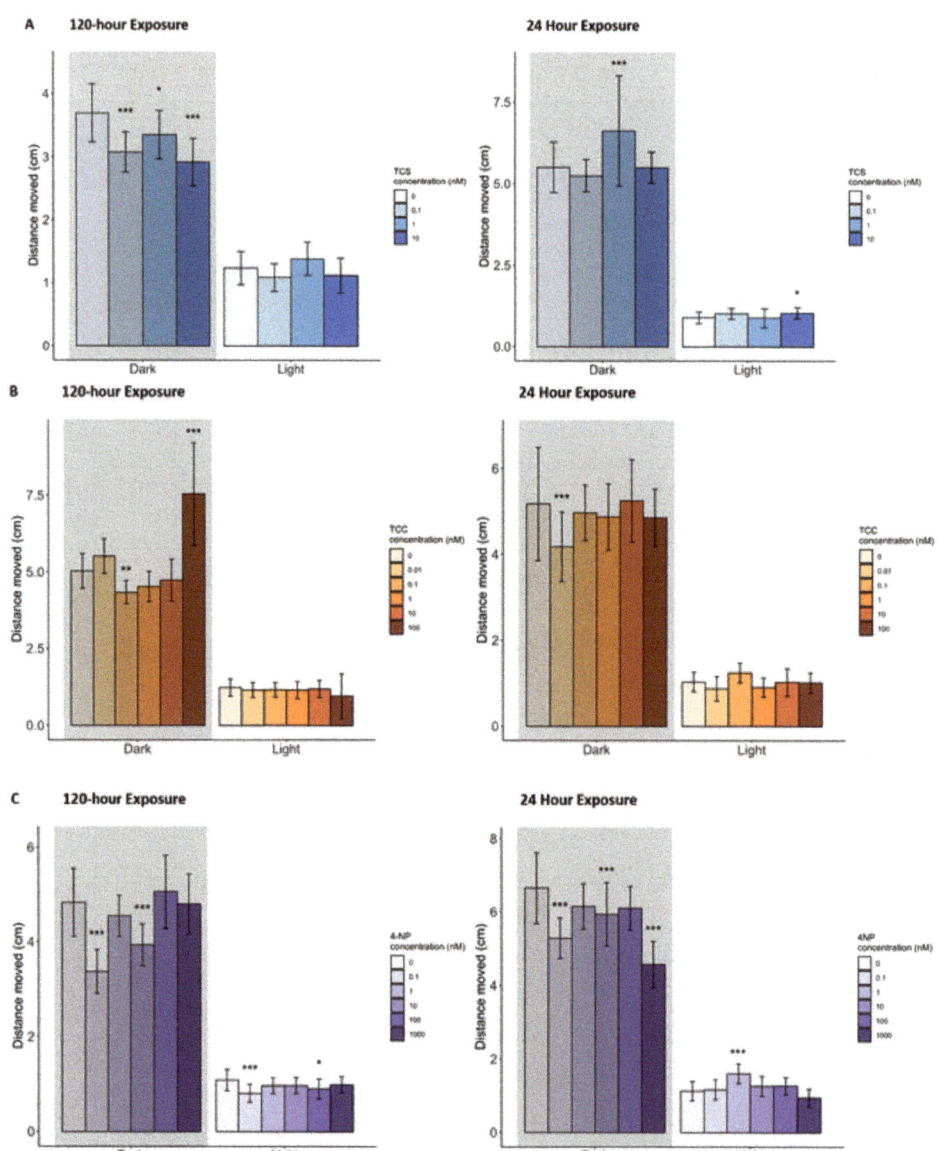

Figure 2. Average distance moved (cm) by larval zebrafish during light and dark cycles following (**A**) triclosan (TCS), (**B**) triclocarbon (TCC), or (**C**) 4-nonylphenol (4NP) exposure starting from 4 h post-fertilization to 5 days post-fertilization (dpf; 120 h exposure) or 4–5 dpf (24 h exposure): * indicates significant difference from control ($p < 0.05$), ** ($p < 0.10$), *** ($p < 0.001$); bars represent standard deviation.

3.1.3. Gene Expression and Pathway Analysis

Table 1 shows the number and direction of change for differentially expressed genes (DEGs) following triclosan exposure. The 120 h exposure group resulted in 31 DEGs with absolute log2-fold changes ≥ 0.75 and adjusted p-values < 0.1, with 17 upregulated and 14 downregulated across all triclosan concentrations. For the 24 h exposure group, there were 45 DEGs across all concentrations with 35 upregulated and 10 downregulated (Table S1). The lowest triclosan concentration (0.1 nM) had the most DEGs regardless of exposure duration, with 19 and 22 DEGs following the 120 and 24 h exposures, respectively, 9 of which were commonly dysregulated by both exposure durations. The significant gene expression profiles were distinct for each exposure concentration following the 24 h exposure, except for hemoglobin, alpha embryonic 1.1 (*hbae1.3*) which was upregulated at 1 and 10 nM. The 120 h exposure was similar, with only one gene, cytochrome P450, family 2, subfamily K, polypeptide 18 (*cyp2k18*), differentially expressed after both 1 and 10 nM exposures.

Table 1. Number of significantly dysregulated genes (significance defined as absolute log2-fold changes ≥ 0.75 and adjusted p-value < 0.1) in zebrafish following extended duration 4NP, TCC or TCS exposure starting at 4 h post-fertilization through 5 days post-fertilization, short term duration starting at 4 dpf through 5 dpf, and total genes dysregulated across all concentrations for extended duration and short term duration. ↓ indicates gene downregulation; ↑ indicates gene upregulation.

	Triclosan			
	0.1 nM	1 nM	10 nM	Total genes
24 h	7 ↓, 15 ↑	2 ↓, 17 ↑	1 ↓, 3 ↑	10 ↓ 35 ↑
120 h	11 ↓, 8 ↑	0 ↓, 1 ↑	3 ↓, 8 ↑	14 ↓ 17 ↑
	Triclocarban			
	0.01 nM	1 nM	100 nM	Total genes
24 h	115 ↓, 54 ↑	26 ↓, 37 ↑	8 ↓, 9 ↑	149 ↓ 100 ↑
120 h	0 ↓, 1 ↑	578 ↓, 258 ↑	478 ↓, 465 ↑	1056 ↓ 724 ↑
	4-nonylphenol			
	0.01 nM	10 nM	1000 nM	Total genes
24 h	0 ↓, 0 ↑	4 ↓, 34 ↑	8 ↓, 6 ↑	12 ↓ 40 ↑
120 h	32 ↓, 7 ↑	1 ↓, 0 ↑	1 ↓, 0 ↑	33 ↓ 7 ↑

Triclosan and 4-nonylphenol shared 8 DEGs (Figure 3) in the following pathways: lipid metabolism, organ development, organ injury and abnormalities, and cancer. The SPINK1 pancreatic cancer pathway was the main pathway expressed following the 24 h triclosan exposure and 120 h 4-nonylphenol exposure, and includes genes such as carboxypeptidase A1 (pancreatic; *cpa1*), which was dysregulated at 1 nM triclosan, and 10 and 1000 nM 4-nonylphenol. For the 24 h duration triclosan exposure, the SP1NK pancreatic cancer pathway had 5 DEGs at 1 nM, while there was only 1 DEG at the 10 nM concentration. Additionally, the 120 h exposure groups for both triclosan and triclocarban shared 14 DEGs (Figure 3) in pathways involving: metabolic processes, including cholesterol biosynthesis, specifically following 1000 nM triclosan and 10 nM triclocarban; xenobiotic processes, specifically following 0.1 and 1 nM triclosan, as well as 1 and 10 nM triclocarban; organ development and morphology, particularly in the cardiovascular and neurological systems, specifically following 0.1, 1, and 1000 nM triclosan and 10 nM triclocarban. Table 2 shows the 14 genes commonly dysregulated by triclosan and triclocarban. There were six DEGs across all three chemicals, including genes related to: cardiovascular system development, such as F-box protein 32 (*fbxo32*) and hemoglobin, alpha embryonic 1.3 (*hbae1.3*); intracellular processes, such as mitochondrial trna (*mt-trna*), si:ch211-153b23.4 (*si:ch211-153b23.4*), and heterogeneous nuclear ribonucleoprotein A0, like (*hnrnpa0l*); and extracellular processes such as pyruvate dehydrogenase kinase 2 (*pdk2b*) (Table 2).

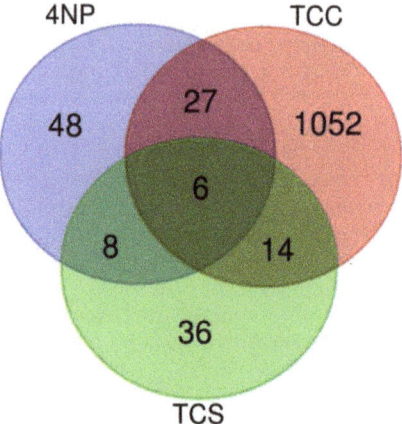

Figure 3. Venn diagram of differentially expressed genes after exposure to triclosan (TCS), triclocarban (TCC) or 4-nonylphenol (4NP) with all exposure concentrations and durations combined. The genes included in this diagram have an adjusted p-value < 0.1 and absolute log2-fold changes >0.75.

IPA analysis of the DEGs following triclosan exposure revealed several pathways of interest. For the 120 h exposure, the top pathways included: xenobiotic metabolism signaling (three genes at both 0.1 and 1 nM); immune system responses, such as NRF2-mediated stress response (two genes at 0.1 and 1 nM, three genes at 1000 nM); metabolic processes, such as cholesterol and glycine synthesis (three genes at 1000 nM); and nervous system organ development and function (five genes at 0.1 nM, two genes at 1 nM) (Table 2). The 24 h triclosan exposure had fewer implicated pathways, but one of the top pathways was organismal injury and abnormalities (33 genes at 0.1 nM, 98 genes at 1 nM, 6 genes at 1000 nM), cancer, specifically the SPINK1 pancreatic cancer pathway, and cardiovascular diseases (16 genes at 1 nM, 1 gene at 1000 nM). Only the lipid metabolism pathway was implicated in both 120 and 24 h triclosan exposure groups. Additionally, there were 10 DEGs expressed in both exposure groups, including: synaptotagmin IV (*syt4*), involved in the nervous system development/signaling pathway; PRELI domain containing 3 (*prelid3b*), involved in the intracellular lipid transport pathway; and perilipin 2 (*plin2*), involved in the lipid metabolism pathway.

Table 2. Differential expression for all genes altered in zebrafish following extended duration 4NP, TCC or TCS exposure starting at 4 h post-fertilization through 5 days post-fertilization and short-term duration starting at 4 dpf through 5 dpf. Significant absolute log2-fold changes (value ≥ 0.75 and adjusted *p*-value < 0.1) in bold. (ND = no difference in expression). Purple background indicates significant differential expression across all 3 chemicals, yellow across 4NP and TCS, green across 4NP and TCC, and blue across TCC and TCS.

	Gene Name	Exposure Concentrations (nM) 0–5 Days												Exposure Concentrations (nM) 4–5 Days											
		4-nonylphenol				Triclosan				Triclocarban				4-nonylphenol				Triclosan				Triclocarban			
		0.1	10	1000	0.1	1	10	0.01	1	100	0.1	10	1000	0.1	1	10	0.01	1	100						
Cardiovascular																										
fbxo32	F-box protein 32	−1.2	−0.8	−0.2	0.4	0.2	1	−0.01	0.5	**1.4**	0.6	0.2	−0.5	**1.1**	0.3	0.1	**1.7**	−1.1	−0.2
tfr1b	transferrin receptor 1b	0.4	0.2	0.3	0.1	−0.1	0.2	−0.03	−0.1	**−1.2**	0.6	**0.8**	−0.2	0.2	0.1	−0.1	−0.4	−0.9	
myl10	myosin, light chain 10, regulatory	−0.1	0.04	−0.3	−0	0.1	−0	−0.2	0.2	−0.1	0.3	0.4	**0.8**	0.2	0.2	−0.1	**−0.8**	−0.7	0.1
mat2aa	methionine adenosyltransferase II, alpha a	−0.1	0.1	0.02	−0	−0	0.4	−0.01	−0.5	**−1**	−0.1	0.2	**0.9**	−0.1	0.2	0.3	−0.1	0.2	0.05
hbae1.1	hemoglobin, alpha embryonic 1.1	−0.1	0.01	−0.3	**−0.8**	−0.3	−0.3	0.1	−0.6	−0.4	−0.4	−0.2	**0.9**	−0.6	**1.1**	**1**	**−0.7**	**−1.6**	−0.2
tcp1	t-complex 1	0.3	−0.2	−0.1	**−0.9**	−0.6	−0.2	0.2	**−1.2**	−0.2	−0.1	−0.3	0.2	**−1**	0.3	0.3	−0.3	−0.1	−0.2
hbbe1.2	hemoglobin beta embryonic- 1.2	0.2	0.3	0.04	ND	−0.3	−0.4	0.04	−0.1	−0.3	−0	−0.2	0.03	−0.1	**1.1**	0.4	−0.3	**−0.8**	**−0.8**
Neurological																										
reep3b	receptor accessory protein 3b	0.5	0.7	0.2	0.6	0.03	0.2	−0.4	0.2	−0.2	0.5	**0.8**	0.2	**0.8**	0.1	0.01	−0.3	−1.4	0.1
syngr3b	synaptogyrin 3b	−0.03	0.7	0.3	**0.9**	0.1	0.2	−0.1	0.3	−0.3	0.5	**1**	0.1	**0.9**	0.1	−0.1	0.3	−0.2	−0.04
agr2	anterior gradient 2	**0.9**	0.4	−0.2	−0.5	−0.4	−0.3	0.2	−0.7	**−0.8**	−0.1	−0.05	0.4	−0.3	0.4	0.05	−0.7	**−2.5**	−0.1
crygm2d18	crystallin, gamma M2d18	−0.2	**−1**	−0.1	0.2	0.1	**−0.8**	0.03	0.04	−0.1	−0.3	0.3	−0.1	**−0.6**	0.6	0.1	−0.2	**−1**	**0.8**
elmo2	engulfment and cell motility 2	−0.2	0.4	−0.1	0.2	−0	0	−0.3	−0	−0.2	0.4	**0.8**	−0.04	0.4	0.5	−0.1	0.5	**1.4**	0.02
arr3b	arrestin 3b, retinal (X-arrestin)	0.1	0.4	−0.02	−0.1	0.2	0.1	−0.3	−0	**−2.6**	0.2	**0.8**	−0.2	0.1	−0.1	−0.2	0.04	−0.4	0.2
opn1sw2	opsin 1 (cone pigments), short-wave-sensitive 2	−0.1	−0.01	0.2	−0.2	−0.3	**−0.8**	−0.1	−0.1	−0.3	−0.3	−0.4	0.2	−0.4	−0.4	−0.3	**−0.8**	**−1.3**	−0.1

Table 2. *Cont.*

Gene Name		Exposure Concentrations (nM) 0–5 Days														Exposure Concentrations (nM) 4–5 Days														
		4-nonylphenol				Triclosan				Triclocarban					4-nonylphenol				Triclosan				Triclocarban				
		0.1	10	1000	0.1	1	10	0.01	1	100	0.1	10	1000	0.1	1	10	0.01	1	100
sii:ch211-153b23.5	si:ch211-153b23.5	0.9	0.3	−0.2	−1	−0.9	−0.6	0.1	−0.9	0.6	0.7	0.5	1	−0.8	0.8	0.1	−0.4	0.4	0.7
mgst3a	microsomal glutathione S- transferase 3a	0.3	−0.1	0.1	−0.9	−0.9	−0.3	0.2	−1.1	−0.5	−0.5	−0.5	−0.1	−1	0.4	0.3	0.02	−0.5	−0.4
rlbp1b	retinaldehyde binding protein 1b	0.1	0.1	0.4	−0	−0	0.2	0.2	0.1	−0.8	0.2	−0.1	0.1	0.4	1.5	0.5	−1.1	−0.5	0.04
Metabolic Processes																			
cel.1	carboxyl ester lipase, tandem duplicate 1	−1.2	−0.7	−0.3	0.6	0.4	0.2	−0.3	0.5	0.4	0.1	−0.3	−0.5	1	0.2	−0.3	0.05	−0.6	−0.5
amy2a	amylase alpha 2A	−0.9	−0.7	−0.1	0.8	0.4	0.2	−0.1	−0.2	0.5	0.1	−0.2	−0.8	0.8	0.4	−0.2	0.4	0.4	−0.6
prss59.2	serine protease 59, tandem duplicate 2	−0.8	−0.4	0.04	0.3	−0.1	0.2	−0.1	−1	0.1	−0	−0.2	−0.4	0.4	1.1	0.2	0.1	−0.3	−0.4
ela2l	elastase 2 like	−0.7	−0.3	0.06	0.3	−0.5	−0.6	0.2	−0.2	−0.1	−0.5	−0.9	−0.4	0.5	0.4	−0.9	0.1	−0.8	−0.3
pip5k1cb	phosphatidylinositol-4-phosphate 5-kinase, type I, gamma b	0.4	0.7	0.4	0.6	0.3	0	−0.4	0	−0.2	0.5	0.9	0.1	0.8	−0.2	−0.3	0.7	−0.3	0.4
fabp10a	fatty acid binding protein 10a, liver basic	−0.9	−0.1	−0.1	0.2	0.01	0.3	−0.1	0.04	−0.1	−0.1	−0.02	−0.7	0.4	0.6	0.2	0.4	−0.9	−0.2
zgc-92590	zgc-92590	−0.9	−0.3	−1.1	−0	−0	0	−0.1	−0.3	−1.6	−0.1	0.2	−0.2	0.1	0.8	0.1	−0.03	−1.1	−0.9
sii:ch211-234p6.10	si:ch211-234p6.10	−0.8	−0.03	0.4	0	−0.2	0.2	−0.04	0.1	−0.3	0.4	0.1	−0.6	0.3	0.6	0.2	0.9	0.3	−0.4
lpin1a	lipin 1	−0.8	−0.4	−0.4	0.1	−0	0	−0.2	0.2	0.9	0.2	−0	−0.3	0.4	0.02	−0.1	0.6	0.2	−0.03
fkbp9	FKBP prolyl isomerase 9	0.8	0.6	−0.02	−0.1	−0.3	0	−0.4	−1.7	−2	−0	1	0.8	−0.2	−0.2	0.04	−0.1	−0.6	−0.1
abcc2	ATP-binding cassette, sub-family C (CFTR/MRP), member 2	−0.4	0.1	−0.1	0.4	1	1	−0.1	0.4	1	0.3	0.2	0.2	0.7	−0.2	0.6	−0.1	0.2	−0.1
plin2	perilipin 2	0.06	−0.2	0.6	−0	0.1	1.1	−0.2	−0.3	1.1	0.8	0.4	0.4	0.3	1	1.1	0.4	−0.6	−0.1

Table 2. *Cont.*

Gene Name		Exposure Concentrations (nM) 0–5 Days									Exposure Concentrations (nM) 4–5 Days								
		4-nonylphenol			Triclosan			Triclocarban			4-nonylphenol			Triclosan			Triclocarban		
		0.1	10	1000	0.1	1	10	0.01	1	100	0.1	10	1000	0.1	1	10	0.01	1	100
fdps	farnesyl diphosphate synthase (farnesyl pyrophosphate synthetase, dimethylallyltranstransferase, geranyltranstransferase)	0.1	0.4	0.1	−0.1	−0.4	0.4	0.02	−0.1	−1.2	0.03	0.4	0.3	0.5	**1**	0.6	0.04	0.3	0.1
Immune System																			
irg1l	immunoresponsive gene 1, like	**1.5**	0.2	0.04	−0.5	−0.2	−0.2	0.2	−0.7	**1.8**	0.8	0.4	0.6	−0.3	0.8	0.2	−0.3	0.6	**1.1**
hsp90b1	heat shock protein 90, beta (grp94), member 1	**1**	0.2	0.3	−0.3	−0.1	−0.2	0.4	−**0.9**	0.2	0.1	−0.1	0.4	−0.4	−0.4	0.2	−0.3	−**0.9**	−0.1
ctsl.1	cathepsin L.1	−0.4	0.2	−0.4	0.7	0.3	−0.1	−0.1	−0.2	0.7	0.6	0.5	−**1.3**	0.8	0.5	−0.5	**1**	−0.1	−0.4
Extracellular Processes																			
si:dkey-14d8.6	si:dkey-14d8.6	−**1.1**	−0.4	0.01	**1**	0.4	0	−0.2	−0.2	0.5	0.2	−0.2	−0.3	0.9	0.4	−0.5	0.5	−0.2	−0.4
pdk2b	Pyruvate dehydrogenase kinase 2b	−**1.1**	−0.5	0.004	−0	0.2	0.9	−0.1	0.3	**1.2**	0.5	−0.2	−0.5	0.3	0.4	0.3	**0.8**	0.5	−0.6
cpa1	carboxypeptidase A1 (pancreatic)	−**1.3**	−0.9	−0.5	0.1	0.02	0.3	−0.1	−0.6	−0.2	−0.1	−0.3	−0.6	0.3	0.4	0.3	**0.9**	−0.8	−0.4
Intracellular Processes																			
si:ch211-153b23.4	si:ch211-153b23.4	**1.2**	−0.2	−0.3	−0.9	−0.5	−0.5	0.2	−0.4	**1.4**	0.8	0.1	0.8	−0.5	**1.2**	0.5	−0.3	0.05	0.5
mt-tRNA	tRNA on mitochondrial genome	−0.002	−0.04	−0.05	−**0.9**	−0.4	−0.6	0.1	0.4	0.5	−0.3	−**1.1**	−0.4	−0.4	0.2	0.1	−**1**	−**1.9**	−0.4
prelid3b	PRELI domain containing 3	0.1	0.4	0.3	**1**	0.4	0.5	−0.4	0.1	−0.5	0.4	0.8	0.1	**1.3**	0.3	−0.1	0.2	−0.6	0.2
hnrnpa0l	heterogeneous nuclear ribonucleoprotein A0, like	0.4	0.2	−0.03	−**0.8**	−0.5	−0.3	0.001	−**1**	−**0.8**	0.2	0.5	**1**	−**0.8**	0.5	0.1	−0.1	0.1	−0.02

Table 2. Cont.

| | Gene Name | Exposure Concentrations (nM) 0–5 Days | | | | | | | | | | | | | | | Exposure Concentrations (nM) 4–5 Days | | | | | | | | | | | | | | | |
| | | 4-nonylphenol | | | | | Triclosan | | | | | Triclocarban | | | | | 4-nonylphenol | | | | | Triclosan | | | | | Triclocarban | | | | |
		0.1	10	1000	0.1	1	10	0.01	1	100	0.1	10	1000	0.1	1	10	0.01	1	100
calcoco1b	calcium binding and coiled-coil domain 1b	−1	−0.7	−0.3	0.1	0.1	0.4	0.1	0.7	2.2	0.4	−0.2	−1.3	0.7	0.7	0.2	0.9	1.1	−0.1
mknk2b	MAPK interacting serine/threonine kinase 2b	−0.9	−0.5	−0.2	0	0.1	0.2	0.03	−0.1	1.1	−0	−0.2	−0.4	0.3	0.1	−0	0.6	−0.1	0.1
trim63a	tripartite motif containing 63a	−0.8	−0.6	−0.2	−0.3	−0.1	0.1	0.2	0.3	0.9	0.04	−0.3	−0.6	0.1	0.2	−0.2	0.5	−1.6	−0.2
si:ch211-207n23.2	si:ch211-207n23.2	0.9	0.3	0.03	−0.5	−0.4	−0.5	0.3	−0.3	1.5	0.8	0.5	0.2	−0.3	0.5	0.1	0.2	0.3	0.4
si:ch211-153b23.3	si:ch211-153b23.3	1.2	−0.02	−0.1	−0.1	−0	−0	0.2	0.3	1.9	0.3	0.3	0.5	−0	0.7	0.3	−0.2	0.8	1.3
sult5a1	sulfotransferase family 5A, member 1	1.7	0.7	0.1	−0.1	−0	−0.1	0.1	−0.2	2.2	0.4	0.3	0.2	−0.1	1	0.1	−0.3	0.9	0.7
mt-atp8	ATP synthase 8, mitochondrial	−0.5	−0.4	−0.7	−0.4	0.2	0	−0.2	0.8	0.1	−0.7	−1	0.2	−0.3	−0.4	0	−0.5	0.2	−0.1
dgcr8	DGCR8 microprocessor complex subunit	0.1	0.4	0.02	0.2	0.2	0.3	−0.2	0.3	−0.8	0.3	0.8	0.1	0.1	−0.02	0.1	0.1	0.6	0.1
bub3	BUB3 mitotic checkpoint protein	0.2	0.3	−0.04	−0.2	−0.1	−0.1	−0.1	−1.3	−1.1	0.3	0.8	0.7	−0.1	−0.1	−0.1	−0.3	−1	0.5
xpot	exportin, tRNA (nuclear export receptor for tRNAs)	−0.3	0.2	−0.3	0.1	−0	−0.1	−0.4	−0.8	−1.5	0.3	0.8	0.4	−0	−0.1	−0.2	0	0.4	0.9
si:ch211-250g4.3	si:ch211-250g4.3	0.1	0.3	0.3	0.7	0.4	0.7	0.1	0.8	1	0.3	0.8	−0.05	0.8	0.5	0.3	0.8	−0.2	−0.02
cebpd	CCAAT enhancer binding protein delta	−0.1	0.03	0.04	−0.4	0.1	0.6	0.2	−0.2	1.2	0.1	−0.5	−0.8	0	0.4	0.4	0.9	0.01	−0.2
hnrnpa0a	heterogeneous nuclear ribonucleoprotein A0a	0.5	0.2	−0.1	−0.5	−0.4	−0.4	−0.1	−0.5	−0.8	0.2	0.4	0.8	−0.8	0	0.02	−0.02	−0.4	−0.3
fkbp5	FKBP prolyl isomerase 5	−0.4	0.1	0.1	0.4	0.7	1	0.2	−0.1	0.2	0.01	−0.5	0.5	0.5	−0.4	ND	0.8	−0.02	−0.03
ddx39ab	DEAD (Asp-Glu-Ala-Asp) box polypeptide 39Ab	0.3	−0.1	−0.1	−1	−0.6	−0.4	0.1	−0.6	−0.9	−0.1	−0.2	0.3	−1.2	0.3	0.3	−0.6	−0.9	−0.2

Table 2. Cont.

	Gene Name	Exposure Concentrations (nM) 0–5 Days									Exposure Concentrations (nM) 4–5 Days								
		4-nonylphenol			Triclosan			Triclocarban			4-nonylphenol			Triclosan			Triclocarban		
		0.1	10	1000	0.1	1	10	0.01	1	100	0.1	10	1000	0.1	1	10	0.01	1	100
tyrp1b	tyrosinase-related protein 1b	0.4	−0.02	0.02	−1	−0.8	−0.4	0.2	−0.2	−1.5	−0.3	−0.4	−0.1	−0.7	0.4	0.2	−1	−0.4	−0.2
hnrnpa0b	heterogeneous nuclear ribonucleoprotein A0b	0.5	0.2	0.3	−0.8	−0.5	0.2	0.2	−0.9	−1.3	−0.2	−0.01	0.2	−0.6	0.3	0.4	0.1	0.6	−0.1
phtf2	putative homeodomain transcription factor 2	−0.4	0.2	0.2	0.5	0.2	0.2	−0.2	0.3	−0	0.3	0.3	−0.1	0.9	0.2	0	−0.2	−2.4	0.2
zmp:0000001081	zmp:0000001081	0.2	0.1	0.3	0.7	0.3	0.6	0.2	0.2	1.4	0.2	0.7	−0.3	0.9	0.5	0.2	0.8	0.2	0.2
sich211-113a14.18	sich211-113a14.18	−0.005	−0.1	0.02	0.1	0.3	0.1	0.1	0.7	0.1	−0.1	−0.2	−0.3	−0.1	−1.4	−0.5	0.2	2.4	−0.1
zgc:113263	zgc:113263	−0.1	−0.2	0.1	0.2	0.01	0	−0.1	0.2	0.4	0.3	0.2	0.2	0.01	−1.2	−0.2	0.2	1.4	0.1
sich211-132b12.7	sich211-132b12.7	0.1	0.4	−0.1	0.2	0	0.7	−0.3	−0.2	−2	−0.2	0.5	−0.1	0.9	1	0.5	0.1	−0.03	−0.1
tm4sf21b	transmembrane 4 L six family member 21b	−0.02	−0.4	−0.2	−0.1	−0	0.2	0.3	−1	−0.4	−0.5	−0.5	0.1	0.1	1.3	0.6	−0.3	0.2	−0.6
Xenobiotic Signaling																			
cyp3a65	cytochrome P450, family 3, subfamily A, polypeptide 65	−1	−0.6	−0.1	0.5	1	0.7	0.1	0.6	2	0	−0.5	−0.8	0.5	0.1	0.1	−0.1	−0.6	−0.4
ucp1	uncoupling protein 1	−0.9	−0.4	0.1	0.5	0.1	0.4	−0.1	0.3	1.4	0.1	−0.2	−0.5	0.7	0.9	0.1	0.7	−0.1	−0.5
cyp2k18	cytochrome P450, family 2, subfamily K, polypeptide 18	0.3	−0.03	0.2	0.2	2	2.4	−0.1	−0.1	1.9	0.2	−0.2	0.6	0.1	−0.2	0.3	−0.2	2.4	−0.5

3.2. Triclocarban

3.2.1. Larval Abnormalities and Mortality

Larval fish exposed to triclocarban for 120 h displayed a significant increase in uninflated swim bladders at the (100 nM) compared to control ($p < 0.005$; Figure 1B). Additionally, total abnormalities were approaching significance ($p = 0.059$), primarily due to the swim bladder abnormalities observed in the 100 nM exposure group. However, fish from the 24 h exposure groups did not experience any significant developmental abnormalities. No significant difference was observed in mortality or the percentage of unhatched eggs at any concentration in either the 120- or 24 h triclocarban exposure.

3.2.2. Behavior

In the 120 h triclocarban exposure, fish exposed to the second-lowest concentration (0.1 nM) showed a decrease in distance moved in the dark ($p < 0.01$), while fish exposed to the highest concentration (100 nM) showed a significant increase in distance moved in the dark ($p < 0.001$) compared to control. The 24 h exposure group exhibited a decrease in distance moved in the dark at the 0.01 nM concentration ($p < 0.001$) (Figure 2B). No significant changes in movement were detected during the light cycle following either the 120- or 24 h exposure.

3.2.3. Gene Expression and Pathway Analysis

Triclocarban exposure resulted in significantly more DEGs compared to the other two chemicals with a total of 2019 DEGs with absolute log2-fold changes ≥ 0.75 and adjusted p-values <0.1. There were 1770 DEGs (724 upregulated and 1056 downregulated, with 10 variably regulated depending on concentration) from the 120 h exposure and 249 DEGs (100 upregulated and 149 downregulated) from the 24 h exposure (Tables 1 and S1). A total of 45 genes were commonly dysregulated following both exposure durations, with two pathways implicated across all exposure concentrations and all exposure durations: cancer and organismal injury and abnormalities, both of which included genes such as tyrosinase-related protein 1b (*tyrp1b*), sequestosome 1 (*sqstm1*), pyruvate dehydrogenase kinase 2 (*pdk2*), and heterogeneous nuclear ribonucleoprotein L (*hnrnpl*). Other affected pathways included endocrine system disorders (expressed following the 1 and 10 nM exposures) and molecular transport and small molecule biochemistry (expressed following the 1 nM exposure).

In addition to the 14 genes commonly dysregulated by triclocarban and triclosan (previously described above), triclocarban and 4-nonylphenol commonly dysregulated 27 genes (Figure 3), with the top affected pathway being lipid metabolism, which was dysregulated following both exposure durations to 0.1 nM 4-nonylphenol, short term exposure to 1000 nM nonylphenol, and long-term exposure to 10 nM triclocarban. The small molecule biochemistry pathway was also dysregulated following both exposure durations to 1 nM triclocarban, long term exposure to 10 nM 4-nonylphenol, and short-term exposure to 0.1 and 1000 nM 4-nonylphenol.

3.3. 4-nonylphenol

3.3.1. Larval Abnormalities and Mortality

4-nonylphenol exposure, regardless of concentration, did not affect mortality rate following either the 120- or 24 h exposure duration. While fish exposed to 4-nonylphenol for 120 h had no significant abnormalities ($p > 0.05$), the 24 h exposure fish showed significant deficiency in swim bladder development and total abnormalities at the 1000 nM concentration compared to control ($p < 0.001$) (Figure 1C). Global Chi-Square analysis showed increased cardiac edema ($p < 0.05$) and skeletal abnormalities ($p < 0.001$) in the 24 h exposure group, but pairwise comparisons showed no significant difference between any specific concentration compared to the control group. However, the 1 nM concentration was approaching significance compared to control for uninflated swim bladder ($p = 0.08$). The 1000 nM concentration for the 24 h exposure had the highest occurrence of abnormalities: skeletal (21%), uninflated swim bladder (21%), and yolk sac edema (10%). The percentage of unhatched eggs was increased in the 120 h exposure for the 0.1 and 100 nM concentrations compared to control, although it was not significant ($p > 0.05$).

3.3.2. Behavior

Larval fish exposed to 4-nonylphenol moved significantly less in the dark at 0.1 and 10 nM after both the 120 and 24 h exposures, and also at 100 nM after the 24 h exposure ($p < 0.001$). In the light, decreased movement was observed in the 120 h exposure at 0.1 and 100 nM ($p < 0.001$ and $p < 0.05$, respectively), while increased movement was observed in the 24 h exposure at 1 nM ($p < 0.001$) (Figure 2C).

3.3.3. Gene Expression and Pathway Analysis

4-nonylphenol exposure resulted in 93 DEGs with absolute log2-fold changes ≥ 0.75 and adjusted p-values < 0.1 across both exposure durations, with 47 upregulated and 46 downregulated (Table 1). The 120 h exposure resulted in 41 DEGs (33 downregulated and 7 upregulated), with 40 dysregulated at 0.1 nM alone, and 1 gene, zgc:92590 (*zgc:92590*), upregulated at both 0.1 and 1000 nM. The 24 h exposure resulted in 52 DEGs, with 34 upregulated and 4 downregulated at 10 nM alone, and the remaining 14 genes dysregulated at 1000 nM (Table S1). The significant gene expression profiles were distinct for each exposure concentration following the 24 h exposure duration. However, three genes were commonly dysregulated following the 120 and 24 h exposure durations: cytochrome P450, family 3, subfamily A, polypeptide 65 (*cyp3a65*), involved in the xenobiotic signaling pathway; amylase alpha 2A (*amy2a*), involved in the metabolic processes pathway; and pyruvate kinase L/R (*pklr*) involved in cardiac development/metal ion binding pathways.

Overall, 4-nonylphenol exposure resulted in fewer differentially expressed pathways compared to triclocarban or triclosan exposures (Table 2). The main pathways implicated following the 120 h exposure were similar to those affected by triclosan and triclocarban exposure, namely lipid metabolism, small molecule biochemistry functions, organismal injury and abnormalities. Cancer was also implicated in the long-term 4-nonylphenol exposure, specifically the SPINK1 pancreatic cancer pathway, with 4 DEGs following 10 nM exposure, including carboxypeptidase A2 (pancreatic; *cpa2*), carboxypeptidase A1 (pancreatic; *cpa1*), carboxypeptidase B1 (tissue; *cpb1*), and chymotrypsinogen B, tandem duplicate 1 (*ctrb2*). In addition to these, another 4 DEGs in the SPINK1 pathway were present after exposure to 0.1 nM 4-nonylphenol: chymotrypsin-like (*ctrl*), chymotrypsin like elastase 2A (*cela2a*), serine protease 2 (*prss2*) and chymotrypsin like elastase 1 (*cela1*) (Table 2). Embryonic development was also an affected pathway following the 24 h exposure duration, and included genes implicated in cardiovascular development, such as transferrin receptor 1b (*tfr1b*), and neurological development, such as synaptogyrin 3b (*syngr3b*), retinal X-arrestin (*arrb3*), and receptor accessory protein 3b (*reep3b*).

4. Discussion

Our results show a wide range of responses to these three EDCs, with notable differences in mortality, morphology, neurobehavior, and gene expression between triclocarban and triclosan, despite their relatively similar functions and chemical structures. Results also varied depending on the timing of exposure, elucidating the importance of examining different windows of developmental exposure. For example, triclocarban led to more significant morphological abnormalities in the 120 h exposure group, whereas none were found with triclosan exposure, but this was likely due to the 100% mortality rate seen at the two highest exposure concentrations (100 and 1000 nM) during both 120 and 24 h exposure periods (Figures 1 and 2). Triclosan-induced mortality generally agrees with existing data, which found a dose-dependent decrease in survival rate with significant reductions in mortality at concentrations < 40 µg/L [27]. However, some studies showed relatively lower or no mortality at concentrations >1000 nM in adult zebrafish [28] and other fish species [29–32], suggesting that developing organisms are more sensitive to triclosan exposure than adults, and different species have variable sensitivity to this contaminant. Some larval zebrafish studies additionally showed relatively lower mortality at equal or higher concentrations to those investigated in the current study, which could potentially be explained by different exposure methods through variations in water renewal or vehicle [33,34]. Our results are also surprising because a recent study predicted that triclocarban would have more adverse effects than triclosan, with a predicted no effect concentration of 0.0147 µg/L versus 0.1757 µg/L for triclosan [35]. Our two highest concentrations of triclosan (100 and 1000 nM) correspond to 43.4 µg/L and 433.9 µg/L respectively, whereas the highest concentration of triclocarban (10 nM) corresponds to 3.2 µg/L, thus potentially explaining why mortality was increased with triclosan, but not triclocarban exposure in this study. 4-nonylphenol mortality followed along the lines of similar studies, which showed no significant mortality in zebrafish embryos exposed from 4 to 168 hpf to concentrations ranging from 0.1–100 µg/L [36].

A decreased percentage of unhatched eggs was another effect seen at 10 nM triclosan, with only 1% of the eggs unhatched at 10 nM compared to 15% for the control group in the 120 h exposure duration. Zebrafish egg hatching is mainly dependent on one enzyme, zebrafish hatching enzyme 1 (*zhe1*) [37]. Although enzymatic activity and therefore zebrafish hatching rate is mainly dependent on development rate and temperature, studies have shown abnormalities in embryo hatching in response to adverse environmental factors, such as glucocorticoids, salinity, and EDCs [38,39]. These outcomes conflict with previous data on triclosan that indicates triclosan has no effect on the hatching rate of zebrafish [40]. While no genes associated with *zhe1* were dysregulated in this study, the gene *cyp2k18*, which is involved in several processes in the body contributing to homeostasis, including exogenous drug catabolic process, organic acid metabolic process, and xenobiotic metabolic process, as well as heme-binding activity, was significantly upregulated after both 1 and 10 nM exposures in the 120 h triclosan exposure group, with an absolute log-fold change of 2.4 in the 10 nM group. Although no literature specifically links this gene to a decreased percentage of unhatched eggs, *cyp2k18* upregulation is considered a marker of toxicity and stress. For example, *cyp2k18* is significantly upregulated due to drug toxicity [41], and *cyp2k18* transgenic zebrafish have been developed in order to assess toxicity to chemotherapy drugs [42]. Finally, upregulation of *cyp* genes has been linked to tumorigenesis in both murine and Japanese medaka (*Oryzias latipes*) models [43]. Therefore, upregulation of *cyp2k18* is likely a marker of triclosan toxicity in the current study and should be explored further as a potential marker of adverse environmental factors.

Triclocarban and 4-nonylphenol both showed a significant increase in uninflated swim bladders at the highest exposure concentration (100 nM), with total abnormalities for triclocarban approaching significance mainly due to the swim bladder abnormalities observed in the 100 nM exposure group. For 4-nonylphenol, there was a trend of swim bladder abnormalities in all concentrations of the 24 h exposure, but only the 1000 nM concentration was significant. Several DEGs expressed across all concentrations in the long term triclocarban exposure group included *sqstm1*, involved in axogenesis and nervous system development, and *pdk2*, which is involved in glucose metabolism in the mitochondria. Both genes are involved in the organismal injury and abnormalities pathway. Although no studies link *sqstm1* upregulation specifically to swim bladder deflation, upregulation has been linked to tumorigenesis in bronchial epithelial cells in humans [44], while *pdk2* upregulation has been linked to the development of pulmonary hypertension [45]. The zebrafish swim bladder shares many developmental [46] and transcriptomic [47] traits with the human pulmonary system, so upregulation of these genes could be contributing to the deflated swim bladders and overall abnormalities seen with the long-term triclocarban exposure. Abnormalities in blood circulation and oxygen delivery have been implicated in uninflated swim bladders in zebrafish [48], and downregulation of *pklr*, which is expressed in red blood cells, occurred at short term 1000 nM 4-nonylphenol exposure.

4-nonylphenol-induced abnormalities were only seen following the 24 h exposure, with the most DEGs following the 24 h exposure at 10 and 1000 nM concentrations

(41 genes), compared to the 120 h exposure period (2 genes). This illustrates that the specific window of susceptibility is particularly important 4-nonylphenol toxicity. Cardiac edema was another abnormality noted after short term 4-nonylphenol exposure; however, it was only significant with global Chi-Square analysis. When pairwise comparison was conducted, no significant difference in cardiac edema between individual concentration levels and the control group occurred. 4-nonylphenol exposure following the 24 h exposure period resulted in the most DEGs related to the cardiovascular system, such as *tfr1b*, which is responsible for hemoglobin biosynthesis, significantly upregulated at 10 nM and approaching significance at 0.1 nM. Resulting hemoglobin defects can lead to increased viscosity and cardiac edema, as seen in a study exposing African catfish (*Clarias gariepinus*) to concentrations of 4-nonylphenol ranging from 250 to 1000 µg/L [49].

Behavioral abnormalities resulted from exposure to each of the three chemicals, with triclosan exposure resulting in the most behavioral changes, followed by 4-nonylphenol, and then triclocarban. Zebrafish are more active in the dark [50] as they search for better lit environments where they can better identify food sources and increase their likelihood of survival [51]. Our triclosan behavioral results for the 120 h exposure are similar to previous findings of hypoactivity in response to triclosan exposure in a variety of aquatic organisms, mice, and humans [40,52–57]. We have expanded upon this previous research by demonstrating that this hypoactivity response is present at lower concentrations (0.1 nM, 1 nM, and 10 nM; Figure 2A) than previously tested (10 µg/L to 0.6 mg/L). This hypoactivity may be related to neurological dysfunction associated with dysregulation of *syt4*, which binds phospholipids in the nervous system, and *prelid3b*, which transports lipids in the nervous and musculature systems, which are both differentially expressed following the long- and short-term triclosan exposures. Conversely, the 24 h triclosan exposure resulted in hyperactivity at only the 1 nM concentration during the dark cycle and at 10 nM during the light cycle. Two existing studies also found hyperactivity in response to triclosan exposure in zebrafish [58] and humans (boys, but not girls) [59]. In both studies, exposure to higher concentrations of triclosan, or a mixture of triclosan and its metabolites, seemed to precipitate the hyperactive behavior, though both exposure periods were much longer than the exposure period in the current study. The need for further research into how triclosan exposure duration and concentration affects behavior is highlighted by the differential effects seen within our study, namely the lower concentrations inducing hypoactive behavior following embryonic exposure, but hyperactivity following larval exposure.

Hypoactivity was the main behavioral change seen in response to 4-nonylphenol in both the 24 and 120 h exposures. 0.1 and 10 nM showed hypoactivity from both exposure time periods, but in the 24 h exposure, the most significant hypoactivity was at the highest concentration (1000 nM) during the dark cycle. Activity in the light phases was generally decreased as well, though only significantly for long term exposure at the 0.1 and 100 nM concentrations. These results agree with previous 4-nonylphenol studies showing some level of hypoactivity in response to exposure in mice and fish [60–65]. In contrast, only the 21 nM exposure resulted in hyperactivity during the light phase, although it is unclear why this occurred.

A non-monotonic response was seen in triclocarban behavior from the 24 h exposure, which only resulted in hypoactivity at 0.01 nM. Conversely, in the 120 h exposure in the dark phase, the second lowest concentration resulted in hypoactivity while the highest concentration resulted in hyperactivity. There have not been many studies on the effects of triclocarban exposure on behavior. One study of fathead minnows (*Pimephales promelas*) found reduced aggression in adult males at 560 and 1576 ng/L and no change in larval behavior [66], while a study of *Gammarus locusta* found reduced activity in females at 500 ng/L [67].

Finally, IPA analysis revealed many pathways common to all chemicals and a few unique to each chemical. The top pathways shared by all chemicals involved metabolic processes (Table 2), such as lipid metabolism, proteolysis, and cholesterol and glycine synthesis. Out of the 49 differentially regulated genes shared in various combinations between triclosan, triclocarban, and 4-nonylphenol, the majority were involved in metabolic processes, specifically with the pancreas, such as *cel.1*, implicated in lipomatosis and diabetes, *amy2a* involved in carbohydrate metabolism, *prss59.2* implicated in pancreatitis, *el2a* involved in proteolysis, and *zgc:92590* involved in proteolysis. The upregulation of these genes is mainly seen in triclosan and 4-nonylphenol, which may be contributing to the pancreatic cancer pathway seen with these chemical exposures. The SPINK1 pancreatic cancer pathway is thought to increase risk of pancreatic cancer by lowering the activation threshold of trypsin [68]. SPINK1 is believed to promote proliferation of cancer cells by inducing EGFR phosphorylation, which results in activation of the mitogen-activated protein kinase (MAPK) pathway [69]. Additionally, the SPINK1 pathway activates the NRF2 pathway, which leads to increased proliferation and decreased apoptosis of cancer cells [70]. Although the SPINK1 gene itself was not dysregulated within our study, genes related to the SPINK1 and the NRF2 pathway, namely abcc2, were dysregulated. The gene *abcc2*, which is implicated in multiple forms of carcinomas, was upregulated in both triclosan and triclocarban exposures, specifically the long-term exposures for triclosan 10 nM and triclocarban 100 nM. Triclosan has been implicated as inducing metabolic acidosis and regressing pancreatic islet cells into pycnotic cells leading to cell death [71], as well as various cancers such as liver and breast tumors [72]. Additionally, chymotrypsin C in humans has also been implicated in pancreatic cancer [73], and the orthologue in zebrafish, *ela2l* was upregulated in the short-term duration exposure for both 4-nonylphenol at 10 nM and triclosan at 10 nM. Endocrine disruptors are commonly known to contribute to endocrine disruption leading to disease processes such as obesity, type 2 diabetes and metabolic syndromes [74], so it is unsurprising that the metabolic pathways listed above were shared by all three chemicals examined within this study.

In conclusion, the findings of the present study expand upon and contribute to the limited studies of developmental-toxicity regarding exposure to 4-nonylphenol, triclosan and triclocarban in embryogenesis and larval zebrafish. Furthermore, our data indicates several future areas of exploration, such as the metabolic effects of chemical exposure, including potential pathways leading to pancreatic cancer, and disorders such as diabetes and hepatitis, as well as the potential impacts of gene differentiation on cardiovascular abnormalities and behavior. Our results indicate that environmentally relevant levels of exposure can disrupt neurologic, behavioral, cardiovascular and metabolic pathways, potentially leading to adverse health outcomes such as cardiac edema and significant mortality and should be explored further. Overall, our results indicate the potential for gene expression changes and population impacts, depending on the time of exposure, caused by these three known endocrine disruptors, and these potential impacts should be studied further not only in aquatic lifeforms, but human health as well.

Supplementary Materials: The following are available online at https://www.mdpi.com/article/10.3390/toxics10020053/s1, Table S1: Significant gene expression changes (Log2 fold change and *p*-value) results of zebrafish embryos following exposure to triclosan, triclocarban or 4 nonylphenol.

Author Contributions: Conceptualization, T.R.B. and D.K.P.; methodology, T.R.B. and J.N.S.; formal analysis, J.P., J.N.S., C.A. and C.-C.W.; investigation, J.N.S., A.S.H., C.A., D.N.M. and Z.S.; resources, T.R.B. ; data curation, J.P., J.N.S., C.A., C.-C.W. and A.S.H.; writing—original draft preparation, J.P., C.-C.W., B.B.B. and T.R.B.; writing—review and editing, J.P., C.-C.W., B.B.B. and T.R.B.; supervision, T.R.B.; project administration, J.P., J.N.S. and T.R.B.; funding acquisition, T.R.B. and D.K.P. All authors have read and agreed to the published version of the manuscript.

Funding: Funding was provided by the Wayne State University Office of Vice President for Research (WSU SEED grant for project development to T.R.B. and D.K.P.; Postdoctoral funding to CW). Additional funding was provided by the National Center for Advancing Translational Sciences [K01 OD01462 to T.R.B.], the WSU Center for Urban Responses to Environmental Stressors [P30 ES020957 to D.N.M. and T.R.B.], the National Institute of Environmental Health Sciences [F31 ES030278 to D.N.M.], and the National Science Foundation [Grant No. 1735038 to C.A.].

Institutional Review Board Statement: Zebrafish use protocols were approved by the Institutional Animal Care and Use Committee at Wayne State University, according to the National Institutes of Health Guide to the Care and Use of Laboratory Animals (Protocol 16-03-054; approved 4 August 2016).

Informed Consent Statement: Not applicable.

Data Availability Statement: The data presented in this article are available within the text and Supplementary Materials.

Acknowledgments: We acknowledge Emily Crofts, Kim Bauman, and all members of the Warrior Aquatic, Translational, and Environmental Research (WATER) lab at Wayne State University for help with zebrafish care and husbandry. We would like to acknowledge the Wayne State University Applied Genomics Technology Center for providing sequencing services and the use of Ingenuity Pathway Analysis Software.

Conflicts of Interest: The authors declare no conflict of interest.

References

1. NIEHS. National Institute of Environmental Health Sciences. Endocrine Disruptors. Available online: https://www.niehs.nih.gov/health/topics/agents/endocrine/index.cfm (accessed on 22 December 2021).
2. Halden, R.U.; Lindeman, A.E.; Aiello, A.E.; Andrews, D.; Arnold, W.A.; Fair, P.; Fuoco, R.E.; Geer, L.A.; Johnson, P.I.; Lohmann, R.; et al. The florence statement on triclosan and triclocarban. *Environ. Health Perspect.* **2017**, *125*, 064501. [CrossRef] [PubMed]
3. Baker, B.B.; Haimbaugh, A.S.; Sperone, F.G.; Johnson, D.M.; Baker, T.R. Persistent contaminants of emerging concern in a great lakes urban-dominant watershed. *J. Great Lakes Res.* **2021**. [CrossRef]
4. Weatherly, L.M.; Gosse, J.A. Triclosan exposure, transformation, and human health effects. *J. Toxicol. Environ. Health. Part B Crit. Rev.* **2017**, *20*, 447–469. [CrossRef] [PubMed]
5. Halden, R.U.; Paull, D.H. Co-occurrence of triclocarban and triclosan in U.S. water resources. *Environ. Sci. Technol.* **2002**, *39*, 1420–1426. [CrossRef] [PubMed]
6. Pycke, B.F.G.; Geer, L.A.; Dalloul, M.; Abulafia, O.; Jenck, A.M.; Halden, R.U. Human fetal exposure to triclosan and triclocarban in an urban population from Brooklyn, New York. *Environ. Sci. Technol.* **2014**, *48*, 8831–8838. [CrossRef] [PubMed]
7. Feng, Y.; Zhang, P.; Zhang, Z.; Shi, J.; Jiao, Z.; Shao, B. Endocrine Disrupting Effects of Triclosan on the Placenta in Pregnant Rats. *PLoS ONE* **2016**, *11*, e0154758. [CrossRef]
8. Radwan, P.; Wielgomas, B.; Radwan, M.; Krasiński, R.; Klimowska, A.; Zajdel, R.; Kaleta, D.; Jurewicz, J. Triclosan exposure and in vitro fertilization treatment outcomes in women undergoing in vitro fertilization. *Environ. Sci. Pollut. Res. Int.* **2021**, *28*, 12993–12999. [CrossRef]
9. James, M.O.; Li, W.; Summerlot, D.P.; Rowland-Faux, L.; Wood, C.E. Triclosan is a potent inhibitor of estradiol and estrone sulfonation in sheep placenta. *Environ. Int.* **2010**, *36*, 942–949. [CrossRef]
10. Vélez, M.P.; Arbuckle, T.E.; Fraser, W.D. Female exposure to phenols and phthalates and time to pregnancy: The Maternal-Infant Research on Environmental Chemicals (MIREC) Study. *Fertil. Steril.* **2015**, *103*, 1011–1020. [CrossRef]
11. Koeppe, E.S.; Ferguson, K.K.; Colacino, J.A.; Meeker, J.D. Relationship between urinary triclosan and paraben concentrations and serum thyroid measures in NHANES 2007–2008. *Sci. Total Environ.* **2013**, *445–446*, 299–305. [CrossRef]
12. Hinther, A.; Bromba, C.M.; Wulff, J.E.; Helbing, C.C. Effects of triclocarban, triclosan, and methyl triclosan on thyroid hormone action and stress in frog and mammalian culture systems. *Environ. Sci. Technol.* **2011**, *45*, 5395–5402. [CrossRef] [PubMed]
13. Christen, V.; Crettaz, P.; Oberli-Schrämmli, A.; Fent, K. Some flame retardants and the antimicrobials triclosan and triclocarban enhance the androgenic activity in vitro. *Chemosphere* **2010**, *81*, 1245–1252. [CrossRef] [PubMed]
14. Villeneuve, D.L.; Jensen, K.M.; Cavallin, J.E.; Durhan, E.J.; Garcia-Reyero, N.; Kahl, M.D.; Ankley, G.T. Effects of the antimicrobial contaminant triclocarban, and co-exposure with the androgen 17β-trenbolone, on reproductive function and ovarian transcriptome of the fathead minnow (Pimephales promelas). *Environ. Toxicol. Chem.* **2017**, *36*, 231–242. [CrossRef] [PubMed]
15. High Priority Chemicals Data System (HPCDS). 2020. Available online: https://hpcds.theic2.org/ (accessed on 14 November 2021).
16. Kourouma, A.; Keita, H.; Duan, P.; Quan, C.; Bilivogui, K.K.; Qi, S.; Yang, K. Effects of 4-nonylphenol on oxidant/antioxidant balance system inducing hepatic steatosis in male rat. *Toxicol. Rep.* **2015**, *2*, 1423–1433. [CrossRef]
17. Bistakova, J.; Forgacs, Z.; Bartos, Z.; Szivosne, M.R.; Jambor, T.; Knazicka, Z.; Lukac, N. Effects of 4-nonylphenol on the steroidogenesis of human adrenocarcinoma cell line (NCI-H295R). *J. Environ. Sci. Health Part A* **2017**, *52*, 221–227. [CrossRef]
18. Duan, P.; Hu, C.; Butler, H.J.; Quan, C.; Chen, W.; Huang, W.; Yang, K. Effects of 4-nonylphenol on spermatogenesis and induction of testicular apoptosis through oxidative stress-related pathways. *Reprod. Toxicol.* **2016**, *62*, 27–38. [CrossRef]
19. Mao, Z.; Zheng, X.; Zhang, Y.; Tao, X.; Li, Y.; Wang, W. Occurrence and Biodegradation of Nonylphenol in the Environment. *Int. J. Mol. Sci.* **2012**, *13*, 491–505. [CrossRef]
20. Hua, W.; Bennett, E.R.; Letcher, R.J. Triclosan in waste and surface waters from the upper Detroit River by liquid chromatography-electrospray-tandem quadrupole mass spectrometry. *Environ. Int.* **2005**, *31*, 621–630. [CrossRef]

21. Shen, J.Y.; Chang, M.S.; Yang, S.; Wu, G.J. Simultaneous determination of triclosan, triclocarban, and transformation products of triclocarban in aqueous samples using solid-phase micro-extraction-HPLC-MS/MS. *J. Sep. Sci.* **2012**, *35*, 2544–2552. [CrossRef]
22. Gilroy, È.A.M.; Muir, D.C.G.; McMaster, M.E.; Darling, C.; Campbell, L.M.; Alaee, M.; Sherry, J.P. Halogenated phenolic compounds in wild fish from Canadian Areas of Concern. *Environ. Toxicol. Chem.* **2017**, *36*, 2266–2273. [CrossRef]
23. Blair, B.D.; Crago, J.P.; Hedman, C.J.; Klaper, R.D. Pharmaceuticals and personal care products found in the Great Lakes above concentrations of environmental concern. *Chemosphere* **2013**, *93*, 2116–2123. [CrossRef] [PubMed]
24. Vimalkumar, K.; Arun, E.; Krishna-Kumar, S.; Poopal, R.K.; Nikhil, N.P.; Subramanian, A.; Babu-Rajendran, R. Occurrence of triclocarban and benzotriazole ultraviolet stabilizers in water, sediment, and fish from Indian rivers. *Sci. Total Environ.* **2018**, *625*, 1351–1360. [CrossRef] [PubMed]
25. Kimmel, C.B.; Ballard, W.W.; Kimmel, S.R.; Ullmann, B.; Schilling, T.F. Stages of embryonic development of the zebrafish. *Dev. Dyn. Off. Publ. Am. Assoc. Anat.* **1995**, *203*, 253–310. [CrossRef] [PubMed]
26. R Core Team. *R: A Language and Environment for Statistical Computing [Computer Software Manual]*; R Core Team: Vienna, Austria, 2016.
27. Wirt, H.; Botka, R.; Perez, K.E.; King-Heiden, T. Embryonic exposure to environmentally relevant concentrations of triclosan impairs foraging efficiency in zebrafish larvae. *Environ. Toxicol. Chem.* **2018**, *37*, 3124–3133. [CrossRef] [PubMed]
28. Oliveira, R.; Domingues, I.; Grisolia, C.K.; Soares, A.M.V.M. Effects of triclosan on zebrafish early-life stages and adults. *Environ. Sci. Pollut. Res.* **2009**, *16*, 679–688. [CrossRef] [PubMed]
29. Silva, D.C.V.R.; Araújo, C.V.M.; López-Doval, J.C.; Neto, M.B.; Silva, F.T.; Paiva, T.C.B.; Pompêo, M.L.M. Potential effects of triclosan on spatial displacement and local population decline of the fish Poecilia reticulata using a non-forced system. *Chemosphere* **2017**, *184*, 329–336. [CrossRef]
30. Horie, Y.; Yamagishi, T.; Takahashi, H.; Iguchi, T.; Tatarazako, N. Effects of triclosan on Japanese medaka (Oryzias latipes) during embryo development, early life stage and reproduction. *J. Appl. Toxicol.* **2018**, *38*, 544–551. [CrossRef]
31. Salierno, J.D.; Lopes, M.; Rivera, M. Latent effects of early life stage exposure to triclosan on survival in fathead minnows, Pimephales promelas. *J. Environ. Sci. Health* **2016**, *51*, 695–702. [CrossRef]
32. Ishibashi, H.; Matsumura, N.; Hirano, M.; Matsuoka, M.; Shiratsuchi, H.; Ishibashi, Y.; Arizono, K. Effects of triclosan on the early life stages and reproduction of medaka Oryzias latipes and induction of hepatic vitellogenin. *Aquat. Toxicol.* **2004**, *67*, 167–179. [CrossRef]
33. Macedo, S.; Torres, T.; Santos, M.M. Methyl-triclosan and triclosan impact embryonic development of Danio rerio and Paracentrotus lividus. *Ecotoxicology* **2017**, *26*, 482–489. [CrossRef]
34. Falisse, E.; Voisin, A.-S.; Silvestre, F. Impacts of triclosan exposure on zebrafish early-life stage: Toxicity and acclimation mechanisms. *Aquat. Toxicol.* **2017**, *189*, 97–107. [CrossRef] [PubMed]
35. Fan, B.; Li, J.; Wang, X.; Gao, X.; Chen, J.; Ai, S.; Li, W.; Huang, Y.; Liu, Z. Study of aquatic life criteria and ecological risk assessment for triclocarban (TCC). *Environ. Pollut.* **2019**, *254*, 112956. [CrossRef] [PubMed]
36. Wu, M.; Xu, H.; Shen, Y.; Qiu, W.; Yang, M. Oxidative stress in zebrafish embryos induced by short-term exposure to bisphenol A, nonylphenol, and their mixture. *Environ. Toxicol. Chem.* **2011**, *30*, 2335–2341. [CrossRef] [PubMed]
37. Sano, K.; Inohaya, K.; Kawaguchi, M.; Yoshizaki, N.; Iuchi, I.; Yasumasu, S. Purification and characterization of zebrafish hatching enzyme—An evolutionary aspect of the mechanism of egg envelope digestion. *FEBS J.* **2008**, *275*, 5934–5946. [CrossRef] [PubMed]
38. Willi, R.A.; Faltermann, S.; Hettich, T.; Fent, K. Active Glucocorticoids Have a Range of Important Adverse Developmental and Physiological Effects on Developing Zebrafish Embryos. *Environ. Sci. Technol.* **2018**, *52*, 877–885. [CrossRef]
39. Liang, X.; Souders, C.L., 2nd; Zhang, J.; Martyniuk, C.J. Tributyltin induces premature hatching and reduces locomotor activity in zebrafish (Danio rerio) embryos/larvae at environmentally relevant levels. *Chemosphere* **2017**, *189*, 498–506. [CrossRef]
40. Pullaguri, N.; Nema, S.; Bhargava, Y.; Bhargava, A. Triclosan alters adult zebrafish behavior and targets acetylcholinesterase activity and expression. *Environ. Toxicol. Pharmacol.* **2020**, *75*, 103311. [CrossRef]
41. Poon, K.L.; Wang, X.; Lee, S.G.P.; Ng, A.S.; Goh, W.H.; Zhao, Z.; Al-Haddawi, M.; Wang, H.; Mathavan, S.; Ingham, P.W.; et al. Editor's Highlight: Transgenic Zebrafish Reporter Lines as Alternative In Vivo Organ Toxicity Models. *Toxicol. Sci.* **2017**, *156*, 133–148. [CrossRef]
42. Van Sebille, Y.Z.; Gibson, R.J.; Wardill, H.R.; Carney, T.J.; Bowen, J.M. Highlight article: Use of zebrafish to model chemotherapy and targeted therapy gastrointestinal toxicity. *Exp. Biol. Med.* **2019**, *244*, 1178–1185. [CrossRef]
43. Lin, C.; Chou, P.; Chen, P. Two azole fungicides (carcinogenic triadimefon and non-carcinogenic myclobutanil) exhibit different hepatic cytochrome P450 activities in medaka fish. *J. Hazard. Mater.* **2014**, *277*, 150–158. [CrossRef]
44. Huang, H.; Zhu, J.; Li, Y.; Zhang, L.; Gu, J.; Xie, Q.; Jin, H.; Che, X.; Li, J.; Huang, C.; et al. Upregulation of SQSTM1/p62 contributes to nickel-induced malignant transformation of human bronchial epithelial cells. *Autophagy* **2016**, *12*, 1687–1703. [CrossRef] [PubMed]
45. Ryan, J.J.; Archer, S.L. Emerging concepts in the molecular basis of pulmonary arterial hypertension (PAH): Part I: Metabolic plasticity and mitochondrial dynamics in the pulmonary circulation and right ventricle in PAH. *Circulation* **2015**, *131*, 1691–1702. [CrossRef] [PubMed]
46. Cass, A.N.; Servetnick, M.D.; McCune, A.R. Expression of a lung developmental cassette in the adult and developing zebrafish swimbladder: Swimbladder gene expression. *Evol. Dev.* **2013**, *15*, 119–132. [CrossRef] [PubMed]

47. Zheng, W.; Wang, Z.; Collins, J.E.; Andrews, R.M.; Stemple, D.; Gong, Z. Comparative transcriptome analyses indicate molecular homology of zebrafish swimbladder and mammalian lung. *PloS One* **2011**, *6*, e24019. [CrossRef]
48. Winata, C.L.; Korzh, S.; Kondrychyn, I.; Korzh, V.; Gong, Z. The role of vasculature and blood circulation in zebrafish swimbladder development. *BMC Dev. Biol.* **2010**, *10*, 3. [CrossRef]
49. Kumaran, S.S.; Kavitha, C.; Ramesh, M.; Grummt, T. Toxicity studies of nonylphenol and octylphenol: Hormonal, hematological and biochemical effects in clarias gariepinus. *J. Appl. Toxicol.* **2011**, *31*, 752–761. [CrossRef]
50. Ogungbemi, A.; Leuthold, D.; Scholz, S.; Küster, E. Hypo- or hyperactivity of zebrafish embryos provoked by neuroactive substances: A review on how experimental parameters impact the predictability of behavior changes. *Environ. Sci. Eur.* **2019**, *31*, 1–26. [CrossRef]
51. Burgess, H.A.; Granato, M. Modulation of locomotor activity in larval zebrafish during light adaptation. *J. Exp. Biol.* **2007**, *210 Pt 14*, 2526–2539. [CrossRef]
52. Tabari, S.A.; Esfahani, M.L.; Hosseini, S.M.; Rahimi, A. Neurobehavioral toxicity of triclosan in mice. *Food Chem. Toxicol.* **2019**, *130*, 154–160. [CrossRef]
53. Sahu, V.K.; Karmakar, S.; Kumar, S.; Shukla, S.P.; Kumar, K. Triclosan toxicity alters behavioral and hematological parameters and vital antioxidant and neurological enzymes in Pangasianodon hypophthalmus (Sauvage, 1878). *Aquat. Toxicol.* **2018**, *202*, 145–152. [CrossRef]
54. Brown, J.; Bernot, M.J.; Bernot, R.J. The influence of TCS on the growth and behavior of the freshwater snail, Physa acuta. *J. Environ. Sci. Health* **2012**, *47*, 1626–1630. [CrossRef] [PubMed]
55. Nassef, M.; Matsumoto, S.; Seki, M.; Khalil, F.; Kang, I.J.; Shimasaki, Y.; Honjo, T. Acute effects of triclosan, diclofenac and carbamazepine on feeding performance of Japanese medaka fish (Oryzias latipes). *Chemosphere* **2010**, *80*, 1095–1100. [CrossRef] [PubMed]
56. Fritsch, E.B.; Connon, R.E.; Werner, I.; Davies, R.E.; Beggel, S.; Feng, W.; Pessah, I.N. Triclosan impairs swimming behavior and alters expression of excitation-contraction coupling proteins in fathead minnow (Pimephales promelas). *Environ. Sci. Technol.* **2013**, *47*, 2008–2017. [CrossRef] [PubMed]
57. Etzel, T.; Muckle, G.; Arbuckle, T.E.; Fraser, W.D.; Ouellet, E.; Seguin, J.R.; Braun, J.M. Prenatal urinary triclosan concentrations and child neurobehavior. *Environ. Int.* **2018**, *114*, 152–159. [CrossRef] [PubMed]
58. Liu, J.; Sun, L.; Zhang, H.; Shi, M.; Dahlgren, R.A.; Wang, X.; Wang, H. Response mechanisms to joint exposure of triclosan and its chlorinated derivatives on zebrafish (*Danio rerio*) behavior. *Chemosphere* **2018**, *193*, 820–832. [CrossRef]
59. Jackson-Browne, M.S.; Papandonatos, G.D.; Chen, A.; Yolton, K.; Lanphear, B.P.; Braun, J.M. Early-life triclosan exposure and parent-reported behavior problems in 8-year-old children. *Environ. Int.* **2019**, *128*, 446–456. [CrossRef]
60. Ward, A.J.W.; Duff, A.J.; Currie, S. The effects of the endocrine disrupter 4-nonylphenol on the behaviour of juvenile rainbow trout (Oncorhynchus mykiss). *Can. J. Fish. Aquat. Sci.* **2006**, *63*, 377–382. [CrossRef]
61. Sharma, M.; Chadha, P.; Borah, M.K. Fish Behaviour and Immune Response as a Potential Indicator of Stress Caused by 4-Nonylphenol. *Am. J. BioScience* **2015**, *3*, 278–283. [CrossRef]
62. Chandrasekar, G.; Arner, A.; Kitambi, S.S.; Dahlman-Wright, K.; Lendahl, M.A. Developmental toxicity of the environmental pollutant 4-nonylphenol in zebrafish. *Neurotoxicology Teratol.* **2011**, *33*, 752–764. [CrossRef]
63. Mao, Z.; Zheng, Y.L.; Zhang, Y.Q. Behavioral impairment and oxidative damage induced by chronic application of nonylphenol. *Int. J. Mol. Sci.* **2011**, *12*, 114–127. [CrossRef]
64. Xia, Y.; Niu, C.; Pei, X. Effects of chronic exposure to nonylphenol on locomotor activity and social behavior in zebrafish (*Danio rerio*). *J. Environ. Sci.* **2010**, *22*, 1435–1440. [CrossRef]
65. Yokota, H.; Seki, M.; Maeda, M.; Oshima, Y.; Tadokoro, H.; Honjo, T.; Kobayashi, K. Life-Cycle Toxicity of 4-Nonylphenol To Medaka (Oryzias Latipes). *Environ. Toxicol. Chem.* **2001**, *20*, 2552–2560. [CrossRef] [PubMed]
66. Schultz, M.M.; Bartell, S.E.; Schoenfuss, H.L. Effects of triclosan and triclocarban, two ubiquitous environmental contaminants, on anatomy, physiology, and behavior of the fathead minnow (Pimephales promelas). *Arch. Environ. Contam. Toxicol.* **2012**, *63*, 114–124. [CrossRef] [PubMed]
67. Barros, S.; Montes, R.; Quintana, J.B.; Rodil, R.; Oliveira, J.M.A.; Santos, M.M.; Neuparth, T. Chronic effects of triclocarban in the amphipod Gammarus locusta: Behavioural and biochemical impairment. *Ecotoxicol. Environ. Saf.* **2017**, *135*, 276–283. [CrossRef]
68. Suzuki, M.; Shimizu, T. Is SPINK1 gene mutation associated with development of pancreatic cancer? New insight from a large retrospective study. *Ebiomedicine* **2019**, *50*, 5–6. [CrossRef]
69. Räsänen, K.; Itkonen, O.; Koistinen, H.; Stenman, U. Emerging roles of SPINK1 in cancer. *Clin. Chem.* **2016**, *62*, 449–457. [CrossRef]
70. Guo, M.; Zhou, X.; Han, X.; Zhang, Y.; Jiang, L. SPINK1 is a prognosis predicting factor of non-small cell lung cancer and regulates redox homeostasis. *Oncol. Lett.* **2019**, *18*, 6899–6908. [CrossRef]
71. Ajao, C.; Andersson, M.A.; Teplova, V.V.; Nagy, S.; Gahmberg, C.G.; Andersson, L.C.; Hautaniemi, M.; Kakasi, B.; Roivainen, M.; Salkinoja-Salonen, M. Mitochondrial toxicity of triclosan on mammalian cells. *Toxicol. Rep.* **2015**, *2*, 624–637. [CrossRef]
72. Dinwiddie, M.T.; Terry, P.D.; Chen, J. Recent evidence regarding triclosan and cancer risk. *Int. J. Environ. Res. Public Health* **2014**, *11*, 2209–2217. [CrossRef] [PubMed]

73. Wang, H.; Sha, W.; Liu, Z.; Chi, C. Effect of chymotrypsin C and related proteins on pancreatic cancer cell migration. *Acta Biochim. Biophys. Sin.* **2011**, *43*, 362–371. [CrossRef]
74. Casals-Casas, C.; Desvergne, B. Endocrine disruptors: From endocrine to metabolic disruption. *Annu. Rev. Physiol.* **2011**, *73*, 135–162. [CrossRef] [PubMed]

Article

Transgenerational Effects of Prenatal Endocrine Disruption on Reproductive and Sociosexual Behaviors in Sprague Dawley Male and Female Rats

Bailey A. Kermath [1], Lindsay M. Thompson [2], Justin R. Jefferson [2], Mary H. B. Ward [2] and Andrea C. Gore [1,2,*]

[1] Institute for Neuroscience, The University of Texas at Austin, Austin, TX 78712, USA; baileykermath@gmail.com

[2] Division of Pharmacology & Toxicology, College of Pharmacy, The University of Texas at Austin, Austin, TX 78712, USA; lindsay.thompson82@utexas.edu (L.M.T.); justin.r.jefferson@gmail.com (J.R.J.); maryhbw@gmail.com (M.H.B.W.)

* Correspondence: andrea.gore@austin.utexas.edu; Tel.: +1-512-471-3669; Fax: +1-512-471-5002

Citation: Kermath, B.A.; Thompson, L.M.; Jefferson, J.R.; Ward, M.H.B.; Gore, A.C. Transgenerational Effects of Prenatal Endocrine Disruption on Reproductive and Sociosexual Behaviors in Sprague Dawley Male and Female Rats. *Toxics* 2022, *10*, 47. https://doi.org/10.3390/toxics10020047

Academic Editors: Tracie Baker and Jessica Plavicki

Received: 7 December 2021
Accepted: 14 January 2022
Published: 20 January 2022

Publisher's Note: MDPI stays neutral with regard to jurisdictional claims in published maps and institutional affiliations.

Copyright: © 2022 by the authors. Licensee MDPI, Basel, Switzerland. This article is an open access article distributed under the terms and conditions of the Creative Commons Attribution (CC BY) license (https://creativecommons.org/licenses/by/4.0/).

Abstract: Endocrine-disrupting chemicals (EDCs) lead to endocrine and neurobehavioral changes, particularly due to developmental exposures during gestation and early life. Moreover, intergenerational and transgenerational phenotypic changes may be induced by germline exposure (F2) and epigenetic germline transmission (F3) generation, respectively. Here, we assessed reproductive and sociosexual behavioral outcomes of prenatal Aroclor 1221 (A1221), a lightly chlorinated mix of PCBs known to have weakly estrogenic mechanisms of action; estradiol benzoate (EB), a positive control; or vehicle (3% DMSO in sesame oil) in F1-, F2-, and F3-generation male and female rats. Treatment with EDCs was given on embryonic day (E) 16 and 18, and F1 offspring monitored for development and adult behavior. F2 offspring were generated by breeding with untreated rats, phenotyping of F2s was performed in adulthood, and the F3 generation were similarly produced and phenotyped. Although no effects of treatment were found on F1 or F3 development and physiology, in the F2 generation, body weight in males and uterine weight in females were increased by A1221. Mating behavior results in F1 and F2 generations showed that F1 A1221 females had a longer latency to lordosis. In males, the F2 generation showed decreased mount frequency in the EB group. In the F3 generation, numbers of ultrasonic vocalizations were decreased by EB in males, and by EB and A1221 when the sexes were combined. Finally, partner preference tests in the F3 generation revealed that naïve females preferred F3-EB over untreated males, and that naïve males preferred untreated over F3-EB or F3-A1221 males. As a whole, these results show that each generation has a unique, sex-specific behavioral phenotype due to direct or ancestral EDC exposure.

Keywords: endocrine-disrupting chemical (EDC); polychlorinated biphenyl (PCB); Aroclor 1221 (A1221); transgenerational; social behavior; mating behavior; paced mating; ultrasonic vocalization (USV); estradiol

1. Introduction

Endocrine-disrupting chemicals (EDCs) interfere with hormone action within an organism [1,2]. These chemicals, or mixture of chemicals, act upon the neuroendocrine systems that govern physiological processes such as reproduction, immune function, metabolism, and sex-typical behaviors in adulthood. Exposure to environmental EDCs during critical periods of development such as gestation can alter the organization of these neuroendocrine systems and predispose organisms towards disease and maladaptive traits. Known as the Developmental Origins of Health and Disease or DOHaD [3], this phenomenon has been well studied for a variety of health outcomes in individuals who experienced direct exposure early in life (F1 generation). Regarding neuroendocrine functions and hormone-dependent behaviors, the focus of this study, exposures to EDCs including bisphenol A

(BPA), phthalates, and persistent organic pollutants such as polychlorinated biphenyls (PCBs) induce adverse phenotypic outcomes in animal studies [4–17], and are associated with increased prevalence of neurobehavioral disorders in epidemiological studies in humans [18–22].

EDCs also exert actions on the F2 generation, exposed as germ cells within the F1 embryo. The F3 generations and beyond can exhibit phenotypic changes in the absence of direct exposure, presumably through germline epigenetic inheritance [23,24]. Although few in number, studies on inter- and transgenerational effects of EDCs have reported sexually dimorphic effects on behaviors, especially those influenced by early life endogenous hormones ([16,25–32]; reviewed in [33]). More research comparing generational effects is needed to better understand how legacy chemicals that are no longer actively manufactured but are still persistent in the environment, such as PCBs, may lead to heritable effects generations later.

The current study aims to build upon previous studies in the lab that identified transgenerational effects of PCBs on physiology, behavior, and hypothalamic gene expression throughout development [29,30,34,35]. Here, we extend these findings by examining mating behavior and sociosexual ultrasonic vocalization and partner preference activity in the F1, F2 and F3 generations to show sex- and generation-specific disruption in adult female and male rats.

2. Materials and Methods

2.1. Experimental Design and Animal Husbandry

All animal protocols were conducted in accordance with NIH and USDA guidelines and were approved by the Institutional Animal Care and Use Committee (IACUC) at The University of Texas at Austin. Sprague Dawley rats were obtained from Harlan Laboratories (Houston, TX, USA), switched to the low-phytoestrogen Harlan-Teklad 2019 Global Diet ad libitum, and housed in same-sex groups (2–3 per cage) under constant humidity and temperature (21–22 °C) and a partially reversed 12:12 L:D cycle (lights on at 2400 h). Virgin females were impregnated in house. The morning after a sperm-positive vaginal smear was termed embryonic day (E) 1. On E16 and E18, during the period of sexual differentiation of the brain, F0 dams were weighed and randomly injected with one of three treatment groups: 1 mg/kg Aroclor 1221 (A1221, an estrogenic PCB mixture, administered intraperitoneally [i.p.]), 50 µg/kg estradiol benzoate (EB; administered subcutaneously [s.c.]), or a negative vehicle control (3% DMSO in sesame oil, injected i.p. or s.c., and combined into one DMSO group). Dosages and routes were selected to be identical to other studies in our lab and to be human relevant [35–39]. F0 litters were spread over 6 cohorts for a total of: DMSO, $n = 14$; EB, $n = 11$; A1221, $n = 12$.

Behavioral and physiological reproductive endpoints were examined after rats reached sexual maturity, using 1 male and female from each litter (Figure 1). F1 males and females were examined for sexual behaviors as young adults (P60) while mated to naïve rats (purchased from Harlan). After behavioral testing, F1 females carried litters to term. F2 offspring were also observed for sexual behavior during mating at P60 and the pregnant F2 dams carried the F3 generation to term. Finally, F3 maternal-maternal lineage females and paternal-paternal lineage males were examined for adult sociosexual behaviors (P60–120). A set of untreated rats (UNT, $n = 6$) were raised in the lab alongside the F3 offspring as an additional negative control group, in which dams were restrained and finger-poked to simulate an injection. Harlan-raised males and females used for F3 sociosexual experiments were received at 2 months of age and allowed to acclimate to the lab for 3–4 weeks before experimentation.

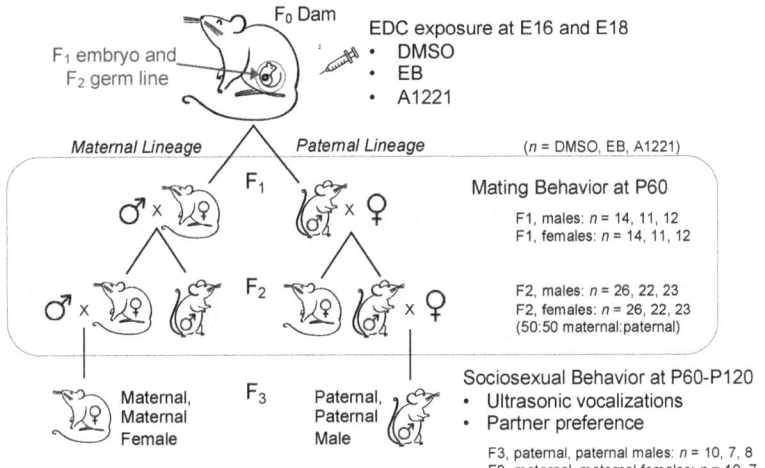

Figure 1. The transgenerational experimental design. Abbreviations: EDC: endocrine-disrupting chemical, E: embryonic day, DMSO: dimethyl sulfoxide, EB: estradiol benzoate, A1221: Aroclor 1221, and P: postnatal day. Gray shading indicates those generations used in the current study for mating behaviors. The F3 generation was used for sociosexual behaviors.

2.2. Tissue Collection

Males and female rats were euthanized between P113–127. For all rats, adrenals and gonads were removed, weighed, and normalized to body weight. Trunk blood was collected from F1 and F2 rats, allowed to clot and spun at $1500 \times g$ for 5 min. Serum was separated and stored at $-80\ ^\circ$C until further analysis.

2.3. Serum Hormone Assays

F1- and F2-generation serum samples were used to investigate the concentrations of circulating testosterone (males) and estradiol (females). Concentrations of serum testosterone were detected in duplicate using an RIA kit, as recommended by the manufacturer (Cat. No. 07189102, MP Biomedicals, Santa Ana, CA, USA). The assay range was 0.1–10 ng/mL, assay sensitivity 0.03 ng/mL and intra-assay variability 1.8%. Serum estradiol samples were run in duplicate using the estradiol RIA kit (Cat. No. DSL-4800, Beckman Coulter, Brea, CA, USA). The assay range was 5–720 pg/mL, assay sensitivity 2.2 pg/mL and intra-assay variability 3.0%.

2.4. Ovariectomy and Hormone Priming for Sociosexual Experiments

Stimulus females used in the ultrasonic vocalization testing were ovariectomized. During surgery, an estradiol Silastic capsule was placed s.c. between the shoulder blades. After recovery, these rats received a s.c. dose of 590 μg progesterone 4 h prior to use to induce receptivity. For the other behaviors, females remained ovarian-intact but were hormone-primed to ensure receptivity during experiments. Ovarian-intact females were given 50 μg estradiol s.c. 52 h, and 590 μg progesterone 4 h, prior to behavioral testing [32]. In all cases, receptivity was confirmed with a sexually experienced male that was otherwise not used in the experiment.

2.5. Reproductive Behavior and Fertility in F1 and F2 Rats

To determine whether prenatal endocrine disruption adversely affects adult reproductive behavior in the F1 and F2 generations, mating trials were conducted at P60 in a non-paced setting. F1 and F2 females were tested on the day of behavioral estrus with a sexually experienced, Harlan-purchased male. F1 and F2 males were tested with sexually naïve, Harlan females in behavioral estrus. Mating trials were performed under dim red light and videotaped for subsequent scoring. Males were acclimated for 10 min to the mating chamber (30 × 38 cm) 5 h before the trial and then returned to the same chamber for 5 min immediately before the trial start at 1600 h. The start time was recorded when the female was placed into the mating chamber. Trials only proceeded if the female was receptive and the male displayed mounting behavior within the first 20 min.

Videos were scored by an experimenter blind to treatment for the following male sexual behaviors: mount frequency, intromission frequency, latencies to mount, intromit, and ejaculate, and the postejaculatory interval (PEI). Because the experimental males were sexually inexperienced and thus slow to display mating behavior, their ejaculation latencies and PEI scores were capped at 30 min after the first mount and 15 min after ejaculation, respectively. Intromission rate was calculated as number of intromissions over the number of mounts with or without penetration. Copulatory rate was calculated as the number of mounts and intromissions from the start time until ejaculation. Female sexual behaviors scored were proceptive (hops and darts only, as ear wiggling could not be scored from the videotape), receptive (lordosis quotient, or the percentage of lordosis responses for the first 10 male mounts, and lordosis intensity score, rating the magnitude of each spinal dorsiflexion from 0 to 3, with 0 representing no spinal dorsiflexion and 3 an exaggerated dorsiflexion and head and rump elevation) and rejection (kicking, boxing, biting, escape, rolling) behaviors for the first 10 male copulatory acts. We further calculated the proceptive rate and rejection rate as the number of acts over the time scored and the latency to display the first lordotic response.

2.6. USV Recording in Sociosexual Context in F3 Rats

USVs were elicited in a sociosexual context for the F3 generation and recorded in a glass chamber (30 × 76 × 45 cm) equipped with an ultrasonic microphone (CM16, Avisoft Bioacoustics, Glienicke/Nordbahn, Germany), as published [6,29]. USVs were sampled at a 250 kHz sampling rate with 16-bit resolution through an A/D card (National Instruments, Austin, TX, USA) using RECORDER NA-DAQ software (v4.2.16, Avisoft Bioacoustics, Glienicke/Nordbahn, Germany). All trials were performed 1–3 h after lights off under dim red light. Experimental rats were sexually naïve, F3 EDC- and control-lineage males and females, aged P60–P120. F3 females were ovarian-intact and hormone-primed to be receptive on the final day of testing. Each experimental rat underwent three separate days of trials, following a previously validated protocol [40]. Days 1 and 2 consisted of a 10-min trial in the recording chamber to habituate the animals and obtain baseline USV recordings. On the final day, a sexually experienced stimulus rat of the opposite sex was placed into the chamber with the experimental rat, separated by a wire mesh partition. They were allowed to interact through the mesh wire for 5 min at which point the stimulus rat was removed from the room and 10 min of USVs were recorded from the experimental rat. Recorded USVs were analyzed with SASlab Pro software (v5.2.07, Avisoft Bioacoustics, Glienicke/Nordbahn, Germany), which automatically measures the number and acoustic parameters of USVs. Sonograms were generated under a 512 FFT-length and 75% overlap frame setup. As flat 50 kHz USVs may have unique communicative properties compared to calls with frequency modulation, USVs were separated into flat and frequency-modulated (FM) calls using an unbiased and replicable technique that categorizes USVs based on their bandwidth, or the maximum peak frequency minus the minimum peak frequency. Calls with a bandwidth of 5 kHz or more were classified as FM and a bandwidth of less than 5 kHz as flats (non-FM) [41,42]. The total number of 50 kHz USVs, number of FM and non-FM calls for the first 5 min of each recording session were analyzed.

2.7. Partner Preference in F3 Rats

F3 EDC- and control-lineage males and females were used after USV testing, approximately 4–7 h after lights off under dim red light. Partner preference trials were conducted as previously described [32]. All rats were sexually naïve and all females remained gonadally intact but were hormone primed to be receptive on the final day of testing. Trials were conducted in a glass arena (122 × 46 cm) and recorded by a video camera connected to ANY-maze software (v4, Stoelting Co., Wood Dale, IL, USA). In order to determine whether F3 EDC- or control-lineage rats could be distinguished from untreated animals in a mating-induced partner preference paradigm, we placed an F3 experimental rat (A1221-, EB-, DMSO-lineage or UNT) opposite a Harlan rat as the Stimulus rats. After Stimulus rats were placed on opposing sides of the arena, a Chooser rat of the opposite sex was allowed to explore and interact with the stimulus rats through wire mesh dividers [32]. Harlan-raised males and females were used as the Choosers and were a separate set from those used as stimulus animals.

Chooser rats were habituated to the empty arena in a 10-min trial on days 1 and 2. On day 3, Stimulus rats were placed behind opposite wire mesh dividers and allowed to acclimate for 5 min. Next, the Chooser rat was placed in the center of the arena and given 10 min to explore and interact with the Stimulus rats across the wire mesh. Trials were repeated up to three times, in which the location and identity of the stimulus rats were exchanged to avoid confounding biases. Behaviors (grooming, rearing, facial investigation, contact with Plexiglas dividers, speed) and total time and total time active (the combination of all scored behaviors) spent in each zone were scored by an experimenter blind to treatment and analyzed by ANY-maze software (v4, Stoelting Co., Wood Dale, IL, USA). Behavior from the area immediately surrounding the wire divider of the stimulus rat (the wire zone) was used for analysis. Data from the Harlan stimulus rat were subtracted from the F3-lineage rat to calculate a preference score in which positive numbers indicate more time spent near the F3-lineage rat.

2.8. Statistical Analysis

Data were analyzed with R 4.1.0 [43], the rstatix [44], the emmeans [45], the lme4 [46], the lmerTest [47], and the ARTool [48–50] packages. Scores over 2.5 standard deviations were considered outliers and removed from the analysis. When outliers were present, only one outlier was detected and removed per group with the one exception of the number of proceptive behaviors in the female F2-DMSO group, in which two outliers were removed. Outliers were distributed evenly across groups. Maternal and paternal lines in the F2 generation were combined for statistical analysis as parental lineage did not significantly impact the endpoints examined. For all somatic (F1, F2 and F3 generations) and mating behavior (F1 and F2 only) outcomes, a one-way analysis of variance (ANOVA) was run for Treatment. Kruskal–Wallis tests were used when data did not meet the Levene's homogeneity of variance or Shapiro–Wilk normality tests. Holm-Sidak or Dunn pairwise post hoc comparisons were run when a significant main effect was found. USV parameters were analyzed with a two-way ANOVA for Sex and Treatment. If data did not meet ANOVA assumptions, even after attempts of data transformation techniques, we used the Aligned Rank Transform (ART) for non-parametric factorial ANOVA [49,50] and the corresponding ART-C pairwise post hoc comparisons [48]. Finally, wire zone preference scores from the partner preference test were run separately for males and females using a linear mixed model with F0 treatment as a Fixed Variable, Animal ID as a Random Variable, and Trial Number as a Repeated Variable. For all data, alpha was set to 0.05.

3. Results

A summary of statistically significant results is provided in Table 1.

Table 1. Summary of significant results.

	Females		Males	
	EB	A1221	EB	A1221
Somatic (F1, F2, F3)				
Body weight	-	-	-	↑ (F2)
Adrenal weight (normalized)	-	-	↑ (trend, F1 and F3)	-
Uterine weight (normalized)	-	A1221 > EB (F2)	-	-
Hormones (F1, F2)—Estradiol (females), Testosterone (males)—n.s.				
Mating Behaviors (F1, F2)				
Mount frequency	↓ (F2)	-	-	-
Lordosis latency	-	↑ (F1)	-	-
Partner Preference Behavior (F3)				
Total time active	↓	↓ (trend)	-	-
Time rearing	↓	↓	↑ (trend)	-
# Rearing bouts	↓ (trend)	-	-	-
Time at Plexiglas	↓	↓ (trend)	↑	-
# Plexiglas bouts	↓	-	-	-
Ultrasonic Vocalizations (F3)				
(combined for the sexes)	EB	A1221		
# Total calls	↓	↓		
# Non-FM calls	↓			

↓ Decreased compared to DMSO (or UNT for Partner Preference). ↑ Increased compared to DMSO (or UNT for Partner Preference). FM: Frequency modulated. n.s. and -: No significant effects. #: Number.

3.1. Transgenerational Somatic Changes

3.1.1. Males

Few somatic changes were detected in the measured outcomes for EDC-lineage rats. A trend was observed for an F0 treatment effect (EB slightly larger than DMSO) in normalized adrenal weights of F1 ($F_{(2,33)} = 2.997$, $p = 0.064$) and F3 ($F_{(2,25)} = 2.856$, $p = 0.076$) males (Table 2). Similarly, no changes were detected in serum testosterone levels (F3 hormones not measured) or normalized testes weight. However, in the F2 generation, we observed a significant effect of treatment on male body weight ($H(2) = 9.054$, $p = 0.011$) with A1221-lineage males having greater average body weights compared to DMSO controls ($p = 0.008$; Figure 2a).

3.1.2. Females

In females, we found an effect of F0 treatment on normalized uterine weight in the F2 generation ($H(2) = 6.434$; $p = 0.040$), in which A1221-lineage females had greater uterine weights compared to EB (Dunn's post hoc, $p = 0.044$; Table 2). No effect was found for female body weight (Figure 2b), normalized adrenal weight, normalized ovarian weight or serum estradiol (F3 hormones not measured). Hormone priming for the sociosexual tests also resulted in an expected increase in normalized ovarian and uterine weights in the F3 generation, regardless of F0 treatment (Table 2).

Table 2. Table of somatic data for each generation.

F1 MALES	DMSO (n = 14)		EB (n = 11)		A1221 (n = 12)		p-Values
	Mean	±SE	Mean	±SE	Mean	±SE	
Body Weight (g)	475.2	(±10.9)	462.9	(±12.0)	461.7	(±11.6)	n.s
Norm Adrenal Weight (mg)	0.105	(±2.2 × 10^{-3})	*0.113*	(±2.6 × 10^{-3})	0.107	(±2.8 × 10^{-3})	$p = 0.064$
Norm Testes Weight (mg)	8.9	(±0.24)	9.0	(±0.07)	9.3	(±0.25)	n.s.
Serum Testosterone (ng/mL)	1.4	(±0.2)	1.2	(±0.1)	1.1	(±0.2)	n.s.

F2 MALES	DMSO (n = 26)		EB (n = 22)		A1221 (n = 23)		
	Mean	±SE	Mean	±SE	Mean	±SE	
Body Weight (g)	449.3	(±5.5)	461.5	(±8.2)	**482.2**	(±11.4)	$p = 0.011$
Norm Adrenal Weight (mg)	0.108	(±1.8 × 10^{-3})	0.111	(±2.3 × 10^{-3})	0.109	(±2.7 × 10^{-3})	n.s.
Norm Testes Weight (mg)	9.3	(±0.009)	9.2	(±0.16)	9.0	(±0.18)	n.s.
Serum Testosterone (ng/mL)	1.1	(±0.1)	1.2	(±0.2)	0.8	(±0.1)	n.s.

F3 MALES	DMSO (n = 10)		EB (n = 9)		A1221 (n = 9)		
	Mean	±SE	Mean	±SE	Mean	±SE	
Body Weight (g)	472.4	(±15.2)	464.0	(±12.9)	491.3	(±7.0)	n.s.
Norm Adrenal Weight (mg)	0.108	(±4.7 × 10^{-3})	*0.114*	(±3.5 × 10^{-3})	0.100	(±2.7 × 10^{-3})	$p = 0.076$
Norm Testes Weight (mg)	9.1	(±0.27)	9.4	(±0.23)	8.8	(±0.17)	n.s.

F1 FEMALES	DMSO (n = 14)		EB (n = 11)		A1221 (n = 12)		
	Mean	±SE	Mean	±SE	Mean	±SE	
Body Weight (g)	286.2	(±6.3)	290.6	(±8.7)	286.0	(±5.4)	n.s.
Norm Adrenal Weight (mg)	0.219	(±9.5 × 10^{-3})	0.205	(±4.3 × 10^{-3})	0.208	(±5.7 × 10^{-3})	n.s.
Norm Ovarian Weight (mg)	0.540	(±2.3 × 10^{-2})	0.528	(±1.9 × 10^{-2})	0.538	(±1.7 × 10^{-2})	n.s.
Norm Uterine Weight (mg)	1.47	(±0.1)	1.99	(±0.4)	1.71	(±0.2)	n.s.
Serum Estradiol (pg/mL)	22.3	(±2.9)	21.8	(±5.4)	19.0	(±2.3)	n.s.

F2 FEMALES	DMSO (n = 26)		EB (n = 22)		A1221 (n = 23)		
	Mean	±SE	Mean	±SE	Mean	±SE	
Body Weight (g)	283.0	(±3.4)	291.4	(±4.6)	289.6	(±3.7)	n.s.
Norm Adrenal Weight (mg)	0.208	(±3.9 × 10^{-3})	0.197	(±4.9 × 10^{-3})	0.203	(±4.0 × 10^{-3})	n.s.
Norm Ovarian Weight (mg)	0.540	(±1.5 × 10^{-2})	0.555	(±1.7 × 10^{-2})	0.547	(±1.3 × 10^{-2})	n.s.
Norm Uterine Weight (mg)	1.63	(±0.1)	1.45	(±0.1)	**1.97**	(±0.2) *	$p = 0.04$
Serum Estradiol (pg/mL)	15.4	(±1.6)	14.4	(±1.3)	22.2	(±3.4)	n.s.

F3 FEMALES	DMSO (n = 11)		EB (n = 7)		A1221 (n = 9)		
	Mean	±SE	Mean	±SE	Mean	±SE	
Body Weight (g)	285.8	(±7.1)	273.3	(±8.4)	287.4	(±8.3)	n.s.
Norm Adrenal Weight (mg)	0.202	(±6.1 × 10^{-3})	0.208	(±8.3 × 10^{-3})	0.207	(±7.4 × 10^{-3})	n.s.
Norm Ovarian Weight (mg)	0.437	(±4.4 × 10^{-2})	0.447	(±1.8 × 10^{-2})	0.475	(±3.1 × 10^{-2})	n.s.
Norm Uterine Weight (mg)	2.97	(±0.5)	2.33	(±0.3)	4.49	(±1.2)	n.s.

Body weights are shown for the day of euthanasia, with adrenal, ovarian, uterine, and testicular weights measured postmortem, and the ANOVA p-value for a main effect of Treatment. Norm: normalized to body weight. SE: standard error of the mean. n.s.: No significant effects. Bold text indicates significantly different ($p < 0.05$) from DMSO, and italicized text indicates a trend ($0.05 < p < 0.1$) from DMSO in post hoc comparisons. *, A1221 significantly different from EB ($p = 0.04$).

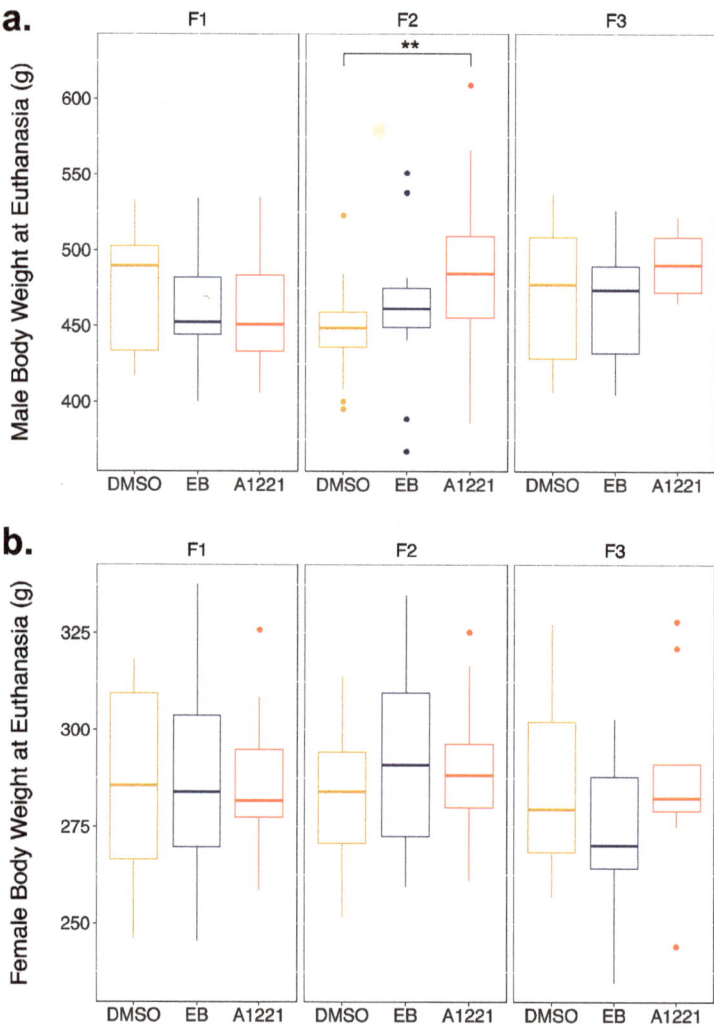

Figure 2. Boxplots of body weight at euthanasia (~P120) for adult (**a**) males and (**b**) females. Data were analyzed by one-way ANOVA or Kruskal–Wallis for effect of F0 treatment, followed by Holm–Sidak or Dunn's pairwise comparisons. ** $p < 0.01$. In this and other boxplot graphs, the line represents the median, the lower and upper outline of the boxes the 25th and 75th percentile, respectively, and the lines the 95th percentile.

3.2. Reproductive Behavior in the F1 and F2 Generations

3.2.1. Males

Overall, few effects of perinatal EDC treatment were found in male mating behavior (Figure 3). In the F2 generation, a significant effect of treatment was found for mount frequency (F(2,66) =3.374; $p = 0.035$). Holm–Sidak post hoc analysis showed that EB-lineage males had a lower mount frequency compared to DMSO ($p = 0.035$; Figure 3e) suggesting that EB males required fewer mounts to reach ejaculation. However, no changes were observed in intromission or ejaculation behaviors (Figure 3). All treatments groups showed long average ejaculation latencies with high variability within the groups, presumably due to male subjects being sexually naïve at the time of testing.

3.2.2. Females

Sexually naïve F1 and F2 females were examined for copulatory, proceptive and receptive behaviors (Figure 4). All females were in behavioral estrus during mating trials and were successfully able to lordose in response to male mounting and intromitting behavior. In F1 females, the latency to display the first lordotic response was affected by treatment (H(33) = 3.83; p = 0.032). Post hoc analysis revealed A1221-exposed females had significantly longer latencies compared to DMSO-exposed females (p = 0.015) despite mounting attempts by a sexually experienced male (Figure 4c). Overall, females displayed high levels of aversive behavior and few proceptive behaviors, likely due to the non-paced setting of the mating trials.

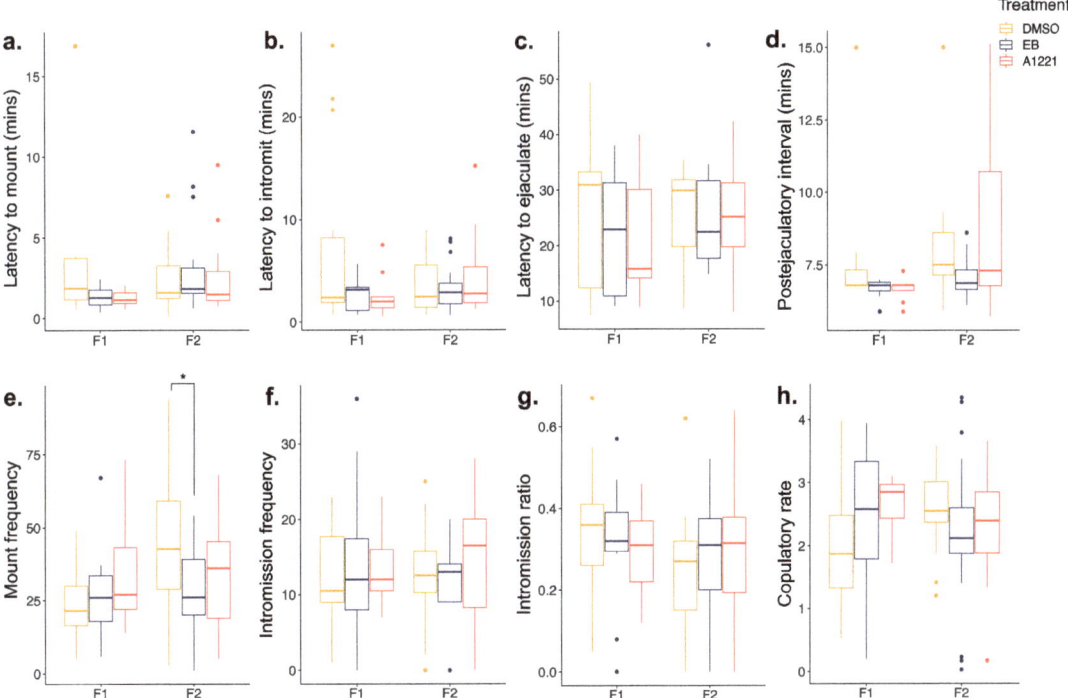

Figure 3. Boxplots of male mating behavior at P60 for F1 and F2 generations. Sexually naïve males were scored for (**a**) latency to mount, (**b**) latency to intromit, (**c**) latency to ejaculate, (**d**) postejaculatory interval, (**e**) mount frequency, (**f**) intromission frequency, (**g**) intromission ratio (calculated as number of intromissions divided by number of mounts), and (**h**) copulatory rate (calculated as the number of mounts and intromissions from the start time until ejaculation). Data were analyzed by one-way ANOVA or Kruskal–Wallis for effect of F0 treatment, followed by Holm–Sidak or Dunn's pairwise comparisons. F1: n = 14 DMSO, 11 EB, 11 A1221; F2: n = 26 DMSO, 22 EB, 22 A1221. * p < 0.05.

3.3. Sociosexual Behaviors in the F3 Generation

3.3.1. Ultrasonic Vocalizations (USVs)

We examined the number and duration of appetitive 50 kHz USVs in F3 adults within a mating context (Figure 5). Experimental females were ovarian-intact but hormone primed to ensure receptivity during testing. To reduce potential variability caused by mixed maternal vs. paternal lineages, only maternal, maternal F3 females and paternal, paternal F3 males were used (see Figure 1). Two-way ANOVA tests revealed significant sex and treatment effects in USVs within the first 5 min of separation from the stimulus rat. In particular, the total call number was significantly affected by sex (F(1,46) = 34.96; $p < 0.001$), F0 treatment (F(3,46) = 6.80; $p < 0.001$) and their interaction (F(3,46) = 5.72; $p = 0.002$; Figure 5c). Post hoc treatment contrasts revealed a trend for EB-lineage males to call less frequently than DMSO controls ($p = 0.065$). When sexes were combined to further examine the treatment main effect, we found that DMSO-lineage controls emitted more USVs than both EB ($p = 0.003$) and A1221 ($p = 0.007$) groups (Figure 5e). There was also a trend for reduced call number between EB-lineage rats and our in-house bred untreated controls (UNT; $p = 0.063$).

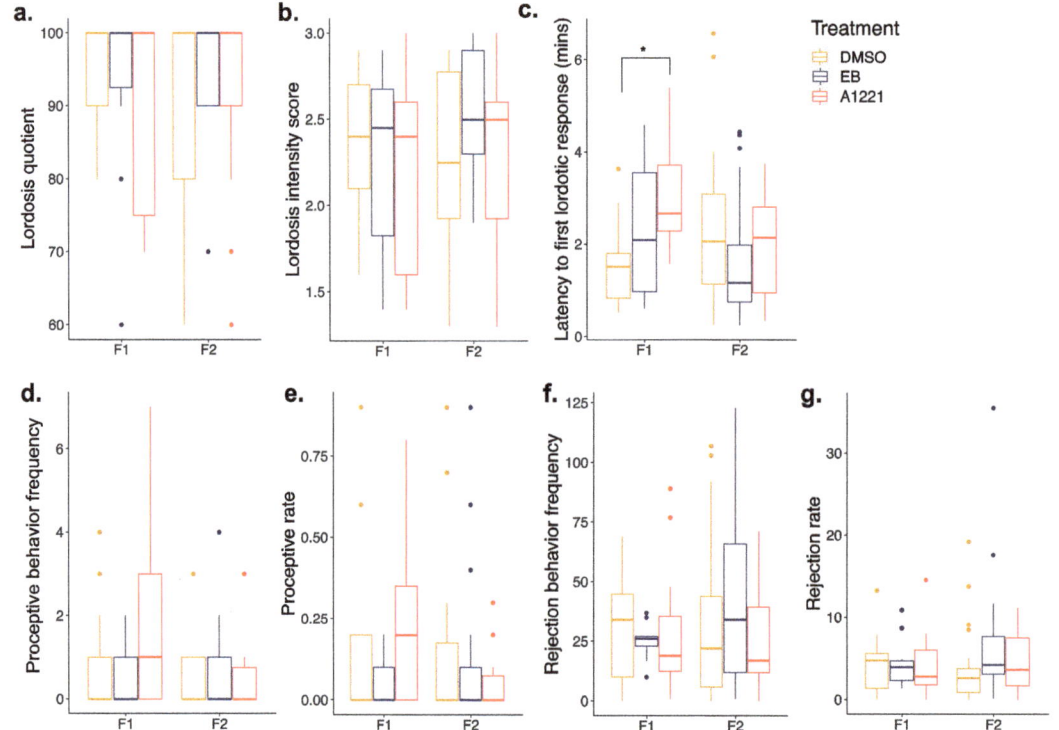

Figure 4. Boxplots of female mating behavior at P60 for F1 and F2 generations. Sexually naïve females were scored for (**a**) lordosis quotient, (**b**) lordosis intensity score, (**c**) latency to first lordosis response, (**d**) proceptive behavior frequency, (**e**) proceptive rate, (**f**) rejection behavior frequency and (**g**) rejection rate. Data were analyzed by one-way ANOVA or Kruskal–Wallis for effect of F0 treatment, followed by Holm–Sidak or Dunn's pairwise comparisons. F1: n = 14 DMSO, 11 EB, 12 A1221; F2: n = 23 DMSO, 22 EB, 23 A1221. * $p < 0.05$.

Similar effects were seen when analyzing two subtypes of USVs: frequency-modulated (FM) and non-FM calls (Table 3). Males emitted both types of calls more frequently (non-FM: F(1,42) = 31.87, $p < 0.001$; FM: F(1,41) = 19.89, $p < 0.001$) and called for longer average durations (F(1,44) = 4.932, $p = 0.032$) than females. The number of non-FM calls was also affected by F0 treatment (F(3,42) = 2.79, $p = 0.024$), with EB males having fewer calls of this subtype than DMSO ($p = 0.031$).

3.3.2. Partner Preference

The partner preference paradigm was used to determine the extent to which F3 EDC- or control-lineage rats would be preferred (or avoided) to untreated animals in a mating context. An F3 experimental rat (A1221-, EB-, DMSO-lineage or UNT) was placed opposite a naïve rat purchased from Harlan as the Stimulus animals. A separate set of naïve, Harlan-purchased Chooser rats (of the opposite sex) interacted with the stimulus rats through wire mesh dividers held in place by Plexiglas. Similar to the USV experiments, females were gonadally intact and hormone primed to be receptive. A blind experimenter scored the Chooser rats' behaviors including grooming, rearing, facial investigation of the stimulus rats through the wire mesh and physical contact with the adjacent Plexiglas. As most behaviors occurred in proximity to the stimulus rats, we focused our analysis to the region adjacent to the wire mesh divider (called the wire zone; Figure 6b). The full set of parameters scored within AnyMaze are listed in Supplemental Table S1.

Table 3. Ultrasonic vocalization parameters for F3 males and females.

MALES	UNT ($n = 5$)		DMSO ($n = 10$)		EB ($n = 6$)		A1221 ($n = 6$)		p-Values	(Sex-Combined)
	Mean	±SE	Mean	±SE	Mean	±SE	Mean	±SE	Treatment	Sex
Number of total calls	101.2	(±23.6)	124.5	(±28.3)	*33.5*	*(±12.3)*	68.8	(±7.8)	**$p < 0.001$**	$p < 0.001$
Number of non-FM calls	48.2	(±10.5)	62.8	(±13.9)	**22.8**	**(±7.5)**	36.5	(±10.1)	**$p = 0.024$**	$p < 0.001$
Number of FM calls	53.0	(±14.5)	47.2	(±11.5)	17.4	(±6.2)	49.5	(±9.4)	n.s.	$p < 0.001$
Percentage of FM calls	50.3	(±4.8)	47.3	(±3.8)	46.2	(±7.1)	59.7	(±4.8)	n.s.	n.s.
Average call duration (ms)	1.44	(±0.22)	1.41	(±0.13)	1.13	(±0.13)	1.61	(±0.19)	n.s.	$p = 0.032$
FEMALES	UNT ($n = 5$)		DMSO ($n = 9$)		EB ($n = 6$)		A1221 ($n = 8$)			
	Mean	±SE	Mean	±SE	Mean	±SE	Mean	±SE		
Number of total calls	10.0	(±2.1)	21.4	(±10.4)	13.8	(±6.7)	7.3	(±4.5)		
Number of non-FM calls	5.8	(±1.5)	10.7	(±4.0)	8.8	(±4.1)	9.6	(±5.0)		
Number of FM calls	4.2	(±0.7)	26.5	(±12.3)	7.8	(±3.6)	4.3	(±2.8)		
Percentage of FM calls	45.0	(±5.7)	40.5	(±10.0)	55.4	(±12.4)	32.0	(±11.8)		
Average call duration (ms)	0.80	(±0.10)	1.06	(±0.22)	1.19	(±0.20)	1.28	(±0.26)		

Two-way ANOVA p-values for a main effect of Treatment and Sex are provided for the sexes combined (shown next to the male data, but applicable to both sexes). Bold text indicates significantly different at $p < 0.05$ from DMSO in post hoc pairwise comparisons within each sex, and italicized text indicates a trend ($0.05 < p < 0.1$). n.s.: No significant effects, FM: frequency modulated.

In trials where naïve female Harlan Choosers were exposed to F3 experimental males, linear mixed modeling (LMM) analysis showed that the females' time spent rearing ($p = 0.044$) and time spent contacting the Plexiglas ($p = 0.020$) were significantly affected by F0 treatment (Figure 6c). Post hoc analysis revealed that Chooser females preferred the EB-lineage males more often than they preferred UNT controls (time rearing, trend $p = 0.082$; time Plexiglas, $p = 0.030$).

When naïve male Harlan Choosers were tested, their total time active ($p = 0.017$), time spent ($p = 0.009$) and number ($p = 0.022$) of rearing bouts, time spent ($p = 0.014$) and number ($p = 0.026$) of bouts contacting the Plexiglas, and number of facial investigation bouts (trend, $p = 0.075$) were affected by F0 treatment (Figure 6d). Post hoc analysis demonstrated that naïve Chooser males avoided EDC-lineage females more frequently than UNT controls (time active UNT vs. A1221 (trend, $p = 0.063$) and UNT vs. EB ($p = 0.049$); rearing number UNT vs. EB (trend, $p = 0.060$); rearing time UNT vs. A1221 ($p = 0.032$) and UNT vs. EB ($p = 0.025$); Plexiglas number UNT vs. EB ($p = 0.045$); Plexiglas time UNT vs. A1221 (trend, $p = 0.056$) and UNT vs. EB (($p = 0.032$); Figure 6d and Supplemental Table S1).

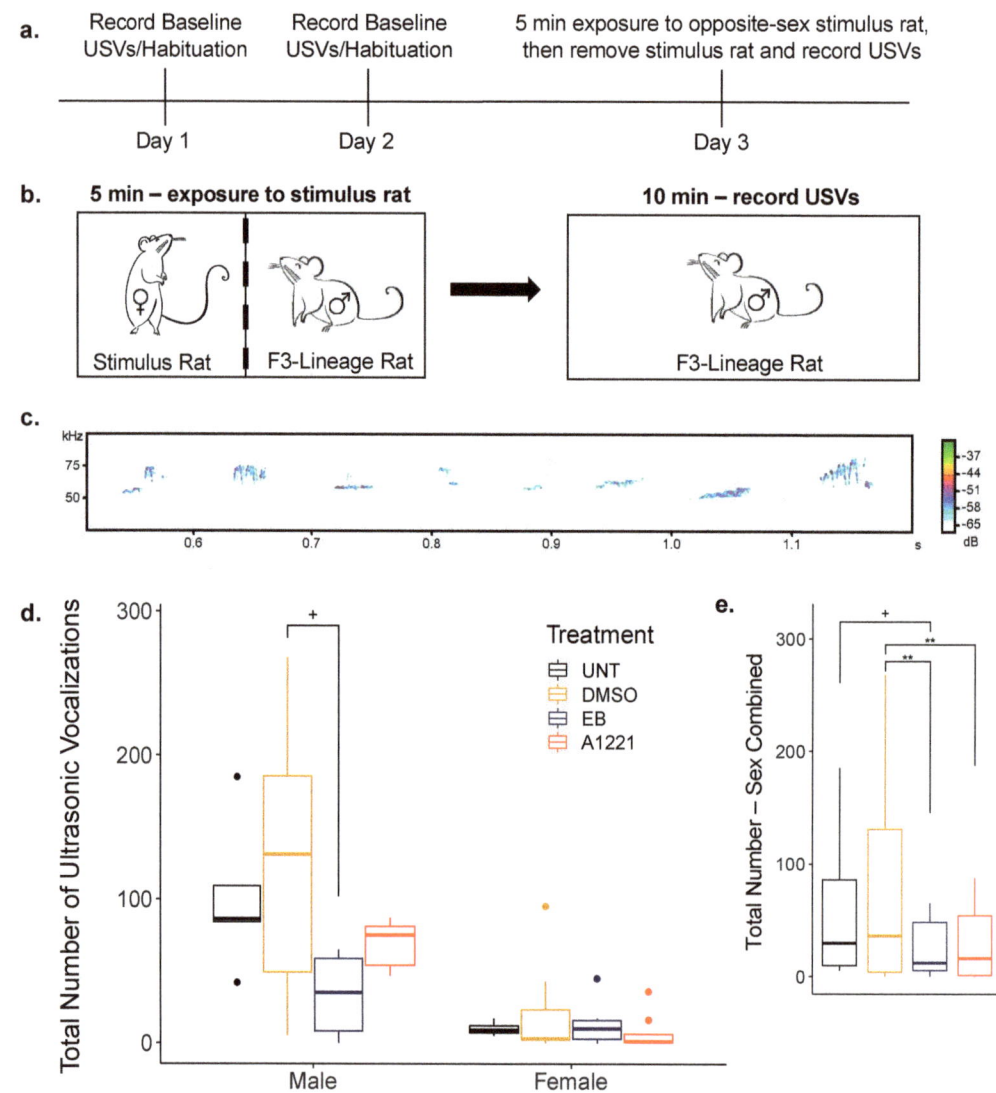

Figure 5. Ultrasonic vocalizations (USVs) emitted by F3−generation males and females (P60−120) in response to an opposite−sex rat. An untreated control group (UNT) was raised across generations in−house alongside the F3 litters. (**a**) Timeline of USV experiment; (**b**) Diagram of experiment on day 3; (**c**) example spectrogram of recorded USVs; (**d**) Boxplots of the total number of USV calls (frequency modulated [FM] and non−FM) during the first 5 min of recording by sex; (**e**) Boxplots of the total number of USV calls with sex combined. Data were analyzed by two−way ANOVA or Aligned Rank Transformation (ART) for effect of F0 treatment and sex, followed by Holm−Sidak or ART−C pairwise comparisons. Males: n = 6 UNT, 10 DMSO, 6 EB, 6 A1221; females: n = 5 UNT, 9 DMSO, 6 EB, 8 A1221. + $p <$ 0.07; ** $p <$ 0.01.

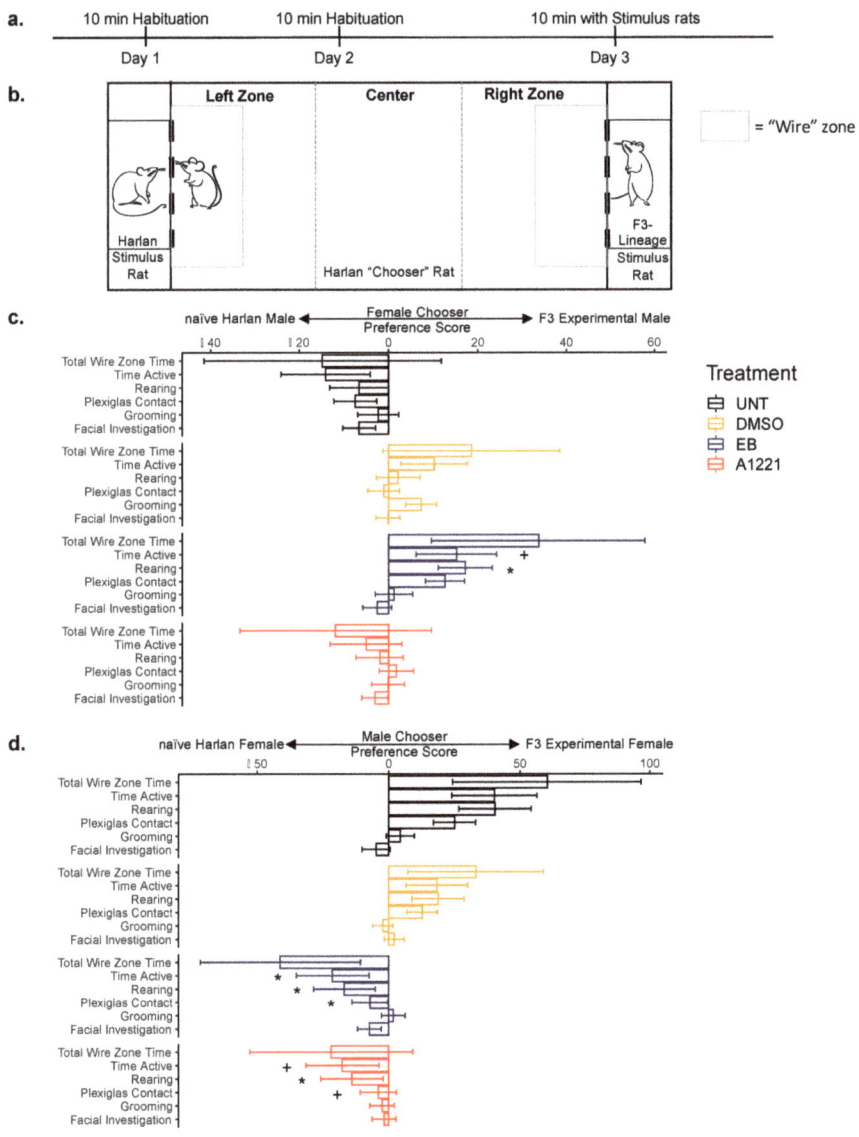

Figure 6. Partner preference (PP) in a mating context by F3-generation males and females (P60–120). The Chooser was a naïve rat purchased from Harlan, given a choice between two opposite-sex rats: an in-lab-generated F3 rat (UNT, DMSO, EB, A1221) and a purchased rat. A preference score was calculated by subtracting time spent with the F3 rat minus time spent with the Harlan rat, in which positive numbers indicate more time spent near the F3-lineage rat and a negative score indicating time towards the Harlan rat. (**a**) Timeline of PP experiment; (**b**) diagram of experiment on day 3, with the wire zone shaded in gray; preference scores from the wire zone for (**c**) naïve female Choosers with F3-lineage males and (**d**) naïve male Choosers with F3-lineage females. Data were analyzed by linear mixed model (LMM) for effect of F0 treatment within each sex followed by Holm–Sidak pairwise comparisons. LMM estimated marginal means and standard errors are graphed. Males: n = 5 UNT, 9 DMSO, 7 EB, 8 A1221; females: n = 5 UNT, 10 DMSO, 7 EB, 7 A1221. * = $p < 0.05$ vs. UNT; + = $p \leq 0.083$ vs. UNT.

4. Discussion

The current study demonstrates that a transient gestational exposure to estrogenic EDCs can significantly alter behaviorally relevant endpoints for at least three generations. Interestingly, this occurred in a sex- and generation-specific manner. We found modest but significant effects on copulatory behavior in F1 females and F2 males. In the F3 generation, EDC treatment decreased appetitive 50 kHz ultrasonic vocalizations in response to a rat of the opposite sex and affected the preference of EDC-lineage males and females for a naïve conspecific in a mating context. Finally, we found few somatic changes in adulthood. It is interesting that multigenerational effects of EDCs are preferentially manifested, at least in this paradigm, in neurobehavioral rather than somatic outcomes, a result that may relate to the exquisite sensitivity of the brain to developmental hormones and its potential for epigenetic programming [51]. However, it is possible that other somatic or biochemical outcomes not examined here are altered by prenatal EDC exposure, as we only measured a few specific endpoints and only at one timepoint in adulthood. For example, previous studies on prenatal PCB exposure found delays in the timing of puberty in males [52] and transgenerational effects on anogenital index and female sex steroid hormone levels at P60 [36].

A1221 has weakly estrogenic activity but also has other mechanistic actions including through thyroid and aromatase-mediated pathways [53,54]. Thus, while A1221 can produce similar effects to EB, it is not a pure estrogen and will often deviate from the EB group due to its non-estrogenic-mediated actions as shown in this and previous studies [6]. Here, EDC exposure was given on days E16 and E18 of gestation during the period of germline epigenetic changes and the beginning of brain sexual differentiation in the rat. Both processes are vulnerable to environmental perturbations and A1221 exposure at this time can cause epimutations that become embedded in the germline, leading to changes in somatic gene expression in later generations [24] and lifelong alterations in sex-typical reproductive physiology and behavior. Studies have found differences in maternal versus paternal lineage transmission of disease phenotypes as there are many sex differences in germline de- and re-methylation dynamics [55]. In this study, both maternal and paternal lineages were investigated in the F2 generation; however, we did not find any significant lineage effects on our endpoints. Finally, due to experimental constraints we were unable to perform experiments on every lineage combination in the F3 generation and instead selected F3 females of maternal, maternal lineage and F3 males of paternal, paternal lineage. This is an important area of future study.

4.1. Transgenerational Somatic Endpoints

Of the somatic changes monitored in the current study, only a few changes were observed in adulthood of EDC-lineage rats, mirroring previous results using this treatment model [36]. Here, we reproduced an increase in body weight at euthanasia in F2-A1221 males. A1221 males had a modest ~7% increase in body weight at euthanasia compared to controls, all given the same ad libitum diet of low-phytoestrogen rat chow. Whether this weight increase is due to increased consumption or a difference in metabolism and energy expenditure between groups should be addressed in future studies. For instance, additional markers of altered metabolism, such as serum insulin or adipokines, could be examined. This finding suggests that PCBs may act according to the "obesogen hypothesis," in which EDC activity can predispose organisms to obesity and metabolic dysfunction. Future research should investigate the extent to which the transgenerational effects of A1221 can synergistically increase weight gain with a high fat diet in adulthood. In the Mennigen et al. (2018) study [36], both F2 and F3 males with A1221 lineage had increased body weight; however, this was primarily driven through the maternal line and our study used only paternal F3 males. Therefore, this discrepancy is likely due to mechanisms of maternal vs. paternal inheritance.

Increased body weight was also previously found in the female F2- and F3-A1221 littermates [36]; however, there are major differences between these subjects and those in the current study. Here, the F2 females carried a litter to term and were euthanized after weaning, and F3 females were euthanized after the completion of all sociosexual experiments at P120. This resulted in females whose age and postpartum status were vastly different from those in the previous study. Similarly, our finding of an increase in the normalized uterine weight of the F2-A1221 females compared to EB, which was not seen previously, could be due to an interaction between EDC-lineage and postpartum status or might be due to differences in cycle status between the groups. Unfortunately, we did not track the cycle status of the females in the present study, as we presumed that females would be roughly distributed throughout the estrous cycle, and this precludes our ability to rule out cycle effects. Finally, our findings agree with previous work showing that treatment of dams with EDCs on gestational days 16 and 18 do not significantly influence serum testosterone or estradiol concentrations in the F1 and F2 generations [36].

4.2. F1 and F2 Generation Adult Mating Behavior

Perturbation of the reproductive axis by estrogenic compounds may affect the expression of sexual behavior in adulthood [56]; thus, we studied the copulatory behavior of the F1 and F2 generations as they were mated to propagate litters for the transgenerational experiment. The timing and setup of the mating trials were designed to replicate the conditions from previous experiments on perinatal EDC treatment in our lab. Therefore, sexually inexperienced EDC-lineage rats were mated to untreated, Harlan-raised rats under non-paced mating conditions.

In the F1 generation, prenatal EDC treatment did not alter male copulatory behavior during their first exposure to sexual experience. While a study using a PCB mixture (PCB 126, 138, 153 and 180) found that prenatal exposure delayed latencies in first and subsequent testing of F1 males [57] our model used a differing PCB mixture that may have differing mechanisms of action. On the other hand, our F1-A1221 females significantly delayed their first lordotic event in response to mounting attempts by a sexually experienced male compared to DMSO. A delay in receptive behavior may indicate a deficiency in copulatory motivation. Similarly, using the same A1221 dose, F1 females in a paced mating paradigm also delayed the pacing of mating encounters and event-return latencies [37] although, in both cases, female lordosis remained intact. Some studies have found EDC effects on lordosis and proceptivity using prenatal endocrine active UV filters [58] or exogenous estradiol [59]. However, other specific PCB mixtures had no influence on female lordosis [57,60] as we found here.

The F2 generation showed a different pattern of results. F2 females had no effects of EDCs in their mating behavior; however, F2 males of EB lineage had a decrease in the mount frequency compared to DMSO. The decrease in the number of mounts did not affect the average intromission ratio, or copulatory efficiency, in which a higher percentage of intromissions to mounts may indicate greater ease to achieve an erection [61]. The decrease is also unlikely to reflect a decreased motivation for sexual activity because the latencies to mount and intromit, better indices of motivation, were not affected. In any case, F2 male mating behavior was not severely impacted by either EDC treatments, at least when comparing the initial sexual event. Future studies should address whether reproductive behavior after repeated sexual experience trials reveals other significant effects.

4.3. F3 Generation Adult Sociosexual Behaviors

Ultrasonic vocalizations are emitted by rodents throughout development and are thought to represent affective states and possibly facilitate communication. In adulthood, rat USVs can be characterized by two main types: 22 kHz calls, associated with aversive stimuli, and those in the 50 kHz or above range, associated with arousal states and positive affect [62]. Rats produce a high rate of 50 kHz calls during positive social interactions such as reproductive behavior, juvenile play and tickling by an experimenter. While the 22 kHz calls are emitted by males after ejaculation, the 50 kHz calls are associated with solicitation and copulatory acts [63]. In this study, we used a well-documented paradigm for inducing 50 kHz calls through a brief exposure to a hormonally receptive rat of the opposite sex [40]. Upon removal of the stimulus animal, rats will reliably produce 50 kHz calls. We also added an additional negative control group of untreated rats (UNT) bred in-house alongside our F3 generation. Both negative control groups, UNT and F3-DMSO, behaved similarly. We found a decrease in 50 kHz USV production with EDC lineage, particularly in males. Unfortunately, due to a low n per group, our study was underpowered. However, when sexes were combined, we were able to see statistically significant decreases in both EB and A1221 groups compared to DMSO control. As 50 kHz calls appear to facilitate mating interactions by signaling a readiness to mate and orienting the activity of the estrous female [63,64], a decrease in USV calls may indicate a deficit in reproductive fitness.

The 50 kHz calls often display variation in subtype and can be roughly separated into frequency-modulated (FM) or non-FM calls. Although the functional implications are not fully understood for these subtypes, FM calls may signal a dopamine-dependent reward state and are preferentially increased in anticipation of cocaine and amphetamine [41,65]. In contrast, flat calls appear to help coordinate social behavior as they are evoked after separation from cage-mates or potential mates and can induce approach behavior in both mating and non-mating environments [66]. Our findings show a decrease in non-FM calls, which would include the flat subtype, with EB lineage. This may suggest a deficit in the coordination of reproductive behavior instead of a decreased motivation to mate. Interestingly, when F3-A1221 pups were separated from their mother, the rate of neonatal USVs were also decreased in paternal-lineage pups [29], so this effect appears to be consistent throughout development.

In this study, we observed notable sex differences in USV calls, with males calling more frequently and for longer call duration than females. While males are known to emit more 50 kHz calls during rough-and-tumble play behavior than females [67], the two sexes generally produce similar call rates during mating encounters [29,68]. Acquisition of sexual experience and hormonal status of both the experimental and stimulus rats can affect the number of vocalizations [68–70]. In naturally cycling females, calls are maximized during proestrus compared to the other cycle states as well as after hormone administration in ovariectomized females [68,71]. The sexually inexperienced females of this study remained ovarian-intact but were supplemented with both estradiol and progesterone to induce the appropriate physiological state. Further, receptivity was confirmed (a lordosis response to an experienced male's mount) prior to the experiment. Unfortunately, this setup failed to induce vocalizations in the females, while males produced calls at a similar rate to that seen in sexually naïve males in the same paradigm [40]. Future studies should assess female USV production during the appropriate stage of their estrous cycle to determine if calls are increased during their natural behavioral estrus. Thus, while the EDC effects appear to be driven solely by paternal-lineage males, our interpretation of F3 female behavior must take into account that this floor effect may mask further decreases in females USV production.

Finally, we investigated whether F3 rats inherited indicators of reproductive deficits from their EDC ancestry. To test this hypothesis, we allowed naïve Chooser rats to select from an F3 experimental rat or a naïve rat (raised at Harlan), using a partner preference paradigm that previously showed a female preference for F3-vehicle males over F3-vinclozolin males (in that study, males showed no preference for either type of female; [32]). Here, we made the surprising observation of a distinction between our in-house negative controls (UNT and F3-DMSO) and the naïve Harlan stimulus rats, especially when males were choosing between females. While males on average tended to prefer UNT and DMSO females compared to naïve Harlan females, they tended to avoid F3-EB and F3-A1221 females. These results emphasize the importance of negative controls, as environmental factors such as rearing environment (in-house vs. Harlan) can affect behavior. Conversely, when naïve females were Choosers, they showed higher preference scores for F3-EB males than for the UNT controls. This study did not attempt to determine the basis for the differences in choice, although this result is particularly interesting in the context of decreased USV production seen in F3-EB males. Other physical stimuli, such as pheromones, and behavioral cues, also play a role in mate choice, and may outweigh any deficits in social USV calls.

5. Conclusions

These results show that prenatal EDC treatment has distinct effects within each generation, in a sexually dimorphic manner, showing the complexity of studying inheritance of EDC exposure. These results extend and complement other data showing transgenerational studies on EDCs as well as other environmental stressors [72] that influence health and disease.

Supplementary Materials: The following supporting information can be downloaded at: https://www.mdpi.com/article/10.3390/toxics10020047/s1, Table S1: Table of estimated marginal means and standard errors for F3 partner preference scores.

Author Contributions: Conceptualization, B.A.K. and A.C.G.; investigation, B.A.K., L.M.T., J.R.J. and M.H.B.W.; methodology, B.A.K., A.C.G. and L.M.T.; data curation, B.A.K. and L.M.T.; formal analysis, B.A.K.; writing—original draft preparation, B.A.K. and A.C.G.; writing—review and editing, A.C.G.; visualization, B.A.K. and A.C.G.; project administration, L.M.T. and B.A.K.; funding acquisition, A.C.G. All authors have read and agreed to the published version of the manuscript.

Funding: This work was supported by NIH (NIEHS) R01 ES023254 and R01 ES029464 grants.

Institutional Review Board Statement: This study was conducted according to the guidelines of the NIH and approved by the IACUC at the University of Texas at Austin (AUP-2016-00029, approved 3 July 2016).

Informed Consent Statement: Not applicable.

Data Availability Statement: Data will be made available upon request.

Acknowledgments: We thank Fay Guarraci and Juan Dominguez for their guidance in scoring female and male reproductive behavior, respectively. We also thank Spurthi Tarugu for scoring the USV data and Mandee Bell for help with animal husbandry. Vector icons were obtained through Vecteezy.com (syringe) and Shutterstock (rats).

Conflicts of Interest: The authors declare no conflict of interest. The funder had no role in the design of the study; in the collection, analyses, or interpretation of data; in the writing of the manuscript, or in the decision to publish the results.

References

1. Zoeller, R.T.; Brown, T.R.; Doan, L.L.; Gore, A.C.; Skakkebaek, N.E.; Soto, A.M.; Woodruff, T.J.; Vom Saal, F.S. Endocrine-Disrupting Chemicals and Public Health Protection: A Statement of Principles from The Endocrine Society. *Endocrinology* **2012**, *153*, 4097–4110. [CrossRef]
2. Gore, A.C.; Chappell, V.A.; Fenton, S.E.; Flaws, J.A.; Nadal, A.; Prins, G.S.; Toppari, J.; Zoeller, R.T. EDC-2: The Endocrine Society's Second Scientific Statement on Endocrine-Disrupting Chemicals. *Endocr. Rev.* **2015**, *36*, E1–E150. [CrossRef] [PubMed]
3. Barker, D.J.P. The developmental origins of adult disease. *Eur. J. Epidemiol.* **2003**, *18*, 733–736. [CrossRef] [PubMed]
4. Reilly, M.P.; Weeks, C.D.; Crews, D.; Gore, A.C. Application of a Novel Social Choice Paradigm to Assess Effects of Prenatal Endocrine-Disrupting Chemical Exposure in Rats (Rattus norvegicus). *J. Comp. Psychol.* **2018**, *132*, 253–267. [CrossRef]
5. Reilly, M.P.; Weeks, C.D.; Topper, V.Y.; Thompson, L.M.; Crews, D.; Gore, A.C. The effects of prenatal PCBs on adult social behavior in rats. *Horm. Behav.* **2015**, *73*, 47–55. [CrossRef]
6. Topper, V.Y.; Reilly, M.P.; Wagner, L.M.; Thompson, L.M.; Gillette, R.; Crews, D.; Gore, A.C. Social and neuromolecular phenotypes are programmed by prenatal exposuRes to endocrine-disrupting chemicals. *Mol. Cell Endocrinol.* **2019**, *479*, 133–146. [CrossRef]
7. Jones, B.A.; Watson, N.V. Perinatal BPA exposure demasculinizes males in measuRes of affect but has no effect on water maze learning in adulthood. *Horm. Behav.* **2012**, *61*, 605–610.
8. Tian, Y.H.; Baek, J.H.; Lee, S.Y.; Jang, C.G. Prenatal and postnatal exposure to bisphenol a induces anxiolytic behaviors and cognitive deficits in mice. *Synapse* **2010**, *64*, 432–439. [CrossRef]
9. Mhaouty-Kodja, S.; Belzunces, L.P.; Canivenc, M.C.; Schroeder, H.; Chevrier, C.; Pasquier, E. Impairment of learning and memory performances induced by BPA: Evidences from the literature of a MoA mediated through an ED. *Mol. Cell Endocrinol.* **2018**, *475*, 54–73. [CrossRef] [PubMed]

10. Jasarevic, E.; Williams, S.A.; Vandas, G.M.; Ellersieck, M.R.; Liao, C.; Kannan, K.; Roberts, R.M.; Geary, D.C.; Rosenfeld, C.S. Sex and dose-dependent effects of developmental exposure to bisphenol A on anxiety and spatial learning in deer mice (Peromyscus maniculatus bairdii) offspring. *Horm. Behav.* **2013**, *63*, 180–189. [CrossRef]
11. Matsuda, S.; Matsuzawa, D.; Ishii, D.; Tomizawa, H.; Sutoh, C.; Nakazawa, K.; Amano, K.; Sajiki, J.; Shimizu, E. Effects of perinatal exposure to low dose of bisphenol A on anxiety like behavior and dopamine metabolites in brain. *Prog. Neuro-Psychopharmacol. Biol. Psychiatry* **2012**, *39*, 273–279. [CrossRef]
12. Patisaul, H.B.; Bateman, H.L. Neonatal exposure to endocrine active compounds or an ERbeta agonist increases adult anxiety and aggression in gonadally intact male rats. *Horm. Behav.* **2008**, *53*, 580–588. [CrossRef] [PubMed]
13. Xu, X.; Hong, X.; Xie, L.; Li, T.; Yang, Y.; Zhang, Q.; Zhang, G.; Liu, X. Gestational and lactational exposure to bisphenol-A affects anxiety- and depression-like behaviors in mice. *Horm. Behav.* **2012**, *62*, 480–490. [CrossRef] [PubMed]
14. Bell, M.R.; Thompson, L.M.; Rodriguez, K.; Gore, A.C. Two-hit exposure to polychlorinated biphenyls at gestational and juvenile life stages: 1. Sexually dimorphic effects on social and anxiety-like behaviors. *Horm. Behav.* **2016**, *78*, 168–177.
15. Gillette, R.; Reilly, M.P.; Topper, V.Y.; Thompson, L.M.; Crews, D.; Gore, A.C. Anxiety-like behaviors in adulthood are altered in male but not female rats exposed to low dosages of polychlorinated biphenyls in utero. *Horm. Behav.* **2017**, *87*, 8–15. [CrossRef]
16. Quinnies, K.M.; Harris, E.P.; Snyder, R.W.; Sumner, S.S.; Rissman, E.F. Direct and transgenerational effects of low doses of perinatal di-(2-ethylhexyl) phthalate (DEHP) on social behaviors in mice. *PLoS ONE* **2017**, *12*, e0171977. [CrossRef]
17. Carbone, S.; Ponzo, O.J.; Gobetto, N.; Samaniego, Y.A.; Reynoso, R.; Scacchi, P.; Moguilevsky, J.A.; Cutrera, R. Antiandrogenic effect of perinatal exposure to the endocrine disruptor di-(2-ethylhexyl) phthalate increases anxiety-like behavior in male rats during sexual maturation. *Horm. Behav.* **2013**, *63*, 692–699. [CrossRef]
18. Kobrosly, R.W.; Evans, S.; Miodovnik, A.; Barrett, E.S.; Thurston, S.W.; Calafat, A.M.; Swan, S.H. Prenatal phthalate exposuRes and neurobehavioral development scoRes in boys and girls at 6–10 years of age. *Environ. Health Perspect.* **2014**, *122*, 521–528.
19. England-Mason, G.; Martin, J.W.; MacDonald, A.; Kinniburgh, D.; Giesbrecht, G.F.; Letourneau, N.; Dewey, D. Similar names, different results: Consistency of the associations between prenatal exposure to phthalates and parent-ratings of behavior problems in preschool children. *Environ. Int.* **2020**, *142*, 105892. [CrossRef]
20. Yoo, S.J.; Joo, H.; Kim, D.; Lim, M.H.; Kim, E.; Ha, M.; Kwon, H.J.; Paik, K.C.; Kim, K.M. Associations between Exposure to Bisphenol A and Behavioral and Cognitive Function in Children with Attention-deficit/Hyperactivity Disorder: A Case-control Study. *Clin. Psychopharmacol. Neurosci.* **2020**, *18*, 261–269. [CrossRef] [PubMed]
21. Perera, F.; Nolte, E.L.R.; Wang, Y.; Margolis, A.E.; Calafat, A.M.; Wang, S.; Garcia, W.; Hoepner, L.A.; Peterson, B.S.; Rauh, V.; et al. Bisphenol A exposure and symptoms of anxiety and depression among inner city children at 10–12 years of age. *Environ. Res.* **2016**, *151*, 195–202. [CrossRef]
22. Vermeir, G.; Covaci, A.; Van Larebeke, N.; Schoeters, G.; Nelen, V.; Koppen, G.; Viaene, M. Neurobehavioural and cognitive effects of prenatal exposure to organochlorine compounds in three year old children. *BMC Pediatr.* **2021**, *21*, 99. [CrossRef] [PubMed]
23. Anway, M.D.; Cupp, A.S.; Uzumcu, M.; Skinner, M.K. Epigenetic transgenerational actions of endocrine disruptors and male fertility. *Science* **2005**, *308*, 1466–1469. [CrossRef]
24. Gillette, R.; Son, M.J.; Ton, L.; Gore, A.C.; Crews, D. Passing experiences on to future generations: Endocrine disruptors and transgenerational inheritance of epimutations in brain and sperm. *Epigenetics* **2018**, *13*, 1106–1126. [CrossRef] [PubMed]
25. Nesan, D.; Feighan, K.M.; Antle, M.C.; Kurrasch, D.M. Gestational low-dose BPA exposure impacts suprachiasmatic nucleus neurogenesis and circadian activity with transgenerational effects. *Sci. Adv.* **2021**, *7*, eabd1159. [CrossRef]
26. Quinnies, K.M.; Doyle, T.J.; Kim, K.H.; Rissman, E.F. Transgenerational Effects of Di-(2-Ethylhexyl) Phthalate (DEHP) on Stress Hormones and Behavior. *Endocrinology* **2015**, *156*, 3077–3083. [CrossRef] [PubMed]
27. Wolstenholme, J.T.; Edwards, M.; Shetty, S.R.; Gatewood, J.D.; Taylor, J.A.; Rissman, E.F.; Connelly, J.J. Gestational exposure to bisphenol A produces transgenerational changes in behaviors and gene expression. *Endocrinology* **2012**, *153*, 3828–3838. [CrossRef]
28. Wolstenholme, J.T.; Goldsby, J.A.; Rissman, E.F. Transgenerational effects of prenatal bisphenol A on social recognition. *Horm. Behav.* **2013**, *64*, 833–839. [CrossRef]
29. Krishnan, K.; Mittal, N.; Thompson Lindsay, M.; Rodriguez-Santiago, M.; Duvauchelle Christine, L.; Crews, D.; Gore Andrea, C. Effects of the endocrine-disrupting chemicals, vinclozolin and polychlorinated biphenyls, on physiological and sociosexual phenotypes in F2 generation Sprague-Dawley rats. *Environ. Health Perspect.* **2018**, *126*, 097005. [CrossRef] [PubMed]
30. Krishnan, K.; Rahman, S.; Hasbum, A.; Morales, D.; Thompson, L.M.; Crews, D.; Gore, A.C. Maternal care modulates transgenerational effects of endocrine-disrupting chemicals on offspring pup vocalizations and adult behaviors. *Horm. Behav.* **2019**, *107*, 96–109. [CrossRef]
31. Crews, D.; Gillette, R.; Scarpino, S.V.; Manikkam, M.; Savenkova, M.I.; Skinner, M.K. Epigenetic transgenerational inheritance of altered stress responses. *Proc. Natl. Acad. Sci. USA* **2012**, *109*, 9143–9148. [CrossRef]
32. Crews, D.; Gore, A.C.; Hsu, T.S.; Dangleben, N.L.; Spinetta, M.; Schallert, T.; Anway, M.D.; Skinner, M.K. Transgenerational epigenetic imprints on mate preference. *Proc. Natl. Acad. Sci. USA* **2007**, *104*, 5942–5946. [CrossRef]
33. Robaire, B.; Delbes, G.; Head, J.A.; Marlatt, V.L.; Martyniuk, C.J.; Reynaud, S.; Trudeau, V.L.; Mennigen, J.A. A cross-species comparative approach to assessing multi- and transgenerational effects of endocrine disrupting chemicals. *Environ. Res.* **2021**, *204*, 112063. [CrossRef] [PubMed]
34. Krishnan, K.; Hasbum, A.; Morales, D.; Thompson, L.M.; Crews, D.; Gore, A.C. Endocrine-disrupting chemicals alter the neuromolecular phenotype in F2 generation adult male rats. *Physiol. Behav.* **2019**, *211*, 112674. [CrossRef] [PubMed]

35. Gore, A.C.; Thompson, L.M.; Bell, M.; Mennigen, J.A. Transgenerational effects of polychlorinated biphenyls: 2. Hypothalamic gene expression in rats †. *Biol. Reprod.* **2021**, *105*, 690–704. [CrossRef] [PubMed]
36. Mennigen, J.A.; Thompson, L.M.; Bell, M.; Tellez Santos, M.; Gore, A.C. Transgenerational effects of polychlorinated biphenyls: 1. Development and physiology across 3 generations of rats. *Environ. Health* **2018**, *17*, 18. [CrossRef]
37. Steinberg, R.M.; Juenger, T.E.; Gore, A.C. The effects of prenatal PCBs on adult female paced mating reproductive behaviors in rats. *Horm. Behav.* **2007**, *51*, 364–372. [CrossRef]
38. Dickerson, S.M.; Cunningham, S.L.; Patisaul, H.B.; Woller, M.J.; Gore, A.C. Endocrine disruption of brain sexual differentiation by developmental PCB exposure. *Endocrinology* **2011**, *152*, 581–594. [CrossRef]
39. Fitzgerald, E.F.; Belanger, E.E.; Gomez, M.I.; Cayo, M.; McCaffrey, R.J.; Seegal, R.F.; Jansing, R.L.; Hwang, S.A. Polychlorinated biphenyl exposure and neuropsychological status among older residents of upper Hudson River communities. *Environ. Health Perspect.* **2008**, *116*, 209–215. [CrossRef]
40. McGinnis, M.Y.; Vakulenko, M. Characterization of 50-kHz ultrasonic vocalizations in male and female rats. *Physiol. Behav.* **2003**, *80*, 81–88. [CrossRef]
41. Simola, N.; Ma, S.T.; Schallert, T. Influence of acute caffeine on 50-kHz ultrasonic vocalizations in male adult rats and relevance to caffeine-mediated psychopharmacological effects. *Int. J. Neuropsychopharmacol.* **2010**, *13*, 123–132. [CrossRef]
42. Wang, H.; Liang, S.; Burgdorf, J.; Wess, J.; Yeomans, J. Ultrasonic vocalizations induced by sex and amphetamine in M2, M4, M5 muscarinic and D2 dopamine receptor knockout mice. *PLoS ONE* **2008**, *3*, e1893. [CrossRef] [PubMed]
43. R Core Team. *R: A Language and Environment for Statistical Computing*; R Foundation for Statistical Computing: Vienna, Austria, 2021.
44. Kassambara, R. Rstatix: Pipe-Friendly Framework for Basic Statistical Tests. Available online: https://CRAN.R-project.org/package=rstatix (accessed on 13 November 2021).
45. Lenth, R. Emmeans: Estimated Marginal Means, AKA least-Square Means. Available online: https://CRAN.R-project.org/package=emmeans (accessed on 13 November 2021).
46. Bates, D.; Mächler, M.; Bolker, B.; Walker, S.K. Fitting Linear Mixed-Effects Models Using Lme4. *J. Stat. Softw.* **2015**, *67*, 1–48. [CrossRef]
47. Kuznetsova, A.; Brockhoff, P.B.; Christensen, R.H.B. LmerTest Package: Tests in Linear Mixed Effects Models. *J. Stat. Softw.* **2017**, *82*, 1–26. [CrossRef]
48. Elkin, L.A.; Kay, M.; Higgins, J.J.; Wobbrock, J.O. An Aligned Rank Transform Procedure for Multifactor Contrast Tests. In Proceedings of the ACM Symposium on User Interface Software and Technology (UIST 2021), Virtual Event, 10–14 October 2021; ACS Press: New York, NY, USA, 2021; pp. 754–768.
49. Kay, M.; Elkin, L.; Higgins, J.J.; Wobbrock, J.O. ARTool: Aligned Rank Transform for Nonparametric Factorial ANOVAs. R package version 0.11.1. Available online: https://github.com/mjskay/ARTool (accessed on 13 November 2021). [CrossRef]
50. Wobbrock, J.O.; Findlater, L.; Gergle, D.; Higgins, J.J. The Aligned Rank Transform for Nonparametric Factorial Analyses Using Only Anova Procedures. In Proceedings of the ACM Conference on Human Factors in Computing Systems (CHI 2011), Vancouver, BC, USA, 7–12 May 2011; ACS Press: New York, NY, USA, 2021; pp. 143–146.
51. Streifer, M.; Gore, A.C. Epigenetics, estrogenic endocrine-disrupting chemicals (EDCs), and the brain. *Adv. Pharmacol.* **2021**, *92*, 73–99. [PubMed]
52. Walker, D.M.; Goetz, B.M.; Gore, A.C. Dynamic postnatal developmental and sex-specific neuroendocrine effects of prenatal polychlorinated biphenyls in rats. *Mol. Endocrinol.* **2014**, *28*, 99–115. [CrossRef]
53. Kilic, N.; Sandal, S.; Colakoglu, N.; Kutlu, S.; Sevran, A.; Yilmaz, B. Endocrine disruptive effects of polychlorinated biphenyls on the thyroid gland in female rats. *Tohoku J. Exp. Med.* **2005**, *206*, 327–332. [CrossRef]
54. Woodhouse, A.J.; Cooke, G.M. Suppression of aromatase activity in vitro by PCBs 28 and 105 and Aroclor 1221. *Toxicol. Lett.* **2004**, *152*, 91–100. [CrossRef]
55. Reik, W.; Dean, W.; Walter, J. Epigenetic reprogramming in mammalian development. *Science* **2001**, *293*, 1089–1093. [CrossRef]
56. Dickerson, S.M.; Gore, A.C. Estrogenic environmental endocrine-disrupting chemical effects on reproductive neuroendocrine function and dysfunction across the life cycle. *Rev. Endocr. Metab. Disord.* **2007**, *8*, 143–159. [CrossRef] [PubMed]
57. Colciago, A.; Casati, L.; Mornati, O.; Vergoni, A.V.; Santagostino, A.; Celotti, F.; Negri-Cesi, P. Chronic treatment with polychlorinated biphenyls (PCB) during pregnancy and lactation in the rat. Part 2: Effects on reproductive parameters, on sex behavior, on memory retention and on hypothalamic expression of aromatase and 5alpha-reductases in the offspring. *Toxicol. Appl. Pharmacol.* **2009**, *239*, 46–54.
58. Faass, O.; Schlumpf, M.; Reolon, S.; Henseler, M.; Maerkel, K.; Durrrer, S.; Lichtensteiger, W. Female sexual behavior, estrous cycle and gene expression in sexually dimorphic brain regions after pre- and postnatal exposure to endocrine active UV filters. *Neurotoxicology* **2009**, *30*, 249–260. [CrossRef]
59. Henley, C.L.; Nunez, A.A.; Clemens, L.G. Estrogen treatment during development alters adult partner preference and reproductive behavior in female laboratory rats. *Horm. Behav.* **2009**, *55*, 68–75. [CrossRef]
60. Cummings, J.A.; Clemens, L.G.; Nunez, A.A. Exposure to PCB 77 affects partner preference but not sexual behavior in the female rat. *Physiol. Behav.* **2008**, *95*, 471–475. [CrossRef]
61. Hull, E.M.; Meisel, R.L.; Sachs, B.D. Male Sexual Behavior. In *Hormones, Brain and Behavior*; Elsevier: Amsterdam, The Netherlands, 2002; pp. 139–214.

62. Portfors, C.V. Types and functions of ultrasonic vocalizations in laboratory rats and mice. *J. Am. Assoc. Lab. Anim. Sci.* **2007**, *46*, 28–34. [PubMed]
63. Barfield, R.J.; Thomas, D.A. The role of ultrasonic vocalizations in the regulation of reproduction in rats. *Ann. N. Y. Acad. Sci.* **1986**, *474*, 33–43. [CrossRef] [PubMed]
64. White, N.R.; Barfield, R.J. Playback of female rat ultrasonic vocalizations during sexual behavior. *Physiol. Behav.* **1989**, *45*, 229–233. [CrossRef]
65. Ahrens, A.M.; Ma, S.T.; Maier, E.Y.; Duvauchelle, C.L.; Schallert, T. Repeated intravenous amphetamine exposure: Rapid and persistent sensitization of 50-kHz ultrasonic trill calls in rats. *Behav. Brain Res.* **2009**, *197*, 205–209. [CrossRef]
66. Wöhr, M.; Houx, B.; Schwarting, R.K.W.; Spruijt, B. Effects of experience and context on 50-kHz vocalizations in rats. *Physiol. Behav.* **2008**, *93*, 766–776. [CrossRef]
67. Kisko, T.M.; Schwarting, R.K.W.; Wöhr, M. Sex differences in the acoustic featuRes. of social play-induced 50-kHz ultrasonic vocalizations: A detailed spectrographic analysis in wild-type Sprague–Dawley and Cacna1c haploinsufficient rats. *Dev. Psychobiol.* **2021**, *63*, 262–276. [CrossRef]
68. McGinnis, M.Y.; Kahn, D.F. Inhibition of male sexual behavior by intracranial implants of the protein synthesis inhibitor anisomycin into the medial preoptic area of the rat. *Horm. Behav.* **1997**, *31*, 15–23. [CrossRef] [PubMed]
69. Bogacki-Rychlik, W.; Rolf, M.; Bialy, M. Anticipatory 50-kHz Precontact Ultrasonic Vocalizations and Sexual Motivation: Characteristic Pattern of Ultrasound Subtypes in an Individual Analyzed Profile. *Front. Behav. Neurosci.* **2021**, *15*, 722456. [CrossRef] [PubMed]
70. Brudzynski, S.M. Biological Functions of Rat Ultrasonic Vocalizations, Arousal Mechanisms, and Call Initiation. *Brain Sci.* **2021**, *11*, 605. [CrossRef] [PubMed]
71. Matochik, J.A.; White, N.R.; Barfield, R.J. Variations in scent marking and ultrasonic vocalizations by Long-Evans rats across the estrous cycle. *Physiol. Behav.* **1992**, *51*, 783–786. [CrossRef]
72. Sobolewski, M.; Abston, K.; Conrad, K.; Marvin, E.; Harvey, K.; Susiarjo, M.; Cory-Slechta, D.A. Lineage- and Sex-Dependent Behavioral and Biochemical Transgenerational Consequences of Developmental Exposure to Lead, Prenatal Stress, and Combined Lead and Prenatal Stress in Mice. *Environ. Health Perspect.* **2020**, *128*, 27001. [CrossRef]

Article

Two Hits of EDCs Three Generations Apart: Effects on Social Behaviors in Rats, and Analysis by Machine Learning

Ross Gillette [1], Michelle Dias [1], Michael P. Reilly [1], Lindsay M. Thompson [1], Norma J. Castillo [1], Erin L. Vasquez [1], David Crews [2] and Andrea C. Gore [1,*]

[1] Division of Pharmacology and Toxicology, College of Pharmacy, The University of Texas at Austin, Austin, TX 78712, USA; rossg@austin.utexas.edu (R.G.); michelledias10@gmail.com (M.D.); michaelreillyp@gmail.com (M.P.R.); lindsay.thompson82@utexas.edu (L.M.T.); norma.j.castillo95@utexas.edu (N.J.C.); erinvasquez6@gmail.com (E.L.V.)

[2] Department of Integrative Biology, The University of Texas at Austin, Austin, TX 78712, USA; crews@mail.utexas.edu

* Correspondence: andrea.gore@austin.utexas.edu

Abstract: All individuals are directly exposed to extant environmental endocrine-disrupting chemicals (EDCs), and indirectly exposed through transgenerational inheritance from our ancestors. Although direct and ancestral exposures can each lead to deficits in behaviors, their interactions are not known. Here we focused on social behaviors based on evidence of their vulnerability to direct or ancestral exposures, together with their importance in reproduction and survival of a species. Using a novel "two hits, three generations apart" experimental rat model, we investigated interactions of two classes of EDCs across six generations. PCBs (a weakly estrogenic mixture Aroclor 1221, 1 mg/kg), Vinclozolin (antiandrogenic, 1 mg/kg) or vehicle (6% DMSO in sesame oil) were administered to pregnant rat dams (F0) to directly expose the F1 generation, with subsequent breeding through paternal or maternal lines. A second EDC hit was given to F3 dams, thereby exposing the F4 generation, with breeding through the F6 generation. Approximately 1200 male and female rats from F1, F3, F4 and F6 generations were run through tests of sociability and social novelty as indices of social preference. We leveraged machine learning using DeepLabCut to analyze nuanced social behaviors such as nose touching with accuracy similar to a human scorer. Surprisingly, social behaviors were affected in ancestrally exposed but not directly exposed individuals, particularly females from a paternally exposed breeding lineage. Effects varied by EDC: Vinclozolin affected aspects of behavior in the F3 generation while PCBs affected both the F3 and F6 generations. Taken together, our data suggest that specific aspects of behavior are particularly vulnerable to heritable ancestral exposure of EDC contamination, that there are sex differences, and that lineage is a key factor in transgenerational outcomes.

Keywords: endocrine-disrupting chemicals (EDC); Aroclor 1221 (A1221); PCBs; vinclozolin; social behavior; sex differences; transgenerational; epigenetic

Citation: Gillette, R.; Dias, M.; Reilly, M.P.; Thompson, L.M.; Castillo, N.J.; Vasquez, E.L.; Crews, D.; Gore, A.C. Two Hits of EDCs Three Generations Apart: Effects on Social Behaviors in Rats, and Analysis by Machine Learning. *Toxics* **2022**, *10*, 30. https://doi.org/10.3390/toxics10010030

Academic Editors: Jessica Plavicki and Jodi Flaws

Received: 19 November 2021
Accepted: 7 January 2022
Published: 11 January 2022

Publisher's Note: MDPI stays neutral with regard to jurisdictional claims in published maps and institutional affiliations.

Copyright: © 2022 by the authors. Licensee MDPI, Basel, Switzerland. This article is an open access article distributed under the terms and conditions of the Creative Commons Attribution (CC BY) license (https://creativecommons.org/licenses/by/4.0/).

1. Introduction

We live in a world that is irreversibly contaminated as a consequence of the chemical revolution that began in the 1940s. The industrial, agricultural, and pharmaceutical industries, to name only a few, have produced hundreds of thousands of chemicals, among which nearly 1000 are now classified as endocrine-disrupting chemicals (EDCs) [1,2]. The consequences of EDC exposure are manifested as endocrine and neurological disorders in individuals directly exposed, especially during sensitive life stages such as fetal development. Furthermore, exposure can cause disease and dysfunction for multiple generations without additional exposure due to heritable epigenetic mechanisms [3,4]. Thus, the complex diseases and dysfunctions associated with EDCs represent the interaction of historical and contemporary exposures. The complexities arising from nearly a century of EDC exposure—about five generations in humans and hundreds of generations in rodents—must be studied in a laboratory setting if there is any hope that we can anticipate similar issues arising in humans.

Among those phenotypes affected by EDCs are social behaviors. These behaviors allow individuals to identify and distinguish others in or outside of their species, serve to establish cohesive social structure and hierarchies, provide cues necessary for parental and sexual behaviors, and are critical for the survival of a species. Several EDC classes including polychlorinated biphenyls (PCBs), vinclozolin, bisphenol A (BPA), phthalates, and chlorpyrifos cause changes in social behaviors in rodents [5–12]. In fact, behaviors, particularly those that are hormone-sensitive such as social behavior, appear to be among the most sensitive to EDCs [13,14]. Beyond these studies on direct exposure are those, although fewer in number, that have demonstrated intergenerational EDC effects on social behaviors [15–19]. However, there has been no work, to our knowledge, on the interactions of ancestral and direct exposures, a gap in understanding of the current real-world dilemma.

Neurobehavioral research must be conducted in the context of sex differences and parental lineage of origin based on strong evidence for sexually dimorphic effects of EDCs, and the importance of maternal vs. paternal exposure on phenotypic outcomes [20]. However, the sheer numbers of animals necessitated by multigenerational breeding and testing of both sexes through parental lineages, and the labor necessary to score nuanced social behaviors such as nose-to-nose interactions of rodent conspecifics, has made large-scale social behavioral experiments prohibitive. Yet, it is these difficult-to-observe behaviors that provide the most salient information about a potential mate that are the most ethologically important [21]. Advances in computer vision and object classification and recognition have established the tools necessary for using machine learning to automate the identification of complex animal behavior [22,23]. These tools were developed and applied to the current analysis.

Here, we used two different EDC classes selected for differences in their historical usage and in their mechanism of action. Polychlorinated biphenyls (PCBs) represent a legacy group of EDCs that were widely used until their ban in the United States and elsewhere in the 1970s. The industrial PCB mixture used herein, Aroclor 1221 (A1221), acts mainly through estrogenic signaling pathways [24]. Vinclozolin (VIN) is in modern use as a fungicide and is primarily antiandrogenic in its action [25,26]. Both VIN [27] and PCBs [28] at high dosages induce overt reproductive toxicity, and at lower concentrations, especially during development, act as EDCs to perturb hormones and their actions [14]. Each has been characterized for its neurobehavioral consequences both for direct and ancestral exposure [1,6,14,19,29–32] but not for their interactions across generations. This is a particularly important but untested concept considering the real-world scenario that humans and animals today were likely exposed to high levels of PCBs 50 years ago, leading to a potentially heritable "imprint," and now their descendants are subjected to exposures to modern classes of EDCs such as VIN.

2. Materials and Methods

2.1. Animals and Treatment

All animal work was conducted using humane procedures that were approved by the Institutional Animal Care and Use Committee at The University of Texas at Austin in accordance with NIH guidelines. Three month old male and female Sprague-Dawley rats were purchased from Envigo (Envigo, Indianapolis, IN, USA) and acclimated to the animal housing facility and light cycle (14:10 dark:light) for two weeks. All rooms were kept at a consistent temperature (22 C) and all rats had ad libitum access to filtered tap water and a low phytoestrogen diet. (Teklad 2019: Envigo, Indianapolis, IN, USA).

The vaginal cytology of virgin breeder females was observed daily the week prior to mating. On the day of proestrus, females (F0) were paired overnight with a sexually experienced male rat and observed under red light for copulatory behaviors. If the female displayed receptive behaviors, the pair was left overnight with food and water. The next morning, the presence of sperm in a vaginal smear was used to confirm pregnancy and marked as embryonic day 1 (E1). Pregnant F0 dams were randomly assigned to one of three treatment groups (Table 1): vehicle (6% DMSO in sesame oil), Aroclor 1221 (A1221, 1 mg/kg), or Vinclozolin (VIN, 1 mg/kg). The dosages and route were selected to match previous work and to fall within ranges of human exposures [33–36]. The investigators were blind to treatments throughout the study, with the code broken only when all experimental work was completed. Pregnant dams were weighed and injected daily via i.p. injection from E8–E18 two hours prior to lights out. This age range was selected to encompass a prenatal period when germline epigenetic marks are established, as well as the beginning of the critical period of brain sexual differentiation [37,38]. The i.p. route also matched prior work, although current studies in the lab have switched to feeding EDCs, with similar results, to better approximate the route of most human exposures. At E18, dams were given nesting material and left undisturbed until birth (referred to as P0). On the day after birth, postnatal day 1 (P1), each litter was culled to 5 males and 5 females per litter based on median anogenital index (anogenital distance divided by the cube root of bodyweight) to maintain an equivalent sex ratio between all litters. At P21, pups were weaned into separate cages with 2–3 same-sex littermates per cage. Of the five males and females retained in each litter, 2–3 of each sex were used for behavioral testing. On P80, one male and one female from each litter that had not been used for behavioral testing was mated with an untreated animal to generate the F2 paternal and maternal lineages, respectively. The breeding protocol is shown in Figure 1.

Table 1. Sample sizes, indicated as # litters/# individual females/# individual males.

First Hit	F1	F3 Maternal	F3 Paternal		
DMSO	10/21/25	11/21/21	10/23/21		
A1221	10/28/24	10/18/24	11/22/18		
VIN	10/28/24	12/19/23	11/18/21		

First Hit/Second Hit	F4 Maternal	F4 Paternal	F6 Maternal	F6 Paternal
DMSO/DMSO	13/22/21	12/25/22	12/22/23	12/21/24
A1221/A1221	9/15/18	8/18/15	9/18/20	8/18/20
A1221/VIN	9/18/19	9/17/18	9/22/18	9/20/20
VIN/A1221	12/21/22	11/21/21	11/19/23	11/22/21
VIN/VIN	13/20/22	12/19/20	12/22/22	10/19/22

Note: The total number of litters, individual females, and individual males used in this study are shown, with treatment, generation, and lineage indicated. On average, 10 litters were used per group and in most cases 2 male and 2 female individuals from each litter were behaviorally characterized, although those groups with fewer litters (due to timeline limitations) occasionally included a third individual per sex.

In the resulting F2 generation, one P80 female from each of the maternal litters was bred with an untreated male to continue the maternal lineage, and one P80 male from each paternal litter was bred with an untreated female to continue the paternal lineage. All untreated male and female breeders were purchased from Envigo, delivered to the lab at P60, and were allowed 2 weeks of habituation before breeding began. This same maternal and paternal breeding paradigm was used until the sixth (F6) generation.

To ascertain interactions of direct and transgenerational EDC exposures, a subset of F3 females from the maternal lineage was bred with untreated males, and a group of naïve females purchased from Envigo was bred with a subset of F3 males from the paternal lineage (Figure 1A). Exposure to EDCs was performed identically as in the F1 generation. Because of the large number of possible combinations, the two hits across three generations (F1/F4) were limited to VEH/VEH, A1221/A1221, A1221/VIN, VIN/VIN, and VIN/A1221 (Figure 1A).

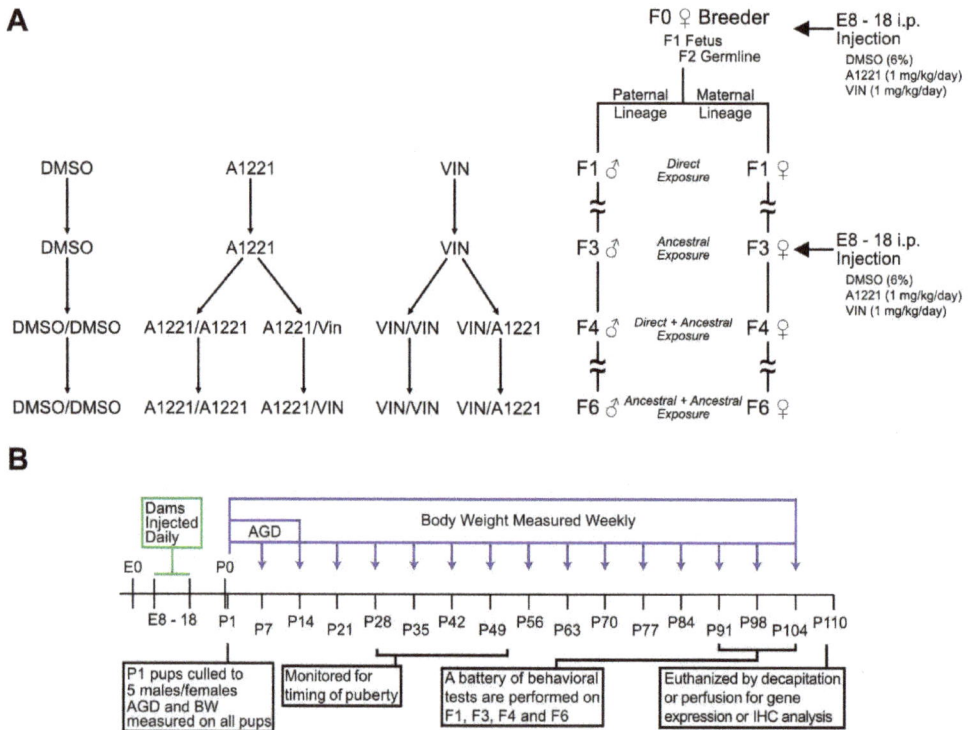

Figure 1. Experimental design and timeline. The experimental design is shown (**A**) including parental lineages, generations, and treatment groups used for analysis. EDC exposure was administered via i.p. injection to F0 and F3 pregnant dams from E8-18. The timeline (**B**) indicates all experimental manipulations and measurements taken. AGD: anogenital distance.

A total of 306 litters were used across all treatment groups with an average N~10 litters per generation/lineage/treatment. The total number of litters used for each generation, lineage, sex, and treatment are shown in Table 1, with approximately 20 individuals per generation/lineage/sex/treatment and a total of 1209 behaviorally characterized animals (Table 1). The litters and resulting behaviorally characterized individuals were spread across 7 cohorts that spanned 4 years.

2.2. Sociability Apparatus and Behavioral Analysis

Each individual animal was run through a battery of behavioral tests beginning on P90: Ultrasonic Vocalizations (USV), and Mate Preference (MP), Open Field (OF), Sociability, Social Novelty, Light/Dark box (LD), Elevated Plus (EP). The order of behavioral tasks was the same for all animals (as listed above) and each behavioral task was separated by 48 h and always occurred between 1 and 4 h after lights off under dim red light. All animals were transported to the behavioral analysis rooms in black-out covered carts as to prevent light pulse exposure. In the current manuscript, data from the sociability and social novelty tasks are presented. One week after completing the behavior tests, animals were euthanized by rapid decapitation (Figure 1B).

A three-chambered sociability chamber (Stoelting, AnyMaze) [39] was used as published and according to the methodology previously published by our lab [6,19,31]. The chamber consists of a 100 cm (wide) × 100 cm (long) Plexiglas square partitioned into 3 equivalent chambers approximately 33 cm (wide) by 100 cm (long), with a door measuring 10 cm × 11 cm leading to the middle chamber (Figure 2). The left and right chambers included a cylindrical stimulus animal enclosure in the bottom corners of the arena that was 15 cm wide and had vertical metal rods separated by 1 cm that allowed facial investigation and nose touching but prevented more extensive interactions.

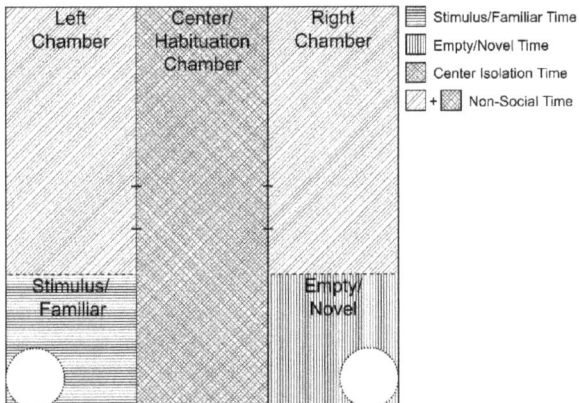

Figure 2. Sociability apparatus. A diagram of the three-chambered sociability apparatus is shown where the fill pattern indicates the various sections of the apparatus that were digitally segmented for analysis. The test occurred in 3 distinct phases. The first (habituation—5 min) occurred with the experimental rat placed in the center chamber (cross-hatched) with the doors to the left and right chambers shut such that the experimental animal could not access either side arena. During the subsequent two stages (sociability—10 min and social novelty—10 min) the doors were opened, allowing the experimental rat free access to the three chambers. The far corners of the side chambers each held a cylindrical holding chamber, with or without a rat contained within. For scoring purposes, the center chamber was scored as social isolation time. Total nonsocial time was calculated by adding the time spent in the distal left and right chambers (diagonal) with the time spent in the center chamber. Time spent near either the stimulus/familiar animal (horizontal) or empty cage/novel animal (vertical) was scored only when the target experimental animal's center of mass was in close proximity (less than approximately one body length) to the respective enclosure.

A sociability trial consisted of 3 distinct stages: habituation (5 min), sociability (10 min), and social novelty (10 min). In the habituation phase, an animal was placed in the center chamber with the entrances blocked such that the target experimental animal could not access the left or right chambers. A same sex- and age-matched stimulus animal was placed in one stimulus enclosure and randomly positioned in either the left- or right-flanking chamber. An empty stimulus enclosure was placed in the opposite flanking chamber. Doors were opened at the beginning of the sociability phase, during which the experimental animal was allowed to freely explore the arena and investigate the stimulus animal or empty stimulus enclosure. At the end of the sociability phase the target experimental animal was temporarily removed from the apparatus and a novel stimulus animal was placed in the empty stimulus enclosure. The placement of the enclosures containing the familiar and the novel stimulus rats on the left or right was random to avoid direction or side biases. During the social novelty phase, the target experimental animal was returned to the center portion of the arena and again allowed to freely explore the arena for 10 min and investigate the familiar and the novel stimulus animal. At the end of the social novelty phase, each animal was returned to their respective home cage. Stimulus rats were used in no more than 3 trials in one day to avoid behavioral changes due to repeated testing. All female experimental and stimulus animals were used for testing while in diestrus.

All behavioral trials were analyzed in real time with AnyMaze software (Stoelting), as published [6,19,31]. The testing arena was digitally segmented into left, center, and right chambers. An additional digital segment was drawn around the stimulus enclosure on each side (approximately 33 cm × 33 cm—Figure 2) to indicate proximity of the target experimental animal to the stimulus enclosures. An animal was considered to enter an area if 80% of the animal's area was inside of that area, which is equivalent to an animal having 4 paws within the area.

2.3. Data Exclusion Criteria

Occupancy plots showing animal position across the duration of a trial were extracted from AnyMaze and visually analyzed for errors in tracking; the latter were marked for retracking in an attempt to rescue the data. After retracking, the occupancy plots were analyzed again. Instances that did not show improved tracking were excluded from analysis. Further trials were removed from analysis for various technical reasons (incomplete habituation period, animals that escaped the arena, and animals that altered the position of stimulus restraint chambers). Raw data were then extracted from AnyMaze and checked for obvious anomalies in the distance traveled to indicate poor tracking fidelity. Trials that had lost more than 20 s of experimental time due to video retracking were also excluded from analysis. Finally, stage 1 of sociability (Stimulus vs. Empty Chamber) was analyzed to determine trials in which the experimental animal did not visit each flanking chamber for at least 10 s. These trials were left as part of the analysis for stage 1 but were removed from stage 2 (Familiar vs. Novel) of sociability because these experimental rats would not have had the opportunity to become familiar with the stimulus animal, making a choice between a "familiar" and a novel rat moot.

2.4. Nose Touch Detection with Machine Learning and DeepLabCut

Facial investigation and nose touches are a critical aspect of rodent social interaction and investigation [6,32,40] but they are also the most labor-intensive part of the analysis, as they involve an investigator iteratively viewing and scoring every recorded trial of the sociability and the social novelty tests from thousands of trials. To automate the detection of nose touches from recorded behavioral trials we used DeepLabCut (version 2.2b [41]) which employs deep residual neural networks to predict the location of individual body parts. A total of 16 trial videos and 20 frames from each video (320 total frames) were used to create a training dataset in which an experimenter manually labeled 8 individual body parts (nose, left and right ears, left and right flank, body center, tail base, and tail end) of the target experimental animal and the nose and tail of the stimulus animals (Figure 3). The manually labeled frames were split into a training set (95%) and test set (5%) and the network was trained for 250K iterations. The resulting body part position data were then used to calculate the distance between the nose of the target experimental animal and the two stimulus animals for every frame of each video. A nose touch was then marked when the distance between the experimental and stimulus animals' noses was below a specified threshold. Nose touch instances and duration were then calculated by employing run-length encoding, in which consecutive video frames in which a nose touch was detected were grouped together as a nose touch instance. A second and separate human-scored validation dataset was then used to optimize the distance between two noses and the time within that distance that most closely represented human scoring. We determined that a distance of 7.5 pixels between the experimental and stimulus animal's noses and 200 consecutive milliseconds (about 7 video frames) within that distance most closely matched what a human scorer considered a nose touch. Finally, video frames in which the machine learning model displayed low certainty ($p < 0.90$) of the position of either the experimental animal's or stimulus animal's nose were excluded from analysis.

Once validated, the trained model was applied to all behavioral videos, nose-touch instances were extracted using the parameters determined above, and the results were analyzed identically to the statistical methods used for the other behavioral metrics described above. A third and separate validation video set (16 videos) was randomly selected from all of the videos processed with the final model, again hand-scored by a blind experimenter, and compared to the machine-learning model to determine the accuracy of the model against a dataset that was entirely removed and separate from the optimization process. This third validation set was used to determine if the model generalized well or was overfit to the datasets used to train the machine learning model or optimize the detection of nose touches.

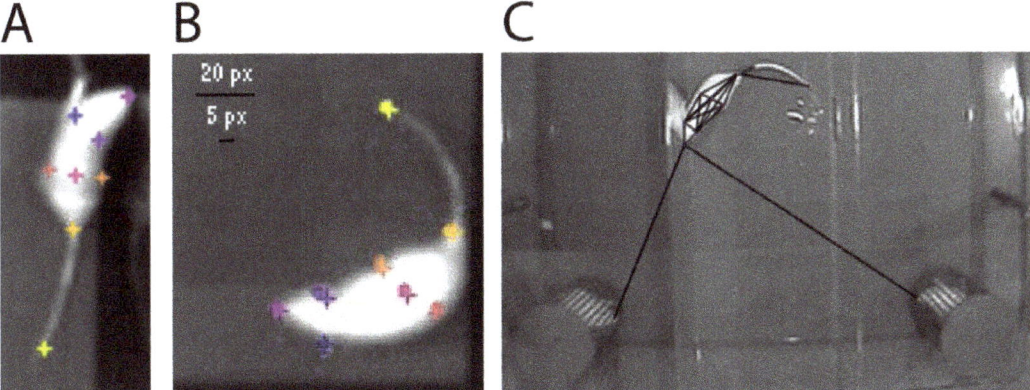

Figure 3. Automated animal tracking for nose-touch detection. Representative images of the training process produced in DeepLabCut are shown. (**A**) The 8 user-labeled body parts (nose, left and right ears, left and right flank, body center, tail base, and tail tip) are indicated on an image (+) used to train the model to identify each individual body part; (**B**) a test image showing the difference in placement between user-labeled body parts (+) and the location of the same body part (•) predicted by the model; (**C**) the final pose-estimation model demonstrating the automated body-part tracking in an experimental video where the noses of two stimulus animals are simultaneously tracked in relation to the body and nose of the experimental animal.

2.5. Statistics

All statistical analyses were performed in R (version 4.0.2—[42]) with the base *stats* package (version 4.0.2). Figures were created with *ggplot2* (version 3.3.3), and were edited only for style with Adobe Illustrator (CS5). Behavioral metrics were checked for normality (Shapiro–Wilk Test [43]) and homogeneity of variance (Bartlett's Test [44]) using R's base *stats* package. As is typical of large behavioral datasets, most of the individual metrics violated these assumptions that are required for the traditional application of parametric statistics. To determine sex differences the raw data were analyzed with a Kruskal–Wallis one-way analysis of variance [45] within generation and parental lineage where all treatments were collapsed into sex. Effect size was calculated for sex differences using Cohen's D [46] and reported with each statistic. To determine effects of EDC treatment, individually optimized Box–Cox power transformations (R—*EnvStats* version 2.4.0 [47]) were applied to all behavioral metrics separately but equally across all generations, lineages, sexes, and treatments within a metric. As large differences were expected due to sex differences in behavioral tests, one-way ANOVAs (R—*car* version 3.0-10 [48]) were applied within sex to determine effects of treatment (e.g., EDC exposure) and considered significant if $p < 0.05$. The effect size, or the amount of variance accounted for by the linear model, was calculated as partial-eta-squared (R—*lsr* version 0.5 [49]) for each significant effect and reported with the corresponding statistic. A partial-eta-squared value of 0.01 is considered small, 0.09 is considered medium, and 0.25 is considered large. If an ANOVA were determined significant, Tukey's Honest Significant Difference [50] pairwise post hoc tests were applied to determine individual group difference and *p*-values were appropriately adjusted for multiple comparisons.

3. Results

3.1. Sociability

3.1.1. Sex Differences

Given the large sample size due to our experimental design, we were afforded the unique opportunity to characterize sex differences inherent to the Sociability test with data collapsed across all other variables (treatment, generation, lineage). Previous studies show that sex differences are large with a bimodal distribution, leading us to use nonparametric tests (Kruskal–Wallis H-test) on an N of ~600 per group. The social behavior data are summarized in Table 2.

Table 2. Sex differences in the sociability and social novelty tests.

	Sociability			
	Females		Males	
Measure	Mean	SEM	Mean	SEM
Distance Traveled (m) *	52.63	0.36	38.58	0.3
Social Preference Score *	0.62	0.01	0.67	0.01
Stimulus Investigation Time (s) *	224.35	2.68	245.90	3.02
Empty Investigation Time (s) *	135.26	2.13	123.31	2.59
Stimulus Animal # Visits *	10.26	0.13	8.68	0.14
Empty Chamber # Visits *	8.42	0.11	6.25	0.09
Stimulus Animal Mean Visit Time (s) *	24.46	0.5	34.76	1.06
Empty Chamber Mean Visit Time (s) *	17.58	0.38	22.78	0.68
Total Nonsocial Time (s) *	240.01	2.79	229.86	2.86
Center Isolation Time (s) *	99.54	1.19	107.72	1.6
Total Nose-Touch Time (s)	20.21	0.40	19.51	0.39
Average Nose-Touch Duration (s) *	0.66	0.01	0.63	0.01
	Social Novelty			
	Females		Males	
Measure	Mean	SEM	Mean	SEM
Distance Traveled (m) *	46.35	0.39	32.44	0.29
Social Novelty Score *	0.62	0.01	0.60	0.01
Familiar Investigation Time (s) *	124.31	2.16	142.47	2.84
Novel Investigation Time (s)	208.61	2.81	215.03	3.17
Familiar Animal # Visits *	8.01	0.11	6.32	0.09
Novel Animal # Visits *	10.09	0.13	7.74	0.11
Familiar Animal Mean Visit Time (s) *	16.89	0.55	24.67	0.6
Novel Animal Mean Visit Time (s) *	22.46	0.44	31.39	0.72
Total Social Time (s) *	332.92	2.96	357.50	3.16
Total Nonsocial Time (s) *	273.37	3.77	249.89	4.04
Center Isolation Time (s) *	125.54	1.78	116.37	1.92
Nose-Touch Novelty Score	0.65	0.01	0.64	0.01
Familiar Animal Nose-Touch Time (s) *	9.33	0.27	8.62	0.30
Novel Animal Nose-Touch Time (s) *	17.77	0.41	15.41	0.38

Note: Sex differences in sociability (top) and social novelty (bottom) are shown for all animals with treatment, generation, and lineage collapsed. Mean and standard error of the mean (SEM) are shown for females (N = 596) and males (N = 601). These summary data provide a definitive comparison of sex differences for the most important measures from the sociability (Top) and social novelty (Bottom) behavioral tasks. Nearly all metrics observed were sexually dimorphic and are indicated by * where $p < 0.001$. Of greatest relevance to the individual tasks, males displayed a stronger social preference score (Top) but females showed a stronger social novelty score (Bottom). #: number of events.

Females were more active and traveled farther in the 10 min trial than did males ($H(1) = 556.17$, $p < 0.0001$, $d = 1.72$). Females showed a reduced social preference score compared to males, ($H(1) = 45.15$, $p < 0.0001$, $d = 0.31$) and spent less time near the stimulus animal ($H(1) = 29.39$, $p < 0.0001$, $d = 0.31$) and more time near an empty stimulus enclosure ($H(1) = 28.52$, $p < 0.0001$, $d = 0.20$). Females visited both the stimulus animal ($H(1) = 60.7$, $p < 0.0001$, $d = 0.47$) and the empty enclosure more often than males ($H(1) = 198.53$, $p < 0.0001$, $d = 0.88$). However, the females' average visit time to both the stimulus animal ($H(1) = 83.79$, $p < 0.0001$, $d = 0.50$) and the empty enclosure were shorter than those in males

($H(1) = 24.69$, $p < 0.0001$, $d = 0.38$). Overall, females spent slightly more time in nonsocial areas of the apparatus ($H(1) = 7.11$, $p = 0.0077$, $d = 0.15$), but males spent more time in the isolated center chamber than did females ($H(1) = 11.15$, $p < 0.0008$, $d = 0.24$).

3.1.2. Effect of EDC Exposure

Prenatal EDC exposure did not affect locomotion (total distance traveled) in any generation, breeding lineage, or sex. We calculated a social preference score as the time spent investigating the stimulus animal divided by the sum of the time spent investigating both the stimulus animal and an empty enclosure to allow a direct comparison of preference that encompassed the two choices presented to the target animal. Direct exposure to prenatal EDCs (F1 generation) did not affect social preference or other more nuanced behavioral metrics (time spent with the stimulus animal, visits, or nonsocial time). In the F3 generation, which had only ancestral and not direct EDC exposure, there was a significant effect of treatment on the social preference score in females of the paternal lineage ($F(2,60) = 3.54$, $p = 0.04$, $\eta p^2 = 0.11$—Figure 4), driven by a decrease in social preference due to ancestral VIN exposure ($p = 0.04$). Other endpoints were unaffected in the paternal lineage. In the F3 maternal lineage, the only metric affected was the average time females spent visiting an empty stimulus chamber ($F(2,60) = 3.54$, $p = 0.04$, $\eta p^2 = 0.11$—data not shown).

In the F4 generation, which represents ancestral exposure (first hit) combined with a second hit of direct fetal exposure three generations later, social preference was not affected by treatment in either sex or lineage. In females, other behaviors were affected, including the number of visits to the arena containing a stimulus animal ($F(4,95) = 2.94$, $p = 0.02$, $\eta p^2 = 0.11$—Figure 5A). This effect was driven primarily by a decreased number of visits by the A1221/VIN females ($p = 0.009$) compared to DMSO vehicle. This seems to have been compensated for by an increased mean visit time to the stimulus animal ($F(4,95) = 3.91$, $p = 0.01$, $\eta p^2 = 0.14$—Figure 5B), again driven by the A1221/VIN group vs. DMSO ($p = 0.01$). An effect on the mean visit time to the empty enclosure was also identified ($F(4,91) = 2.80$, $p = 0.03$, $\eta p^2 = 0.1$—Figure 5C), in this case driven by an increase in visits by the A1221/A1221 females compared to DMSO vehicle ($p < 0.01$). The same A1221/A1221 females showed a decrease in total nonsocial time ($F(4,91) = 4.81$, $p = 0.001$, $\eta p^2 = 0.17$—Figure 5D) compared to DMSO ($p < 0.01$) and VIN/A1221 ($p = 0.02$).

Males from the F4 paternal lineage showed few effects with the sole exception of the VIN/VIN treatment group compared to vehicle, as seen for mean visit time to the stimulus animal ($F(4,91) = 2.49$, $p = 0.05$, $\eta p^2 = 0.10$—Figure 5B) and the empty enclosure ($F(4,91) = 3.00$, $p = 0.02$, $\eta p^2 = 0.12$—Figure 5C) in which the VIN/VIN males showed increase duration visits to both ($p = 0.05$ and $p = 0.03$), respectively compared to DMSO vehicle. This increase in mean visit time was accompanied by a decrease in total nonsocial time ($F(4,91) = 2.66$, $p = 0.04$, $\eta p^2 = 0.10$—Figure 5D) by VIN/VIN compared to vehicle males ($p = 0.02$).

A single effect of treatment was identified in the maternal lineage females of the F4 generation, for which total nonsocial time ($F(4,91) = 3.76$, $p = 0.007$, $\eta p^2 = 0.14$—Figure 5E) was reduced in the A1221/VIN ($p = 0.04$), VIN/A1221 ($p < 0.01$), and VIN/VIN ($p = 0.03$) groups compared to DMSO.

Treatment effects on the social preference in the F6 generation, which represents two cumulative ancestral exposures three generations apart, were exclusive to females from the paternal lineage ($F(4,95) = 3.07$, $p = 0.02$, $\eta p^2 = 0.11$—Figure 4). The A1221/VIN group showed a marginally increased preference for social affiliation compared to controls ($p = 0.08$) and this score was significantly greater in A1221/VIN than the VIN/VIN group ($p = 0.04$). This effect was accompanied by an inverse relationship in time investigating the empty enclosure (A1221/VIN < VIN/VIN; ($F(4,95) = 3.05$, $p = 0.02$, $\eta p^2 = 0.11$—Figure 5F).

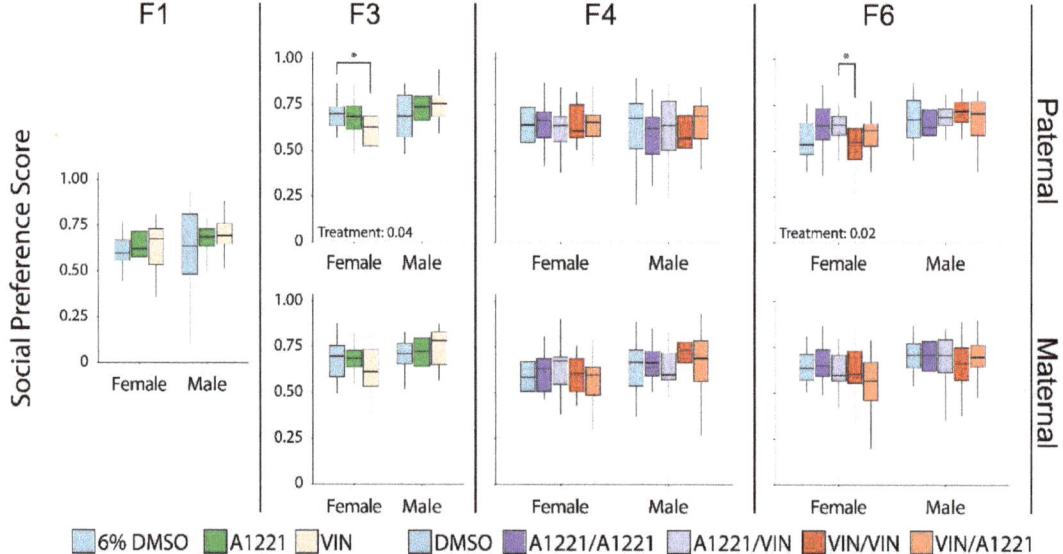

Figure 4. In the sociability test, the social preference score (time spent investigating the stimulus animal divided by the sum of the time spent investigating both the stimulus animal and an empty enclosure) for sociability is graphed as box and whisker plots (minimum, 25% quartile, median, 75% quartile, and maximum) shown separately by generation (F1 to F6 from left to right), lineage (paternal—top and maternal—bottom), and sex (indicated on x-axis). Lineage does not apply to the F1 generation. The social preference score was calculated as the time spent near the stimulus animal divided by the time spent near the stimulus animal plus the time spent near the empty enclosure. Scores above 0.5 indicate a preference for socializing with the stimulus animal. All group means were above the 0.5 threshold. Significant post hoc tests as determined by Tukey HSD are indicated with bars connecting the respective groups (* <0.05).

3.2. Social Novelty

3.2.1. Sex Differences

We again leveraged the large sample size from our experiments to establish a definitive sex differences profile characteristic of social novelty (Table 2). Females traveled a greater distance in the 10 min trial than males ($H(1) = 533.72, p < 0.0001, d = 1.67$). Females also showed a modest increase in social novelty score over males ($H(1) = 4.28, p = 0.039, d = 0.14$), spent less time with the familiar stimulus animal ($H(1) = 18.04\ p < 0.0001, d = 0.29$), but there was no difference in the time spent with the novel stimulus animal ($H(1) = 0.70, p = 0.403, d = 0.09$). Females displayed more visits to both the familiar stimulus animal ($H(1) = 122.81, p < 0.0001, d = 0.68$) and the novel stimulus animal ($H(1) = 164.05, p < 0.0001, d = 0.81$) but spent less average time per visit with the familiar ($H(1) = 144.24, p < 0.0001, d = 0.55$) and novel stimulus animals than males ($H(1) = 104.46, p < 0.0001, d = 0.61$). Overall, females had less total social time, the time associating with either the novel or familiar stimulus animal ($H(1) = 31.89, p < 0.0001, d = 0.33$), more nonsocial time ($H(1) = 24.03, p < 0.0001, d = 0.25$), and more time isolated in the center chamber of the apparatus ($H(1) = 19.07, p < 0.0001, d = 0.20$).

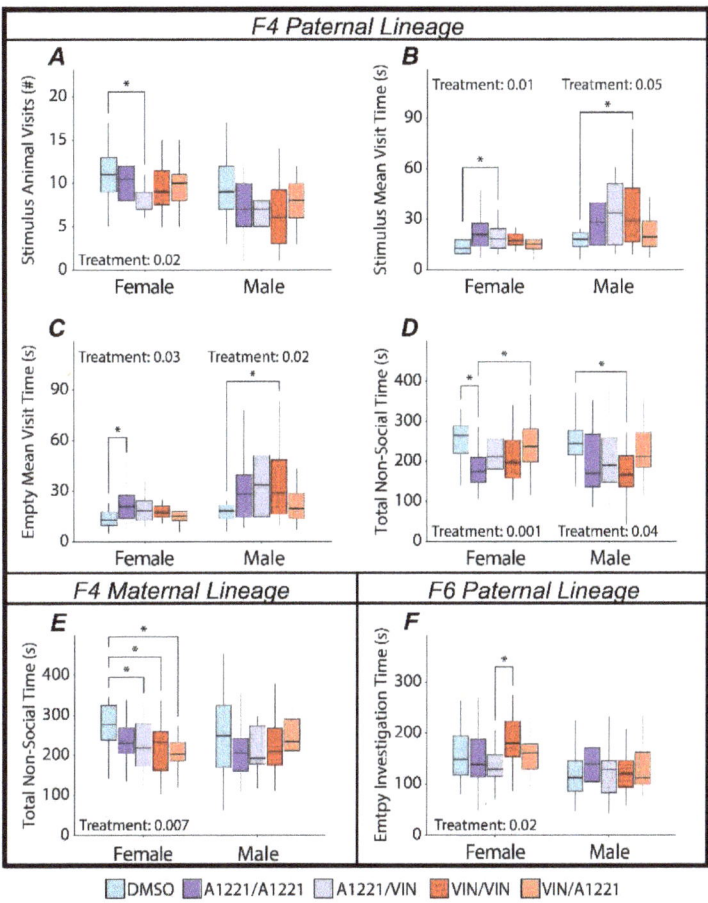

Figure 5. Aspects of social interaction dynamics in the test for sociability are graphed as box and whisker plots (minimum, 25% quartile, median, 75% quartile, and maximum). Main effects of treatment were determined by a one-way ANOVA within each sex and lineage, and indicated for $p < 0.05$. Significant post hoc results as determined by Tukey HSD and appropriately adjusted for multiple comparisons are indicated with bars connecting the respective groups (* <0.05). Shown are significant results for the F4 Paternal lineage (**A–D**), the F4 maternal lineage (**E**), and the F6 Paternal Lineage (**F**).

3.2.2. Effect of EDC Exposure

In the F1 generation there were no effects of EDC exposure on the social novelty score, defined as the time spent with the novel animal divided by the sum of the time spent with both the novel and familiar stimulus animals. In the F3 generation, the social novelty score was affected by ancestral EDC exposure in females from the paternal lineage ($F(2,58) = 3.49$, $p = 0.04$, $\eta p^2 = 0.11$—Figure 6), which was largely driven by an increase in the A1221 group compared to DMSO ($p = 0.03$). This change was associated with an effect of treatment in the amount of time spent with the familiar animal ($F(2,58) = 4.08$, $p = 0.02$, $\eta p^2 = 0.12$—Figure 7A), attributable to a decrease in the A1221 group vs. DMSO ($p = 0.02$). Time spent with the novel animal was unaffected. While the number of entries to the familiar stimulus animal chamber was not changed, familiar mean visit time was affected ($F(2,58) = 4.38$, $p = 0.02$, $\eta p^2 = 0.13$—Figure 7B), with a decrease in the A1221 group compared to both DMSO ($p = 0.03$) and VIN ($p = 0.04$) ancestral exposure. In the maternal lineage, social novelty score was unaffected, whereas aspects of social interaction were changed in both maternal lineage males and females. Treatment affected the number of visits to the familiar animal in both females ($F(2,56) = 4.63$, $p = 0.01$, $\eta p^2 = 0.14$—Figure 7C) and males ($F(2,66) = 5.60$, $p = 0.01$, $\eta p^2 = 0.15$—Figure 7C). In females this was driven by a decrease in visits of the A1221 group compared to the VIN group ($p = 0.01$). In males, both A1221 ($p = 0.05$) and VIN ($p = 0.01$) exposed individuals showed an increase in total visits to the stimulus animal compared to DMSO males. Mean visit time of males from the maternal lineage was decreased ($F(2,66) = 3.19$, $p = 0.05$, $\eta p^2 = 0.09$—Figure 7D) and driven primarily by a decrease in the VIN group compared to DMSO ($p = 0.04$). The social novelty score was not affected in either sex or breeding lineage in the F4 or F6 generations (Figure 6), nor was there any change in aspects of social interaction.

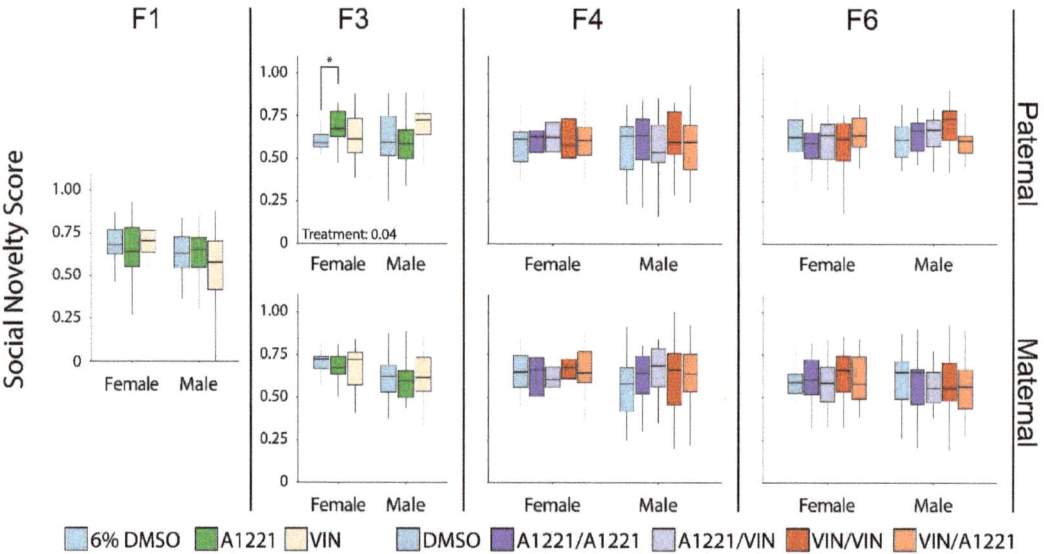

Figure 6. In the social novelty test, the social novelty score (time spent with the novel animal divided by the sum of the time spent with both the novel and familiar stimulus animals) is graphed as box and whisker plots (minimum, 25% quartile, median, 75% quartile, and maximum) shown separately by generation (F1 to F6 from left to right), lineage (paternal—top and maternal—bottom), and sex (indicated on x-axis). Lineage does not apply to the F1 generation. The social novelty score was calculated as the time spent near the novel stimulus animal divided by the time spent near the novel animal plus the time spent near the familiar stimulus animal. Scores above 0.50 indicate a preference for socializing with the novel animal. All group means were above the 0.50 threshold. Significant post hoc tests as determined by Tukey's HSD are indicated with bars connecting the respective groups (* <0.05).

In F6 paternal lineage males there was an effect of treatment on the number of visits to the familiar animal in males ($F(4,99) = 3.34$, $p = 0.01$, $\eta p^2 = 0.12$—Figure 7E). In the F6 maternal lineage, there was an effect on locomotion (distance traveled) in males ($F(4,98) = 3.21$, $p = 0.02$, $\eta p^2 = 0.12$—not shown), for which the A1221/VIN treatment group ($p = 0.01$) was increased compared to the VIN/VIN group. This was the only effect of EDC exposure on locomotion in any sex, generation, treatment, or lineage in the social novelty test. In females from the F6 maternal lineage, the only behavioral metric affected was total social time ($F(4,102) = 2.83$, $p = 0.03$, $\eta p^2 = 0.10$—Figure 7F) where the A1221/A1221 group spent more time associating with either the familiar or novel animal compared to DMSO ($p = 0.05$) and VIN/A1221 ($p = 0.03$).

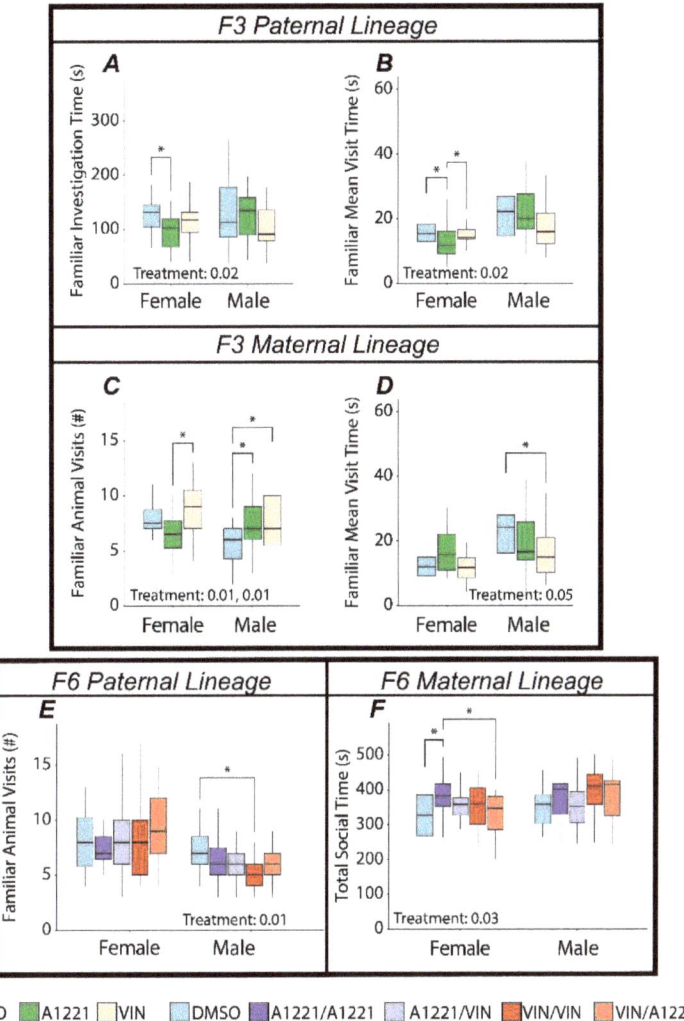

Figure 7. Aspects of social interaction dynamics in the test for social novelty are graphed as box and whisker plots (minimum, 25% quartile, median, 75% quartile, and maximum). Main effects of treatment within sex and lineage were determined by a one-way ANOVA within each sex and lineage, and indicated for $p < 0.05$. Significant post hoc results as determined by Tukey HSD and appropriately adjusted for multiple comparisons are indicated with bars connecting the respective groups (* <0.05). Shown are significant effects for the F3 paternal (**A,B**) and maternal (**C,D**) lineages, and F6 paternal (**E**) and maternal (**F**) lineages.

3.3. Nose-Touching Behaviors

3.3.1. Validation of Machine Learning Accuracy

Three independent validation datasets were used to train, optimize, and verify the automated detection of nose-touch instances with machine learning. The first was used to train the machine learning model to accurately detect and track individual body parts on the experimental animal and the stimulus animals. The accuracy of the model was determined by comparing the distance in pixels between the coordinate location of a manually indicated body part (e.g., a rat's nose) and the coordinate location of a body part predicted by the model (Figure 3B). Our final model showed a root mean squared error (RMSE) of 6.08 pixels. For comparison, a rat's nose is approximately 5 pixels long and 5.5 pixels wide. The second validation dataset was used to optimize the parameters that determined what should constitute a nose touch. The same dataset was manually scored for nose touches three separate times by the same blind experimenter and once by the algorithm used to extract nose-touch instances. The human-variability (RMSE) in the first two manually scored sets was 4.98 and 3.54 s when compared to the third. Variability between the third human scored dataset and the final computer model was 4.31 s. Because this second dataset was itself used to optimize the parameters for which nose touches were detected, and therefore could be subject to overfitting, we generated a third manually scored dataset that was also scored by the computer model and compared for accuracy (RMSE = 3.63 s). We found that our method to automate the detection of nose touches was as accurate as a human scorer and generalized well to the entire dataset.

3.3.2. Sex Differences in Nose Touching

Nose touch data were analyzed for sex differences in time spent nose touching, average duration of nose touches, and the longest-duration nose touch (Table 2). A nose touch score was also calculated for social novelty as the amount of time nose touching with the novel animal divided by the sum of the time nose touching with both the familiar and novel stimulus animals. In the Sociability test, males and females spent similar time nose-touching with a stimulus animal ($H(1) = 1.08$, $p = 0.30$, $d = 0.07$) but the average duration of nose touches was longer in females than males ($H(1) = 9$, $p = 0.003$, $d = 0.07$). In social novelty, males and females did not show a difference in nose-touch novelty score ($H(1) = 0.04$, $p = 0.84$, $d = 0.04$) but females spent more total time nose touching with both the familiar ($H(1) = 8.29$, $p = 0.004$, $d = 0.09$) and novel ($H(1) = 18.71$, $p < 0.0001$, $d = 0.22$) stimulus animals.

3.3.3. EDC Effects on Nose Touching

In the Sociability test, none of the metrics analyzed for nose touches were found to be affected by treatment in either sex or in any of the generations or lineages. In the social novelty test, the effects identified were exclusive to the F6 paternal lineage. In females, the only metric affected by treatment was total nose-touch time ($F(4,115) = 2.65$, $p = 0.04$, $\eta p^2 = 0.08$—Figure 8A) in which the VIN/VIN group showed an increase compared to DMSO ($p = 0.02$). In males, the total amount of time spent nose touching with the novel animal was affected by treatment ($F(4,113) = 3.11$, $p = 0.02$, $\eta p^2 = 0.10$—Figure 8B); the VIN/VIN group was increased compared to A1221/A1221 ($p = 0.04$) and VIN/A1221 ($p = 0.02$). A more nuanced metric of nose touching (longest-duration nose touch) was also affected ($F(4,113) = 2.80$, $p = 0.03$, $\eta p^2 = 0.10$—Figure 8C) with VIN/VIN higher than VIN/A1221 ($p = 0.01$).

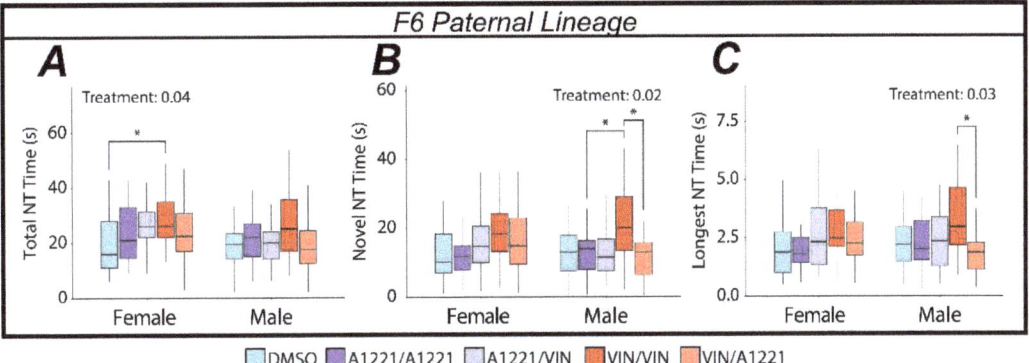

Figure 8. Metrics of nose-touching (NT) in the social novelty test as determined by leveraging machine learning techniques are graphed as box and whisker plots (minimum, 25% quartile, median, 75% quartile, and maximum). Main effects of treatment were determined by a one-way ANOVA within each sex and lineage, and indicated for $p < 0.05$. Significant post hoc results as determined by Tukey's HSD and appropriately adjusted for multiple comparisons are indicated with bars connecting the respective groups (* <0.05). (**A**) The only nose-touch metric affected in females was identified in the F6 paternal lineage where a combination of an ancestral and direct exposure of VIN increased total nose-touch time (the sum of familiar and stimulus NT time). (**B**,**C**) The time spent nose-touching with the novel stimulus animal and the longest nose-touch interaction were both increased in males from the F6 paternal lineage exposed to an ancestral exposure and a direct exposure of VIN. Taken together, these data suggest ancestral VIN exposure may influence social identification or discrimination.

4. Discussion

Our model of two hits of EDCs given three generations apart enabled us to begin to decipher the combinatorial effects of multigenerational exposures to legacy and contemporary chemicals for the first time. Built into our design, and evident in the results, were concepts that are critical to research on EDCs. First, the sexes respond differently to direct exposures to environmental toxicants, especially during critical developmental periods when hormone release and actions differ between the sexes. Second, epigenetic programming of the germline by EDCs is both sex-specific and dependent upon paternal and maternal lineage. Third, exposures to EDCs within and across generations may have unexpected outcomes.

Another novel aspect of our study was developing and applying machine learning with DeepLabCut to analyze nose-touching behavior in rats. This type of endpoint is important because social recognition happens through species-specific cues that may be obvious to a conspecific but not to a human observer. In rats, this involves close-in facial investigation and the assessment of pheromonal and olfactory cues, alterations of which by EDCs may change the dynamics of social interactions [21,51]. EDCs were observed to change nose-touch behavior and facial investigation between same sex-conspecifics in adulthood [6,8].

Finally, independent of any EDC or lineage effects, our massive behavioral dataset enabled us to thoroughly characterize and directly compare male and female rats with a sample size of ~600 animals per sex. We verified the well-known increased locomotor activity of female over male rats. Previous work showing that males spend more time investigating stimulus animals than females [6,52–54] was confirmed here; our males spent more time associating with stimulus animals in both the sociability and social novelty tasks. While males spent more time with a stimulus animal, females made more visits to each stimulus animal but for shorter duration bouts than males. Despite spending less time with stimulus rats, females had a stronger preference for social novelty than males. These comparisons provide a strong baseline for other studies on sex differences in social behaviors.

4.1. Direct Developmental Exposure to EDCs (F1 Generation) Have Few Effects on Social Behaviors

The effects of direct EDC exposure on social behavior have been reported in several studies, with outcomes dependent on the compound used, the timing of exposure, the age at testing, and the endpoints measured. To date, effects of direct EDC exposure on social behaviors have been reported for PCBs [5,6], BPA [7,8], atrazine [7], phthalates [9,10], chlorpyrifos [11], and vinclozolin [12,55]. These experiments show that EDC exposures alter subsets of behavior, and that expected sexual dimorphisms of behavior are sometimes diminished [14].

Our current study did not find any effects of direct A1221 or VIN exposure on social behaviors in the F1 generation. These results were surprising based on this prior literature, but the timing and doses of treatment used here differed from previous work. EDCs often exert nonmonotonic dose-response curves [56], and these effects are further influenced by the developmental stage. The fact that we did not observe direct EDC effects on social behaviors, but that effects were found in subsequent generations, suggests that the selected doses and timing may not be adequate to causes direct developmental changes but were still able to induce heritable epigenetic profiles through actions on the germline, as demonstrated previously [57] with the same dose and exposure paradigm used here.

4.2. EDCs Affect Social Behavior in Ancestrally Exposed Individuals

Studies on effects of ancestral exposure to EDCs on social behavior have demonstrated perturbations in the F3 generation. Transgenerational BPA increased social behavior and impaired dishabituation of social novelty in females [15,16]. Transgenerational exposure to antiandrogenic EDCs such as phthalates [17,18] and VIN [19] reduced social behaviors in males. In all of these studies, different doses and behavioral paradigms makes comparing the results between them and discerning the differential impact of EDC classes difficult. However, it is particularly noteworthy that the doses used here (1 mg/kg/day A1221 or VIN) are much lower than our previous work (e.g., 100 mg/kg [19,31]), better represent real-world exposures, and still produce robust heritable transgenerational phenotypes. Here, we directly compared two different classes of EDCs, and their combination, across generations to determine how social behavior was affected.

4.2.1. The Paternal Lineage Females Are Most Vulnerable to Ancestral Exposure

It has been proposed the male germline is particularly susceptible to environmental input and insult [58] and there are numerous demonstrations that the male germline is directly affected by EDC exposure (reviewed in [4]). Evidence for the female germline is more limited, probably due to the much greater ease in isolating and purifying sperm compared to ova. This has led to bias in work considering transgenerational endpoints in paternal descendants compared to studies on the maternal lineage. Our current study is the most comprehensive in its consideration of lineage of origin and sex due to EDC exposure. We found that ancestral EDC exposure effects on social behavior were more frequent in the paternal lineage, but there were also some maternal lineage effects.

Among the behaviors we analyzed, rats' preference for a conspecific over an empty cage (Sociability), and the preference for a novel over a familiar rat (social novelty), are most relevant to social decision-making in rodents and are discussed here. We found a main effect of treatment in three comparisons, all of which were in females from the paternal lineage: ancestral VIN exposure decreased sociability in the F3 and F6 generations while ancestral A1221 exposure increased the preference for social novelty in the F3 generation. There are three primary points to take away from these results. First, female social behavior is particularly susceptible to EDC exposure, with EDC class (estrogenic vs. antiandrogenic) playing a role. Second, altered social behavior emerges in the F3 generation, presumably due to direct exposure of the F2 germline. Finally, the paternal germline is more susceptible than the maternal to EDC exposure in the context of social behavior, although we emphasize that this result should not be extrapolated to other endpoints affected by EDCs.

It is notable that those rats of the F4 generation receiving a second hit of the same EDC (i.e., A1221/A1221 or VIN/VIN) differed in their behaviors from those receiving a single hit. There are several possible interpretations, including that germline perturbations are corrected or diminished, or that there is an interaction between ancestral exposure and direct exposure whereby one mitigates the other. This remains to be determined. There were, however, aspects of social interaction dynamics altered in F4 rats, in which A1221 decreased the number of visits to an animal but increased the duration of those visits in the paternal lineage.

4.2.2. PCBs Are an Underappreciated Agent of Transgenerational Perturbations

A recent review on the transgenerational impact of EDC exposure on transgenerational endpoints included 43 primary research articles, of which 17 were focused on VIN and only one (from our research group) reported results on PCBs [4]. Our lab subsequently published a second article on transgenerational A1221 effects [59]. PCBs are well described for their endocrine disrupting actions when exposure occurs during development [1] but the literature on the transgenerational effects of PCBs on behavioral or molecular endpoints is limited. We previously showed that A1221 induced heritable epimutations in both F3 sperm and brain [57]; increased body weight, circulating progesterone, and estradiol [60], and caused aberrant gene expression in the hypothalamus in the F3 generation [59]. The current study adds to the transgenerational literature with results showing that A1221 increased preference for social novelty in paternal F3 females, an effect that was not present in the F1 generation, nor did it persist to the F4 or F6 generations after an additional hit of A1221. Two hits of A1221 also did not change the overall preference for social novelty but induced changes in social interaction dynamics (nonsocial time and average visit time) in the F4 generation. These effects again did not persist to the F6 generation; they also did not occur in groups where the first hit or the second hit were different EDCs. These effects demonstrate that A1221 is an agent of heritable behavioral change, at least in the context of social behavior.

4.2.3. The Order of EDC Exposure Is Important

A1221 represents an EDC class (PCBs) that is no longer manufactured, but persists in our environment. VIN represents an EDC that was introduced about five decades after A1221 and is still in agricultural use. Based on evidence that A1221 and VIN alter unique subsets of differentially methylated regions in F3 sperm [57] we hypothesized that associated behavioral phenotypes would differ. The present data confirm that hypothesis. A number of effects were dependent on the order of the EDC exposure; for example, A1221/VIN females from the paternal F6 generation had increased preference for social affiliation when compared to VIN/VIN females. While neither of these groups were determined to be different from the control group, these data show that the changes caused by specific EDCs can set a trajectory that can either amplify or diminish further exposures.

We are not aware of any experimental precedence to these findings, so putting these results into context is difficult. However, we strongly believe that future experiments considering heritable transgenerational phenotypes should try to include historically relevant exposure models that might also include complex EDC mixtures so that we can more realistically model human and wildlife exposures.

Author Contributions: Conceptualization: R.G., D.C. and A.C.G.; methodology: R.G., M.D., M.P.R., L.M.T., D.C. and A.C.G.; investigation: M.P.R., L.M.T., N.J.C. and E.L.V.; formal analysis: R.G. and M.D.; writing—original draft: R.G.; writing—review and editing: R.G., D.C. and A.C.G. All authors have read and agreed to the published version of the manuscript.

Funding: This research was funded by the National Institutes of health RO1 ES023254 to D.C. and A.C.G., RO1 ES029464 to A.C.G and a PhRMA Postdoctoral Fellowship to R.G.

Institutional Review Board Statement: The study was conducted according to the guidelines of the National Institutes of Health, and approved by the Institutional Animal Care and Use Committee at The University of Texas at Austin (AUP-2016-00029, approved 3 July 2016, and AUP-2018-00171, approved on 8 June 2018).

Data Availability Statement: Data will be made available upon request.

Acknowledgments: The authors recognize Mandee Bell for assistance with animal husbandry and Connor D. Weeks and Andrew Zentay for their assistance performing behavioral tasks presented in this study.

Conflicts of Interest: The authors declare no conflict of interest. The funders had no role in the design of the study; in the collection, analyses, or interpretation of data; in the writing of the manuscript, or in the decision to publish the results.

References

1. Gore, A.C.; Chappell, V.A.; Fenton, S.E.; Flaws, J.A.; Nadal, A.; Prins, G.S.; Toppari, J.; Zoeller, R.T. EDC-2: The Endocrine Society's Second Scientific Statement on Endocrine-Disrupting Chemicals. *Endocr. Rev.* **2015**, *36*, E1–E150. [CrossRef]
2. Ribeiro, E.; Ladeira, C.; Viegas, S. EDCs Mixtures: A Stealthy Hazard for Human Health? *Toxics* **2017**, *5*, 5. [CrossRef]
3. Rissman, E.F.; Adli, M. Minireview: Transgenerational Epigenetic Inheritance: Focus on Endocrine Disrupting Compounds. *Endocrinology* **2014**, *155*, 2770–2780. [CrossRef]
4. Van Cauwenbergh, O.; Di Serafino, A.; Tytgat, J.; Soubry, A. Transgenerational epigenetic effects from male exposure to endocrine-disrupting compounds: A systematic review on research in mammals. *Clin. Epigenetics* **2020**, *12*, 1–23. [CrossRef]
5. Jolous-Jamshidi, B.; Cromwell, H.C.; McFarland, A.M.; Meserve, L.A. Perinatal exposure to polychlorinated biphenyls alters social behaviors in rats. *Toxicol. Lett.* **2010**, *199*, 136–143. [CrossRef]
6. Reilly, M.P.; Weeks, C.D.; Topper, V.Y.; Thompson, L.M.; Crews, D.; Gore, A.C. The effects of prenatal PCBs on adult social behavior in rats. *Horm. Behav.* **2015**, *73*, 47–55. [CrossRef]
7. Porrini, S.; Belloni, V.; Della Seta, D.; Farabollini, F.; Giannelli, G.; Dessì-Fulgheri, F. Early exposure to a low dose of bisphenol A affects socio-sexual behavior of juvenile female rats. *Brain Res. Bull.* **2005**, *65*, 261–266. [CrossRef]
8. Wolstenholme, J.T.; Taylor, J.A.; Shetty, S.R.J.; Edwards, M.; Connelly, J.J.; Rissman, E.F. Gestational Exposure to Low Dose Bisphenol A Alters Social Behavior in Juvenile Mice. *PLoS ONE* **2011**, *6*, e25448. [CrossRef]
9. Kougias, D.G.; Cortes, L.; Moody, L.; Rhoads, S.; Pan, Y.-X.; Juraska, J.M. Effects of Perinatal Exposure to Phthalates and a High-Fat Diet on Maternal Behavior and Pup Development and Social Play. *Endocrinology* **2018**, *159*, 1088–1105. [CrossRef]
10. Lee, K.-I.; Chiang, C.-W.; Lin, H.-C.; Zhao, J.-F.; Li, C.-T.; Shyue, S.-K.; Lee, T.-S. Maternal exposure to di-(2-ethylhexyl) phthalate exposure deregulates blood pressure, adiposity, cholesterol metabolism and social interaction in mouse offspring. *Arch. Toxicol.* **2016**, *90*, 1211–1224. [CrossRef]
11. Venerosi, A.; Ricceri, L.; Tait, S.; Calamandrei, G. Sex dimorphic behaviors as markers of neuroendocrine disruption by environmental chemicals: The case of chlorpyrifos. *Neuro Toxicol.* **2012**, *33*, 1420–1426. [CrossRef]
12. Colbert, N.K.W.; Pelletier, N.C.; Cote, J.M.; Concannon, J.B.; Jurdak, N.A.; Minott, S.B.; Markowski, V.P. Perinatal Exposure to Low Levels of the Environmental Antiandrogen Vinclozolin Alters Sex-Differentiated Social Play and Sexual Behaviors in the Rat. *Environ. Health Perspect.* **2005**, *113*, 700–707. [CrossRef]
13. Gore, A.C.; Patisaul, H.B. Neuroendocrine disruption: Historical roots, current progress, questions for the future. *Front. Neuroendocr.* **2010**, *31*, 395–399. [CrossRef]
14. Gore, A.C.; Krishnan, K.; Reilly, M.P. Endocrine-disrupting chemicals: Effects on neuroendocrine systems and the neurobiology of social behavior. *Horm. Behav.* **2019**, *111*, 7–22. [CrossRef]
15. Wolstenholme, J.T.; Edwards, M.; Shetty, S.R.J.; Gatewood, J.D.; Taylor, J.A.; Rissman, E.F.; Connelly, J.J. Gestational Exposure to Bisphenol A Produces Transgenerational Changes in Behaviors and Gene Expression. *Endocrinology* **2012**, *153*, 3828–3838. [CrossRef]
16. Wolstenholme, J.T.; Goldsby, J.A.; Rissman, E.F. Transgenerational effects of prenatal bisphenol a on social recognition. *Horm. Behav.* **2013**, *64*, 833–839. [CrossRef]
17. Quinnies, K.M.; Harris, E.P.; Snyder, R.W.; Sumner, S.S.; Rissman, E.F. Direct and transgenerational effects of low doses of perinatal di-(2-ethylhexyl) phthalate (DEHP) on social behaviors in mice. *PLoS ONE* **2017**, *12*, e0171977. [CrossRef]
18. Quinnies, K.M.; Doyle, T.J.; Kim, K.H.; Rissman, E.F. Transgenerational Effects of Di-(2-Ethylhexyl) Phthalate (DEHP) on Stress Hormones and Behavior. *Endocrinology* **2015**, *156*, 3077–3083. [CrossRef]
19. Crews, D.; Gillette, R.; Scarpino, S.V.; Manikkam, M.; Savenkova, M.I.; Skinner, M.K. Epigenetic transgenerational inheritance of altered stress responses. *Proc. Natl. Acad. Sci. USA* **2012**, *109*, 9143–9148. [CrossRef]

20. Robaire, B.; Delbes, G.; Head, J.A.; Marlatt, V.L.; Martyniuk, C.J.; Reynaud, S.; Trudeau, V.L.; Mennigen, J.A. A cross-species comparative approach to assessing multi- and transgenerational effects of endocrine disrupting chemicals. *Environ. Res.* **2022**, *204*, 112063. [CrossRef]
21. Crews, D.; Gore, A.; Hsu, T.S.; Dangleben, N.L.; Spinetta, M.; Schallert, T.; Anway, M.D.; Skinner, M.K. Transgenerational epigenetic imprints on mate preference. *Proc. Natl. Acad. Sci. USA* **2007**, *104*, 5942–5946. [CrossRef] [PubMed]
22. Krizhevsky, A.; Sutskever, I.; Hinton, G.E. ImageNet Classification with Deep Convolutional Neural Networks. *Commun. Acm.* **2017**, *60*, 84–90. [CrossRef]
23. Luo, H.; Xiong, C.; Fang, W.; Love, P.; Zhang, B.; Ouyang, X. Convolutional neural networks: Computer vision-based workforce activity assessment in construction. *Autom. Constr.* **2018**, *94*, 282–289. [CrossRef]
24. Connor, K.; Ramamoorthy, K.; Moore, M.; Mustain, M.; Chen, I.; Safe, S.; Zacharewski, T.; Gillesby, B.; Joyeux, A.; Balaguer, P. Hydroxylated Polychlorinated Biphenyls (PCBs) as Estrogens and Antiestrogens: Structure–Activity Relationships. *Toxicol. Appl. Pharmacol.* **1997**, *145*, 111–123. [CrossRef]
25. Kelce, W.R.; Monosson, E.; Gamcsik, M.P.; Laws, S.C.; Gray, L.E. Environmental Hormone Disruptors: Evidence That Vinclozolin Developmental Toxicity Is Mediated by Antiandrogenic Metabolites. *Toxicol. Appl. Pharmacol.* **1994**, *126*, 276–285. [CrossRef]
26. Haith, D.A.; Rossi, F.S. Risk Assessment of Pesticide Runoff from Turf. *J. Environ. Qual.* **2003**, *32*, 447–455. [CrossRef]
27. Gray, L.E.; Ostby, J.; Furr, J.; Wolf, C.J.; Lambright, C.; Parks, L.; Veeramachaneni, D.N.; Wilson, V.; Price, M.; Hotchkiss, A.; et al. Effects of environmental antiandrogens on reproductive development in experimental animals. *Hum. Reprod. Updat.* **2001**, *7*, 248–264. [CrossRef]
28. Park, H.-Y.; Hertz-Picciotto, I.; Sovcikova, E.; Kocan, A.; Drobna, B.; Trnovec, T. Neurodevelopmental toxicity of prenatal polychlorinated biphenyls (PCBs) by chemical structure and activity: A birth cohort study. *Environ. Health* **2010**, *9*, 51. [CrossRef]
29. Streifer, M.; Gore, A.C. Epigenetics, Estrogenic Endocrine-Disrupting Chemicals (EDCs), and the Brain. *Adv. Pharmacol.* **2021**, *92*, 73–99. [CrossRef]
30. Hernandez, M.E.; Gore, A.C. Chemical Contaminants—A Toxic Mixture for Neurodevelopment. *Nat. Rev. Endocrinol.* **2017**, *13*, 322–323. [CrossRef]
31. Gillette, R.; Miller-Crews, I.; Nilsson, E.E.; Skinner, M.K.; Gore, A.C.; Crews, D. Sexually Dimorphic Effects of Ancestral Exposure to Vinclozolin on Stress Reactivity in Rats. *Endocrinology* **2014**, *155*, 3853–3866. [CrossRef]
32. Topper, V.Y.; Reilly, M.P.; Wagner, L.M.; Thompson, L.M.; Gillette, R.; Crews, D.; Gore, A.C. Social and neuromolecular phenotypes are programmed by prenatal exposures to endocrine-disrupting chemicals. *Mol. Cell. Endocrinol.* **2019**, *479*, 133–146. [CrossRef]
33. Dickerson, S.M.; Cunningham, S.L.; Patisaul, H.B.; Woller, M.J.; Gore, A.C. Endocrine Disruption of Brain Sexual Differentiation by Developmental PCB Exposure. *Endocrinology* **2011**, *152*, 581–594. [CrossRef]
34. Fitzgerald, E.F.; Belanger, E.E.; Gomez, M.I.; Cayo, M.; McCaffrey, R.J.; Seegal, R.F.; Jansing, R.L.; Hwang, S.-A.; Hicks, H.E. Polychlorinated Biphenyl Exposure and Neuropsychological Status among Older Residents of Upper Hudson River Communities. *Environ. Health Perspect.* **2008**, *116*, 209–215. [CrossRef]
35. Alyea, R.A.; Gollapudi, B.B.; Rasoulpour, R.J. Are we ready to consider transgenerational epigenetic effects in human health risk assessment? *Environ. Mol. Mutagen.* **2013**, *55*, 292–298. [CrossRef]
36. Steinberg, R.M.; Juenger, T.; Gore, A.C. The effects of prenatal PCBs on adult female paced mating reproductive behaviors in rats. *Horm. Behav.* **2007**, *51*, 364–372. [CrossRef]
37. Arnold, A.P.; Gorski, R.A. Gonadal Steroid Induction of Structural Sex Differences in the Central Nervous System. *Annu. Rev. Neurosci.* **1984**, *7*, 413–442. [CrossRef]
38. Lee, H.J.; Hore, T.A.; Reik, W. Reprogramming the Methylome: Erasing Memory and Creating Diversity. *Cell Stem Cell* **2014**, *14*, 710–719. [CrossRef]
39. Nadler, J.J.; Moy, S.S.; Dold, G.; Simmons, N.; Perez, A.; Young, N.B.; Barbaro, R.P.; Piven, J.; Magnuson, T.R.; Crawley, J.N. Automated apparatus for quantitation of social approach behaviors in mice. *Genes Brain Behav.* **2004**, *3*, 303–314. [CrossRef]
40. Reilly, M.P.; Weeks, C.D.; Crews, D.; Gore, A.C. Application of a novel social choice paradigm to assess effects of prenatal endocrine-disrupting chemical exposure in rats (Rattus norvegicus). *J. Comp. Psychol.* **2018**, *132*, 253–267. [CrossRef]
41. Mathis, A.; Mamidanna, P.; Cury, K.M.; Abe, T.; Murthy, V.N.; Mathis, M.W.; Bethge, M. DeepLabCut: Markerless pose estimation of user-defined body parts with deep learning. *Nat. Neurosci.* **2018**, *21*, 1281–1289. [CrossRef] [PubMed]
42. R Core Team. Available online: https://www.r-project.org/ (accessed on 10 November 2021).
43. Royston, J.P. Algorithm AS 181: The W Test for Normality. *J. R. Stat. Soc. Ser. C Appl. Stat.* **1982**, *31*, 176. [CrossRef]
44. Bartlett, M.S. Properties of sufficiency and statistical tests. *Proc. R. Soc. Lond. Ser. A Math. Phys. Sci.* **1937**, *160*, 268–282. [CrossRef]
45. Kruskal, W.H.; Wallis, W.A. Use of Ranks in One-Criterion Variance Analysis. *J. Am. Stat. Assoc.* **1952**, *47*, 583–621. [CrossRef]
46. Diener, M.J. *Corsini Encyclopedia of Psychology*; Wiley: Hoboken, NJ, USA, 2010. [CrossRef]
47. Millard, S. *EnvStats: An R Package for Environmental Statistics*; Springer: New York, NY, USA, 2013; ISBN 978-1-4614-8455-4.
48. Fox, J.; Weisberg, S. *An R Companion to Applied Regression*, 3rd ed.; Sage: Thousand Oaks, CA, USA, 2019.
49. Navarro, D. *Learning Statistics with R: A Tutorial for Psychology Students and other Beginners*; University of Adelaide: Adelaide, Australia, 2015.
50. Tukey, J.W. *Exploratory Data Analysis*; Addison-Wesley: Reading, MA, USA, 1977; Volume 2.
51. Hernandez Scudder, M.E.; Weinberg, A.; Thompson, L.; Crews, D.; Gore, A.C. Prenatal EDCs Impair Mate and Odor Preference and Activation of the VMN in Male and Female Rats. *Endocrinology* **2020**, *161*, bqaa124. [CrossRef] [PubMed]

52. Johnston, A.L.; File, S.E. Sex differences in animal tests of anxiety. *Physiol. Behav.* **1991**, *49*, 245–250. [CrossRef]
53. Holmes, M.M.; Niel, L.; Anyan, J.J.; Griffith, A.T.; Monks, D.A.; Forger, N.G. Effects of Bax gene deletion on social behaviors and neural response to olfactory cues in mice. *Eur. J. Neurosci.* **2011**, *34*, 1492–1499. [CrossRef]
54. Karlsson, S.A.; Haziri, K.; Hansson, E.; Kettunen, P.; Westberg, L. Effects of sex and gonadectomy on social investigation and social recognition in mice. *BMC Neurosci.* **2015**, *16*, 83. [CrossRef]
55. Hotchkiss, A.; Ostby, J.; Vandenbergh, J.; Gray, L. An environmental antiandrogen, vinclozolin, alters the organization of play behavior. *Physiol. Behav.* **2003**, *79*, 151–156. [CrossRef]
56. Vandenberg, L.N.; Colborn, T.; Hayes, T.B.; Heindel, J.J.; Jacobs, D.R., Jr.; Lee, D.-H.; Shioda, T.; Soto, A.M.; vom Saal, F.S.; Welshons, W.V.; et al. Hormones and Endocrine-Disrupting Chemicals: Low-Dose Effects and Nonmonotonic Dose Responses. *Endocr. Rev.* **2012**, *33*, 378–455. [CrossRef]
57. Gillette, R.; Son, M.J.; Ton, L.; Gore, A.C.; Crews, D. Passing experiences on to future generations: Endocrine disruptors and transgenerational inheritance of epimutations in brain and sperm. *Epigenetics* **2018**, *13*, 1106–1126. [CrossRef] [PubMed]
58. Soubry, A.; Hoyo, C.; Jirtle, R.L.; Murphy, S.K. A paternal environmental legacy: Evidence for epigenetic inheritance through the male germ line. *BioEssays* **2014**, *36*, 359–371. [CrossRef] [PubMed]
59. Gore, A.C.; Thompson, L.M.; Bell, M.; A Mennigen, J. Transgenerational effects of polychlorinated biphenyls: Hypothalamic gene expression in rats. *Biol. Reprod.* **2021**, *105*, 690–704. [CrossRef] [PubMed]
60. Mennigen, J.A.; Thompson, L.M.; Bell, M.; Santos, M.T.; Gore, A.C. Transgenerational effects of polychlorinated biphenyls: Development and physiology across 3 generations of rats. *Environ. Health* **2018**, *17*, 18. [CrossRef]

Article

Effect of Low and High Doses of Two Selective Serotonin Reuptake Inhibitors on Pregnancy Outcomes and Neonatal Mortality

Rafael R. Domingues [1,2,†], Hannah P. Fricke [1,2,†], Celeste M. Sheftel [1,3], Autumn M. Bell [1], Luma C. Sartori [1], Robbie S. J. Manuel [1], Chandler J. Krajco [1], Milo C. Wiltbank [1,2] and Laura L. Hernandez [1,2,3,*]

[1] Department of Animal and Dairy Sciences, University of Wisconsin-Madison, Madison, WI 53706, USA; reisdomingue@wisc.edu (R.R.D.); hfricke@wisc.edu (H.P.F.); underriner@wisc.edu (C.M.S.); ambell3@wisc.edu (A.M.B.); canavessisar@wisc.edu (L.C.S.); rmanuel@wisc.edu (R.S.J.M.); ckrajco@wisc.edu (C.J.K.); wiltbank@wisc.edu (M.C.W.)
[2] Endocrinology and Reproductive Physiology Program, University of Wisconsin-Madison, Madison, WI 53706, USA
[3] Molecular and Cellular Pharmacology Program, University of Wisconsin-Madison, Madison, WI 53706, USA
* Correspondence: llhernan@wisc.edu
† These authors contributed equally to this manuscript.

Abstract: Selective serotonin reuptake inhibitors (SSRI) are the most common antidepressant used by pregnant women; however, they have been associated with adverse pregnancy outcomes and perinatal morbidity in pregnant women and animal models. We investigated the effects of two SSRI, fluoxetine and sertraline, on pregnancy and neonatal outcomes in mice. Wild-type mice were treated daily with low and high doses of fluoxetine (2 and 20 mg/kg) and sertraline (10 and 20 mg/kg) from the day of detection of a vaginal plug until the end of lactation (21 days postpartum). Pregnancy rate was decreased only in the high dose of fluoxetine group. Maternal weight gain was reduced in the groups receiving the high dose of each drug. Number of pups born was decreased in the high dose of fluoxetine and low and high doses of sertraline while the number of pups weaned was decreased in all SSRI-treated groups corresponding to increased neonatal mortality in all SSRI-treated groups. In conclusion, there was a dose-dependent effect of SSRI on pregnancy and neonatal outcomes in a non-depressed mouse model. However, the distinct placental transfer of each drug suggests that the effects of SSRI on pup mortality may be mediated by SSRI-induced placental insufficiency rather than a direct toxic effect on neonatal development and mortality.

Keywords: selective serotonin reuptake inhibitor; perinatal mortality; neonatal morbidity; fluoxetine; sertraline

Citation: Domingues, R.R.; Fricke, H.P.; Sheftel, C.M.; Bell, A.M.; Sartori, L.C.; Manuel, R.S.J.; Krajco, C.J.; Wiltbank, M.C.; Hernandez, L.L. Effect of Low and High Doses of Two Selective Serotonin Reuptake Inhibitors on Pregnancy Outcomes and Neonatal Mortality. *Toxics* **2022**, *10*, 11. https://doi.org/10.3390/toxics10010011

Academic Editors: Kimberly Keil Stietz, Tracie Baker and Jessica Plavicki

Received: 1 December 2021
Accepted: 22 December 2021
Published: 1 January 2022

Publisher's Note: MDPI stays neutral with regard to jurisdictional claims in published maps and institutional affiliations.

Copyright: © 2022 by the authors. Licensee MDPI, Basel, Switzerland. This article is an open access article distributed under the terms and conditions of the Creative Commons Attribution (CC BY) license (https://creativecommons.org/licenses/by/4.0/).

1. Introduction

Psychotropic medications that are taken during pregnancy can pose risks of toxic effects to both mother and fetus [1,2]. About 8–12% of pregnant women take antidepressants [2–4] and selective serotonin reuptake inhibitors (SSRI) are the most commonly used antidepressant [4,5]. Among SSRI, fluoxetine was the first clinically available and remains one of the most popular while sertraline is currently the most prescribed SSRI to pregnant women [6]. Although teratogenic effects of some SSRI (i.e., paroxetine) are well recognized [7], other SSRI (sertraline, citalopram, fluoxetine) continue to be commonly prescribed to pregnant women [6]. Nevertheless, in the past decades multiple studies highlighted the association between SSRI use during gestation and adverse maternal, fetal, and neonatal health outcomes including decreased birthweight, preterm birth, and increased perinatal morbidity and mortality [3,6,8,9].

In addition to its role as a neurotransmitter, serotonin is a hormone with vasoactive properties [10] so that increased serotonin signaling selectively increases vascular resistance in the uterus causing reduced uterine vascular perfusion [11]. SSRI increase free (plasma) serotonin content by inhibiting serotonin uptake into platelets [12], thereby also decreasing uterine vascular perfusion [13]. Reduced uteroplacental blood flow leads to placental dysfunction/insufficiency, the main cause of fetal growth restriction [14,15]. The role of serotonin and SSRI on fetal growth restriction have been reviewed [10,15–18]. Fetal growth restriction is an important cause of prematurity, perinatal morbidity, and lifelong health impairment in addition to being the second leading cause of perinatal mortality [19,20].

Although fluoxetine and sertraline inhibit the serotonin transporter (SERT), their pharmacokinetics differ quite markedly [21]. Following oral ingestion, fluoxetine exhibits greater bioavailability compared to sertraline (80% vs. 44%, respectively) [21]. Plasma concentrations of sertraline follow linear kinetics (increased dose promotes proportional increase in systemic concentrations of the drug) [22]. However, fluoxetine follows nonlinear kinetics resulting in a disproportional increase in systemic concentrations after dose augmentation. Additionally, while sertraline has a half-life of 22 to 36 h, fluoxetine has a half-life of 1–6 days [22]. Furthermore, fluoxetine metabolism produces an active metabolite, norfluoxetine, which has a longer half-life than fluoxetine itself (8–15 days) while sertraline's metabolites are essentially inactive. Lastly, placental transfer of fluoxetine is greater compared to sertraline (70% vs. 25%) [23,24]. It is unclear whether these two popular SSRI with distinct kinetics similarly affect pregnancy outcomes and neonatal morbidity/mortality, particularly given their distinct placental transfer.

Because of the widespread use of SSRI during gestation and their possible detrimental effects on pregnancy and neonatal outcomes, we aimed to compare the effects of low and high doses of fluoxetine and sertraline on pregnancy and neonatal outcomes. Pregnant mice were treated with SSRI during the second half to pregnancy. Altogether, both low and high doses of sertraline and fluoxetine adversely affected neonatal outcomes; however, only the high dose of fluoxetine resulted in decreased pregnancy rate.

2. Materials and Methods

2.1. Animals

All experiments were approved by the Research Animal Care and Use Committee at the University of Wisconsin-Madison and were performed under protocol number A005789-A01. Mice were housed in a controlled environmental facility for biological research in the Biochemistry Department vivarium (fluoxetine study) and the Animal and Dairy Sciences Department vivarium (sertraline study) at the University of Wisconsin-Madison. Animal facilities were maintained at a temperature of 25 °C and a humidity of 50% to 60%, with a 12:12 h light-dark cycle with ad libitum water and food (LabDiet 5015, TestDiet, Richmond, IN, USA). Wild-type C57BL/6J mice were obtained from Jackson Laboratories (stock # 000664, Jackson Laboratories, Bar Harbor, ME, USA). Females included in our study either originated from Jackson Laboratories or were F1 offspring from our breeding colony. Beginning at 6 weeks of age, female mice were bred with a male overnight.

2.2. Experimental Design

After detection of a vaginal plug (day post coitum [DPC] 0.5), dams were individually housed and randomly assigned to treatment groups. All mice received a daily intraperitoneal injection between the hours of 0800 and 0900 of either vehicle, low and high dose of SSRI (either fluoxetine or sertraline) from DPC 0.5 until the end of lactation (21 days postpartum). Virgin females (unmated; 2–6 per group) were treated with vehicle, low and high dose of SSRI (either of fluoxetine or sertraline) for evaluation of the effect of SSRI on weight in nonpregnant mice. Virgin mice received seven treatments, equivalent to pregnant dams treated from DPC0.5 to 6.5. All mice were weighed daily at time of injection. A successful pregnancy was determined by weight gain between DPC 0.5 and 7.5 [25] and confirmed by parturition. Pregnancy length and number of pups born were recorded on the day of parturition. The number of live pups was recorded daily during lactation. Litters were not standardized.

For the fluoxetine study, fluoxetine hydrochloride (F312; Sigma-Aldrich, St. Louis, MO, USA) was reconstituted in saline. Mice were treated with vehicle (saline; n = 28), low dose of fluoxetine (2 mg/kg; n = 32), or high dose of fluoxetine (20 mg/kg; n = 127).

For the sertraline study, sertraline hydrochloride (S6319; Sigma-Aldrich, St. Louis, MO, USA) was reconstituted in 8.3% dimethyl sulfoxide (DMSO) diluted in saline for the low dose group (10 mg/kg; n = 32) and in 15% DMSO diluted in saline for the high dose group (20 mg/kg; n = 16). To account for the different concentrations of DMSO, we had two vehicle groups: 8.3% DMSO diluted in saline (vehicle group for the low dose sertraline; n = 32) and in 15% DMSO diluted in saline (vehicle group for the high dose sertraline; n = 11).

The high dose of fluoxetine (20 mg/kg) has been extensively used in rodent studies [26,27], including reports from our laboratory [28]. However, the systemic concentrations of the drug in mice from previous studies from our laboratory were higher than in humans. Therefore, we selected a low dose (2 mg/kg) that we anticipated would be within expected systemic concentrations of fluoxetine in humans. The low dose of sertraline (10 mg/kg) was based on other reports and is expected to be within normal range of human systemic concentrations [29]. However, we had issues with drug solubility to develop a model with the higher dose of sertraline. Because we did not want to dramatically increase DMSO concentrations nor alter injection volume between studies, we were only able to treat mice at 20 mg/kg.

2.3. Blood Collection and Fluoxetine Assay

Blood samples were collected from the dams on DPC 0.75 and DPC 17.75. Mice were fasted for 6–8 h after the morning treatment until blood collection. Blood was collected from the submandibular vein using a 5.5 mm lancet, placed on ice for 20 min, and centrifuged at 846 g for 20 min at 4 °C; serum was stored at -80 °C until assayed. Serum fluoxetine and norfluoxetine concentrations were measured with a forensic fluoxetine ELISA kit (catalog no. 107619; Neogen, Lexington, KY, USA) according to manufacturer's instructions, samples were diluted 1:100. A displacement curve was prepared from fluoxetine hydrochloride (S6319; Sigma-Aldrich, St. Louis, MO, USA) to create a standard curve for quantification. The cross-reactivity is 100% for fluoxetine and 67% for norfluoxetine; therefore, data is presented as fluoxetine plus norfluoxetine concentrations.

2.4. Statistical Analysis

All statistical analyses were performed on SAS (version 9.4; SAS Institute Inc., Cary, NC, USA). Data were analyzed with PROC MIXED procedure using one-way ANOVA and two-way ANOVA for repeated measures. Tukey HSH was used for post hoc comparisons. Studentized residuals with deviations from assumptions of normality and/or homogeneity of variance were transformed into square root, logarithms, or ranks. Survival analysis was done with PROC LIFETEST using Wilcoxon test. For the fluoxetine study, comparisons were performed among all groups. For the sertraline studies, comparisons between vehicle and treated group were performed separately for the low and high dose. A probability of ≤ 0.05 indicated a difference was significant and a probability between >0.05 and ≤ 0.1 indicated significance was approached. Data are presented as the mean $\pm$ standard error of mean (SEM).

3. Results

3.1. Systemic Fluoxetine Concentrations

Fluoxetine was undetected in vehicle-treated mice (Table 1). In fluoxetine-treated mice, there was a dose-dependent effect of treatment on systemic concentrations of fluoxetine. At 6 h after the first treatment, fluoxetine concentrations were 24-fold greater in mice treated with the high than low dose. At 6 h after the 18th treatment, fluoxetine concentrations had increased 2.4-fold with the low dose and 4.2-fold with the high dose, producing a 43-fold greater fluoxetine in the high than low dose animals.

Table 1. Serum concentrations of fluoxetine + norfluoxetine in pregnant mice on DPC 0.75 and 17.75 (6 h after the first and 18th treatments, respectively).

	Vehicle	Low Dose	High Dose	p Value
DPC 0.75, ng/mL (range)	0 Undetectable	195.4 ± 14.6 [b] (169.5–219.9)	4724.7 ± 354.0 [a] (2756.3–5828.4)	<0.0001
DPC 17.75, ng/mL (range)	0 Undetectable	466.8 ± 61.6 [b] (336.8–746.6)	20,059.2 ± 2176.5 [a] (13,477.8–29,815.6)	<0.0001

[a,b] indicate significant differences among groups.

3.2. Maternal Weight

Weight gain after the onset of treatment was evaluated between DPC 0.5 to 6.5 and DPC 7.5 to 18.5. Until DPC 6.5 there is little to no effect of embryonic weight on total maternal weight [25]; therefore, pregnant, nonpregnant, and virgin mice were included in the analysis for more robust analysis of the effect of SSRI on mouse weight. In the fluoxetine study, overall weight gain during DPC 0.5 to 6.5 was decreased in the high dose group (Figure 1). Although all groups in the fluoxetine study lost weight after the first day of treatment, the high dose of fluoxetine caused greater weight loss after a single treatment compared to vehicle and low dose. Additionally, the group receiving the high dose of fluoxetine did not recuperate weight to pretreatment levels until DPC 6.5 while the vehicle and low dose groups reached pretreatment weight on DPC 2.5. Maternal weight gain was overall reduced in the high dose of fluoxetine from DPC 7.5 to 18.5. Maternal weight on the day before parturition was greatest ($p = 0.005$) in the vehicle group (32.7 ± 0.5 g), intermediate in the low dose (32.1 ± 0.7 g), and lowest in the high dose group (30.5 ± 0.4 g).

In the sertraline study, the high dose group had greater weight loss after onset of treatment and overall weight gain between DPC 0.5 and 6.5 and between DPC 7.5 and 18.5 was lower in the high dose group than vehicle. However, maternal weight on the day before parturition was not different between groups ($p = 0.2$, 32.8 ± 0.8 vs. 31.6 ± 0.7 g for the low dose; $p = 0.6$, 34.2 ± 0.5 vs. 33.5 ± 1.1 g for the high dose).

3.3. Pregnancy Establishment and Maintenance

In the fluoxetine study, the high dose significantly reduced pregnancy establishment (pregnancy per plug; Table 2). Sertraline treatment (low and high doses) had no significant effect on the number of pregnant mice. Gestation length was not affected by fluoxetine treatment, but sertraline extended the mean gestation length.

Table 2. Effect of low and high doses of fluoxetine and sertraline on pregnancy outcomes.

	Vehicle	Low Dose	High Dose	p Value
Fluoxetine				
Vaginal plug, n	28	32	127	-
Pregnant dams, n	24	25	24	-
Pregnancy per plug, %	85.7 [a]	78.1 [a]	18.9 [b]	<0.0001
Gestation length, day	19.1 ± 0.1	19.1 ± 0.1	18.9 ± 0.1	0.17
Sertraline low dose				
Vaginal plug, n	32	32	N/A	-
Pregnant dams, n	22	23	N/A	-
Pregnancy per plug, %	68.8	71.9	N/A	0.99
Gestation length, day	18.8 ± 0.1 [B]	19.0 ± 0.1 [A]	N/A	0.096
Sertraline high dose				
Vaginal plug, n	11	N/A	16	-
Pregnant dams, n	7	N/A	11	-
Pregnancy per plug, %	63.6	N/A	68.7	0.9
Gestation length, day	18.9 ± 0.3 [b]	N/A	19.6 ± 0.2 [a]	0.01

[a,b] Indicate significant difference among groups. [A,B] Indicate significance was approached.

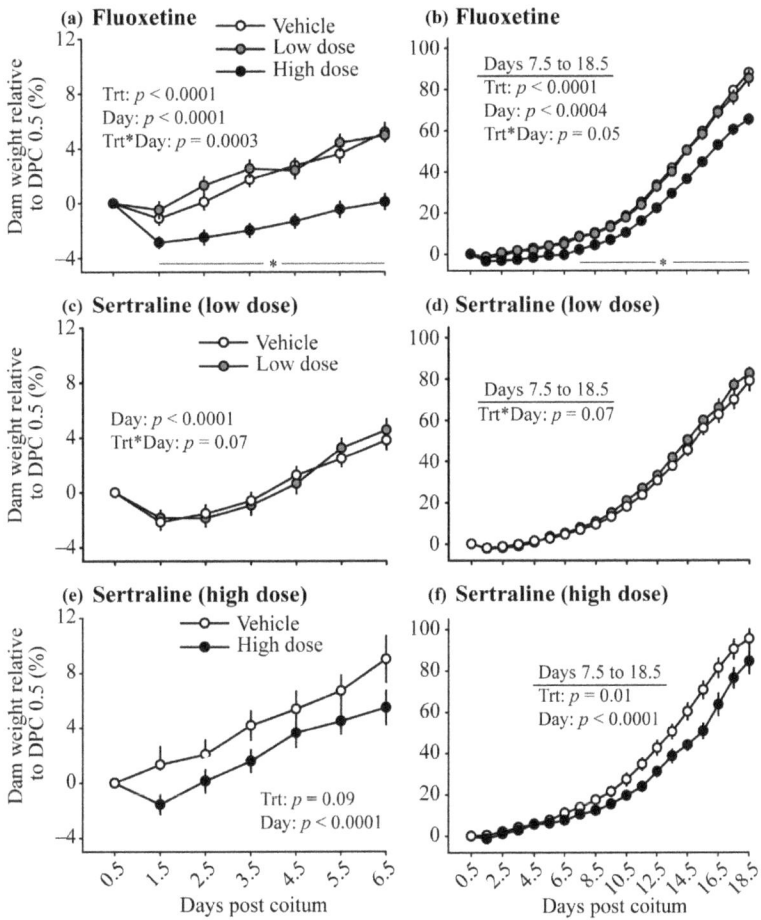

Figure 1. Effect of fluoxetine (**a,b**) and sertraline (**c–f**) on maternal weight gain between days post coitum (DPC) 0.5 to 6.5 (**a,c,e**) using data from pregnant (before the effect of fetal weight on maternal weight), nonpregnant, and virgin mice. Maternal weight gain (**b,d,f**) during entire gestation (DPC 0.5 to 18.5) used data from only pregnant mice with data analyzed for DPC 7.5 to 18.5. * Indicates significantly decreased weight in the high dose of fluoxetine group.

3.4. Neonatal Outcomes and Pup Survival

The low dose of fluoxetine did not significantly affect the number of pups born compared to the vehicle; however, the low dose of sertraline and the high dose of both fluoxetine and sertraline caused a reduction in the number of pups born (Figure 2).

In the fluoxetine study, there was a dose-dependent increase in pup mortality during the 21 days postpartum resulting in less pups weaned per litter (Figure 2). The percentage of litters in which all pups died was greater ($p = 0.0008$) for the high dose fluoxetine group (62.5%) compared to the control (20.8%) and low dose (16.0%) groups. The mean number of pups weaned per litter was also reduced by sertraline treatment (low and high doses). However, the number of litters in which all pups died were not different between groups ($p = 0.6$, 5.9 vs. 13.6% for the low dose; $p = 0.6$, 28.6 vs. 72.7% for the high dose).

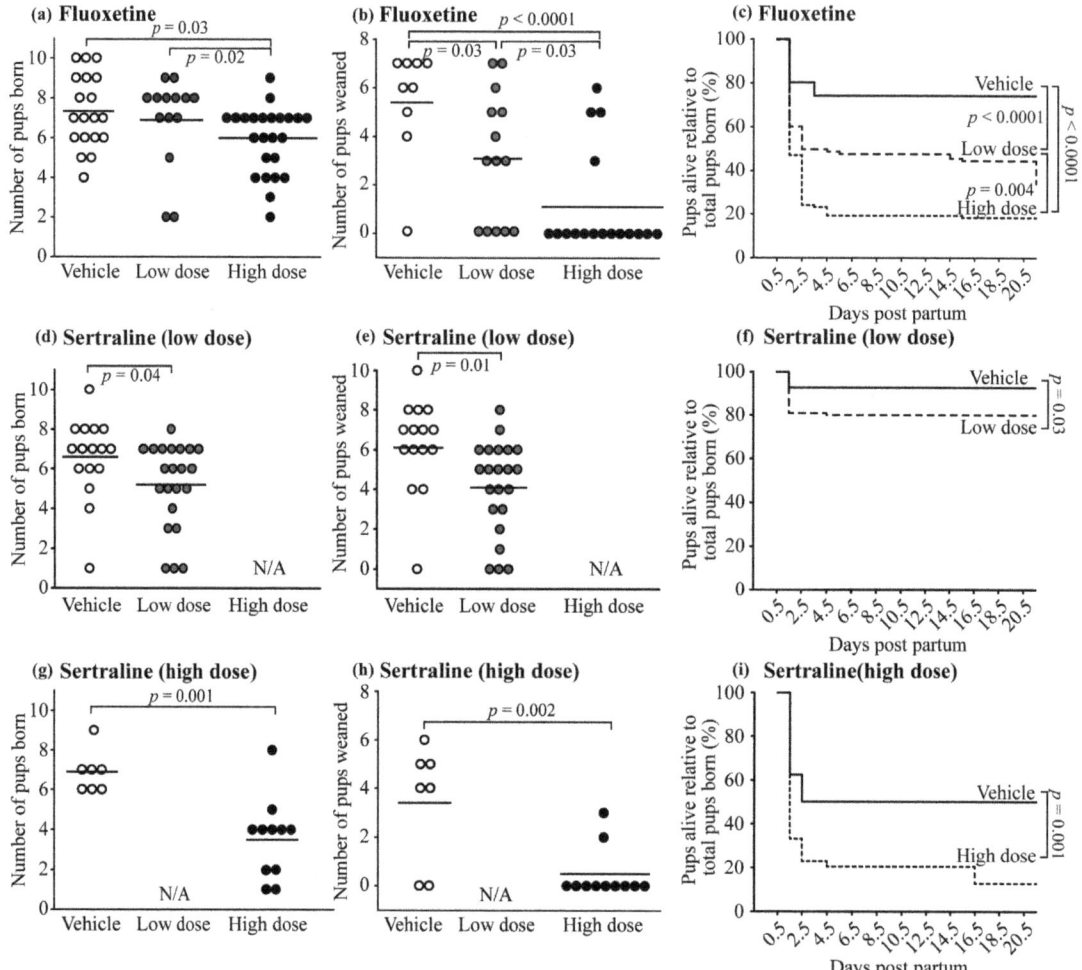

Figure 2. Effect of fluoxetine (**a–c**) and sertraline (**d–i**) on neonatal outcomes: Number of pups born per litter (**a,d,g**); number of pups weaned per litter (**b,e,h**). Survival analysis of pup mortality (**c,f,i**) during lactation (days postpartum 0.5 to 21.5).

To further investigate the effect of in utero SSRI exposure on neonatal mortality, we analyzed pup mortality using survival curves. Besides a clear effect of SSRI on neonatal survival, pup mortality occurred primarily before DPP 4.5 independent of treatment (vehicle vs. SSRI) and dose.

3.5. Pregnancy Complications

Some of the SSRI-treated dams that gave birth to an unusually reduced number of pups appeared to still have unborn pups due to visually large abdominal size. A similar finding was not observed in vehicle-treated dams. Four SSRI-treated dams that had all pups die a few days postpartum were euthanized for necropsy on postpartum days 2.5 to 5.5. These dams had 3 to 5 fully developed dead pups still in the uterus. Additionally, one dam from the high dose sertraline group euthanized on postpartum day 21.5 had 3 dead pups in the uterus that appeared to be mummified. Because our experimental design did not anticipate these issues, only a few (4) dams were euthanized and inspected for unborn pups. Consequently, precise quantification of this finding was not possible. However, reporting this finding is important for designing future studies. To gain insight into the incidence of unborn pups in SSRI-treated dams in the present study we examined the maternal weight change between the day before and the day of parturition (Figure 3). Because maternal weight was not different between groups on the day before parturition (except for high dose fluoxetine), a reduced weight loss suggests that fewer pups were born and is indicative of unborn pups. The high dose of both fluoxetine and sertraline had overall less mean weight loss between the last day of pregnancy and the day of parturition. Based on individual maternal weight loss, it seems likely that other SSRI-treated dams with unusually small litter sizes that were not necropsied also had unborn pups after parturition.

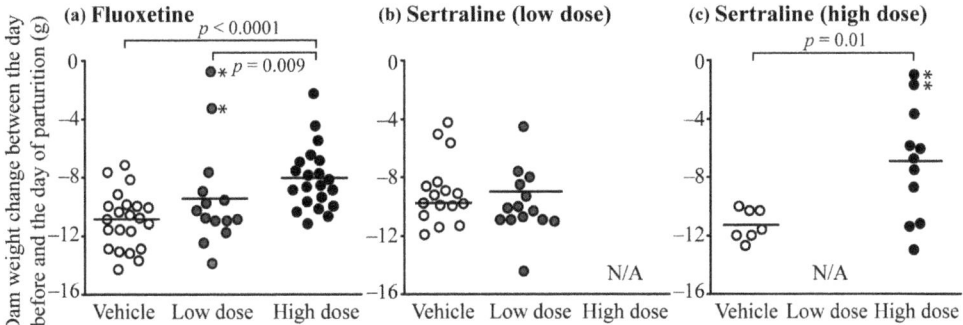

Figure 3. Maternal weight change between the last day of pregnancy and the day of parturition in the fluoxetine (**a**) and sertraline (**b**,**c**) studies. * Denotes dams that were euthanized after parturition and had fully developed dead pups in the uterus.

4. Discussion

Understanding the effects of maternal medication on pregnancy complications and neonatal outcomes is vital to comprehensively assess the risk of perinatal exposure to psychotropic medication on maternal and newborn wellbeing. Herein, we report the effects of two popular antidepressants on pregnancy and neonatal outcomes in a mouse model highlighting a dose-dependent effect of SSRI, particularly fluoxetine, on neonatal outcomes. Interestingly, fluoxetine and sertraline caused comparable reductions in the number of pups born and pup survival despite the distinct placental transfer of each drug; therefore, exposing fetuses to distinct amounts of each drug. This suggests that these adverse neonatal outcomes are likely to be related to the effect of SSRI on the dam and placenta rather than a direct toxic effect of each drug on pup development.

Because SSRI treatments began on DPC 0.5 in the present study, ovulation and fertilization were expected to take place before the onset of treatments and, therefore, to be similar among vehicle and SSRI-treated groups. Additionally, fluoxetine has little to no effect on embryo development in vitro [30] so a direct effect of fluoxetine on embryo development is unlikely. Therefore, the decreased pregnancy per plug in the high dose fluoxetine group (initially observed on DPC 7.5) is likely due to implantation failure. The high dose of fluoxetine could cause implantation failure via multiple mechanisms including: (1) Induction of maternal weight loss that could alter ovarian function [31] disrupting the endocrine environment required for embryo implantation; (2) direct or indirect modulation of estrogen signaling [32–34] with consequent disruption of uterine receptivity; or (3) decreased uterine vascular perfusion [10,11,13,18] disrupting uterine vascular remodeling [35]. Previous studies have also indicated an effect of SSRI on embryo implantation and early pregnancy loss in humans [36,37] and animal models [38,39], although the mechanism remains to be elucidated. Since none of the other doses of SSRI had an effect on embryo implantation, the decreased pregnancy rate in the group receiving the high dose of fluoxetine in the present study may be related to a toxic effect of fluoxetine, as discussed later, rather than an expected effect of the drug at therapeutic concentrations in humans. Further studies are needed to confirm this finding and to define the mechanisms and critical period of fluoxetine exposure on impaired embryo implantation to establish the safety of fluoxetine in early pregnancy development.

The reduced maternal weight gain after DPC 7.5 in the high dose of fluoxetine and sertraline-treated groups suggests smaller litter size, reduced embryonic/fetal growth, or both. Indeed, maternal SSRI treatment during gestation has been linked to intrauterine growth restriction in humans [3,6,8,17] and in animal models [38,40]. The mechanisms of SSRI-induced fetal growth restriction have been a prominent area of research worldwide [10]. The SSRI-induced increase in serotonin signaling has been associated with decreased uterine vascular perfusion [13] and vascular lesions on the maternal and fetal sides of the placenta in women [18]. Therefore, maternal exposure to SSRI may compromise placenta function leading to inadequate nutrient exchange between mother and fetus which can result in fetal growth restriction [10]. On the other hand, the number of pups born was reduced in dams exposed to the high dose of fluoxetine and the low and high doses of sertraline suggesting a role for SSRI on embryo implantation and embryonic/fetal survival. However, the unexpected finding of fully developed dead pups still in the uterus days after parturition in SSRI-treated mice further clouds our interpretation of the effects of SSRI on pregnancy establishment and fetal survival. Furthermore, it is not known whether the intrauterine pup death was a cause or a consequence of fetal retention. Although dystocia has not been reported in women taking SSRI during gestation and fluoxetine does not affect uterine contractions [41], it was an unexpected but critical finding in our study and warrants further investigation. In light of this finding, reduced litter size in rodent models treated with SSRI in previous reports and future studies should be interpreted with caution.

Neonatal mortality, primarily during early postnatal period, was increased in all groups exposed to SSRI in the present study. Previous rodent studies have also shown that fluoxetine and sertraline exposure during the perinatal period increase neonatal mortality [23,40,42]. Fetal developmental malformations (primarily cardiac, respiratory, and neurodevelopmental disorders) have been reported as possible causes of neonatal mortality associated with perinatal SSRI exposure [3,23,43–45]. However, neonatal mortality may be due to placental insufficiency caused by SSRI disruption of uterine/placental vascular perfusion and structure [3,10,13,18] rather than a direct effect of SSRI on fetal development. Placental insufficiency is generally regarded as the major cause of fetal growth restriction which is associated with perinatal morbidity and mortality. Fetal growth restriction (unrelated to SSRI exposure) leads to several fetal adaptations to restricted nutrient availability including morphological heart changes, increased cardiac workload, and cardiac function issues resembling dilated cardiomyopathy [14]. Interestingly, rodents exposed to SSRI perinatally have altered cardiac morphology [43,46] and dilated cardiomyopathy [23]. Because both drugs promote comparable neonatal mortality and increase serotonin concentrations in the maternal side of the placenta altering placental homeostasis [18,47] but have distinct placental transfer (70% fluoxetine [23] vs. 25% sertraline [24]), placental insufficiency is likely to be the underlying mechanism of SSRI-related pup mortality rather than a direct toxic effect of SSRI on fetal development. Nevertheless, these results do not exclude a direct role of fluoxetine (highest placental transfer) on fetal organogenesis.

The placenta regulates maternal and fetal serotonin homeostasis during gestation [3,48–50]. Since embryonic production of serotonin is limited until day 14.5 of gestation in mice, extraembryonic sources of serotonin are required to maintain fetal brain development [48]. SERT located on the apical region of syncytiotrophoblast (maternal side of the placenta) transports maternal-derived serotonin into the placenta during early and mid gestation regulating serotonin content (signaling) on the maternal side of the placenta and providing maternal-derived serotonin needed for fetal development [48,49]. However, during late pregnany serotonin is no longer transported from mother to fetus; instead, organic cation transporter 3 (OCT3) located on the fetal side of the placenta transports fetal serotonin into trophoblasts for degradation [47,49,50]. SSRI inhibition of placental SERT prevents the transport of maternal serotonin into the placenta increasing serotonin signaling on the maternal side of the placenta [47]. Interestingly, OCT3 is inhbited by glucocorticoids [51] and exogenous drugs such as SSRI [47,50]. However, fluoxetine, but not sertraline, decreases the capacity of OCT3 to transport serotonin from fetal circulation into the placenta in an in situ model [47]. Therefore, added to the decreased placental transfer of sertraline [23,24] resulting in lower concentrations of the drug in the fetal circulation, it seems likely that the similar effects of the two drugs on neonatal outcomes are mediated by their common capacity to inhibit SERT on the maternal side of the placenta [47]. This likely leads to increased serotonin signaling with a consequent compromise of placental vascular perfusion and function [11,15].

Systemic concentrations of fluoxetine + norfluoxetine increase after onset of treatment ranging from 160 to 560 ng/mL in humans [21]. In overdosed patients, fluoxetine + norfluoxetine concentrations may reach 1490 ng/mL [52]. In rodent models, although the dose and route of administration of fluoxetine vary among studies, the dose of 20 mg/kg/day via intraperitoneal injection has been widely used [26,27]. Our results clearly demonstrate that this dose produces systemic concentrations that are many fold greater than clinically relevant doses in humans. On the contrary, the low dose of fluoxetine used in our experiment (2 mg/kg/day) resulted in systemic concentrations similar to expected concentrations in humans. Unfortunately, we were unable to measure the sertraline concentration in our study. The overall effects of sertraline (low and high doses) were intermediate compared to the low and high doses of fluoxetine. However, the limitation of sertraline solubility that required increased DMSO concentration and the lack of systemic drug concentrations clouded our full interpretation of the effects of the high dose of sertraline on pregnancy and neonatal outcomes. Although we recognize that extreme dosage treatments are often important for delineating physiologic pathways and investigating possible toxic effects of drugs in animal models, our study highlights the importance of using therapeutic dosages to more accurately evaluate the risk of maternal drug exposure on pregnancy and neonatal outcomes, particularly in translational studies for more direct relevance to human medicine.

Most mice (vehicle and SSRI-treated) in our studies lost weight in the first 24 h after the first treatment. This is expected due to the stress of handling and changing cages for mating, individual housing, and initiation of treatments. However, mice exposed to the high dose of each drug experienced greater weight loss. Nevertheless, mice receiving the high dose of sertraline and low dose of fluoxetine reestablished pretreatment weight by day 2 of treatment (similar to vehicle) while mice receiving the high dose of fluoxetine took 6 days to reestablish pretreatment weight. In animal models, SSRI-induced weight loss has been reported [31,39,40,53]. In humans, short-term fluoxetine treatment is also known to cause weight loss [54]. However, with prolonged treatment, weight gain is most commonly observed. Although the SSRI-induced weight loss in mice may resemble the weight loss observed in short-term fluoxetine treatment in humans, the weight loss in rodents exposed to high doses of SSRI seems to be due to drug overdose because it has been associated with amenorrhea [31], digestive disorders, and death [39,53].

5. Conclusions

Overall, our results demonstrate a dose-dependent effect of SSRI exposure during gestation and lactation on pregnancy outcomes and perinatal pup mortality. The comparable neonatal outcomes during treatment with these two drugs that have distinct placental transfer properties make it likely that SSRI-induced placental insufficiency and fetal growth restriction lead to the observed neonatal morbidity/mortality rather than a direct toxic effect of each drug on perinatal mortality. Lastly, we highlight that some effects of treatments with excessive doses of psychotropic medication in animal models (as for the highest dose of fluoxetine in the present study) may not reflect expected effects in humans due to extreme systemic concentrations of the drug, and therefore, should be interpreted with caution.

Author Contributions: Conceptualization, R.R.D., H.P.F., C.M.S., M.C.W. and L.L.H.; methodology, R.R.D., H.P.F., C.M.S., M.C.W. and L.L.H.; formal analysis, R.R.D.; investigation, R.R.D., H.P.F., C.M.S., A.M.B., L.C.S., R.S.J.M. and C.J.K.; resources, M.C.W. and L.L.H.; data curation, R.R.D., H.P.F., C.M.S., L.C.S., R.S.J.M. and C.J.K.; writing—original draft preparation, R.R.D. and H.P.F.; writing—review and editing, R.R.D., H.P.F., C.M.S., M.C.W., L.L.H., A.M.B., L.C.S., R.S.J.M. and C.J.K.; supervision, M.C.W. and L.L.H.; project administration, R.R.D., H.P.F. and C.M.S.; funding acquisition, M.C.W. and L.L.H. All authors have read and agreed to the published version of the manuscript. R.R.D. and H.P.F. contributed equally to this paper.

Funding: This research was funded by the National Institute of Health grant number R01HD094759 to L.L. Hernandez and Molecular and Cellular Pharmacology T32 training grant (NIH: GM008688-16) to CMS, Metabolism and Nutrition Training Program T32 training grant (NIH: DK007665) to CMS, Endocrine and Reproductive Physiology T32 Training grant (NIH: HD041291) to HPF.

Institutional Review Board Statement: The study was approved by the Animal Care and Use Committee of the College of Agriculture and Life Sciences at the University of Wisconsin-Madison (protocol A005789-A01, date of approval: 21 December 2021).

Informed Consent Statement: Informed consent was obtained from all subjects in the study.

Data Availability Statement: The data presented in this study are available on request from the corresponding author.

Conflicts of Interest: The authors declare no conflict of interest. The funders had no role in the design of the study; in the collection, analyses, or interpretation of data; in the writing of the manuscript, or in the decision to publish the results.

References

1. Brajcich, M.R.; Palau, M.A.; Messer, R.D.; Murphy, M.E.; Marks, J. Why the Maternal Medication List Matters: Neonatal Toxicity From Combined Serotonergic Exposures. *Pediatrics* **2021**, *147*, e2250. [CrossRef]
2. Belik, J. Fetal and Neonatal Effects of Maternal Drug Treatment for Depression. *Semin. Perinatol.* **2008**, *32*, 350–354. [CrossRef] [PubMed]
3. Velasquez, J.C.; Goeden, N.; Bonnin, A. Placental serotonin: Implications for the developmental effects of SSRIs and maternal depression. *Front. Cell. Neurosci.* **2013**, *7*, 1–7. [CrossRef]
4. Oberlander, T.F.; Warburton, W.; Misri, S.; Aghajanian, J.; Hertzman, C. Neonatal outcomes after prenatal exposure to selective serotonin reuptake inhibitor antidepressants and maternal depression using population-based linked health data. *Arch. Gen. Psychiatry* **2006**, *63*, 898–906. [CrossRef]
5. Tran, H.; Robb, A.S. SSRI use during pregnancy. *Semin. Perinatol.* **2015**, *39*, 545–547. [CrossRef]
6. Bandoli, G.; Chambers, C.D.; Wells, A.; Palmsten, K. Prenatal Antidepressant Use and Risk of Adverse Neonatal Outcomes. *Pediatrics* **2020**, *146*, e2493. [CrossRef]
7. Berard, A.; Iessa, N.; Chaabane, S.; Muanda, F.T.; Boukhris, T.; Zhao, J.P. The risk of major cardiac malformations associated with paroxetine use during the first trimester of pregnancy: A systematic review and meta-analysis. *Br. J. Clin. Pharmacol.* **2016**, *81*, 589–604. [CrossRef]
8. Alwan, S.; Friedman, J.M.; Chambers, C. Safety of Selective Serotonin Reuptake Inhibitors in Pregnancy: A Review of Current Evidence. *CNS Drugs* **2016**, *30*, 499–515. [CrossRef]
9. Zhao, X.F.; Liu, Q.; Cao, S.X.; Pang, J.Y.; Zhang, H.J.; Feng, T.T.; Deng, Y.J.; Yao, J.; Li, H.F. A meta-analysis of selective serotonin reuptake inhibitors (SSRIs) use during prenatal depression and risk of low birth weight and small for gestational age. *J. Affect. Disord.* **2018**, *241*, 563–570. [CrossRef] [PubMed]
10. Rosenfeld, C.S. Placental serotonin signaling, pregnancy outcomes, and regulation of fetal brain development. *Biol. Reprod.* **2020**, *102*, 532–538. [CrossRef] [PubMed]
11. Lang, U.; Prada, J.; Clark, K.E. Systemic and uterine vascular response to serotonin in third trimester pregnant ewes. *Eur. J. Obstet. Gynecol. Reprod. Biol.* **1993**, *51*, 131–138. [CrossRef]
12. Blardi, P.; De Lalla, A.; Leo, A.; Auteri, A.; Iapichino, S.; Di Muro, A.; Dell'Erba, A.; Castrogiovanni, P. Serotonin and fluoxetine levels in plasma and platelets after fluoxetine treatment in depressive patients. *J. Clin. Psychopharmacol.* **2002**, *22*, 131–136. [CrossRef]
13. Morrison, J.L.; Chien, C.; Riggs, K.W.; Gruber, N.; Rurak, D. Effect of maternal fluoxetine administration on uterine blood flow, fetal blood gas status, and growth. *Pediatr. Res.* **2002**, *51*, 433–442. [CrossRef] [PubMed]
14. Malhotra, A.; Allison, B.J.; Castillo-Melendez, M.; Jenkin, G.; Polglase, G.R.; Miller, S.L. Neonatal Morbidities of Fetal Growth Restriction: Pathophysiology and Impact. *Front. Endocrinol.* **2019**, *10*, 55. [CrossRef]
15. Ranzil, S.; Walker, D.W.; Borg, A.J.; Wallace, E.M.; Ebeling, P.R.; Murthi, P. The relationship between the placental serotonin pathway and fetal growth restriction. *Biochimie* **2019**, *161*, 80–87. [CrossRef]

16. Ranzil, S.; Ellery, S.; Walker, D.W.; Vaillancourt, C.; Alfaidy, N.; Bonnin, A.; Borg, A.; Wallace, E.M.; Ebeling, P.R.; Erwich, J.J.; et al. Disrupted placental serotonin synthetic pathway and increased placental serotonin: Potential implications in the pathogenesis of human fetal growth restriction. *Placenta* **2019**, *84*, 74–83. [CrossRef]
17. Zullino, S.; Simoncini, T. Impact of selective serotonin reuptake inhibitors (SSRIs) during pregnancy and lactation: A focus on short and long-term vascular effects. *Vasc. Pharmacol.* **2018**, *108*, 74–76. [CrossRef]
18. Levy, M.; Kovo, M.; Miremberg, H.; Anchel, N.; Herman, H.G.; Bar, J.; Schreiber, L.; Weiner, E. Maternal use of selective serotonin reuptake inhibitors (SSRI) during pregnancy-neonatal outcomes in correlation with placental histopathology. *J. Perinatol.* **2020**, *40*, 1017–1024. [CrossRef] [PubMed]
19. Gagnon, R. Placental insufficiency and its consequences. *Eur. J. Obstet. Gynecol. Reprod. Biol.* **2003**, *110*, S99–S107. [CrossRef]
20. Nardozza, L.M.M.; Caetano, A.C.R.; Zamarian, A.C.P.; Mazzola, J.B.; Silva, C.P.; Marcal, V.M.G.; Lobo, T.F.; Peixoto, A.B.; Araujo, E. Fetal growth restriction: Current knowledge. *Arch. Gynecol. Obstet.* **2017**, *295*, 1061–1077. [CrossRef] [PubMed]
21. DeVane, C.L. Metabolism and pharmacokinetics of selective serotonin reuptake inhibitors. *Cell. Mol. Neurobiol.* **1999**, *19*, 443–466. [CrossRef]
22. DeVane, C.L.; Liston, H.L.; Markowitz, J.S. Clinical pharmacokinetics of sertraline. *Clin. Pharmacokinet.* **2002**, *41*, 1247–1266. [CrossRef] [PubMed]
23. Noorlander, C.W.; Ververs, F.F.T.; Nikkels, P.G.J.; van Echteld, C.J.A.; Visser, G.H.A.; Smidt, M.P. Modulation of Serotonin Transporter Function during Fetal Development Causes Dilated Heart Cardiomyopathy and Lifelong Behavioral Abnormalities. *PLoS ONE* **2008**, *3*, e2782. [CrossRef] [PubMed]
24. Heinonen, E.; Blennow, M.; Blomdahl-Wetterholm, M.; Hovstadius, M.; Nasiell, J.; Pohanka, A.; Gustafsson, L.L.; Wide, K. Sertraline concentrations in pregnant women are steady and the drug transfer to their infants is low. *Eur. J. Clin. Pharmacol.* **2021**, *77*, 1323–1331. [CrossRef]
25. Heyne, G.W.; Plisch, E.H.; Melberg, C.G.; Sandgren, E.P.; Peter, J.A.; Lipinski, R.J. A Simple and Reliable Method for Early Pregnancy Detection in Inbred Mice. *J. Am. Assoc. Lab. Anim. Sci.* **2015**, *54*, 368–371. [PubMed]
26. Walia, V.; Gilhotra, N. GABAergic influence in the antidepressant effect of fluoxetine in unstressed and stressed mice. *J. Appl. Pharm. Sci.* **2017**, *7*, 141–147.
27. Ma, L.; Tang, J.Y.; Zhou, J.Y.; Zhu, C.; Zhang, X.; Zhou, P.; Yu, Q.; Wang, Y.; Gu, X.J. Fluoxetine, a selective serotonin reuptake inhibitor used clinically, improves bladder function in a mouse model of moderate spinal cord injury. *Neural Regen. Res.* **2021**, *16*, 2093–2098. [CrossRef] [PubMed]
28. Weaver, S.R.; Fricke, H.P.; Xie, C.; Lipinski, R.J.; Vezina, C.M.; Charles, J.F.; Hernandez, L.L. Peripartum Fluoxetine Reduces Maternal Trabecular Bone After Weaning and Elevates Mammary Gland Serotonin and PTHrP. *Endocrinology* **2018**, *159*, 2850–2862. [CrossRef] [PubMed]
29. Wang, J.S.; DeVane, C.L.; Gibson, B.B.; Donovan, J.L.; Markowitz, J.L.; Zhu, H.J. Population pharmacokinetic analysis of drug-drug interactions among risperidone, bupropion, and sertraline in CF1 mice. *Psychopharmacology* **2006**, *183*, 490–499. [CrossRef] [PubMed]
30. Kaihola, H.; Yaldir, F.G.; Hreinsson, J.; Hörnaeus, K.; Bergquist, J.; Olivier, J.D.; Åkerud, H.; Sundström-Poromaa, I. Effects of fluoxetine on human embryo development. *Front. Cell. Neurosci.* **2016**, *10*, 160. [CrossRef]
31. Uphouse, L.; Hensler, J.G.; Sarkar, J.; Grossie, B. Fluoxetine disrupts food intake and estrous cyclicity in Fischer female rats. *Brain Res.* **2006**, *1072*, 79–90. [CrossRef]
32. Hansen, C.H.; Larsen, L.W.; Sorensen, A.M.; Halling-Sorensen, B.; Styrishave, B. The six most widely used selective serotonin reuptake inhibitors decrease androgens and increase estrogens in the H295R cell line. *Toxicol. In Vitro* **2017**, *41*, 1–11. [CrossRef] [PubMed]
33. Jacobsen, N.W.; Hansen, C.H.; Nellemann, C.; Styrishave, B.; Halling-Sorensen, B. Effects of selective serotonin reuptake inhibitors on three sex steroids in two versions of the aromatase enzyme inhibition assay and in the H295R cell assay. *Toxicol. In Vitro* **2015**, *29*, 1729–1735. [CrossRef]
34. Muller, J.C.; Imazaki, P.H.; Boareto, A.C.; Lourenco, E.L.B.; Golin, M.; Vechi, M.F.; Lombardi, N.F.; Minatovicz, B.C.; Scippo, M.L.; Martino-Andrade, A.J.; et al. In vivo and in vitro estrogenic activity of the antidepressant fluoxetine. *Reprod. Toxicol.* **2012**, *34*, 80–85. [CrossRef]
35. Cartwright, J.E.; Fraser, R.; Leslie, K.; Wallace, A.E.; James, J.L. Remodelling at the maternal-fetal interface: Relevance to human pregnancy disorders. *Reproduction* **2010**, *140*, 803–813. [CrossRef]
36. Kjaersgaard, M.I.S.; Parner, E.T.; Vestergaard, M.; Sorensen, M.J.; Olsen, J.; Christensen, J.; Bech, B.H.; Pedersen, L.H. Prenatal Antidepressant Exposure and Risk of Spontaneous Abortion—A Population-Based Study. *PLoS ONE* **2013**, *8*, e72095. [CrossRef]
37. Almeida, N.D.; Basso, O.; Abrahamowicz, M.; Gagnon, R.; Tamblyn, R. Risk of Miscarriage in Women Receiving Antidepressants in Early Pregnancy, Correcting for Induced Abortions. *Epidemiology* **2016**, *27*, 538–546. [CrossRef] [PubMed]
38. Bauer, S.; Monk, C.; Ansorge, M.; Gyamfi, C.; Myers, M. Impact of antenatal selective serotonin reuptake inhibitor exposure on pregnancy outcomes in mice. *Am. J. Obstet. Gynecol.* **2010**, *203*, 375.e1–375.e4. [CrossRef] [PubMed]
39. Cabrera, R.M.; Lin, Y.L.; Law, E.; Kim, J.; Wlodarczyk, B.J. The teratogenic effects of sertraline in mice. *Birth Defects Res.* **2020**, *112*, 1014–1024. [CrossRef]

40. Muller, J.C.; Boareto, A.C.; Lourenco, E.L.B.; Zaia, R.M.; Kienast, M.F.; Spercoski, K.M.; Morais, R.N.; Martino-Andrade, A.J.; Dalsenter, P.R. In Utero and Lactational Exposure to Fluoxetine in Wistar Rats: Pregnancy Outcomes and Sexual Development. *Basic Clin. Pharmacol. Toxicol.* **2013**, *113*, 132–140. [CrossRef]
41. Vedernikov, Y.; Bolanos, S.; Bytautiene, E.; Fulep, E.; Saade, G.R.; Garfield, R.E. Effect of fluoxetine on contractile activity of pregnant rat uterine rings. *Am. J. Obstet. Gynecol.* **2000**, *182*, 296–299. [CrossRef]
42. Sparenborg, S. Mortality in neonatal rats is increased by moderate prenatal exposure to some monoamine reuptake inhibitors—A brief review. In *Cocaine: Effects on the Developing Brain*; Harvey, J.A., Kosofsky, B.E., Eds.; Academy of Sciences: New York, NY, USA, 1998; Volume 846, pp. 423–426.
43. Haskell, S.E.; Hermann, G.M.; Reinking, B.E.; Volk, K.A.; Peotta, V.A.; Zhu, V.; Roghair, R.D. Sertraline exposure leads to small left heart syndrome in adult mice. *Pediatr. Res.* **2013**, *73*, 286–293. [CrossRef]
44. Marchand, G.J.; Meassick, K.; Wolf, H.; Hopewell, S.K.; Sainz, K.; Anderson, S.M.; Ware, K.; Vallejo, J.; King, A.; Ruther, S.; et al. Respiratory depression in a neonate born to mother on maximum dose sertraline: A case report. *J. Med. Case Rep.* **2021**, *15*, 1–5. [CrossRef]
45. Velasquez, J.C.; Bonnin, A. Placental Transport and Metabolism: Implications for the Developmental Effects of Selective Serotonin Reuptake Inhibitors (SSRI) Antidepressants. In *Prenatal and Postnatal Determinants of Development*; Walker, D.W., Ed.; Humana Press Inc.: Totowa, NJ, USA, 2016; Volume 109, pp. 245–262.
46. Haskell, S.E.; Lo, C.; Kent, M.E.; Eggleston, T.M.; Volk, K.A.; Reinking, B.E.; Roghair, R.D. Cardiac Outcomes After Perinatal Sertraline Exposure in Mice. *J. Cardiovasc. Pharmacol.* **2017**, *70*, 119–127. [CrossRef] [PubMed]
47. Horackova, H.; Karahoda, R.; Cerveny, L.; Vachalova, V.; Ebner, R.; Abad, C.; Staud, F. Effect of Selected Antidepressants on Placental Homeostasis of Serotonin: Maternal and Fetal Perspectives. *Pharmaceutics* **2021**, *13*, 1306. [CrossRef]
48. Bonnin, A.; Goeden, N.; Chen, K.; Wilson, M.L.; King, J.; Shih, J.C.; Blakely, R.D.; Deneris, E.S.; Levitt, P. A transient placental source of serotonin for the fetal forebrain. *Nature* **2011**, *472*, 347–352. [CrossRef]
49. Kliman, H.J.; Quaratella, S.B.; Setaro, A.C.; Siegman, E.C.; Subha, Z.T.; Tal, R.; Milano, K.M.; Steck, T.L. Pathway of Maternal Serotonin to the Human Embryo and Fetus. *Endocrinology* **2018**, *159*, 1609–1629. [CrossRef]
50. Karahoda, R.; Horackova, H.; Kastner, P.; Matthios, A.; Cerveny, L.; Kucera, J.; Kacerovsky, M.; Tebbens, J.D.; Bonnin, A.; Abad, C.; et al. Serotonin homeostasis in the materno-foetal interface at term: Role of transporters (SERT/SLC6A4 and OCT3/SLC22A3) and monoamine oxidase A (MAO-A) in uptake and degradation of serotonin by human and rat term placenta. *Acta Physiol.* **2020**, *229*, e13478. [CrossRef]
51. Gasser, P.J.; Lowry, C.A. Organic cation transporter 3: A cellular mechanism underlying rapid, non-genomic glucocorticoid regulation of monoaminergic neurotransmission, physiology, and behavior. *Horm. Behav.* **2018**, *104*, 173–182. [CrossRef] [PubMed]
52. Sabbioni, C.; Bugamelli, F.; Varani, G.; Mercolini, L.; Musenga, A.; Saracino, M.A.; Fanali, S.; Raggi, M.A. A rapid HPLC-DAD method for the analysis of fluoxetine and norfluoxetine in plasma from overdose patients. *J. Pharm. Biomed. Anal.* **2004**, *36*, 351–356. [CrossRef] [PubMed]
53. Aggarwal, A.; Jethani, S.L.; Rohatgi, R.K.; Kalra, J. Selective Serotonin Re-uptake Inhibitors (SSRIs) Induced Weight Changes: A Dose and Duration Dependent Study on Albino Rats. *J. Clin. Diagn. Res.* **2016**, *10*, AF1–AF3. [CrossRef] [PubMed]
54. Ferguson, J.M. SSRI antidepressant medications: Adverse effects and tolerability. *Prim. Care Companion J. Clin. Psychiatry* **2001**, *3*, 22. [CrossRef] [PubMed]

Article

A Preconception Paternal Fish Oil Diet Prevents Toxicant-Driven New Bronchopulmonary Dysplasia in Neonatal Mice

Jelonia T. Rumph [1,2,3], Kayla J. Rayford [2], Victoria R. Stephens [1,4], Sharareh Ameli [1,4], Pius N. Nde [2], Kevin G. Osteen [1,4,5] and Kaylon L. Bruner-Tran [1,*]

[1] Women's Reproductive Health Research Center, Department of Obstetrics and Gynecology, Vanderbilt University School of Medicine, 1161 21st Ave S, MCN B-1100, Nashville, TN 37232, USA; jrumph19@email.mmc.edu (J.T.R.); victoria.r.stephens@vanderbilt.edu (V.R.S.); s.ameli@vanderbilt.edu (S.A.); Kevin.osteen@vanderbilt.edu (K.G.O.)

[2] Department of Microbiology, Immunology and Physiology, Meharry Medical College, Nashville, TN 37208, USA; krayford@mmc.edu (K.J.R.); pnde@mmc.edu (P.N.N.)

[3] Department of Pharmacology, Vanderbilt University, Nashville, TN 37208, USA

[4] Department of Pathology, Microbiology and Immunology, Vanderbilt University School of Medicine, Nashville, TN 37208, USA

[5] VA Tennessee Valley Healthcare System, Nashville, TN 37208, USA

* Correspondence: kaylon.bruner-tran@vanderbilt.edu

Abstract: New bronchopulmonary dysplasia is a developmental lung disease associated with placental dysfunction and impaired alveolarization. Risk factors for new BPD include prematurity, delayed postnatal growth, the dysregulation of epithelial-to-mesenchymal transition (EMT), and parental exposure to toxicants. Our group previously reported that a history of paternal toxicant exposure increased the risk of prematurity and low birth weight in offspring. A history of paternal toxicant exposure also increased the offspring's risk of new BPD and disease severity was increased in offspring who additionally received a supplemental formula diet, which has also been linked to poor lung development. Risk factors associated with new BPD are well-defined, but it is unclear whether the disease can be prevented. Herein, we assessed whether a paternal fish oil diet could attenuate the development of new BPD in the offspring of toxicant exposed mice, with and without neonatal formula feeding. We investigated the impact of a paternal fish oil diet preconception because we previously reported that this intervention reduces the risk of TCDD associated placental dysfunction, prematurity, and low birth weight. We found that a paternal fish oil diet significantly reduced the risk of new BPD in neonatal mice with a history of paternal toxicant exposure regardless of neonatal diet. Furthermore, our evidence suggests that the protective effects of a paternal fish oil diet are mediated in part by the modulation of small molecules involved in EMT.

Keywords: multigenerational; toxicants; bronchopulmonary dysplasia; therapeutics; lung development

Citation: Rumph, J.T.; Rayford, K.J.; Stephens, V.R.; Ameli, S.; Nde, P.N.; Osteen, K.G.; Bruner-Tran, K.L. A Preconception Paternal Fish Oil Diet Prevents Toxicant-Driven New Bronchopulmonary Dysplasia in Neonatal Mice. *Toxics* **2022**, *10*, 7. https://doi.org/10.3390/toxics10010007

Academic Editors: Kimberly Keil Stietz, Tracie Baker and Jessica Plavicki

Received: 12 October 2021
Accepted: 22 December 2021
Published: 27 December 2021

Publisher's Note: MDPI stays neutral with regard to jurisdictional claims in published maps and institutional affiliations.

Copyright: © 2021 by the authors. Licensee MDPI, Basel, Switzerland. This article is an open access article distributed under the terms and conditions of the Creative Commons Attribution (CC BY) license (https://creativecommons.org/licenses/by/4.0/).

1. Introduction

The Developmental Origin of Health and Disease (DOHaD) concept was first proposed by Dr. David Barker who originally recognized the relationship between maternal malnutrition and the risk of metabolic syndrome in their adult children [1,2]. In recent years, the DOHaD concept has expanded and now recognizes a wide variety of factors that impact (positively or negatively) the fetal environment and, by extension, the child's adult health [3–5]. Although originally focused on maternal factors relevant to the current study (diet, stress, environmental exposures), it now recognizes that the health of the paternal parent can also significantly impact the fetal environment and offspring health.

The father's biological contribution to pregnancy is contained within his seminal fluid at the time of copulation. In addition to the spermatozoa, seminal fluid contains a variety of nutrients (fructose, citric acid), microbes, and proteolytic enzymes that are necessary for sperm survival and the successful fertilization of the oocyte [6–8]. Numerous studies have demonstrated that the quality of the seminal fluid can influence embryonic implantation, the microbial composition of the intrauterine environment, and placental function [9–13]. Importantly, the placenta is largely a paternally derived organ and is critical for pregnancy maintenance and fetal development [14]. For this reason, poor seminal fluid quality has the potential to not only negatively impact the development and function of the placenta but also to negatively impact fetal development and long-term child health.

Smoking and exposure to pollution are factors that reduce the quality of the seminal fluid and contribute to the DOHaD. Our laboratory previously reported that in utero exposure to 2,3,7,8-tetrachlorodibenzo-p-dioxin (TCDD)—a byproduct of smoke and a contributor to pollution—subsequently reduced the number and quality of sperm in male mice of reproductive age [10]. Furthermore, the offspring of these mice were susceptible to premature birth and intrauterine growth restriction (IUGR) [15,16]. We later demonstrated that these complications were associated with TCDD associated placental dysfunction that is characterized by alterations in the size of the placenta as well as the expression of placental progesterone receptor and toll-like receptor-4, which each play integral roles in placental function and pregnancy outcomes [10].

Paternal diet has also been shown to influence the DOHaD by modulating the quality of seminal fluid [17]. Our group and others have reported that a paternal fish oil diet improves the quality of seminal fluid. We found that providing a preconception supplemental fish oil diet to male mice that were exposed to TCDD in utero markedly improved both sperm density and motility and was associated with a significant increase in fertility. The improvement in sperm quality also translated to enhanced placental function, marked by a reduced expression of toll-like receptor 4 and increased expression of the progesterone receptor. Intervening with a paternal fish oil diet also reduced the risk of premature birth and eliminated the incidence of offspring IUGR [18].

In human neonates, premature birth and IUGR are each associated with an increased risk of developing new BPD, a developmental lung disease that is characterized by impaired alveolarization [19,20]. Several studies have suggested that placental dysfunction precedes (and contributes to) the development of new BPD [21–24]. The onset and severity of developmental lung diseases, such as new BPD, have also been linked to exposure to pollution in humans and experimental animal models [25–29]. Our laboratory found that the offspring of male mice that were exposed to TCDD in utero were also susceptible to new BPD [29]. Herein, we examined the efficacy of a paternal fish oil diet to prevent this potentially fatal neonatal disease. Our study revealed markedly improved lung development and reduced incidence of new BPD in offspring sired by a father with a history of TCDD exposure that was also provided a fish oil diet preconception.

Finally, exposure to components of fish oil and TCDD have been found to influence proteins that help to regulate epithelial-to-mesenchymal transition (EMT) within the lung; a process that is linked to the development of new BPD [30–33]. Therefore, we investigated whether a paternal fish oil diet attenuated the development of new BPD by modulating the small molecules involved in EMT. We also observed how neonatal diet impacted these parameters because maternal milk is associated with better infant health outcomes than formula feeding, which has been linked to poor postnatal lung development [34,35]. We found that the reduced risk of new BPD in the offspring of fish oil-supplemented males was associated with significantly reduced beta-catenin gene expression—a molecule involved in EMT and the development of new BPD [36–39].

2. Materials and Methods

2.1. Animals

Adult (10–12 weeks) and neonatal C57BL/6 mice were used in this study. Adult mice were obtained from Envigo (Indianapolis, IN, USA) or born in-house. All neonatal mice were born in-house. The animals were housed in the Barrier Animal Care Facility at the Vanderbilt University Medical Center, which is free from common mouse pathogens. Adult mice were provided food and water ad libitum. Animal room temperatures were maintained between 22–24 °C with a relative humidity of 40–50% on a 12-h light–dark schedule. Experiments described in this study were approved by Vanderbilt University's Institutional Animal Care and Use Committee (IACUC) as per the Animal Welfare Act protocol #M2000098. Approval date: 1 December 2019.

2.2. Chemicals

TCDD (99% in nonane #ED-908) was purchased from Cambridge Isotope Laboratories (Andover, MA, USA). Esbilac Puppy Milk Replacer Powder was purchased from Pet-Ag, Inc (Hampshire, IL, USA). All other chemicals were obtained from Sigma-Aldrich (St. Louis, MO, USA) unless otherwise stated.

2.3. TCDD Exposure and Mating Scheme

Virgin 10- to 12-week-old C57BL/6 females were mated with intact males of a similar age. The females were weighed daily and monitored for the presence of a vaginal semen plug, denoting that copulation had occurred. The morning a vaginal plug was identified, the dam was considered pregnant (denoted as embryonic day $10^{0.5}$ and moved to a new cage. Following the confirmation of pregnancy, dams were exposed to TCDD (10 µg/kg) in corn oil or corn oil vehicle alone by gavage on $10^{15.5}$ at 1100 h CST. The dams provided with vehicle only were used as unexposed controls while dams receiving TCDD were designated F0 mice (the founding generation).

Although the selected dose of TCDD is higher than typical human exposures, this dose reflects the more rapid clearance of this toxicant in mice compared to humans. This dose is well below the LD50 for adult C57BL/6 mice (230 µg/kg) [40] and is not overtly teratogenic or abortogenic. In our hands, parturition typically occurred on E20 for both control and F0 pregnancies. Finally, since the half-life of TCDD is 11 days in C57BL/6 mice, the offspring of the F0 dams (F1 pups) were directly exposed to TCDD in utero and during lactation [40]. Germ cells present within F1 fetuses were also directly exposed to TCDD, and these cells had the potential to become the F2 generation.

2.4. Diet and Mating Scheme for the F1 Generation

Purina Mills (TestDiet division) provided the 5% Menhaden fish oil diet, which also contained 1.5% corn oil to prevent the depletion of omega-6 fatty acids. Menhaden fish oil, (OmegaProtein, Houston, TX, USA) has an established fatty acid profile (~40% omega-3 fatty acids) and was processed by the manufacturer to remove dioxins and polychlorinated biphenyls. The fish oil diet is a modification of Purina's low phytoestrogen rodent chow 5VR5, which was used as the control (standard) diet. Protein, total fat, and energy content was the same for each diet. The fish oil diet was maintained in vacuum-sealed bags at −20 °C until use and once provided to mice, it was replaced every 3 days.

After weaning, F1 and control males were maintained on a standard or fish oil diet for at least 7 weeks (one full cycle of spermatogenesis) and mated at 10–12 weeks of age with age-matched unexposed C57BL/6 females. Once a vaginal semen plug was identified, dams were singly housed until parturition. Offspring sired by control males were denoted as CT whereas the offspring of F1 males were denoted F2TCDD.

2.5. Formula Feeding

Beginning on postnatal day 7 (PND7), CT and F2TCDD pups were sexed, weighed, and randomized to a strict maternal milk diet or a supplemental formula diet. Pups were bottle-fed 30 µL of formula three times a day over the course of four days using a small nipple attached to a 1ml syringe (Miracle Nipple Mini for Pets and Wildlife). Each 30 µL dose was provided in two aliquots of 15 µL, each 10 min apart. All pups remained with dams for the duration of the study and were allowed to nurse ad libitum. Pups were weighed daily and monitored for macroscopic signs of new BPD (e.g., labored breathing and difficulty feeding) from PND 7–10.

2.6. Euthanasia and Collection of Tissue

To assess the development of new BPD, we performed necropsies on PND 11 as our previous studies revealed that the disease was prevalent by this time in F2TCDD pups [29]. On PND 11 at 1100 h local time, CT and F2TCDD pups were weighed and observed for external signs of new BPD (labored breathing and/or delayed growth), then euthanized by decapitation performed under deep anesthesia as per the AAALAC guidelines. Following euthanasia, the peritoneal cavity and rib cage were opened, and the lungs and trachea were isolated and weighed. The lungs were perfused and inflated using an intratracheal injection of $1\times$ phosphate buffered saline (PBS) as previously described [41,42]. A 1mL syringe filled with PBS was inserted into the trachea and used to inflate the left lobe of the lung. A small string was tied around the right lobe to prevent inflation. The non-inflated lobe was stored at $-80\ °C$ for qPCR and immunoblot analyses and the inflated lobe was fixed in 10% buffered formalin. The fixed tissues were processed and paraffin-embedded, and slides containing 4 μm sections were prepared by Vanderbilt University's Translational Pathology Shared Resource Center (TPSR).

2.7. Hematoxylin and Eosin (H and E) Staining

Formalin-fixed, paraffin-embedded tissue sections were mounted on slides, then deparaffinized in xylene and rehydrated in ethanol. Slides were then incubated in Hematoxylin, washed, and incubated in Eosin. The slides were dehydrated in ethanol and xylene, then coverslipped using Cytoseal XYL (Thermo Scientific, Waltham, MA, USA; Cat# 8312-4).

2.8. Alveolus Diameter and Radial Alveoli Count

The measurements of alveoli diameter and radial alveolar count were conducted as previously described by others [43–45]. The linear ruler on Image J was used to assess the horizontal length of the individual alveoli that developed in the alveolar space of each group. Measurements were taken using the horizontal length of the alveoli to account for irregularities in alveolar shape/branching. H and E-stained slides from at least six pups from six separate litters were used to determine the average alveolus diameter for each group. At least 10 measurements were taken per $100\times$ image to obtain the average alveolus diameter per pup. The manual measurements of the pulmonary alveolar space were confirmed using the mean linear intercept, a semi-quantitative method for lung morphology that has been used previously [46]. For the radial alveolar count, a line was drawn from the surface epithelium to the nearest bronchiole. Alveoli that intersected the line were counted and the average was taken for each group in the same manner as described above.

2.9. Histological Determination of BPD

We developed an unbiased lung scoring system to identify mice that did or did not have new BPD. The diagnosis of new BPD was determined using a lung injury scale that accounted for alveoli diameter, radial alveolar count, and the presence of red blood cells (RBCs) in the lung. We measured these parameters because reduced alveolus diameter and radial alveolar count are established markers for reduced lung alveolarization and impaired lung function [47]. Furthermore, the altered distribution of RBCs in the lung is related to respiratory distress and hypertension, which are associated with new BPD [48–51]. We carefully assessed multiple sections from 10 non-littermate CT pups to determine the normal range of these parameters.

Healthy pups had an average alveolus diameter of 25 ± 5 μM and an average radial alveolar count of 5–7. Healthy pups also exhibited little to no RBCs infiltrating the alveolar space. A lung injury score of 1–4 was indicative of a healthy lung, marked by the formation of distinct alveoli. Pups were considered to have mild BPD when they exhibited an average alveolus diameter of 20 ± 4 µM, an average radial alveolar count of 3–4, and the presence of RBCs in 20–50% of the alveolar space. Pups with mild BPD received a lung injury score of 5–7. Pups diagnosed with severe BPD exhibited an average alveolus diameter of 16 µM or less, an average radial alveolar count of 3 or less, and the presence of RBCs in more than 50% of the alveolar space. Pups with severe BPD were given a lung injury score of 8 or 9. To date, pups have not exhibited overlapping pulmonary phenotypes.

2.10. qRT-PCR

Lung tissue was lysed using the Trizol reagent (Invitrogen, Carlsbad, CA, USA) and the total RNA was purified from tissue lysates using the RNeasy Mini Kit (Qiagen, Valencia, CA, USA). An iScript cDNA synthesis kit was used to generate cDNA from 1 µg of the total RNA (Bio-Rad) using random decamer primers. The same thermal cycling program was applied to all primers: 95 °C for 30 s, 40 cycles of 95 °C for 5 s, and 60 °C for 5 s using a Bio-Rad CFX96 real-time thermocycler. The melting curve was analyzed to confirm product purity. All reactions were performed in triplicate. A *Ribosomal 18s* transcript was used as a housekeeping gene to normalize the transcript levels of E-cadherin and β-catenin for all samples. The results were evaluated using the delta-delta Ct method as previously described [15,18]

2.11. Immunoblot Assays

Lung tissue was homogenized in a RIPA buffer (Life Technologies, cat# R0278; CA, USA), containing a protease inhibitor cocktail set III at 1:100 (Calbiochem, Gibbstown, NJ, USA) and phosphatase inhibitor cocktails 2 and 3 at 1:100 each (Sigma Aldrich, St. Louis, MO, USA), and using a Tissue Tearor (United Lab Plastics, St. Louis, MO, USA). Lysates (20 µg/well) were separated by SDS-PAGE using 4–15% gradient polyacrylamide gels and transferred onto nitrocellulose membranes (Life Technologies, Carlsbad, CA, USA). The membranes were blocked in Intercept TBS Blocking Buffer (LI-COR Biosciences, Lincoln, NE, USA) and then incubated with an appropriate primary antibody diluted at 1:1000 in the blocking buffer at 4 °C overnight on a shaker: rabbit anti-beta-Catenin antibody (Cell Signaling Technology, Danvers, MA, USA; Cat# 8480); mouse anti-e-cadherin (Cell Signaling Technology, Danvers, MA, USA; Cat# 3195). Blots were incubated in a mouse anti-beta-actin monoclonal antibody (Sigma Aldrich, MO, USA; Cat# C7207) for 1 h at room temperature. The blots were washed and incubated with the IRDye 680RD Donkey anti-Mouse or IRDye 800 CW Goat anti-Rabbit (Licor, Lincoln, NE; cat# 926-68072 and 925-3221) secondary antibody in the blocking buffer containing 0.01% Tween 20 for 1 h at room temperature. The blots were washed with 1X TBS 0.01% Tween 20 and the bound antibody bands were scanned using the infrared fluorescence detection Odyssey Imaging System (LI-COR Biosciences). The housekeeping beta-actin signal was used for the normalization of the data. Each experiment was conducted in biological triplicates and the quantitation of band intensity was performed by densitometry using Image J.

2.12. Statistical Analysis

The alveolar measurements, qRT-PCR, and immunoblot data were analyzed using GraphPad Prism's one-way ANOVA and the Tukey post-hoc test. For all experiments, six non-littermates were used to obtain the average for each group. The presented images are representative of each group. The data are represented as the mean ± standard deviation. $p < 0.05$ were considered significant. Significance was determined by comparing each treatment group to maternal milk-fed CT pups. All experiments were repeated twice using different non-littermates. In each group, approximately half of the pups were male and the other half were female. The majority of the pups were male in groups with uneven samples sizes.

3. Results

3.1. A Paternal Fish Oil Diet Preconception Improves Postnatal Growth in Pups with a History of TCDD Exposure

Infants diagnosed with new bronchopulmonary dysplasia (BPD) typically exhibit low birth weight, delayed postnatal growth, and lung hypoplasia [52–54]. To determine the efficacy of a preconception paternal fish oil diet intervention in eliminating factors associated with new BPD, we monitored pup growth from postnatal day (PND) 7–10 and assessed lung-to-body weight ratios on PND 11. Control (CT) pups had an average body weight of 3.3 g by PND 7, which increased to an average of 5.3 g by PND 10. Seven-day-old pups sired by fathers exposed to 2,3,7,8-tetrachlorodibenzo-p-dioxin in utero (F2TCDD pups) who were maintained on a standard diet preconception had a similar average body weight to CT pups. By PND 7, F2TCDD pups on a maternal milk diet had an average body weight of 3.3 g, which was reduced by 0.3 g in formula-supplemented pups. The average body weight of F2TCDD pups receiving maternal milk plateaued at 4 g by PND 8, but the average body weight of pups who received supplemental formula gradually increased to 5 g by PND 10. Regardless of their postnatal diet, F2TCDD pups sired by a father on a fish oil diet had an average body weight of 4.5 g by PND 7, which gradually increased to 6 g by PND 10. However, these trends in postnatal growth did not reach statistical significance (Figure 1A), as demonstrated in Table S1.

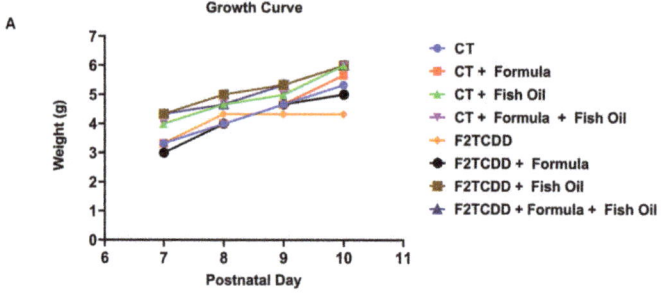

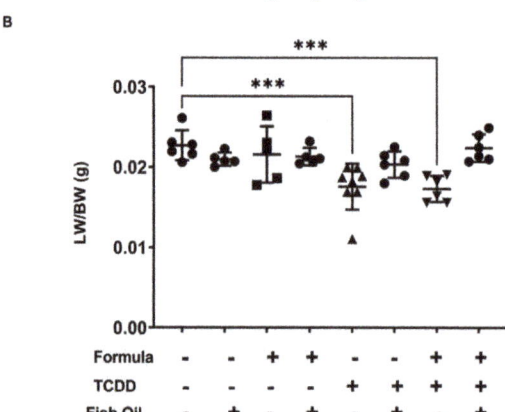

Figure 1. A dietary fish oil intervention improves postnatal development in pups: CT and F2TCDD pup body weight was monitored from postnatal day 7–10 (**A**), lung hypoplasia was determined by measuring lung-to-body weight ratios in CT and F2TCDD pups (**B**). Growth curve data represents the mean value of 4–5 non-littermate pups. Lung-to-body weight ratio data represents individual values of 4–5 non-littermate pups. Standard deviation is shown. *** $p \leq 0.001$.

Lung hypoplasia manifests as a lung-to-body weight ratio of 0.0115, or 1.15%, or less [55,56]. On PND 11, CT pups displayed an average lung-to-body weight ratio of 0.022 g, or 2.2%. F2TCDD pups sired by a standard diet father exhibited an average lung-to-body weight ratio of 0.018, or 1.8%, when they received a strict maternal milk diet and 0.017, or 1.7%, when they received supplemental formula. F2TCDD pups were not diagnosed with lung hypoplasia, but they displayed a significant reduction in lung-to-body weight ratios compared to CT pups ($p < 0.0005$). Intervening with a paternal fish oil diet preconception normalized lung-to-body weight ratios to 0.020, or 2%, in maternal milk-fed and 0.021, or 2.1%, in formula-supplemented F2TCDD pups (Figure 1B). These results confirm our previous finding that a paternal fish oil diet intervention influences the offspring's postnatal development [18]. Herein, we demonstrated that a paternal fish oil diet intervention improved postnatal growth rates and lung-to-body weight ratios in pups with a history of paternal TCDD exposure, regardless of their postnatal diet.

3.2. A Paternal Fish Oil Diet Mitigates Delayed Lung Development in Pups with a History of TCDD Exposure

To confirm that a paternal fish oil diet preconception improves lung development in pups, we observed pulmonary histology using hematoxylin- and eosin-stained (H and E) lung slides. CT pups exhibited normal lung development and alveolarization, marked by the distinct formation of alveoli (Figure 2A,E). Supplementing the CT pup diet with formula led to a visible reduction in alveolar space and the thickening of alveolar walls (Figure 2B,F). CT pups sired by a father who received a fish oil diet preconception exhibited improved lung alveolarization, regardless of their postnatal diet (Figure 2C,D,G,H).

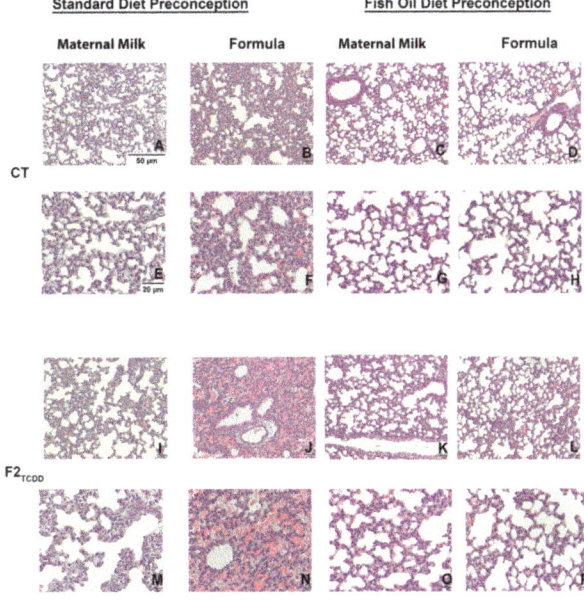

Figure 2. A paternal dietary fish oil intervention improves lung development in pups: Representative images of Hematoxylin- and Eosin-stained perfused lung tissue of PND11 CT pups sired by a standard or fish oil diet father following a maternal milk diet or supplemental formula at a magnification of 100× (**A–D**) and 400× (**E–H**); F2TCDD pups sired by a standard or fish oil diet father following a maternal milk diet or formula supplementation at a magnification of 100× (**I–L**) and 400× (**M–P**).

Lung H and Es confirmed that F2TCDD pups sired by fathers on a standard diet exhibited an impaired development of distinct alveoli and thickening of alveolar walls (Figure 2I,M). When provided supplemental formula, many F2TCDD pups experienced interalveolar red blood cell (RBC) infiltration (Figure 2J,N). A paternal fish oil diet preconception improved lung alveolarization and reduced the risk of interalveolar RBC infiltration in F2TCDD pups, regardless of postnatal diet (Figure 2K,L,O,P).

H- and E-stained slides were used to assess lung morphology by manually measuring the pulmonary space and radial alveoli count as previously described [43,45]. The pulmonary alveolar space measurements were verified using mean line intercept (MLI), an automated semi-quantitative lung morphology analysis [46]. CT pups sired by a father on a standard diet preconception had an average alveolar diameter of 25 ± 5 μm. Supplementing CT pups with formula led to a significant reduction in the average alveolar diameter ($p = 0.0027$). A paternal fish oil diet intervention normalized the average alveolar diameter of formula-supplemented CT pups. F2TCDD pups sired by a standard diet father also exhibited a significant reduction in the diameter of their alveolar space, regardless of postnatal diet ($p < 0.0001$). Intervening with a paternal fish oil diet in F2TCDD pups normalized their alveolar diameter to that of CT pups (Figure 3A).

We confirmed that formula-fed CT pups sired by a standard diet father also exhibited a significant decrease in lung MLI ($p < 0.0228$). F2TCDD pups also exhibited a significant reduction in lung MLI, independent of postnatal diet ($p = 0.0001$) (Figure 3B). All CT pups exhibited an average number of alveoli of 6 ± 1, independent of their postnatal diet. F2TCDD pups sired by a father on a standard diet displayed a significant reduction in their average number of alveoli compared to CT pups ($p < 0.0001$). Intervening with a paternal fish oil diet preconception in F2TCDD pups normalized their average number of alveoli to that of CT pups (Figure 3C). Our results confirm formula-fed CT pups and formula-fed/maternal milk-fed F2TCDD pups exhibit poor postnatal lung development, marked by a reduction in alveolar diameter, lung MLI, and radial alveolar count. We also confirmed the hypothesis that intervening with a paternal fish oil diet preconception in these groups improves lung development by increasing the average diameter of alveolar space, lung MLI, and number of alveoli in the offspring.

3.3. A Paternal Fish Oil Diet Reduces the Incidence of New BPD in Pups with a History of TCDD Exposure

To determine the incidence of new BPD among pups, we used a scoring system based on relevant histologic markers, as detailed in the methods and demonstrated in Figure 4B–G. Regardless of their postnatal diet, F2TCDD pups sired by a father receiving a standard diet preconception displayed a significant increase in the incidence of new BPD compared to all CT pups ($p < 0.0001$) (Figure 4A). This translated to 85% of maternal milk-fed and 71% of formula-supplemented F2TCDD pups developing new BPD. Intervening with a paternal fish oil diet preconception reduced the incidence of new BPD to 25% in maternal milk-fed and 10% in formula-supplemented F2TCDD pups (Table 1). These findings support our hypothesis that a paternal fish oil diet preconception mitigates the development of new BPD in pups with a history of paternal TCDD exposure.

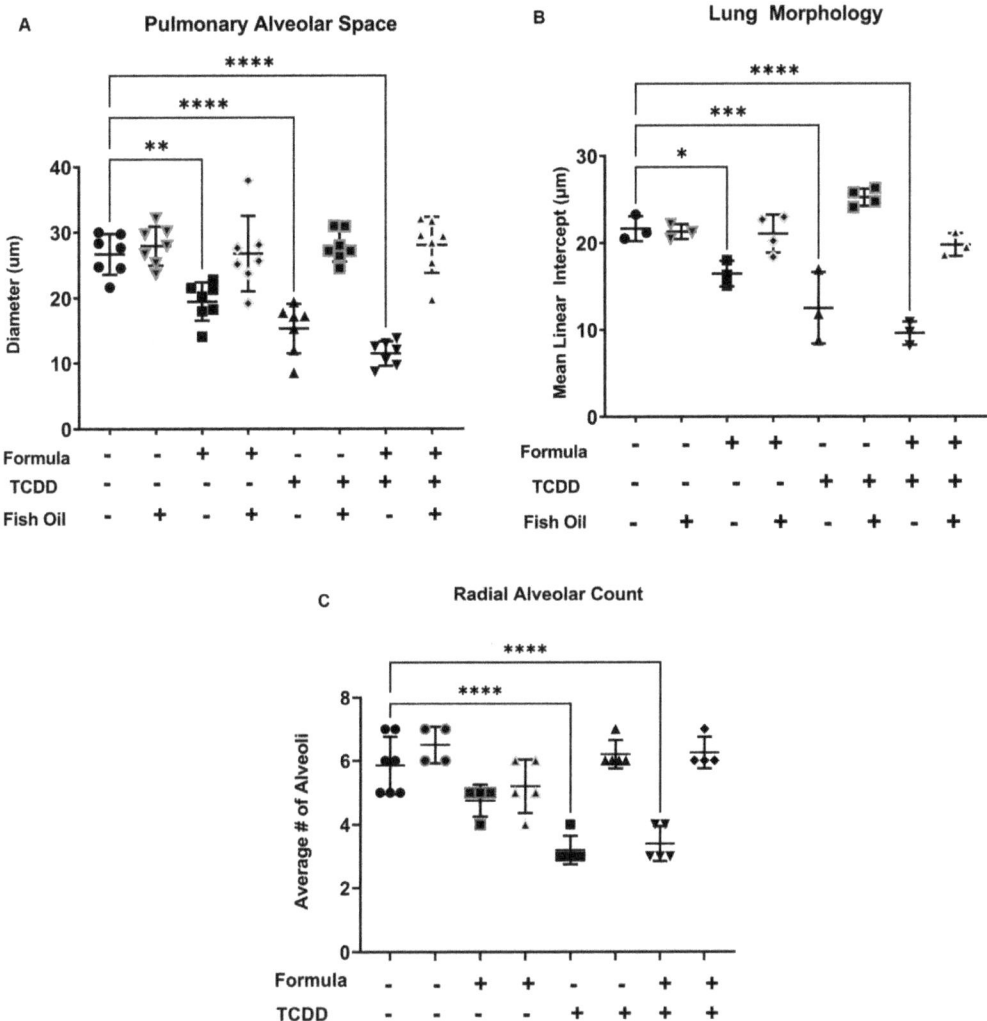

Figure 3. A paternal dietary fish oil intervention improves the alveolarization of pup lungs: Pulmonary alveolar space (**A**), mean linear intercept (**B**), and radial alveolar count (**C**) of CT and F2TCDD pups ± a fish oil intervention and/or supplemental formula was measured on PND11. Groups used for manual determination of pulmonary alveolar space and radial alveolar count contained 6–10 non-littermates. Data points represent the mean values from individual pups. Standard deviation is shown. Groups used for automated mean linear intercept contained 3–4 non-littermates.* $p \leq 0.05$; ** $p \leq 0.01$;*** $p \leq 0.001$;**** $p \leq 0.0001$.

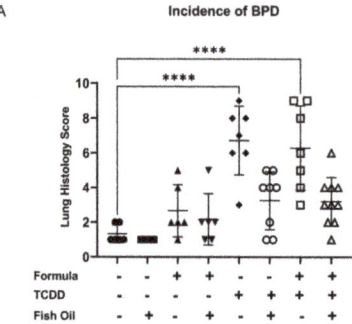
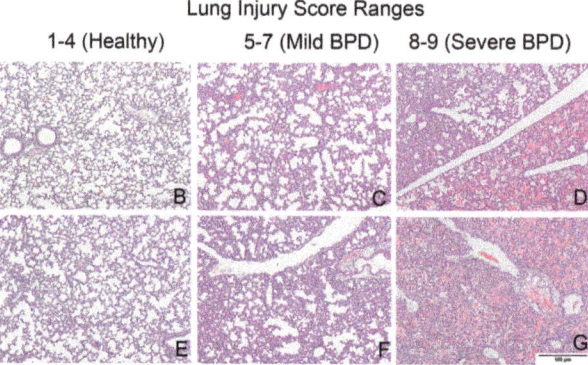

Figure 4. A dietary fish oil intervention reduces the incidence of BPD in TCDD-exposed pups: Incidence of BPD was determined in CT and F2TCDD pups ± a fish oil intervention and/or formula supplementation (**A**), the incidence of BPD was determined using a novel scale based on lung histology (**B–G**). Data points represent the individual lung injury scores of 6–10 non-littermates from each group. Standard deviation is shown. **** $p \leq 0.0001$.

Table 1. Incidence of pups with BPD across all groups. CT, control; FO, fish oil.

Exposure Group	Incidence of New BPD	Average Lung Injury Score
CT	0/7 = 0%	1
CT + FO	0/7 = 0%	1
CT + FORMULA	1/16 = 16%	3
CT + FO + FORMULA	1/6 = 16%	2
F2TCDD	6/7 = 85%	6
F2TCDD + FO	2/8 = 25%	3
F2TCDD + FORMULA	5/7 = 71%	7
F2TCDD + FO + FORMULA	1/10 = 10%	3

3.4. Diet and History of TCDD Exposure Influences Pup Pulmonary Beta-Catenin and E-Cadherin Expression

The small molecules beta-catenin and E-cadherin play integral roles in lung development. These proteins are involved in epithelial-to-mesenchymal transition (EMT)—a process that is dysregulated during the development and progression of new BPD [30]. Since EMT contributes to normal lung development and the development of lung diseases, we aimed to determine whether a paternal fish oil diet reduced the incidence of new BPD in offspring by modulating the small molecules associated with EMT. We examined pup beta-catenin and E-cadherin expression to observe whether a paternal fish oil diet intervention improves new BPD outcomes by influencing these small molecules.

CT pups displayed no differences in their expression of beta-catenin at the transcript level, regardless of paternal and postnatal diet. F2TCDD pups on a maternal milk diet ($p = 0.0234$) and supplemental formula diet ($p = 0.0091$) displayed a significant increase in beta-catenin gene expression when their father received a standard diet preconception. Intervening with a paternal fish oil diet preconception reduced beta-catenin gene expression in F2TCDD pups, regardless of postnatal diet (Figure 5A).

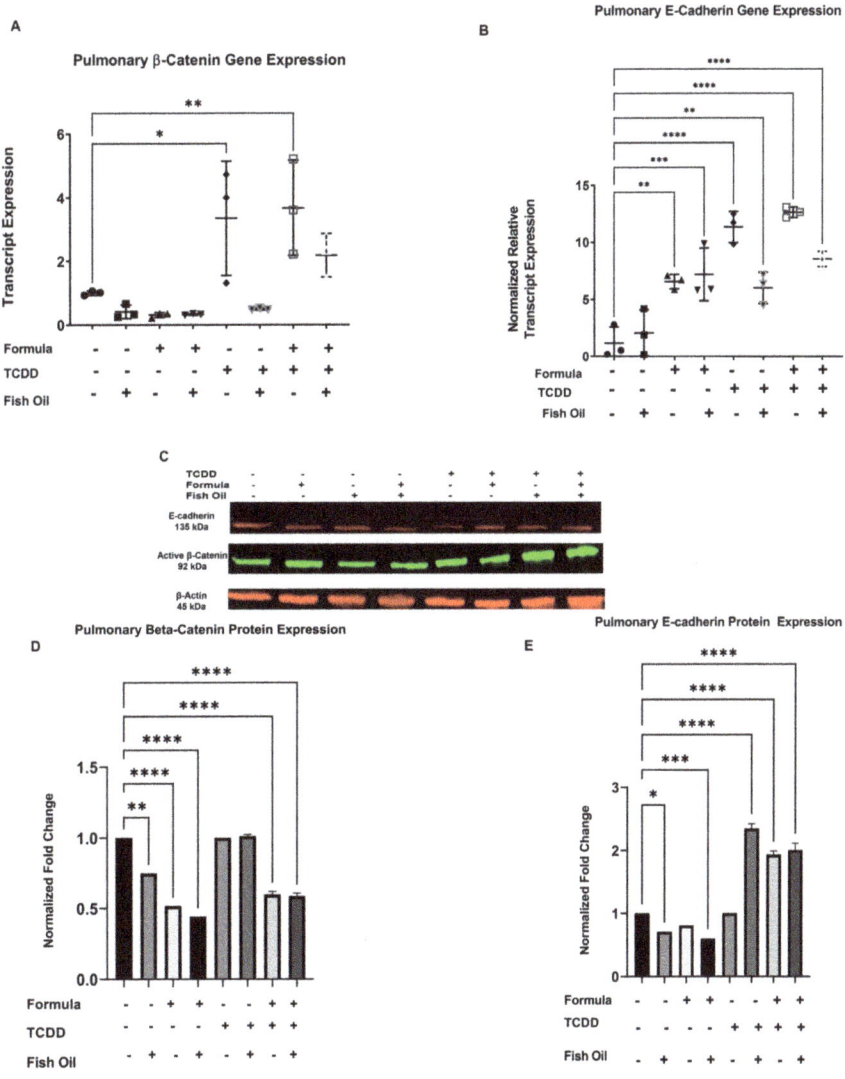

Figure 5. Diet and history of TCDD exposure influence beta-catenin and E-cadherin expression: RNA isolated from the lungs of PND 11 CT and F2TCDD pups ± a fish oil intervention and/or supplemental formula was used to quantify the expression of beta-catenin and E-cadherin gene expression (A,B), beta-catenin and E-cadherin protein expression were measured through immunoblotting (C), and quantified using densitometry (D,E). * $p \leq 0.05$; ** $p \leq 0.01$; *** $p \leq 0.001$; **** $p \leq 0.0001$.

Significantly, compared to CT pups sired by a standard diet father and only provided maternal milk, E-cadherin gene expression was increased in CT pups receiving supplemental formula who were sired by fathers on a standard diet preconception ($p = 0.0017$). A paternal fish oil diet further increased E-cadherin gene expression in formula-supplemented CT pups ($p = 0.0006$). F2TCDD pups sired by standard diet fathers exhibited significant increases in E-cadherin gene expression, independent of postnatal diet ($p < 0.0001$). A paternal fish oil diet intervention in maternal milk-fed ($p = 0.0044$) and formula-supplemented F2TCDD pups ($p < 0.0001$) also significantly increased E-cadherin gene expression (Figure 5B).

We also quantified the protein expression of pulmonary beta-catenin and E-cadherin in pups through immunoblotting (Figure 5C and Figure S1). A paternal fish oil diet intervention significantly decreased beta-catenin protein expression in CT pups ($p = 0.0020$). Beta-catenin protein expression was further reduced in CT pups receiving supplemental formula, regardless of paternal diet preconception ($p < 0.0001$). Independent of paternal diet preconception, beta-catenin protein expression in F2TCDD pups was similar to maternal milk-fed CT pups sired by standard diet fathers. The formula supplementation of F2TCDD pups significantly reduced beta-catenin protein expression, regardless of paternal diet ($p < 0.0001$) (Figure 5D).

Significantly, CT pups sired by a father on a fish oil diet preconception exhibited a reduction in E-cadherin protein expression following a strict maternal milk diet ($p = 0.0108$) and a supplemental formula diet ($p < 0.0001$). F2TCDD pups sired by a standard diet father expressed similar levels of E-cadherin protein expression to maternal milk-fed CT pups sired by a standard diet father. Regardless of offspring postnatal diet, a paternal fish oil diet intervention in F2TCDD pups led to increased E-cadherin protein expression ($p < 0.0001$). The formula supplementation of F2TCDD pups sired by standard diet fathers also led to a significant increase in E-cadherin protein expression ($p < 0.0001$) (Figure 5E).

4. Discussion

We previously reported that a paternal history of TCDD exposure increased the offspring's risk of premature birth and IUGR due to impaired placental function [18]. A paternal history of TCDD exposure also increased the offspring's risk of new BPD, a disease previously associated with prematurity, IUGR, and placental dysfunction [19,21,22,57]. The severity of new BPD worsened in neonatal mice that received supplemental formula, although this dietary intervention is commonly provided to premature human infants to enhance their nutritional intake and promote lung development [29,58,59]. Intervening with a paternal fish oil diet preconception, following his own in utero TCDD exposure, reduced the risk of premature birth and IUGR in offspring in association with improved placental function [18]. Herein, we investigated the efficacy of a paternal fish oil diet preconception in preventing the development of new BPD in offspring with a history of paternal TCDD exposure. We also observed whether the protection offered by a paternal fish oil diet persisted independently of the offspring's postnatal diet (maternal milk vs. supplemental formula).

Herein we confirmed that a paternal history of TCDD exposure impaired postnatal growth in offspring from PND 7 to 10. A paternal fish oil diet preconception improved offspring growth independent of postnatal diet; however, this trend did not reach significance (Figure 1A and Table S1). As demonstrated by Figure 1B, a history of paternal TCDD exposure also impaired lung development, denoted by an increased risk of lung hypoplasia, which is associated with poor lung health outcomes [60]. Intervening with a paternal fish oil diet preconception in males with a history of in utero TCDD exposure reduced the offspring's risk of lung hypoplasia, independent of postnatal diet.

We also observed the lung histology of pups from each group and found that CT pups exhibited normal alveolarization when they received a postnatal maternal milk diet (Figure 2A,E). Our results confirmed that postnatal formula supplementation may be associated with impaired lung development [34,61], marked by the thickening of the alveolar walls in CT pups who received postnatal formula supplementation (Figure 2B,F). A paternal fish oil diet preconception also improved lung histology in CT pups provided with supplemental formula. The paternal history of TCDD exposure preceding a standard diet preconception impaired lung development in pups who received a maternal milk diet (Figure 2I,M) and a supplemental formula diet (Figure 2J,N). Additionally, formula supplementation led to interalveolar RBC infiltration—a sign of pulmonary hypertension, which is associated with the development of new BPD in human neonates [62,63]. Intervening with a paternal fish oil diet preconception in males with a history of TCDD exposure improved lung development in their offspring, independent of postnatal diet (Figure 2K–P). This intervention also reduced the risk of interalveolar RBC infiltration in formula-supplemented F2TCDD pups.

We used the previously described methods to assess pulmonary alveolar space, lung MLI, and radial alveolar count, which are altered in developmental lung diseases [64]. We confirmed that formula supplementation has a negative effect on the diameter of pulmonary alveolar space, as it was significantly reduced in CT pups who received supplemental formula (Figure 3A). We also observed that a history of paternal TCDD exposure reduced pulmonary alveolar space in offspring, regardless of their postnatal diet; however, a paternal fish oil diet normalized the pulmonary alveolar space of these pups (Figure 3A).

We used lung MLI to confirm that postnatal formula supplementation and paternal history of TCDD exposure independently reduced alveolar space. Formula-fed CT pups sired by standard diet fathers displayed a significant decrease in lung MLI; this trend persisted in maternal milk- and formula-fed F2TCDD pups sired by standard diet fathers (Figure 3B). CT pups did not exhibit significant changes in radial alveolar count. However, a paternal history of TCDD exposure led to a significant reduction in the radial alveolar count, which was normalized in pups whose father received a supplemental fish oil diet preconception (Figure 3C). These results confirm that historical exposure to pollution negatively impacts lung development, however, components of fish oil may mitigate this effect [65,66].

Using a novel lung injury score system, we confirmed our previous report that a history of paternal TCDD exposure increased the offspring's risk of new BPD and that formula supplementation increased disease severity (Figure 4). These data support our hypothesis that a paternal fish oil diet preconception can reduce the risk of new BPD in offspring with a history of ancestral TCDD exposure. Our results also show that postnatal formula supplementation is associated with poor lung health outcomes [67] and a non-significant increase in the incidence of new BPD in CT mice (Figure 4A). Overall, our results support the theory that formula supplementation and paternal history of toxicant exposure are independent risk factors for the development and severity of new BPD (Table 1).

Studies have suggested that the pathophysiology of new BPD involves EMT. Therefore, to explore the potential mechanisms associated with the protective effects of fish oil against the development of new BPD, we examined E-cadherin and beta-catenin expression. These small molecules are influenced by TCDD and components of fish oil [32,33] and have each been shown to be involved in EMT. Additionally, aberrant expression of beta-catenin has been linked to new BPD [36,68,69]. Therefore, we investigated the efficacy of a paternal fish oil diet preconception in attenuating the development of new BPD by modulating the small molecules involved in EMT. We confirmed that maternal milk- and formula-fed pups with a history of paternal TCDD exposure exhibited an increased gene expression of beta-catenin (Figure 5A). We found that formula supplementation increased the gene expression of E-cadherin, independent of a paternal history of TCDD exposure. However, a paternal fish oil diet insignificantly reduced the gene expression of E-cadherin in offspring with a history of TCDD exposure (Figure 5B). Pups with a history of TCDD exposure displayed a significant increase in the co-expression of beta-catenin and E-cadherin at the gene level—a potential marker for EMT [70,71]. However, beta-catenin and E-cadherin protein expression were aberrant between groups (Figure 5D,E). Our results suggest that new BPD in offspring with a history of TCDD exposure may be associated with increased beta-catenin gene expression, supporting previous findings by others that this small molecule is dysregulated in new

BPD. Surprisingly, our results also suggest that the protein expression of beta-catenin and E-cadherin may play a less significant role in the attenuation of new BPD.

Overall, our study suggests that a preconception fish oil diet in males, following in utero TCDD exposure, reduces the offspring's risk of developing new BPD and that this effect is mediated in part through the modulation of beta-catenin gene expression, a small molecule involved in EMT. Although E-cadherin is also involved in EMT, its aberrant expression between groups suggests that it does not play a major role in new BPD outcomes. We theorize that a paternal fish oil diet preconception reduces the risk of new BPD in offspring with a history of TCDD exposure by improving placental function, which eliminates the risk of delayed postnatal growth in offspring and subsequently improves lung development. It is also likely that a paternal fish oil diet increases the levels of fish oil components (e.g., Docosahexaenoic acid (DHA) and Eicosapentaenoic acid (EPA)) in seminal fluid [72], which may, in turn, contribute to the fatty acids present in the intrauterine environment. DHA and EPA are critical to infant health and the development of lungs and other organs; however, PUFA stores are often depleted after birth following changes in the nutritional content of an infant's postnatal diet [73,74]. Relevant to public health, this study suggests that a paternal fish oil diet may be an efficacious preventative measure in attenuating the risk of new BPD in the offspring of fathers who have been exposed to toxicants via smoking or as a consequence of their occupation.

Supplementary Materials: The following are available online at https://www.mdpi.com/article/10.3390/toxics10010007/s1: Figure S1, Full immunoblot. Table S1, *p*-values representing differences in growth curve measurements between groups. CT, Control; FO, Fish oil.

Author Contributions: Conceptualization, J.T.R. and K.L.B.-T.; methodology, J.T.R. and K.J.R.; validation, J.T.R. and K.L.B.-T.; formal analysis, J.T.R.; investigation, J.T.R., V.R.S. and S.A.; resources, K.L.B.-T., K.G.O. and P.N.N.; data curation, J.T.R. and K.J.R.; writing—original draft preparation, J.T.R.; writing—review and editing, J.T.R., K.L.B.-T. and P.N.N.; visualization, J.T.R.; supervision, K.L.B.-T. and P.N.N.; project administration, J.T.R. and K.L.B.-T.; funding acquisition, K.L.B.-T. and K.G.O. All authors have read and agreed to the published version of the manuscript.

Funding: This research was funded by the National Institute of Environmental Health Sciences, Grant/Award Number TOX T32 ES007028; the National Institute of General Medical Sciences of the National Institutes of Health, Award Numbers T32GM007628, 1SC1AI127352, 5R25GM059994, 1F31AI67579, and U54MD007586; and the Veterans Administration, Grant/Award Number I01BX002583.

Institutional Review Board Statement: This study was conducted according to the guidelines of the Declaration of Helsinki, and approved by the Institutional Review Board of Vanderbilt University (protocol code: M200098 and date of approval: 12 January 2019).

Informed Consent Statement: Not Applicable.

Data Availability Statement: All relevant data is within the manuscript.

Acknowledgments: The authors would like to acknowledge the assistance of Lou Ann Brown, Paula Austin, Ryan Doster, and Tianbing Ding. The authors would also like to acknowledge Vanderbilt University's Translational Pathology Shared Resource supported by the NCI/NIH Cancer Center Support Grant 2P30 CA068485-14, as well as Meharry Medical College's Consolidated Research instrumentation, Informatics, Statistics, and Learning Integration Suite.

Conflicts of Interest: The authors declare no conflict of interest.

References

1. Barker, D.J. The origins of the developmental origins theory. *J. Intern. Med.* **2007**, *261*, 412–417. [CrossRef] [PubMed]
2. Roseboom, T.J. Epidemiological evidence for the developmental origins of health and disease: Effects of prenatal undernutrition in humans. *J. Endocrinol.* **2019**, *242*, T135–T144. [CrossRef] [PubMed]
3. Heindel, J.J.; Lawler, C.; Gluckman, P.D.; Hanson, M.A. Developmental Origins of Health and Disease: Role of exposure to environmental chemicals in developmental origins of health and disease. *Endocrinology* **2015**, *156*, 3416–3421. [CrossRef] [PubMed]
4. Hsu, C.-N.; Tain, Y.-L. The Good, the Bad, and the Ugly of Pregnancy Nutrients and Developmental Programming of Adult Disease. *Nutrients* **2019**, *11*, 894. [CrossRef] [PubMed]

5. Codagnone, M.G.; Spichak, S.; O'Mahony, S.M.; O'Leary, O.F.; Clarke, G.; Stanton, C.; Dinan, T.G.; Cryan, J.F. Programming Bugs: Microbiota and the Developmental Origins of Brain Health and Disease. *Biol. Psychiatry* **2019**, *85*, 150–163. [CrossRef]
6. Moretti, E.; Capitani, S.; Figura, N.; Pammolli, A.; Federico, M.G.; Giannerini, V.; Collodel, G. The presence of bacteria species in semen and sperm quality. *J. Assist. Reprod. Genet.* **2009**, *26*, 47–56. [CrossRef]
7. Singer, R.; Sagiv, M.; Barnet, M.; Levinsky, H. Semen volume and fructose content of human semen. Survey of the years 1980–1989. *Acta Eur. Fertil.* **1990**, *21*, 205–206.
8. Singer, R.; Landau, B.; Joshua, H.; Zukerman, Z.; Pick, I.; Sigienriech, E.; Chowers, I. Protein content of human seminal plasma and spermatozoa in relation to sperm counts. *Acta Eur. Fertil.* **1976**, *7*, 281–284.
9. Schjenken, J.E.; Sharkey, D.J.; Green, E.S.; Chan, H.Y.; Matias, R.A.; Moldenhauer, L.M.; Robertson, S.A. Sperm modulate uterine immune parameters relevant to embryo implantation and reproductive success in mice. *Commun. Biol.* **2021**, *4*, 572. [CrossRef]
10. Ding, T.; Mokshagundam, S.; Rinaudo, P.F.; Osteen, K.G.; Bruner-Tran, K.L. Paternal developmental toxicant exposure is associated with epigenetic modulation of sperm and placental Pgr and Igf2 in a mouse model. *Biol. Reprod.* **2018**, *99*, 864–876. [CrossRef]
11. Di Mascio, D.; Saccone, G.; Bellussi, F.; Vitagliano, A.; Berghella, V. Type of paternal sperm exposure before pregnancy and the risk of preeclampsia: A systematic review. *Eur. J. Obstet. Gynecol. Reprod. Biol.* **2020**, *251*, 246–253. [CrossRef]
12. Robertson, S.A.; Sharkey, D.J. Seminal fluid and fertility in women. *Fertil. Steril.* **2016**, *106*, 511–519. [CrossRef]
13. Bromfield, J.J. Seminal fluid and reproduction: Much more than previously thought. *J. Assist. Reprod. Genet.* **2014**, *31*, 627–636. [CrossRef]
14. Wang, X.; Miller, D.C.; Harman, R.; Antczak, D.F.; Clark, A.G. Paternally expressed genes predominate in the placenta. *Proc. Natl. Acad. Sci. USA* **2013**, *110*, 10705–10710. [CrossRef] [PubMed]
15. Ding, T.; McConaha, M.; Boyd, K.L.; Osteen, K.G.; Bruner-Tran, K.L. Developmental dioxin exposure of either parent is associated with an increased risk of preterm birth in adult mice. *Reprod. Toxicol.* **2011**, *31*, 351–358. [CrossRef] [PubMed]
16. Bruner-Tran, K.L.; Osteen, K.G. Developmental exposure to TCDD reduces fertility and negatively affects pregnancy outcomes across multiple generations. *Reprod. Toxicol.* **2011**, *31*, 344–350. [CrossRef] [PubMed]
17. Schagdarsurengin, U.; Steger, K. Epigenetics in male reproduction: Effect of paternal diet on sperm quality and offspring health. *Nat. Rev. Urol.* **2016**, *13*, 584–595. [CrossRef]
18. McConaha, M.E.; Ding, T.; Lucas, J.A.; Arosh, J.A.; Osteen, K.G.; Bruner-Tran, K.L. Preconception omega-3 fatty acid supplementation of adult male mice with a history of developmental 2,3,7,8-tetrachlorodibenzo-p-dioxin exposure prevents preterm birth in unexposed female partners. *Reproduction* **2011**, *142*, 235–241. [CrossRef]
19. Mestan, K.K.; Steinhorn, R.H. Fetal origins of neonatal lung disease: Understanding the pathogenesis of bronchopulmonary dysplasia. *Am. J. Physiol. Lung Cell Mol. Physiol.* **2011**, *301*, L858–L859. [CrossRef]
20. Thébaud, B.; Goss, K.N.; Laughon, M.; Whitsett, J.A.; Abman, S.H.; Steinhorn, R.H.; Aschner, J.L.; Davis, P.G.; McGrath-Morrow, S.A.; Soll, R.F.; et al. Bronchopulmonary dysplasia. *Nat. Rev. Dis. Primers* **2019**, *5*, 78. [CrossRef]
21. Young, K.; Sosenko, I.; Claure, N. Placental dysfunction and impaired fetal growth: A relationship with bronchopulmonary dysplasia and pulmonary hypertension. *Thorax* 2021. [CrossRef]
22. Torchin, H.; Ancel, P.-Y.; Goffinet, F.; Hascoët, J.-M.; Truffert, P.; Tran, D.; Lebeaux, C.; Jarreau, P.-H. Placental Complications and Bronchopulmonary Dysplasia: EPIPAGE-2 Cohort Study. *Pediatrics* **2016**, *137*, e20152163. [CrossRef] [PubMed]
23. Mir, I.N.; Chalak, L.F.; Brown, L.S.; Johnson-Welch, S.; Heyne, R.; Rosenfeld, C.R.; Kapadia, V.S. Impact of multiple placental pathologies on neonatal death, bronchopulmonary dysplasia, and neurodevelopmental impairment in preterm infants. *Pediatr. Res.* **2020**, *87*, 885–891. [CrossRef] [PubMed]
24. Redline, R.W.; Wilson-Costello, D.; Hack, M. Placental and Other Perinatal Risk Factors for Chronic Lung Disease in Very Low Birth Weight Infants. *Pediatr. Res.* **2002**, *52*, 713–719. [CrossRef] [PubMed]
25. Rice, J.L.; McGrath-Morrow, S.A.; Collaco, J.M. Indoor Air Pollution Sources and Respiratory Symptoms in Bronchopulmonary Dysplasia. *J. Pediatr.* **2020**, *222*, 85–90.e82. [CrossRef]
26. Collaco, J.M.; Morrow, M.; Rice, J.L.; McGrath-Morrow, S.A. Impact of road proximity on infants and children with bronchopulmonary dysplasia. *Pediatr. Pulmonol.* **2020**, *55*, 369–375. [CrossRef]
27. Collaco, J.M.; Aoyama, B.C.; Rice, J.L.; McGrath-Morrow, S.A. Influences of environmental exposures on preterm lung disease. *Expert Rev. Respir. Med.* **2021**, *15*, 1271–1279. [CrossRef] [PubMed]
28. Latzin, P.; Röösli, M.; Huss, A.; Kuehni, C.E.; Frey, U. Air pollution during pregnancy and lung function in newborns: A birth cohort study. *Eur. Respir. J.* **2009**, *33*, 594–603. [CrossRef]
29. Mokshagundam, S.; Ding, T.; Rumph, J.T.; Dallas, M.; Stephens, V.R.; Osteen, K.G.; Bruner-Tran, K.L. Developmental 2,3,7,8-tetrachlorodibenzo-p-dioxin exposure of either parent enhances the risk of necrotizing enterocolitis in neonatal mice. *Birth Defects Res.* **2020**, *112*, 1209–1223. [CrossRef]
30. Yang, H.; Fu, J.; Xue, X.; Yao, L.; Qiao, L.; Hou, A.; Jin, L.; Xing, Y. Epithelial-mesenchymal transitions in bronchopulmonary dysplasia of newborn rats. *Pediatr. Pulmonol.* **2014**, *49*, 1112–1123. [CrossRef]
31. Bartis, D.; Mise, N.; Mahida, R.Y.; Eickelberg, O.; Thickett, D.R. Epithelial–mesenchymal transition in lung development and disease: Does it exist and is it important? *Thorax* **2014**, *69*, 760–765. [CrossRef]
32. Sung, N.J.; Kim, N.H.; Bae, N.Y.; Jo, H.S.; Park, S.-A. DHA inhibits Gremlin-1-induced epithelial-to-mesenchymal transition via ERK suppression in human breast cancer cells. *Biosci. Rep.* **2020**, *40*, BSR20200164. [CrossRef]

33. Gao, Z.; Bu, Y.; Liu, X.; Wang, X.; Zhang, G.; Wang, E.; Ding, S.; Liu, Y.; Shi, R.; Li, Q.; et al. TCDD promoted EMT of hFPECs via AhR, which involved the activation of EGFR/ERK signaling. *Toxicol. Appl. Pharm.* **2016**, *298*, 48–55. [CrossRef]
34. Villamor-Martínez, E.; Pierro, M.; Cavallaro, G.; Mosca, F.; Villamor, E. Mother's Own Milk and Bronchopulmonary Dysplasia: A Systematic Review and Meta-Analysis. *Front. Pediatr.* **2019**, *7*, 224. [CrossRef]
35. Schlosser-Brandenburg, J.; Ebner, F.; Klopfleisch, R.; Kühl, A.A.; Zentek, J.; Pieper, R.; Hartmann, S. Influence of Nutrition and Maternal Bonding on Postnatal Lung Development in the Newborn Pig. *Front. Immunol.* **2021**, *12*, 734153. [CrossRef] [PubMed]
36. Alapati, D.; Rong, M.; Chen, S.; Hehre, D.; Hummler, S.C.; Wu, S. Inhibition of β-catenin signaling improves alveolarization and reduces pulmonary hypertension in experimental bronchopulmonary dysplasia. *Am. J. Respir. Cell Mol. Biol.* **2014**, *51*, 104–113. [CrossRef] [PubMed]
37. Lecarpentier, Y.; Gourrier, E.; Gobert, V.; Vallée, A. Bronchopulmonary Dysplasia: Crosstalk Between PPARγ, WNT/β-Catenin and TGF-β Pathways; The Potential Therapeutic Role of PPARγ Agonists. *Front. Pediatr.* **2019**, *7*, 176. [CrossRef]
38. Sucre, J.M.; Vijayaraj, P.; Aros, C.J.; Wilkinson, D.; Paul, M.; Dunn, B.; Guttentag, S.H.; Gomperts, B.N. Posttranslational modification of β-catenin is associated with pathogenic fibroblastic changes in bronchopulmonary dysplasia. *Am. J. Physiol. Lung Cell Mol. Physiol.* **2017**, *312*, L186–L195. [CrossRef] [PubMed]
39. Sucre, J.M.S.; Deutsch, G.H.; Jetter, C.S.; Ambalavanan, N.; Benjamin, J.T.; Gleaves, L.A.; Millis, B.A.; Young, L.R.; Blackwell, T.S.; Kropski, J.A.; et al. A Shared Pattern of β-Catenin Activation in Bronchopulmonary Dysplasia and Idiopathic Pulmonary Fibrosis. *Am. J. Pathol.* **2018**, *188*, 853–862. [CrossRef] [PubMed]
40. Vogel, C.F.; Zhao, Y.; Wong, P.; Young, N.F.; Matsumura, F. The use of c-src knockout mice for the identification of the main toxic signaling pathway of TCDD to induce wasting syndrome. *J. Biochem. Mol. Toxicol.* **2003**, *17*, 305–315. [CrossRef]
41. Davenport, M.L.; Sherrill, T.P.; Blackwell, T.S.; Edmonds, M.D. Perfusion and Inflation of the Mouse Lung for Tumor Histology. *J. Vis. Exp.* **2020**, *162*, e60605. [CrossRef] [PubMed]
42. Karasutani, K.; Baskoro, H.; Sato, T.; Arano, N.; Suzuki, Y.; Mitsui, A.; Shimada, N.; Kodama, Y.; Seyama, K.; Fukuchi, Y.; et al. Lung Fixation under Constant Pressure for Evaluation of Emphysema in Mice. *J. Vis. Exp.* **2019**, *151*, e58197. [CrossRef]
43. Cooney, T.P.; Thurlbeck, W.M. The radial alveolar count method of Emery and Mithal: A reappraisal 1-postnatal lung growth. *Thorax* **1982**, *37*, 572–579. [CrossRef] [PubMed]
44. Dechelotte, P.; Labbé, A.; Caux, O.; Vanlieferinghen, P.; Raynaud, E.J. Defect in pulmonary growth. Comparative study of 3 diagnostic criteria. *Arch. Fr. Pediatr.* **1987**, *44*, 255–261. [PubMed]
45. Klein, A.W.; Becker, R.F.; Bryson, M.R. A method for estimating the distribution of alveolar sizes from histological lung sections. *Trans. Am. Microsc. Soc.* **1972**, *91*, 195–208. [CrossRef]
46. Crowley, G.; Kwon, S.; Caraher, E.J.; Haider, S.H.; Lam, R.; Batra, P.; Melles, D.; Liu, M.; Nolan, A. Quantitative lung morphology: Semi-automated measurement of mean linear intercept. *BMC Pulm. Med.* **2019**, *19*, 206. [CrossRef]
47. Baker, C.D.; Alvira, C.M. Disrupted lung development and bronchopulmonary dysplasia: Opportunities for lung repair and regeneration. *Curr. Opin. Pediatr.* **2014**, *26*, 306–314. [CrossRef] [PubMed]
48. Tóth, S.; Pingorová, S.; Jonecová, Z.; Morochovic, R.; Pomfy, M.; Veselá, J. Adult Respiratory Distress Syndrome and alveolar epithelium apoptosis: An histopathological and immunohistochemical study. *Folia Histochem. Cytobiol.* **2009**, *47*, 431–434. [CrossRef]
49. Go, H.; Ohto, H.; Nollet, K.E.; Sato, K.; Ichikawa, H.; Kume, Y.; Kanai, Y.; Maeda, H.; Kashiwabara, N.; Ogasawara, K.; et al. Red cell distribution width as a predictor for bronchopulmonary dysplasia in premature infants. *Sci. Rep.* **2021**, *11*, 7221. [CrossRef]
50. Janz, D.R.; Ware, L.B. The role of red blood cells and cell-free hemoglobin in the pathogenesis of ARDS. *J. Intensive Care* **2015**, *3*, 20. [CrossRef]
51. Hansmann, G.; Sallmon, H.; Roehr, C.C.; Kourembanas, S.; Austin, E.D.; Koestenberger, M. Pulmonary hypertension in bronchopulmonary dysplasia. *Pediatr. Res.* **2021**, *89*, 446–455. [CrossRef] [PubMed]
52. Bos, A.P.; Hussain, S.M.; Hazebroek, F.W.; Tibboel, D.; Meradji, M.; Molenaar, J.C. Radiographic evidence of bronchopulmonary dysplasia in high-risk congenital diaphragmatic hernia survivors. *Pediatr. Pulmonol.* **1993**, *15*, 231–234. [CrossRef]
53. Singer, L.T.; Davillier, M.; Preuss, L.; Szekely, L.; Hawkins, S.; Yamashita, T.; Baley, J. Feeding interactions in infants with very low birth weight and bronchopulmonary dysplasia. *J. Dev. Behav. Pediatr. JDBP* **1996**, *17*, 69–76. [CrossRef]
54. Dassios, T.; Williams, E.E.; Hickey, A.; Bunce, C.; Greenough, A. Bronchopulmonary dysplasia and postnatal growth following extremely preterm birth. *Arch. Dis. Child. Fet. Neonatal Ed.* **2021**, *106*, 386–391. [CrossRef]
55. Laberge, J.-M.; Puligandla, P. Chapter 64-Congenital Malformations of the Lungs and Airways. In *Pediatric Respiratory Medicine*, 2nd ed.; Taussig, L.M., Landau, L.I., Eds.; Mosby: Philadelphia, PA, USA, 2008; pp. 907–941. [CrossRef]
56. Askenazi, S.S.; Perlman, M. Pulmonary hypoplasia: Lung weight and radial alveolar count as criteria of diagnosis. *Arch. Dis. Child.* **1979**, *54*, 614–618. [CrossRef]
57. D'Angio, C.T.; Maniscalco, W.M. Bronchopulmonary dysplasia in preterm infants: Pathophysiology and management strategies. *Paediatr. Drugs* **2004**, *6*, 303–330. [CrossRef]
58. Pereira, G.R.; Baumgart, S.; Bennett, M.J.; Stallings, V.A.; Georgieff, M.K.; Hamosh, M.; Ellis, L. Use of high-fat formula for premature infants with bronchopulmonary dysplasia: Metabolic, pulmonary, and nutritional studies. *J. Pediatr.* **1994**, *124*, 605–611. [CrossRef]
59. Young, T.E. Nutritional support and bronchopulmonary dysplasia. *J. Perinatol.* **2007**, *27*, S75–S78. [CrossRef]

60. Delgado-Peña, Y.P.; Torrent-Vernetta, A.; Sacoto, G.; de Mir-Messa, I.; Rovira-Amigo, S.; Gartner, S.; Moreno-Galdó, A.; Molino-Gahete, J.A.; Castillo-Salinas, F. Pulmonary hypoplasia: An analysis of cases over a 20-year period. *Pediatria* **2016**, *85*, 70–76. [CrossRef]
61. Huang, J.; Zhang, L.; Tang, J.; Shi, J.; Qu, Y.; Xiong, T.; Mu, D. Human milk as a protective factor for bronchopulmonary dysplasia: A systematic review and meta-analysis. *Arch. Dis. Child. Fet. Neonatal Ed.* **2019**, *104*, F128–F136. [CrossRef]
62. Smukowska-Gorynia, A.; Tomaszewska, I.; Malaczynska-Rajpold, K.; Marcinkowska, J.; Komosa, A.; Janus, M.; Olasinska-Wisniewska, A.; Slawek, S.; Araszkiewicz, A.; Jankiewicz, S.; et al. Red Blood Cells Distribution Width as a Potential Prognostic Biomarker in Patients With Pulmonary Arterial Hypertension and Chronic Thromboembolic Pulmonary Hypertension. *Heart Lung Circ.* **2018**, *27*, 842–848. [CrossRef]
63. Berkelhamer, S.K.; Mestan, K.K.; Steinhorn, R.H. Pulmonary hypertension in bronchopulmonary dysplasia. *Semin. Perinatol.* **2013**, *37*, 124–131. [CrossRef] [PubMed]
64. Ramani, M.; Bradley, W.E.; Dell'Italia, L.J.; Ambalavanan, N. Early exposure to hyperoxia or hypoxia adversely impacts cardiopulmonary development. *Am. J. Respir. Cell Mol. Biol.* **2015**, *52*, 594–602. [CrossRef]
65. Harris, W.S.; Baack, M.L. Beyond building better brains: Bridging the docosahexaenoic acid (DHA) gap of prematurity. *J. Perinatol.* **2015**, *35*, 1–7. [CrossRef]
66. Voynow, J.A.; Auten, R. Environmental Pollution and the Developing Lung. *Clin. Pulm. Med.* **2015**, *22*, 177–184. [CrossRef] [PubMed]
67. Guilbert, T.W.; Stern, D.A.; Morgan, W.J.; Martinez, F.D.; Wright, A.L. Effect of breastfeeding on lung function in childhood and modulation by maternal asthma and atopy. *Am. J. Respir. Crit. Care Med.* **2007**, *176*, 843–848. [CrossRef] [PubMed]
68. Loh, C.Y.; Chai, J.Y.; Tang, T.F.; Wong, W.F.; Sethi, G.; Shanmugam, M.K.; Chong, P.P.; Looi, C.Y. The E-Cadherin and N-Cadherin Switch in Epithelial-to-Mesenchymal Transition: Signaling, Therapeutic Implications, and Challenges. *Cells* **2019**, *8*, 1118. [CrossRef]
69. Zhu, Y.; Tan, J.; Xie, H.; Wang, J.; Meng, X.; Wang, R. HIF-1α regulates EMT via the Snail and β-catenin pathways in paraquat poisoning-induced early pulmonary fibrosis. *J. Cell Mol. Med.* **2016**, *20*, 688–697. [CrossRef]
70. Zhu, G.J.; Song, P.P.; Zhou, H.; Shen, X.H.; Wang, J.G.; Ma, X.F.; Gu, Y.J.; Liu, D.D.; Feng, A.N.; Qian, X.Y.; et al. Role of epithelial-mesenchymal transition markers E-cadherin, N-cadherin, β-catenin and ZEB2 in laryngeal squamous cell carcinoma. *Oncol. Lett.* **2018**, *15*, 3472–3481. [CrossRef] [PubMed]
71. Tian, X.; Liu, Z.; Niu, B.; Zhang, J.; Tan, T.K.; Lee, S.R.; Zhao, Y.; Harris, D.C.H.; Zheng, G. E-Cadherin/β-Catenin Complex and the Epithelial Barrier. *J. Biomed. Biotechnol.* **2011**, *2011*, 567305. [CrossRef] [PubMed]
72. Conquer, J.A.; Martin, J.B.; Tummon, I.; Watson, L.; Tekpetey, F. Effect of DHA supplementation on DHA status and sperm motility in asthenozoospermic males. *Lipids* **2000**, *35*, 149–154. [CrossRef] [PubMed]
73. Haggarty, P. Effect of placental function on fatty acid requirements during pregnancy. *Eur. J. Clin. Nutr.* **2004**, *58*, 1559–1570. [CrossRef] [PubMed]
74. Smith, S.L.; Rouse, C.A. Docosahexaenoic acid and the preterm infant. *Matern. Health Neonatol. Perinatol.* **2017**, *3*, 22. [CrossRef] [PubMed]

Article

The Bladder Is a Novel Target of Developmental Polychlorinated Biphenyl Exposure Linked to Increased Inflammatory Cells in the Bladder of Young Mice

Conner L. Kennedy †, Audrey Spiegelhoff †, Kathy Wang, Thomas Lavery, Alexandra Nunez, Robbie Manuel, Lauren Hillers-Ziemer, Lisa M. Arendt and Kimberly P. Keil Stietz *

Department of Comparative Biosciences, School of Veterinary Medicine, University of Wisconsin-Madison, Madison, WI 53706, USA; clkennedy3@wisc.edu (C.L.K.); aspiegelhoff@wisc.edu (A.S.); kwang399@wisc.edu (K.W.); tlavery@wisc.edu (T.L.); asnunez@wisc.edu (A.N.); rmanuel@wisc.edu (R.M.); lauren.ziemer@nih.gov (L.H.-Z.); lmarendt@wisc.edu (L.M.A.)
* Correspondence: kkeil@wisc.edu
† These authors contributed equally.

Abstract: Bladder inflammation is associated with several lower urinary tract symptoms that greatly reduce quality of life, yet contributing factors are not completely understood. Environmental chemicals are plausible mediators of inflammatory reactions within the bladder. Here, we examine whether developmental exposure to polychlorinated biphenyls (PCBs) leads to changes in immune cells within the bladder of young mice. Female mice were exposed to an environmentally relevant mixture of PCBs through gestation and lactation, and bladders were collected from offspring at postnatal day (P) 28–31. We identify several dose- and sex-dependent PCB effects in the bladder. The lowest concentration of PCB (0.1 mg/kg/d) increased CD45+ hematolymphoid immune cells in both sexes. While PCBs had no effect on CD79b+ B cells or CD3+ T cells, PCBs (0.1 mg/kg/d) did increase F4/80+ macrophages particularly in female bladder. Collagen density was also examined to determine whether inflammatory events coincide with changes in the stromal extracellular matrix. PCBs (0.1 mg/kg/d) decreased collagen density in female bladder compared to control. PCBs also increased the number of cells undergoing cell division predominantly in male bladder. These results implicate perturbations to the immune system in relation to PCB effects on the bladder. Future study to define the underlying mechanisms could help understand how environmental factors can be risk factors for lower urinary tract symptoms.

Keywords: lower urinary tract; bladder; inflammation; POPs; developmental basis of adult disease

1. Introduction

Lower urinary tract symptoms (LUTS) greatly impact quality of life. Patients seeking medical attention for these symptoms represent a significant health care cost [1,2]. LUTS encompass a diverse range of both storage and voiding dysfunction; symptoms can range from obstruction, weak stream or difficulty urinating to increased frequency, urgency, incontinence, and overactive bladder. LUTS are often treated symptomatically because the underlying etiology is not completely understood and likely multifactorial. A better understanding of causative agents may lead to more beneficial therapies and improvement in quality of life for patients.

Bladder inflammation, also termed cystitis, is one major cause of bladder dysfunction [3]. There are known factors which can give rise to bladder inflammation including infection (typically acute inflammation), genetics, autoimmunity and dietary influences [3,4]. Consequences of these factors can lead to states of chronic inflammation, which can induce fibrosis and continuation or worsening of symptoms such as pain, urgency and frequency [3,5,6]. However, a clear factor contributing to bladder inflammation is not always evident and could reside in the environment. Chemicals are capable of eliciting bladder inflammation, for example, cyclophosphamide is commonly used in rodent models to study acute and chronic interstitial cystitis phenotypes [7]. Despite the use of chemicals to model bladder inflammation in rodents, whether exposure to ubiquitous environmental chemicals such as polychlorinated biphenyls (PCBs) can contribute to a state of chronic bladder inflammation, fibrosis and bladder dysfunction is understudied.

PCBs are a class of persistent organic pollutants that continue to pose a risk to human health. PCBs are implicated as risk factors for developing neurodevelopmental disorders (NDDs), which often have comorbid symptoms of bladder dysfunction [8]. PCBs have also been linked to deleterious changes in other health outcomes such as reproductive and immune function [9–18]. Detectable levels of PCBs are present in serum and tissue samples from humans [9,19] and livestock [20], as well as commercial milk [20,21], air [22,23] and water [24]. While the manufacture of PCBs is banned, exposure continues due to persistence of PCBs in the environment (legacy sources), as well as release of contemporary PCBs as unintentional byproducts during manufacturing processes such as paint pigment production [25,26]. Contemporary PCBs have been found to make up a large portion of detected PCBs in recent samples from humans and livestock [20,27,28]. Contemporary PCBs tend to be lower-chlorinated congeners and were not necessarily produced as part of manufactured (Aroclor) PCB mixtures prior to the ban [29,30]. Since some of these contemporary PCBs were not part of legacy PCB mixtures, less is known about their effects on long term health outcomes. Here, we use the Markers of Autism Risk in Babies Learning Early Signs (MARBLES) PCB mixture which mimics the proportion of the top PCB congeners detected in the serum of women at risk of having a child with a neurodevelopmental disorder [28,31]. This mixture not only replicates an environmentally relevant mixture found in the human population, but also consists of legacy higher-chlorinated PCBs in addition to contemporary lower chlorinated PCBs such as PCB 11 [28,32].

PCBs are known to have deleterious effects on the proper function of the immune system. Several studies indicate deficits in the adaptive immune response and immunotoxicity in adults and adolescents [15,16,33–37]. There is also evidence that innate immunity is influenced by PCB exposure, with increases in inflammatory cytokines often observed as indices of immune activation [17,38,39]. In a follow up of Yusho patients who experienced high exposure to PCBs via contaminated food in 1968, plasma concentrations of several cytokines were higher in exposed patients compared to controls more than 30 years after the initial incident [40]. In line with epidemiological data, in animal models, PCBs have been shown to alter proinflammatory and profibrotic markers in serum, liver, and brain [41–43]. Recent animal studies have modeled environmentally relevant concentrations of PCBs using the MARBLES PCB mixture. In mice, developmental exposure to MARBLES PCB results in offspring which display alterations in inflammatory markers in the intestine which coincide with changes in intestinal physiology and the microbiota [44]. In addition, developmental exposure to MARBLES PCB leads to elevated levels of serum cytokines and chemokines in juvenile offspring [45]. While PCB exposures are linked to changes in circulating cytokines, less is known regarding the inflammatory signature within tissues such as the bladder.

To expand upon the evidence that PCBs can increase serum cytokine in juvenile mice following developmental exposure, we sought to determine whether PCBs alter inflammation in other organs of interest such as the bladder. We focus on the bladder as we have previously shown that PCBs are not only detected in mouse bladder following developmental exposure, but that PCBs lead to changes in neuromorphology which are correlated with increased mast cell numbers [32]. Here, we test the hypothesis that developmental exposure to the environmentally relevant MARBLES PCB mixture results in increased immune cells within the bladder of juvenile male and female mice and test whether these cells are associated with changes to the extracellular matrix collagen density.

2. Materials and Methods

2.1. Animals

All procedures involving animals were conducted in accordance with the NIH Guide for the Care and Use of Laboratory Animals and were approved by the University of California-Davis Animal Care and Use Committee (#18853 and 20584; date of approval: 8-5-15 and 8-3-18 respectively). Wild-type mice of 75% C57BL/6J/25% SVJ129 genetic background (Jackson Labs, Sacramento, CA, USA), were used in this study and were collected as part of a larger study as described previously [32,45]. All mice were housed in clear plastic cages containing corn cob bedding and maintained on a 12 h light and dark cycle at 22 ± 2 °C. Feed (Diet 5058, LabDiet, St. Louis, MO, USA) and water were available ad libitum.

2.2. Developmental PCB Exposures

We have previously described the MARBLES PCB mixture and doses selected, 0.1, 1 and 6 mg/kg body weight/day, which result in PCB levels in offspring tissues within ranges observed in humans and do not interfere with reproductive outcomes of the dam [32,45,46]. The proportion of individual PCB congeners in the MARBLES PCB mixture mimics the proportion of the top 12 PCB congeners identified in pregnant women at risk of having a child with a neurodevelopmental disorder [18,28,31]. PCBs were synthesized and authenticated by the Synthesis Core of the University of Iowa Superfund Research Program with >99% purity as reported previously [28]. PCBs were dissolved in organic peanut oil (Spectrum Organic Products, LLC, Melville, NY, USA), and mixed into organic peanut butter (Trader Joe's, Monrovia, CA, USA). PCB doses were measured on a weigh boat and delivered to mouse cages for oral consumption. Organic peanut oil dissolved in peanut butter and without PCBs served as the dosing control (0 mg/kg). Female mice were dosed daily beginning two weeks prior to start of mating and continuing through pregnancy and lactation (through postnatal day 21). Adult male breeders were paired with female mice until a visible copulation plug was seen or until mice steadily gained weight indicative of pregnancy. Male and female offspring were weaned at P21 and group housed with same sex and dose littermates. Mice were euthanized via CO2 prior to collection of tissues at P28–31.

2.3. Immunohistochemistry

Bladders were processed for immunohistochemistry as described previously [32]. Immunofluorescence was performed on bladder sections essentially as described [47]. Briefly, slides were deparaffinized in xylene and rehydrated through a series of graded ethanols. Antigen retrieval was performed in citrate buffer (0.01 M, pH 6.00, 20 min at 50% microwave power). Sections were blocked for 1 h in blocking buffer containing 1% blocking solution, 5% goat serum (16210064, Fisher, Waltham, MA, USA) and 1% BSA fraction V (80055-682, VWR, Radnor, PA) in tris-buffered saline (TBS, 0.2 M Tris-HCl (Fisher BP153-1), 1.5 M NaCl (Fisher BP358-212). Blocking solution was prepared as a stock solution consisting of 10% Blocking reagent (501003304, Fisher) dissolved in 100 mM maleic acid (S25415, Fisher) and 150 mM NaCl (721016, Fisher) pH 7.5. Sections were incubated with primary antibodies listed in Table 1 overnight at 4 °C. Following TBS washes, secondary antibodies listed in Table 1 were applied to tissues for 1 h at room temperature. Secondary antibody was removed and 4′,6-diamidino-2-phenylindole, dilactate solution 300 nM (DAPI) (IC15757401, VWR) was applied for 5 min to counterstain nuclei. Slides were mounted in anti-fade mounting medium (90% glycerol (G33500, Fisher), 0.2% n-propyl gallate (AAA1087722, Fisher), and phosphate-buffered saline (SH3001304, Fisher)) and cover-slipped. For non-fluorescent immunohistochemistry the same protocol as above was followed with the addition of incubating slides in 0.5% hydrogen peroxide (H325500, Fisher) for 20 min prior to antigen retrieval, and for F4/80, sections were blocked in 5% fish gelatin (G7765, Sigma-Aldrich, St. Louis, MO, USA) for 1 h at room temperature. For CD3 and F4/80, after removal of secondary antibody, sections were incubated in

ABC reagent (Vectastain Elite HRP ABC kit, PK-6100, Vector Laboratories) followed by DAB peroxidase substrate kit (SK-4100) for development according to manufacturer's instructions. For CD79b, following removal of primary antibody, the ImmPRESS® HRO Horse Anti-Rabbit IgG Polymer Detection Kit, Peroxidase (MP-7401, Vector Laboratories) was used per manufacturer's instructions. Sections were counterstained for 30 sec with hematoxylin (6765001, Fisher), dehydrated in graded ethanol followed by xylene and mounted with Permount (SP15100, Fisher). Slides were imaged using an Eclipse E600 or Eclipse Ci compound microscope (Nikon Instruments Inc., Melville, NY, USA) with a Photometrics Dyno CCD camera or DS Ri2 camera (Nikon Instruments Inc.) interfaced to NIS elements imaging software (Nikon Instruments Inc.) or a semi-automated BZ-X710 digital microscope and stitching software (Keyence, Itasca, IL, USA). Cell counts were performed in Image J (Version 1.52a) using the cell counter tool in a blinded manner. Cell counts were either normalized to total cells, total stromal cells, or total epithelial cells, or normalized to total bladder area as indicated in each figure. An n = 4–6 bladders per treatment group were used in all analyses.

Table 1. List of Antibodies.

Primary Antibodies	Catalog #	Company	Source	Dilution	Pairing
CD79b	ab134147	Abcam	Rabbit	1:250	
CD3	ab11089	Abcam	Rat	1:250	
CD45 (PTPRC)	ab10558	Abcam	Rabbit	1:750	
E-Cadherin (CDH1)	610181	BD Transduction Labs (via Fisher)	Mouse	1:250	
F4/80 (EMR1; Adgre1)	123102	Biolegend	Rat	1:50	
Ki67	ab15580	Abcam	Rabbit	1:200	
Keratin 5	905901	Biolegend	Chicken	1:500	
Secondary Antibodies/Detection Kits					
Anti-Mouse Alexa Fluor 594	715-545-150	Jackson ImmunoResearch	Donkey	1:250	Cdh1
Anti-Rabbit Alexa Fluor 488	711-545-152	Jackson ImmunoResearch	Donkey	1:250	CD45
Anti-Rabbit Alexa Fluor 594	111-585-144	Jackson ImmunoResearch	Goat	1:250	Ki67
Anti-Rat Biotinylated	112-066-003	Jackson ImmunoResearch	Goat	1:250	CD3, F4/80
Anti-Chicken Alexa 488	703-546-155	Jackson ImmunoResearch	Donkey	1:250	Krt5
ImmPRESS® HRO Horse Anti-Rabbit IgG Polymer Detection Kit, Peroxidase	MP-7401	Vector Laboratories	Horse	Per manufacturer's instructions	CD79b
VECTASTAIN ELITE HRP ABC KIT paired with DAB substrate Kit, Peroxidase	PK-6100, SK-4100	Vector Laboratories		Per manufacturer's instructions	CD3, F4/80

2.4. Collagen Quantification

Picrosirius red staining (PSR) staining was used to determine collagen density as described previously [48]. Images were captured from the red (PSR) and green (autofluorescent) channels and ImageJ was used to process images. Briefly, the green channel (autofluorescence) was subtracted from the red channel (collagen) using the Image Calculator feature of ImageJ. The image was then made binary, using the freehand selection tool a region of interest was drawn around the bladder stroma by an individual blinded to treatment conditions. The total area of this region of interest was determined using the Analyze-Measure feature. Once total area was established, the analyze particles feature was used to determine collagen pixel density within the region. Collagen area/total area was used to determine collagen density.

2.5. Statistics

Normality of data was assessed using the Kolmogorov–Smirnov or D'Agostino–Pearson omnibus (K2) test within GraphPad Prism 8. If data failed to pass normality, transformation (log or square root) was applied to restore normality. Two-way ANOVA followed by Tukey's multiple comparisons tests were used to determine differences between or among treatment groups with p values ≤ 0.05 considered significant. N values for each endpoint are indicated in figure legends. Significant main effects of two-way ANOVAs are indicated in each figure in addition to post hoc comparisons indicated by asterisks and bars.

3. Results

3.1. Developmental PCB Exposure Increases Inflammatory Cells in the Bladder

Since PCBs cause inflammation in many tissues [38,44,49,50], and bladder inflammation can be a driver of LUTS [3], we tested and confirmed that mice developmentally (in utero and via lactation) exposed to MARBLES PCB develop low-grade bladder inflammation as young adults. We first chose to examine hematolymphoid cells defined as those positive for protein tyrosine phosphate receptor type C (PTPRC, CD45). There was a significant overall main effect of PCB dose on the percentage of CD45-positive hematolymphoid immune cells within the bladder, driven by an increase in CD45-positive cells at the 0.1 mg/kg treatment group versus control (Figure 1A–C). This PCB dose-dependent increase was observed in both the epithelium and the stroma (Figure 1D,E). In addition, within the epithelium, there was an increase in CD45-positive cells at the 1 mg/kg treatment group compared to control (Figure 1D).

We next tested whether PCB exposure activates innate or adaptive immune responses within the bladder. We chose to focus on three immune cell populations which are dysregulated by PCBs in other tissues and are commonly dysregulated in patients with bladder dysfunction: B cells (adaptive), T cells (adaptive) [51,52] and macrophages (innate) [53,54]. B cells were immunolabeled with an antibody targeting CD79b (Figure 2A,B). PCB exposure did not significantly change the bladder abundance of CD79b-positive B cells; however, we identified a surprising overall main effect of sex with increases in total CD79b-positive cells in female compared to male bladder tissue (Figure 2C). PCB exposure did not change the number of CD79b-positive B cells within the epithelium or stromal cell types independently (Figure 2D,E). We visualized T cells using an antibody targeting CD3 (Figure 3A,B) and did not identify any significant sex- or dose-dependent differences within the bladder (Figure 3C–E). These results suggest that developmental PCB exposure does not activate the adaptive immune response within the bladder in juvenile mice.

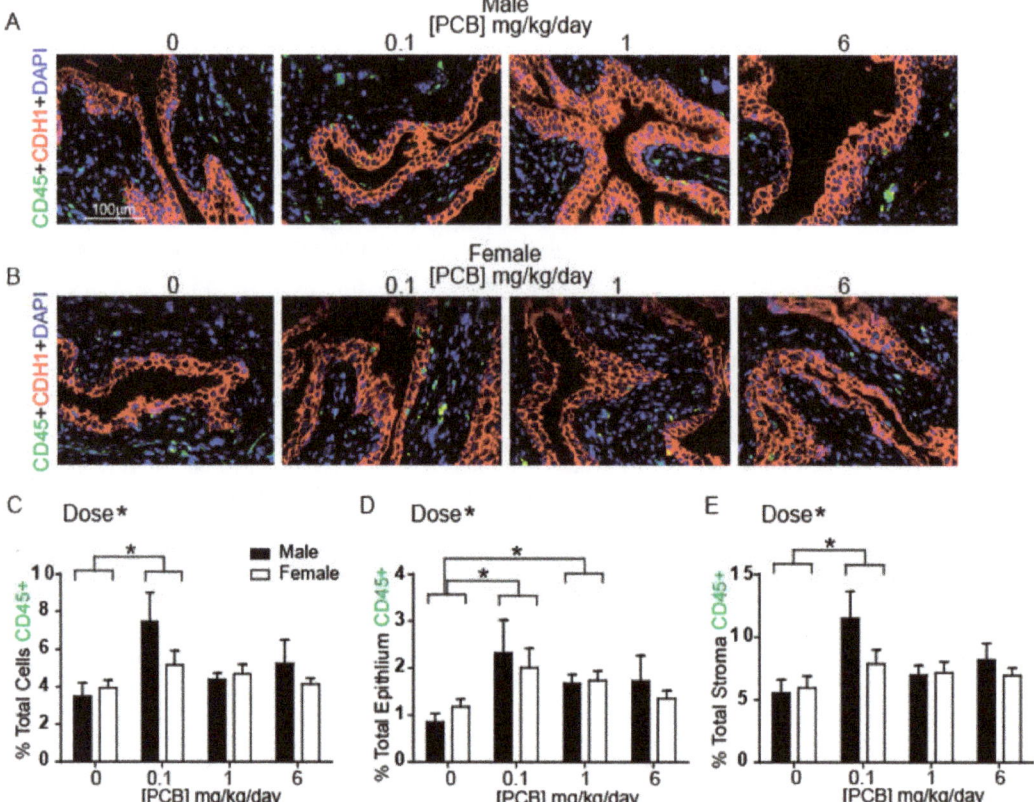

Figure 1. PCBs increase CD45-positive immune cells in bladder of developmentally exposed mice. Mice were exposed to PCBs via maternal diet through gestation and lactation and bladders collected from young male and female offspring at postnatal day (P) 28–31 for immunohistochemistry. Representative images of (**A**) male and (**B**) female mouse bladders at each PCB treatment group incubated with antibodies targeting CD45 (green) to label immune cells, e-cadherin (CDH1, red) to label all epithelium and DAPI (blue) to stain nuclei. Quantification of (**C**) the percent of total cells CD45 positive, (**D**) the percent of total epithelial cells CD45 positive and (**E**) the percent of total stromal cells CD45 positive. Results are the mean ± SEM, n = 4–6 bladders per treatment group, up to 3 images per bladder were averaged for final value. Significant differences at $p < 0.05$ are indicated by asterisk or asterisk and bars, as determined using two-way ANOVA followed by Tukey's multiple comparisons tests.

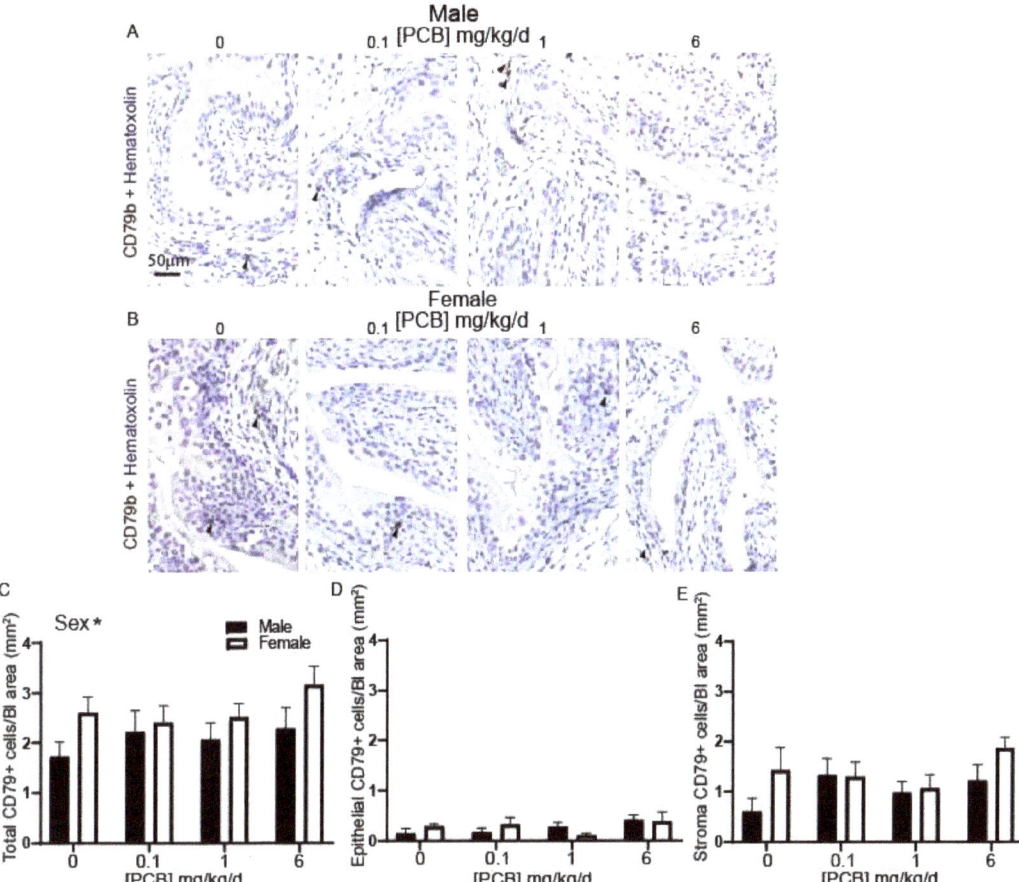

Figure 2. Sex influences CD79b-positive immune cells in bladders of developmentally exposed mice. Mice were exposed to PCBs via maternal diet through gestation and lactation. Bladders were collected from young male and female offspring at postnatal day (P) 28–31 for immunohistochemistry. Representative images of (**A**) male and (**B**) female mouse bladders at each PCB treatment group incubated with antibodies targeting CD79b (brown) to label immune cells and hematoxylin (purple) to label all nuclei. Quantification of (**C**) total cells CD79b positive within bladder, (**D**) epithelial cells CD79b positive within bladder, and (**E**) stromal cells CD79b positive within the bladder. Results are the mean ± SEM, n = 4–6 bladders per treatment group. * indicates significant differences at $p < 0.05$, as determined using two-way ANOVA.

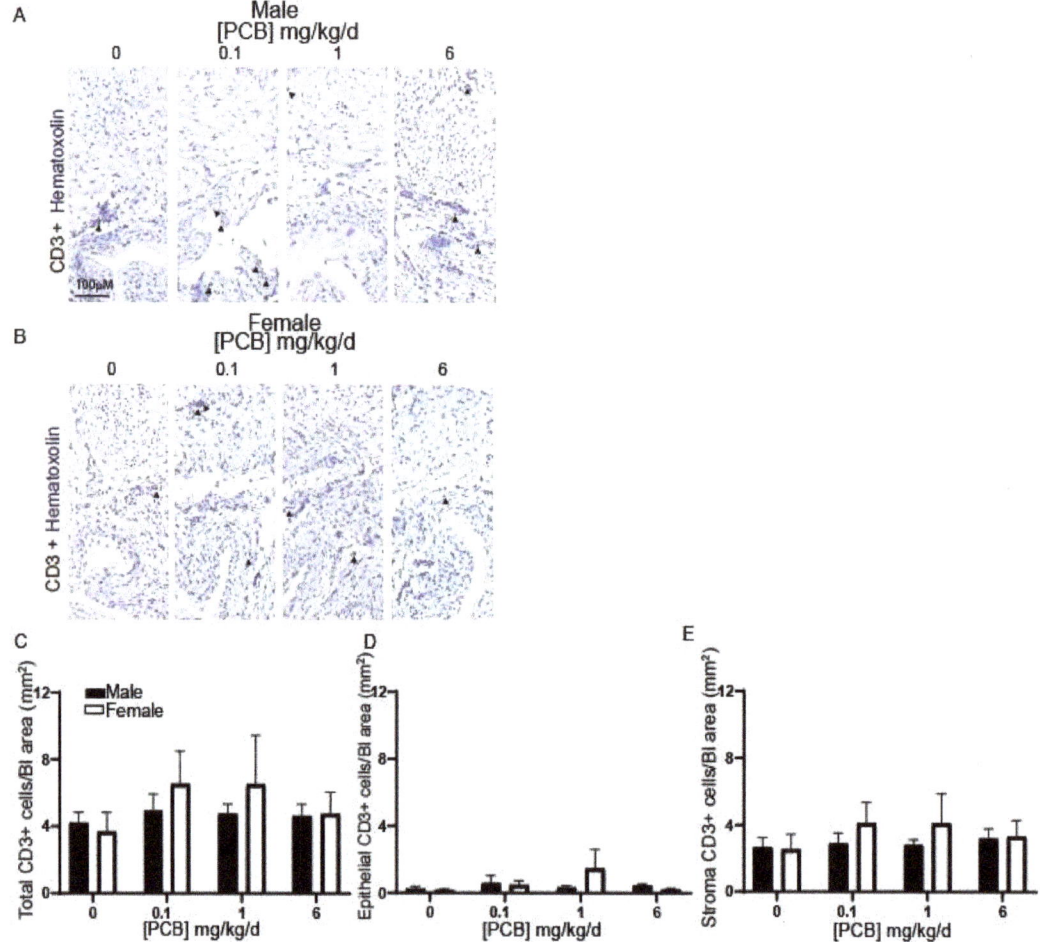

Figure 3. PCBs have no effect on CD3-positive immune cells in bladders of developmentally exposed mice. Mice were exposed to PCBs via maternal diet through gestation and lactation. Bladders were collected from young male and female offspring at postnatal day (P) 28–31 for immunohistochemistry. Representative images of (**A**) male and (**B**) female mouse bladders at each PCB treatment group incubated with antibodies targeting CD3 (brown) to label immune cells and hematoxylin (purple) to label all nuclei. Quantification of (**C**) the percent of total cells CD3 positive, (**D**) the percent of epithelial cells CD3 positive, and (**E**) the percent of stromal cells CD3 positive. Results are the mean ± SEM, n = 4–6 bladders per treatment group. No significant differences as determined by two-way ANOVA.

Macrophages contribute to PCB-mediated inflammation in other tissues [53,55,56] and are mediators of the innate immune response. To test whether developmental PCBs altered this endpoint in our model, we examined F4/80-positive cells within the bladder as a marker of total macrophages (Figure 4A,B). There was a significant overall main effect of dose in the percentage of F4/80-positive macrophages, driven by a specific increase in the 0.1 mg/kg PCB treatment group compared to all other groups (Figure 4C). There was also an overall main effect of sex, with more macrophages in female than male bladders (Figure 4C). We did not observe a PCB-dependent change in the percentage of F4/80-positive cells within bladder epithelium alone (Figure 4D). However, in bladder stroma, there was a significant main effect of sex, dose and an interaction between sex and dose with F4/80-positive cells greater at the 0.1 mg/kg PCB treatment group compared to all other doses. This result was driven by a significant increase in F4/80-positive stromal cells in the female 0.1 mg/kg PCB treatment group compared to all other female treatment groups as well as between male and female at the 0.1 mg/kg PCB group (Figure 4E). These results indicate that the innate immune response can be activated by developmental PCB exposure which is especially evident at the lowest 0.1 mg/kg PCB group in female mice.

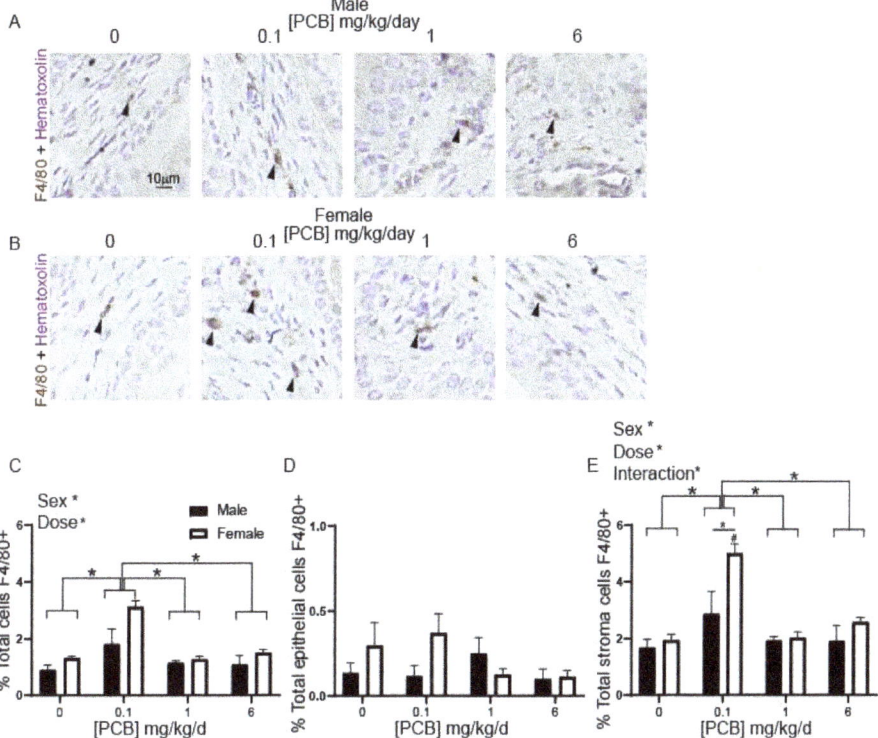

Figure 4. PCBs increase F4/80-positive cells in bladder of developmentally exposed female mice. Mice were exposed to PCBs via maternal diet through gestation and lactation and bladders collected from young male and female offspring at postnatal day (P) 28–31 for immunohistochemistry. Representative images of (**A**) male and (**B**) female mouse bladders at each PCB treatment group incubated with antibodies targeting F4/80 (brown) to label macrophages, nuclei were counterstained with hematoxylin (purple). Quantification of (**C**) percent total cells F4/80 positive, (**D**) percent total epithelial cells F4/80 positive, (**E**) percent total stromal cells F4/80 positive. Results are the mean ± SEM, n = 4–6 bladders per treatment group. Significant differences at $p < 0.05$ indicated by asterisk, asterisk and bars or by a # which indicates significant difference from all other same sex treatment groups as determined using two-way ANOVA followed by Tukey's multiple comparisons tests.

3.2. Developmental PCB Exposure Decreases Collagen Density within Female Bladder Stroma

Chronic or acute inflammation can change collagen density [57] and PCBs have been linked to changes in collagen mRNA abundance and fiber density [58–60]. To test whether developmental PCB exposure changes collagen density in young mice, bladders were stained using picrosirius red (PSR) and stained fibers visualized under a Texas red filter (Figure 5A,B). There was a significant decrease in collagen density within bladder stroma in female mice in the 0.1 mg/kg/d PCB treatment group compared to control (Figure 5C). These results indicate that PCBs are capable of altering the extracellular matrix via changes in collagen density.

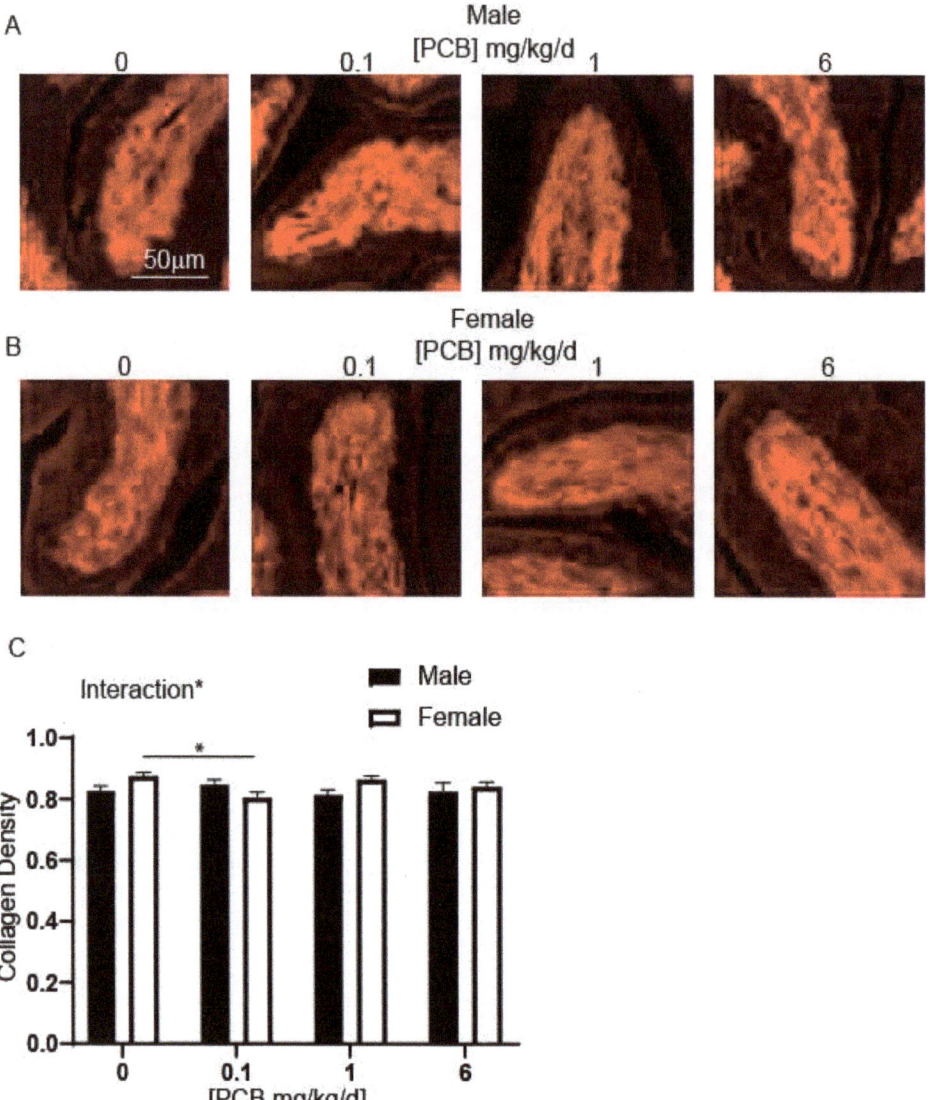

Figure 5. PCBs increase collagen density in bladder of developmentally exposed female mice. Mice were exposed to PCBs via maternal diet through gestation and lactation and bladders collected from young male and female offspring at postnatal day (P) 28–31. Slides were stained with picrosirius red to visualize collagen. Representative images of (**A**) male and (**B**) female mouse bladders at each PCB treatment group stained with picrosirius red to visualize collagen (red). Quantification of (**C**) ratio of collagen per unit area. Results are the mean ± SEM, n = 4–6 bladders per treatment group. Significant differences at $p < 0.05$ indicated by asterisk and asterisk and bar as determined using two-way ANOVA followed by Tukey's multiple comparisons tests.

3.3. Developmental PCB Exposure Increases Proliferation in a Sex- and Dose-Specific Manner

The PCB-mediated (0.1 mg/kg) increase in the number of CD45+ hematolymphoid cells within bladder could indicate ongoing low-grade inflammation or a lingering response to a previous injury. Since proliferation is key to rapid recovery after injury to restore function [61,62], we used an antibody against Ki-67 to visualize cells in the active phase of the cell cycle to test whether PCB exposure at the 0.1 mg/kg dose increases bladder cell proliferation, which would be consistent with ongoing inflammation (Figure 6A,B). There was a significant overall main effect of dose and sex such that the percentage of Ki-67-positive bladder cells was greater in the 0.1 mg/kg PCB treatment group compared to control and was greater in male mice versus female mice (Figure 6C). When split into epithelial and stromal compartments significant effects were only observed in the epithelium (Figure 6D,E) where there was a significant main effect of sex with the percentage of Ki-67-positive cells greater in male versus female bladders (Figure 6D). Together, these results indicate that the 0.1 mg/kg PCB dose increases the number of cells in the active phase of the cell cycle most prominently in male mice.

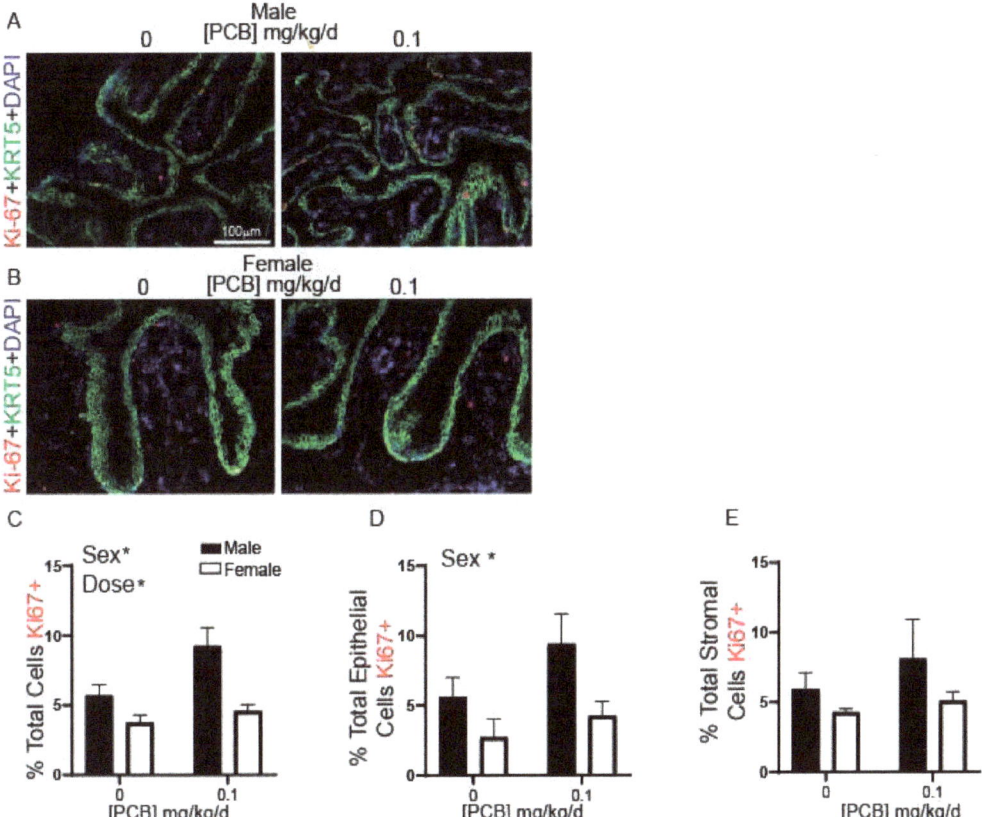

Figure 6. PCBs increase Ki-67-positive cells in male bladder. Mice were exposed to PCBs via maternal diet through gestation and lactation and bladders collected from young male and female offspring at postnatal day (P) 28–31 for immunohistochemistry. Representative images of (A) male and (B) female mouse bladders at each PCB treatment group incubated with antibodies targeting Ki-67 (red) to label proliferating cells, keratin 5 (KRT5, green) to label basal epithelium and DAPI (blue) to stain nuclei. Quantification of (C) the percent of total cells Ki-67 positive, (D) the percent of total epithelial cells Ki-67 positive and (E) the percent of total stromal cells Ki-67 positive. Results are the mean ± SEM, n = 4–6 bladders per treatment group, up to 3 images per bladder were averaged for final value. Significant differences at $p < 0.05$ indicated by asterisk as determined using two-way ANOVA followed by Tukey's multiple comparisons tests.

4. Discussion

PCB exposure during in utero development and via lactation causes low-grade inflammation in the bladder and changes the cellular and molecular composition of the bladder in a dose- and sex-specific manner. The 0.1 mg/kg/d PCB dose increased the bladder concentration of CD45-positive hematolymphoid cells, macrophages and Ki67-positive cells and decreased collagen density in a sex-specific manner. These data suggest that innate immune responses can be evoked by low-dose PCB exposure during bladder development. We also found in female mice that PCBs increase the bladder's density of macrophages, coincident with PCB-mediated changes to the bladder extracellular matrix. The impact of PCB-induced bladder inflammation on urinary function remains to be determined and is an area of future study. In addition, studies are needed to establish the timeline of inflammatory cell recruitment to the bladder of PCB exposed mice and the persistence of these cells into adulthood.

Increasing evidence suggests that exposure to environmental toxicants can contribute to lower urinary tract function in adulthood. Developmental exposure to environmental toxicant, dioxin (TCDD) via the dam, has been shown to decrease void intervals in adult male mouse offspring [63] and further exacerbates voiding dysfunction in susceptible mice [64,65]. There is also evidence that remodeling of collagen is one pathway by which environmental chemicals can impact the lower urinary tract. In rhesus monkeys, developmental TCDD exposure increases inflammatory cells and fibrosis in the prostate of offspring years after exposure [66]. In mice genetically susceptible to prostate neoplasia, developmental exposure to TCDD exacerbates hormone induced changes in collagen fiber size/distribution in the prostate and the bladder [65]. While these studies did not examine inflammation or presence of immune cells within the bladder, evidence that collagen fiber size and distribution were impacted by TCDD are in line with the results obtained here. We found a decrease in collagen density within female bladder of the 0.1 mg/kg/d PCB treatment group compared to controls, the same group which had an increase in F4/80-positive cells in the bladder. One possible explanation for this observation is that PCB exposure leads to increased degradation of the extracellular matrix and collagen, which is characteristic of some inflammatory disorders, such as asthma [57]. This could be due to the activity of macrophages, which produce matrix metalloproteinases (MMPs) that degrade extracellular proteins such as collagens [67]. Alternatively, PCBs may prevent the proper formation of collagen architecture within the bladder at this specific timepoint. This explanation is also supported by studies that found IL-1, an immune mediator released in response to stimuli such as allergens, suppresses the formation of collagen fibers [57], and that PCB contamination can lead to increased expression of mediators such as IL-1 [53]. Whether the observed decrease in collagen density is primarily due to collagen degradation or suppression of formation will require further research.

One factor which makes determining whether benign urologic diseases may have a developmental origin complex is the fact that dose-dependent effects are often seen with environmental chemical exposures. Non-monotonic PCB dose effects are commonly observed in the central nervous system [68,69]. Similarly, we saw increases in immune cells in the bladder in the lowest (0.1 mg/kg) dose group but not in higher dose groups. It can be difficult to compare dose effects across studies when different ages and dosing paradigms are used, nonetheless our findings are consistent with other PCB immune studies. For example, only a low single dose (20 mg/kg) exposure to Aroclor 1260 in adult male mice elevated serum IL-6 in response to a high-fat diet weeks later; higher concentrations of Aroclor 1260 (200 mg/kg) did not [70]. This suggests a dose-specific effect of PCBs on inflammatory markers, as well as the ability of PCBs to interact with other factors such as diet [70]. One explanation for PCB dose effects could be that higher doses of PCBs lead to toxicity and cell death. However, apoptosis is not a likely factor in the lack of response observed in the higher PCB dose groups here. Our previous study in mice dosed in the same manner and of the same age, had detectable levels of PCBs within bladder tissue, but did not display any changes in cleaved caspase 3 apoptotic cells in the bladder, changes in epithelial composition or thickness, or any changes in mouse body mass or urine creatinine concentration [32]. A second explanation for dose effects observed is that low versus high doses of PCBs could produce a different trajectory of changes within the bladder. It is possible that the higher PCB doses elicited an inflammatory response at an earlier timepoint than was examined. On the other hand, it is also possible that each dose triggers a different set of receptors/pathways which contribute to bladder phenotypes. We have previously observed PCB induced increases in nerve fiber density within the male bladder only in the higher 6 mg/kg PCB group [32]. Whether this increase in nerve density was preceded by inflammation or arose independently of an inflammatory event remains to be determined. Dose effects could also be due to differences in metabolizing enzymes present, the effects of the metabolites themselves on the tissue, or differences in receptor targets.

We identified an increase in macrophages within bladders of mice exposed to the lowest PCB dose which was prevalent in female offspring. This finding is consistent with previous studies which have examined effects of PCBs on serum cytokine and chemokine expression [45]. Macrophages are capable of secreting various cytokines including tumor necrosis factor (TNF), interleukin (IL)-1, IL-6, IL-8, IL-12, IL-23 as well as chemokines such as CXCL1 and 2, CCL5, CXCL8-11 [71]. In a study using the same dosing paradigm used here, PCBs elicited dose-dependent increases in several serum cytokines and chemokines including TNFa, IL-1a, IL-1b, IL-10, GM-CSF, IL-17, IL-12, IFN-g, IL-4, IL-9, IL-13, CCL3, and CXCL1 [45]. While not all chemokines were examined, the PCB effect on increasing serum TNFa, IL-1, IL-12 and CXCL1 are consistent with the hypothesis that macrophages may be involved in PCB induced inflammatory responses in the bladder and perhaps other tissues. In addition to serum cytokines, another study using the same dosing paradigm used here reported increases in IL-6 and IL-1b in intestine of developmentally exposed mice [44]. Exposure to PCB 126 (50–500 nM) in macrophage cell lines induced expression of proinflammatory cytokines such as monocyte chemoattractant protein-1 (MCP-1) which is involved in macrophage recruitment [53]. Epidemiological data from Inuits with high PCB consumption report a significant increase in serum levels of inflammatory markers, YLK-40 and hsCRP [17]. Elevation of YKL-40 is of particular interest as it can be secreted by macrophages [72]. Together, these results are in line with increased macrophages in PCB exposed bladder observed here and suggest that macrophages may play an important role in PCB-induced changes in inflammation, especially in female bladder.

Macrophage polarization can lead to differences in response to stimuli and can contribute to sex differences in inflammatory responses. For example, in a myocarditis rat model, males tend to exhibit a M1 driven pro-inflammatory response while females predominantly exhibit an M2 anti-inflammatory response [73]. These states of macrophage polarization also correspond to severity of fibrosis. Males had increased expression of collagen gene transcripts and fibrosis while females had upregulated expression of anti-fibrotic genes and downregulation of pro-fibrotic genes [73]. It is possible that the observation here of increased macrophages in female bladder in the 0.1 mg/kg PCB treatment group that also had decreased collagen density, is due to a mechanism related to macrophage polarization. Whether PCBs alter the polarization of macrophages, and whether the macrophages present in the female bladder were M1 or M2 and linked to changes in expression of genes which are pro- or anti-fibrotic in nature, remains an area of future study.

The presence of increased CD45-positive cells in the bladder suggests that one possible mechanism of action for developmental PCB exposure on the bladder is setting up a state of low-grade inflammation. This may persist or be cleared, but it raises the possibility that PCBs could act as a stressor causing a low-grade inflammatory reaction ongoing in the bladder. This could lead to increased sensitivity to other stressors throughout life especially if those stressors also trigger inflammation. The observed increase in macrophages in female bladder is also intriguing. In humans, macrophages are known to play a role in response to urinary tract infection which is more prevalent in women compared to men [74–76]. Women are also more commonly affected by interstitial cystitis/bladder pain syndrome than men, and it is thought that a component of this diagnosis is underlying bladder inflammation [3,77]. Whether environmental exposures to PCBs can contribute to the etiology or sex biases of these disorders is unknown, but of potential clinical interest as steps to reduce exposure, especially during development, could help reduce risk or severity of these lower urinary tract symptoms later in life.

Author Contributions: Conceptualization, C.L.K., L.M.A. and K.P.K.S.; formal analysis, C.L.K., A.S., K.W., T.L., A.N., L.H.-Z. and K.P.K.S.; funding acquisition, C.L.K., L.M.A. and K.P.K.S.; investigation, C.L.K., A.S., K.W., A.N., R.M., L.H.-Z. and K.P.K.S.; methodology, C.L.K., R.M., L.H.-Z., L.M.A. and K.P.K.S.; project administration, L.M.A. and K.P.K.S.; resources, L.H.-Z., L.M.A. and K.P.K.S.; supervision, L.M.A. and K.P.K.S.; validation, C.L.K., A.S., K.W., T.L., A.N. and K.P.K.S.; visualization, C.L.K., A.S., K.W. and K.P.K.S.; writing—original draft, C.L.K., A.S., K.W., T.L. and K.P.K.S.; writing—review and editing, C.L.K., A.S., K.W., T.L., A.N., R.M., L.H.-Z., L.M.A. and K.P.K.S. All authors have read and agreed to the published version of the manuscript.

Funding: This work was supported by the National Institutes of Health NIEHS R00 ES029537 to KPKS and T32 ES007015 to CLK. This work was also supported by core facilities including the Wisconsin O'Brien Center for Benign Urologic Research U54 DK104310. The synthesis of the MARBLES mix was supported by the Iowa Superfund Research Program at The University of Iowa P42 ES013661. Animal studies were supported by ES014901 and ES011269 and the United States Environmental Protection Agency R833292. The content is solely the responsibility of the authors and does not necessarily represent the official views of the NIH or the USEPA. Further, the NIH and USEPA did not endorse the purchase of any commercial products or services mentioned in the publication.

Institutional Review Board Statement: All procedures involving animals were conducted in accordance with the NIH Guide for the Care and Use of Laboratory Animals and were approved by the University of California-Davis Animal Care and Use Committee (#18853 and 20584).

Informed Consent Statement: Not applicable.

Data Availability Statement: Available upon request to corresponding author.

Acknowledgments: We would like to acknowledge Xueshu Li and Hans-Joachim Lehmler from the University of Iowa for the synthesis of the PCBs used in this study. We would also like to acknowledge Pamela Lein (University of California Davis) and Dale Bjorling (University of Wisconsin Madison) for technical and financial support and expertise pertaining to these studies.

Conflicts of Interest: The authors have no conflict of interest to declare.

References

1. Rubin, E.B.; Buehler, A.E.; Halpern, S.D. States worse than death among hospitalized patients with serious illnesses. *JAMA Intern. Med.* **2016**, *176*, 1557–1559. [CrossRef] [PubMed]
2. Onukwugha, E.; Zuckerman, I.H.; McNally, D.; Coyne, K.S.; Vats, V.; Mullins, C.D. The total economic burden of overactive bladder in the united states: A disease-specific approach. *Am. J. Manag. Care* **2009**, *15*, S90–S97.
3. Grover, S.; Srivastava, A.; Lee, R.; Tewari, A.K.; Te, A.E. Role of inflammation in bladder function and interstitial cystitis. *Ther. Adv. Urol.* **2011**, *3*, 19–33. [CrossRef]
4. Warren, J.W.; Brown, J.; Tracy, J.K.; Langenberg, P.; Wesselmann, U.; Greenberg, P. Evidence-based criteria for pain of interstitial cystitis/painful bladder syndrome in women. *Urology* **2008**, *71*, 444–448. [CrossRef]
5. Ryu, C.M.; Shin, J.H.; Yu, H.Y.; Ju, H.; Kim, S.; Lim, J.; Heo, J.; Lee, S.; Shin, D.M.; Choo, M.S. N-acetylcysteine prevents bladder tissue fibrosis in a lipopolysaccharide-induced cystitis rat model. *Sci. Rep.* **2019**, *9*, 8134. [CrossRef] [PubMed]
6. Wynn, T.A.; Ramalingam, T.R. Mechanisms of fibrosis: Therapeutic translation for fibrotic disease. *Nat. Med.* **2012**, *18*, 1028–1040. [CrossRef]
7. Augé, C.; Gamé, X.; Vergnolle, N.; Lluel, P.; Chabot, S. Characterization and validation of a chronic model of cyclophosphamide-induced interstitial cystitis/bladder pain syndrome in rats. *Front. Pharmacol.* **2020**, *11*, 1305. [CrossRef] [PubMed]
8. Panesar, H.K.; Kennedy, C.L.; Keil Stietz, K.P.; Lein, P.J. Polychlorinated biphenyls (pcbs): Risk factors for autism spectrum disorder? *Toxics* **2020**, *8*, 70. [CrossRef]
9. Lyall, K.; Croen, L.A.; Sjödin, A.; Yoshida, C.K.; Zerbo, O.; Kharrazi, M.; Windham, G.C. Polychlorinated biphenyl and organochlorine pesticide concentrations in maternal mid-pregnancy serum samples: Association with autism spectrum disorder and intellectual disability. *Environ. Health. Perspect.* **2017**, *125*, 474–480. [CrossRef]
10. Eubig, P.A.; Aguiar, A.; Schantz, S.L. Lead and pcbs as risk factors for attention deficit/hyperactivity disorder. *Environ. Health Perspect.* **2010**, *118*, 1654–1667. [CrossRef]

11. Schantz, S.L.; Widholm, J.J.; Rice, D.C. Effects of pcb exposure on neuropsychological function in children. *Environ. Health Perspect.* **2003**, *111*, 357–576. [CrossRef]
12. Tsai, M.S.; Chen, M.H.; Lin, C.C.; Ng, S.; Hsieh, C.J.; Liu, C.Y.; Hsieh, W.S.; Chen, P.C. Children's environmental health based on birth cohort studies of asia. *Sci. Total. Environ.* **2017**, *609*, 396–409. [CrossRef]
13. Neblett, M.F.; Curtis, S.W.; Gerkowicz, S.A.; Spencer, J.B.; Terrell, M.L.; Jiang, V.S.; Marder, M.E.; Barr, D.B.; Marcus, M.; Smith, A.K. Examining reproductive health outcomes in females exposed to polychlorinated biphenyl and polybrominated biphenyl. *Sci. Rep.* **2020**, *10*, 3314. [CrossRef]
14. Meeker, J.D.; Hauser, R. Exposure to polychlorinated biphenyls (pcbs) and male reproduction. *Syst. Biol. Reprod. Med.* **2010**, *56*, 122–131. [CrossRef] [PubMed]
15. Leijs, M.M.; Koppe, J.G.; Olie, K.; van Aalderen, W.M.; de Voogt, P.; ten Tusscher, G.W. Effects of dioxins, pcbs, and pbdes on immunology and hematology in adolescents. *Environ. Sci. Technol.* **2009**, *43*, 7946–7951. [CrossRef]
16. Heilmann, C.; Grandjean, P.; Weihe, P.; Nielsen, F.; Budtz-Jørgensen, E. Reduced antibody responses to vaccinations in children exposed to polychlorinated biphenyls. *PLoS Med.* **2006**, *3*, e311. [CrossRef]
17. Schæbel, L.K.; Bonefeld-Jørgensen, E.C.; Vestergaard, H.; Andersen, S. The influence of persistent organic pollutants in the traditional inuit diet on markers of inflammation. *PLoS ONE* **2017**, *12*, e0177781. [CrossRef]
18. Granillo, L.; Sethi, S.; Keil, K.P.; Lin, Y.; Ozonoff, S.; Iosif, A.M.; Puschner, B.; Schmidt, R.J. Polychlorinated biphenyls influence on autism spectrum disorder risk in the marbles cohort. *Environ. Res.* **2019**, *171*, 177–184. [CrossRef]
19. Schantz, S.L.; Gardiner, J.C.; Aguiar, A.; Tang, X.; Gasior, D.M.; Sweeney, A.M.; Peck, J.D.; Gillard, D.; Kostyniak, P.J. Contaminant profiles in southeast asian immigrants consuming fish from polluted waters in northeastern wisconsin. *Environ. Res.* **2010**, *110*, 33–39. [CrossRef]
20. Sethi, S.; Chen, X.; Kass, P.H.; Puschner, B. Polychlorinated biphenyl and polybrominated diphenyl ether profiles in serum from cattle, sheep, and goats across california. *Chemosphere* **2017**, *181*, 63–73. [CrossRef]
21. Chen, X.; Lin, Y.; Dang, K.; Puschner, B. Quantification of polychlorinated biphenyls and polybrominated diphenyl ethers in commercial cows' milk from california by gas chromatography-triple quadruple mass spectrometry. *PLoS ONE* **2017**, *12*, e0170129. [CrossRef]
22. Heiger-Bernays, W.J.; Tomsho, K.S.; Basra, K.; Petropoulos, Z.E.; Crawford, K.; Martinez, A.; Hornbuckle, K.C.; Scammell, M.K. Human health risks due to airborne polychlorinated biphenyls are highest in new bedford harbor communities living closest to the harbor. *Sci. Total Environ.* **2020**, *710*, 135576. [CrossRef]
23. Bräuner, E.V.; Andersen, Z.J.; Frederiksen, M.; Specht, I.O.; Hougaard, K.S.; Ebbehøj, N.; Bailey, J.; Giwercman, A.; Steenland, K.; Longnecker, M.P.; et al. Health effects of pcbs in residences and schools (hesperus): Pcb-health cohort profile. *Sci. Rep.* **2016**, *6*, 24571. [CrossRef]
24. Howell, N.L.; Suarez, M.P.; Rifai, H.S.; Koenig, L. Concentrations of polychlorinated biphenyls (pcbs) in water, sediment, and aquatic biota in the houston ship channel, texas. *Chemosphere* **2008**, *70*, 593–606. [CrossRef] [PubMed]
25. Hu, D.; Hornbuckle, K.C. Inadvertent polychlorinated biphenyls in commercial paint pigments. *Environ. Sci. Technol.* **2010**, *44*, 2822–2827. [CrossRef] [PubMed]
26. Jahnke, J.C.; Hornbuckle, K.C. Pcb emissions from paint colorants. *Environ. Sci. Technol.* **2019**, *53*, 5187–5194. [CrossRef] [PubMed]
27. Sethi, S.; Keil, K.P.; Chen, H.; Hayakawa, K.; Li, X.; Lin, Y.; Lehmler, H.J.; Puschner, B.; Lein, P.J. Detection of 3,3'-dichlorobiphenyl in human maternal plasma and its effects on axonal and dendritic growth in primary rat neurons. *Toxicol. Sci.* **2017**, *158*, 401–411. [CrossRef] [PubMed]
28. Sethi, S.; Morgan, R.K.; Feng, W.; Lin, Y.; Li, X.; Luna, C.; Koch, M.; Bansal, R.; Duffel, M.W.; Puschner, B.; et al. Comparative analyses of the 12 most abundant pcb congeners detected in human maternal serum for activity at the thyroid hormone receptor and ryanodine receptor. *Environ. Sci. Technol.* **2019**, *53*, 3948–3958. [CrossRef]
29. Klocke, C.; Sethi, S.; Lein, P.J. The developmental neurotoxicity of legacy vs. Contemporary polychlorinated biphenyls (pcbs): Similarities and differences. *Environ. Sci. Pollut. Res. Int.* **2020**, *27*, 8885–8896. [CrossRef]
30. Kodavanti, P.R.; Kannan, N.; Yamashita, N.; Derr-Yellin, E.C.; Ward, T.R.; Burgin, D.E.; Tilson, H.A.; Birnbaum, L.S. Differential effects of two lots of aroclor 1254: Congener-specific analysis and neurochemical end points. *Environ. Health Perspect.* **2001**, *109*, 1153–1161. [CrossRef]
31. Hertz-Picciotto, I.; Schmidt, R.J.; Walker, C.K.; Bennett, D.H.; Oliver, M.; Shedd-Wise, K.M.; LaSalle, J.M.; Giulivi, C.; Puschner, B.; Thomas, J.; et al. A prospective study of environmental exposures and early biomarkers in autism spectrum disorder: Design, protocols, and preliminary data from the marbles study. *Environ. Health. Perspect.* **2018**, *126*, 117004. [CrossRef]
32. Keil Stietz, K.P.; Kennedy, C.L.; Sethi, S.; Valenzuela, A.; Nunez, A.; Wang, K.; Wang, Z.; Wang, P.; Spiegelhoff, A.; Puschner, B.; et al. In utero and lactational pcb exposure drives anatomic changes in the juvenile mouse bladder. *Curr. Res. Toxicol.* **2021**, *2*, 1–18. [CrossRef]
33. Lü, Y.C.; Wu, Y.C. Clinical findings and immunological abnormalities in yu-cheng patients. *Environ. Health Perspect.* **1985**, *59*, 17–29. [CrossRef]
34. Svensson, B.G.; Hallberg, T.; Nilsson, A.; Schütz, A.; Hagmar, L. Parameters of immunological competence in subjects with high consumption of fish contaminated with persistent organochlorine compounds. *Int. Arch. Occup. Environ. Health* **1994**, *65*, 351–358. [CrossRef]

35. Dietert, R.R. Developmental immunotoxicity, perinatal programming, and noncommunicable diseases: Focus on human studies. *Adv. Med.* **2014**, *2014*, 867805. [CrossRef]
36. Stølevik, S.B.; Nygaard, U.C.; Namork, E.; Haugen, M.; Meltzer, H.M.; Alexander, J.; Knutsen, H.K.; Aaberge, I.; Vainio, K.; van Loveren, H.; et al. Prenatal exposure to polychlorinated biphenyls and dioxins from the maternal diet may be associated with immunosuppressive effects that persist into early childhood. *Food Chem. Toxicol.* **2013**, *51*, 165–172. [CrossRef]
37. Heilmann, C.; Budtz-Jørgensen, E.; Nielsen, F.; Heinzow, B.; Weihe, P.; Grandjean, P. Serum concentrations of antibodies against vaccine toxoids in children exposed perinatally to immunotoxicants. *Environ. Health Perspect.* **2010**, *118*, 1434–1438. [CrossRef]
38. Peinado, F.M.; Artacho-Cordón, F.; Barrios-Rodríguez, R.; Arrebola, J.P. Influence of polychlorinated biphenyls and organochlorine pesticides on the inflammatory milieu. A systematic review of in vitro, in vivo and epidemiological studies. *Environ. Res.* **2020**, *186*, 109561. [CrossRef] [PubMed]
39. Imbeault, P.; Findlay, C.S.; Robidoux, M.A.; Haman, F.; Blais, J.M.; Tremblay, A.; Springthorpe, S.; Pal, S.; Seabert, T.; Krümmel, E.M.; et al. Dysregulation of cytokine response in canadian first nations communities: Is there an association with persistent organic pollutant levels? *PLoS ONE* **2012**, *7*, e39931. [CrossRef]
40. Kuwatsuka, Y.; Shimizu, K.; Akiyama, Y.; Koike, Y.; Ogawa, F.; Furue, M.; Utani, A. Yusho patients show increased serum il-17, il-23, il-1β, and tnfα levels more than 40 years after accidental polychlorinated biphenyl poisoning. *J. Immunotoxicol.* **2014**, *11*, 246–249. [CrossRef]
41. Xu, L.; Guo, X.; Li, N.; Pan, Q.; Ma, Y.Z. Effects of quercetin on aroclor 1254-induced expression of cyp. *J. Immunotoxicol.* **2019**, *16*, 140–148. [CrossRef]
42. Ahmed, R.G.; El-Gareib, A.W.; Shaker, H.M. Gestational 3,3′,4,4′,5-pentachlorobiphenyl (pcb 126) exposure disrupts fetoplacental unit: Fetal thyroid-cytokines dysfunction. *Life. Sci.* **2018**, *192*, 213–220. [CrossRef]
43. Bell, M.R.; Dryden, A.; Will, R.; Gore, A.C. Sex differences in effects of gestational polychlorinated biphenyl exposure on hypothalamic neuroimmune and neuromodulator systems in neonatal rats. *Toxicol. Appl. pharmacol.* **2018**, *353*, 55–66. [CrossRef] [PubMed]
44. Rude, K.M.; Pusceddu, M.M.; Keogh, C.E.; Sladek, J.A.; Rabasa, G.; Miller, E.N.; Sethi, S.; Keil, K.P.; Pessah, I.N.; Lein, P.J.; et al. Developmental exposure to polychlorinated biphenyls (pcbs) in the maternal diet causes host-microbe defects in weanling offspring mice. *Environ. Pollut.* **2019**, *253*, 708–721. [CrossRef] [PubMed]
45. Matelski, L.; Keil Stietz, K.P.; Sethi, S.; Taylor, S.L.; Van de Water, J.; Lein, P.J. The influence of sex, genotype, and dose on serum and hippocampal cytokine levels in juvenile mice developmentally exposed to a human-relevant mixture of polychlorinated biphenyls. *Curr. Res. Toxicol.* **2020**, *1*, 85–103. [CrossRef] [PubMed]
46. Klocke, C.; Lein, P.J. Evidence implicating non-dioxin-like congeners as the key mediators of polychlorinated biphenyl (pcb) developmental neurotoxicity. *Int. J. Mol. Sci.* **2020**, *21*, 1013. [CrossRef] [PubMed]
47. Abler, L.L.; Keil, K.P.; Mehta, V.; Joshi, P.S.; Schmitz, C.T.; Vezina, C.M. A high-resolution molecular atlas of the fetal mouse lower urogenital tract. *Dev. Dyn.* **2011**, *240*, 2364–2377. [CrossRef]
48. Wegner, K.A.; Keikhosravi, A.; Eliceiri, K.W.; Vezina, C.M. Fluorescence of picrosirius red multiplexed with immunohistochemistry for the quantitative assessment of collagen in tissue sections. *J. Histochem. Cytochem.* **2017**, *65*, 479–490. [CrossRef] [PubMed]
49. Petriello, M.C.; Brandon, J.A.; Hoffman, J.; Wang, C.; Tripathi, H.; Abdel-Latif, A.; Ye, X.; Li, X.; Yang, L.; Lee, E.; et al. Dioxin-like pcb 126 increases systemic inflammation and accelerates atherosclerosis in lean ldl receptor-deficient mice. *Toxicol. Sci.* **2018**, *162*, 548–558. [CrossRef]
50. Hennig, B.; Meerarani, P.; Slim, R.; Toborek, M.; Daugherty, A.; Silverstone, A.E.; Robertson, L.W. Proinflammatory properties of coplanar pcbs: In vitro and in vivo evidence. *Toxicol. Appl. Pharmacol.* **2002**, *181*, 174–183. [CrossRef]
51. Akiyama, Y.; Yao, J.R.; Kreder, K.J.; O'Donnell, M.A.; Lutgendorf, S.K.; Lyu, D.; Maeda, D.; Kume, H.; Homma, Y.; Luo, Y. Autoimmunity to urothelial antigen causes bladder inflammation, pelvic pain, and voiding dysfunction: A novel animal model for hunner-type interstitial cystitis. *Am. J. Physiol. Renal. Physiol.* **2021**, *320*, F174–F182. [CrossRef]
52. Horváthová, M.; Jahnová, E.; Palkovičová, L.; Trnovec, T.; Hertz-Picciotto, I. Dynamics of lymphocyte subsets in children living in an area polluted by polychlorinated biphenyls. *J. Immunotoxicol.* **2011**, *8*, 333–345. [CrossRef]
53. Wang, C.; Petriello, M.C.; Zhu, B.; Hennig, B. Pcb 126 induces monocyte/macrophage polarization and inflammation through ahr and nf-kappab pathways. *Toxicol. Appl. Pharmacol.* **2019**, *367*, 71–81. [CrossRef] [PubMed]
54. Cartwright, R.; Franklin, L.; Tikkinen, K.A.O.; Kalliala, I.; Miotla, P.; Rechberger, T.; Offiah, I.; McMahon, S.; O'Reilly, B.; Lince, S.; et al. Genome wide association study identifies two novel loci associated with female stress and urgency urinary incontinence. *J. Urol.* **2021**, 101097JU0000000000001822.
55. Ferrante, M.C.; Mattace Raso, G.; Esposito, E.; Bianco, G.; Iacono, A.; Clausi, M.T.; Amero, P.; Santoro, A.; Simeoli, R.; Autore, G.; et al. Effects of non-dioxin-like polychlorinated biphenyl congeners (pcb 101, pcb 153 and pcb 180) alone or mixed on j774a.1 macrophage cell line: Modification of apoptotic pathway. *Toxicol. Lett.* **2011**, *202*, 61–68. [CrossRef]
56. Yang, B.; Wang, Y.; Qin, Q.; Xia, X.; Liu, Z.; Song, E.; Song, Y. Polychlorinated biphenyl quinone promotes macrophage-derived foam cell formation. *Chem. Res. Toxicol.* **2019**, *32*, 2422–2432. [CrossRef]
57. Osei, E.T.; Mostaço-Guidolin, L.B.; Hsieh, A.; Warner, S.M.; Al-Fouadi, M.; Wang, M.; Cole, D.J.; Maksym, G.N.; Hallstrand, T.S.; Timens, W.; et al. Epithelial-interleukin-1 inhibits collagen formation by airway fibroblasts: Implications for asthma. *Sci. Rep.* **2020**, *10*, 8721. [CrossRef] [PubMed]

58. Jin, J.; Wahlang, B.; Shi, H.; Hardesty, J.E.; Falkner, K.C.; Head, K.Z.; Srivastava, S.; Merchant, M.L.; Rai, S.N.; Cave, M.C.; et al. Dioxin-like and non-dioxin-like pcbs differentially regulate the hepatic proteome and modify diet-induced nonalcoholic fatty liver disease severity. *Med. Chem. Res.* **2020**, *29*, 1247–1263. [CrossRef] [PubMed]
59. Ramajayam, G.; Sridhar, M.; Karthikeyan, S.; Lavanya, R.; Veni, S.; Vignesh, R.C.; Ilangovan, R.; Djody, S.S.; Gopalakrishnan, V.; Arunakaran, J.; et al. Effects of aroclor 1254 on femoral bone metabolism in adult male wistar rats. *Toxicology* **2007**, *241*, 99–105. [CrossRef]
60. Lind, P.M.; Larsson, S.; Oxlund, H.; Hâkansson, H.; Nyberg, K.; Eklund, T.; Orberg, J. Change of bone tissue composition and impaired bone strength in rats exposed to 3,3',4,4',5-pentachlorobiphenyl (pcb126). *Toxicology* **2000**, *150*, 41–51. [CrossRef]
61. Kullmann, F.A.; Clayton, D.R.; Ruiz, W.G.; Wolf-Johnston, A.; Gauthier, C.; Kanai, A.; Birder, L.A.; Apodaca, G. Urothelial proliferation and regeneration after spinal cord injury. *Am. J. Physiol. Renal. Physiol.* **2017**, *313*, F85–F102. [CrossRef] [PubMed]
62. Golubeva, A.V.; Zhdanov, A.V.; Mallel, G.; Dinan, T.G.; Cryan, J.F. The mouse cyclophosphamide model of bladder pain syndrome: Tissue characterization, immune profiling, and relationship to metabotropic glutamate receptors. *Physiol. Rep.* **2014**, *2*, e00260. [CrossRef]
63. Turco, A.E.; Oakes, S.R.; Stietz, K.P.K.; Dunham, C.L.; Joseph, D.B.; Chathurvedula, T.S.; Girardi, N.M.; Schneider, A.J.; Gawdzik, J.; Sheftel, C.M.; et al. A neuroanatomical mechanism linking perinatal tcdd exposure to lower urinary tract dysfunction in adulthood. *Dis. Model. Mech.* **2021**, *14*, dmm049068. [CrossRef]
64. Turco, A.E.; Thomas, S.; Crawford, L.K.; Tang, W.; Peterson, R.E.; Li, L.; Ricke, W.A.; Vezina, C.M. In utero and lactational 2,3,7,8-tetrachlorodibenzo-p-dioxin (tcdd) exposure exacerbates urinary dysfunction in hormone-treated c57bl/6j mice through a non-malignant mechanism involving proteomic changes in the prostate that differ from those elicited by testosterone and estradiol. *Am. J. Clin. Exp. Urol.* **2020**, *8*, 59–72. [PubMed]
65. Ricke, W.A.; Lee, C.W.; Clapper, T.R.; Schneider, A.J.; Moore, R.W.; Keil, K.P.; Abler, L.L.; Wynder, J.L.; López Alvarado, A.; Beaubrun, I.; et al. In utero and lactational tcdd exposure increases susceptibility to lower urinary tract dysfunction in adulthood. *Toxicol. Sci.* **2016**, *150*, 429–440. [CrossRef] [PubMed]
66. Arima, A.; Kato, H.; Ise, R.; Ooshima, Y.; Inoue, A.; Muneoka, A.; Kamimura, T.; Fukusato, T.; Kubota, S.; Sumida, H.; et al. In utero and lactational exposure to 2,3,7,8-tetrachlorodibenzo-p-dioxin (tcdd) induces disruption of glands of the prostate and fibrosis in rhesus monkeys. *Reprod. Toxicol.* **2010**, *29*, 317–322. [CrossRef] [PubMed]
67. Galis, Z.S.; Khatri, J.J. Matrix metalloproteinases in vascular remodeling and atherogenesis: The good, the bad, and the ugly. *Circ. Res.* **2002**, *90*, 251–262. [CrossRef]
68. Yang, D.; Kim, K.H.; Phimister, A.; Bachstetter, A.D.; Ward, T.R.; Stackman, R.W.; Mervis, R.F.; Wisniewski, A.B.; Klein, S.L.; Kodavanti, P.R.; et al. Developmental exposure to polychlorinated biphenyls interferes with experience-dependent dendritic plasticity and ryanodine receptor expression in weanling rats. *Environ. Health Perspect.* **2009**, *117*, 426–435. [CrossRef] [PubMed]
69. Wayman, G.A.; Bose, D.D.; Yang, D.; Lesiak, A.; Bruun, D.; Impey, S.; Ledoux, V.; Pessah, I.N.; Lein, P.J. Pcb-95 modulates the calcium-dependent signaling pathway responsible for activity-dependent dendritic growth. *Environ. Health Perspect.* **2012**, *120*, 1003–1009. [CrossRef]
70. Wahlang, B.; Song, M.; Beier, J.I.; Cameron Falkner, K.; Al-Eryani, L.; Clair, H.B.; Prough, R.A.; Osborne, T.S.; Malarkey, D.E.; States, J.C.; et al. Evaluation of aroclor 1260 exposure in a mouse model of diet-induced obesity and non-alcoholic fatty liver disease. *Toxicol. Appl. Pharmacol.* **2014**, *279*, 380–390. [CrossRef]
71. Arango Duque, G.; Descoteaux, A. Macrophage cytokines: Involvement in immunity and infectious diseases. *Front. Immunol.* **2014**, *5*, 491. [CrossRef]
72. Rathcke, C.N.; Vestergaard, H. Ykl-40, a new inflammatory marker with relation to insulin resistance and with a role in endothelial dysfunction and atherosclerosis. *Inflamm. Res.* **2006**, *55*, 221–227. [CrossRef] [PubMed]
73. Barcena, M.L.; Jeuthe, S.; Niehues, M.H.; Pozdniakova, S.; Haritonow, N.; Kühl, A.A.; Messroghli, D.R.; Regitz-Zagrosek, V. Sex-specific differences of the inflammatory state in experimental autoimmune myocarditis. *Front. Immunol.* **2021**, *12*, 686384. [CrossRef] [PubMed]
74. Schwab, S.; Jobin, K.; Kurts, C. Urinary tract infection: Recent insight into the evolutionary arms race between uropathogenic escherichia coli and our immune system. *Nephrol. Dial. Transplant.* **2017**, *32*, 1977–1983. [CrossRef]
75. Lacerda Mariano, L.; Ingersoll, M.A. Bladder resident macrophages: Mucosal sentinels. *Cell Immunol.* **2018**, *330*, 136–141. [CrossRef] [PubMed]
76. Lacerda Mariano, L.; Rousseau, M.; Varet, H.; Legendre, R.; Gentek, R.; Saenz Coronilla, J.; Bajenoff, M.; Gomez Perdiguero, E.; Ingersoll, M.A. Functionally distinct resident macrophage subsets differentially shape responses to infection in the bladder. *Sci. Adv.* **2020**, *6*, eabc5739. [CrossRef]
77. Kim, J.; De Hoedt, A.; Wiggins, E.; Haywood, K.; Jin, P.; Greenwood, B.; Narain, N.R.; Tolstikov, V.; Bussberg, V.; Barbour, K.E.; et al. Diagnostic utility of serum and urinary metabolite analysis in patients with interstitial cystitis/painful bladder syndrome. *Urology* **2021**. [CrossRef]

 *toxics*

Communication

Early Low-Level Arsenic Exposure Impacts Post-Synaptic Hippocampal Function in Juvenile Mice

Karl F. W. Foley [1,*], Daniel Barnett [2], Deborah A. Cory-Slechta [3] and Houhui Xia [1,2]

1. Department of Neuroscience, University of Rochester Medical Center, Rochester, NY 14642, USA; Houhui_Xia@URMC.Rochester.edu
2. Department of Pharmacology & Physiology, University of Rochester Medical Center, Rochester, NY 14642, USA; dmb4001@med.cornell.edu
3. Department of Environmental Medicine, University of Rochester Medical Center, Rochester, NY 14642, USA; Deborah_Cory-slechta@urmc.rochester.edu
* Correspondence: karl_foley@urmc.rochester.edu

Abstract: Arsenic is a well-established carcinogen known to increase mortality, but its effects on the central nervous system are less well understood. Epidemiological studies suggest that early life exposure is associated with learning deficits and behavioral changes. Studies in arsenic-exposed rodents have begun to shed light on potential mechanistic underpinnings, including changes in synaptic transmission and plasticity. However, previous studies relied on extended exposure into adulthood, and little is known about the effect of arsenic exposure in early development. Here, we studied the effects of early developmental arsenic exposure in juvenile mice on synaptic transmission and plasticity in the hippocampus. C57BL/6J females were exposed to arsenic (0, 50 ppb, 36 ppm) via drinking water two weeks prior to mating, with continued exposure throughout gestation and parturition. Electrophysiological recordings were then performed on juvenile offspring prior to weaning. In this paradigm, the offspring are exposed to arsenic indirectly, via the mother. We found that high (36 ppm) and relatively low (50 ppb) arsenic exposure both decreased basal synaptic transmission. A compensatory increase in pre-synaptic vesicular release was only observed in the high-exposure group. These results suggest that indirect, ecologically relevant arsenic exposure in early development impacts hippocampal synaptic transmission and plasticity that could underlie learning deficits reported in epidemiological studies.

Keywords: arsenic; synaptic transmission; long-term potentiation; hippocampus; development

1. Introduction

Early life exposure to toxic chemicals and environmental pollutants is associated with learning deficits and behavioral changes [1–3]. An estimated 200 million people worldwide are exposed to arsenic concentrations in drinking water that exceed the World Health Organization's recommended limit, 10 parts per billion (ppb) [4]. Exposure to concerning levels of arsenic is not limited to toxic waste sites. Rather, arsenic levels commonly exceed 10 ppb in domestic wells throughout the United States, especially in the southwest. While arsenic levels are kept below 10 ppb in municipal water supplies, private wells are unregulated and arsenic levels exceed 10 ppb in 20 out of 37 principal aquifers in the United States [5]. Even mild increases in arsenic exposure are of concern, as exposure is associated with numerous adverse health outcomes and increased mortality from a variety of conditions, including cardiovascular disease and cancer, as well as increased infant mortality [4]. Further, recent studies suggest the consequences of arsenic exposure can span across generations [6–8].

In January 2006, the maximum contaminant level (MCL) of arsenic in public water systems was lowered from 50 to 10 ppb, in compliance with a previous United States Environmental Protection Agency (EPA) ruling [9]. This change was enacted due to several epidemiological studies demonstrating an increased risk of cancer. While acute, high-level arsenic exposure was known to be associated with peripheral neuropathy, relatively little was known about the neurological consequences of chronic, low-level arsenic exposure [National Research Council (NRC) [10]]. However, more recent epidemiological studies have demonstrated that arsenic exposure is associated with deficits in cognitive and motor functions in children and adults [11–18]. Additionally, a recent study suggests inequalities in arsenic exposure reductions following the 2006 change in the MCL, such that there was a higher concentration of arsenic in public water systems serving Hispanic and tribal communities, small rural communities, and southwestern U.S. communities [19]. Given the relatively recent change in the arsenic MCL in public water in the U.S., the continued high exposure in many regions worldwide and the known vulnerability of the developing brain to toxicants and pollutants underscores the critical need to understand the effects of early life exposure to arsenic.

Electrophysiological studies in arsenic-exposed rodent models have begun to shed light on the potential mechanistic underpinnings of the associated cognitive deficits. Rodents exposed to high arsenic concentrations throughout early development and adulthood demonstrate a decrease in synaptic transmission and long-term potentiation (LTP) [20] in the hippocampus that may be secondary to altered glutamate transport [21]. The findings suggest that both pre- and post-synaptic functions are altered due to compensatory changes, since basal synaptic transmission decreased following arsenic exposure, whereas the neurotransmitter release probability increased. These changes are also reflected by changes in the expression of excitatory receptors and other regulators of synaptic signaling [22–26]. Similar changes in synaptic transmission and plasticity have been demonstrated by the ex vivo exposure of hippocampal slices to arsenite metabolites [27,28]. However, the electrophysiological effects of early developmental arsenic exposure have not been distinguished from chronic adulthood exposure. As arsenic can cross the placenta [29], we reasoned that early developmental arsenic exposure could cause changes in synaptic transmission and plasticity even without adulthood exposure. Specifically, given the epidemiological evidence of neurobehavioral deficits in children, we hypothesized that early developmental arsenic exposure alone reduces synaptic transmission and plasticity. Still, the effects of continued adulthood exposure on hippocampal synaptic function could differ from the effects of early developmental exposure alone. For example, whereas acute ex vivo exposure to arsenic metabolites attenuates LTP in hippocampal slices from adult rats, it facilitated LTP in young rats [30]. Further, there are pre- and post-synaptic changes present in adult mice following chronic exposure that suggest compensatory changes in the hippocampal circuit, with a decrease in basal synaptic transmission but an increase in neurotransmitter release probability [20]. However, the directionality of these synaptic changes is unclear, e.g., does a decrease in post-synaptic receptors cause a compensatory increase in pre-synaptic neurotransmitter release, or vice versa? Here, we studied the effects of in vivo arsenic exposure during gestation and early development by exposing dams to a high level (36 ppm) or a low level of arsenic (50 ppb), the MCL for public drinking water in the United States prior to 2006. Of note, in this paradigm, the dam is exposed to arsenic directly via drinking water, whereas the offspring are exposed indirectly via maternal transmission. Surprisingly, we found that the maternal exposure to even low levels of arsenic, i.e., 50 ppb, impairs synaptic transmission in the hippocampus of the offspring. Additionally, we observed different effects of arsenic exposure in our juvenile mice than what has been previously reported in adult mice (Tables 1 and 2).

Table 1. Timeline of arsenic exposure in current and previous electrophysiology studies.

Study	Exposure Type *	Age of Rodents Tested [†]	Exposure Onset	Exposure End
Nelson-Mora et al. (2018)	In vivo	Adult	Gestation	Adulthood
Current Study	In vivo	Juvenile	Gestation	Early development
Kruger et al. (2006)	Ex vivo	Juvenile/Adult	Following brain slice preparation	NA
Kruger et al. (2007)	Ex vivo	Juvenile/Adult		
Kruger et al. (2009)	Ex vivo	Juvenile/Adult		

* In vivo exposure: rodents exposed to arsenic through drinking water; Ex vivo exposure: brain tissue exposed to arsenic metabolites after its removal from the rodent. [†] Juvenile mice: 14 to 23 days old; Adult mice: 2–4 months of age.

Table 2. Effect of arsenic exposure on synaptic function varies by exposure paradigm.

Study	Exposure Level	Basal Transmission	Paired-Pulse Facilitation (PPF)	Long-Term Potentiation (LTP)
		In vivo exposure		
Nelson-Mora et al. (2018)	20 ppm	Decrease	Decrease *	Decrease
Current Study	36 ppm	Decrease	Decrease	Increase
Current study	50 ppb	Decrease	No change	No change [†]
		Acute, ex vivo exposure: Juvenile rodents		
Kruger et al. (2006)	1–100 µM	Decrease [‡]	Not tested	No change
Kruger et al. (2007)	1–100 µM	Decrease	Not tested	Decrease [‡]
Kruger et al. (2009)	1–100 µM	Decrease [‡]	Not tested	Increase [‡]
		Acute, ex vivo exposure: Adult rodents		
Kruger et al. (2006)	0.1–100 µM	Decrease [‡]	No change	Decrease
Kruger et al. (2007)	1–100 µM	Decrease	Not tested	Decrease [‡]
Kruger et al. (2009)	10–100 µM	Decrease [‡]	No change	Decrease

* Statistical test not performed. No change observed at short inter-pulse intervals (10–20 ms); possible decrease at other intervals. [†] Trend observed; warrants further study. [‡] Change not observed for all concentrations or metabolites tested.

2. Materials and Methods

2.1. Arsenic Exposure

All experimental protocols were approved by the Institutional Animal Care and Use Committee of the University of Rochester and carried out in compliance with ARRIVE guidelines. Given that arsenic(V) acid salt (arsenate) is the most common form of arsenic in groundwater (Cullen and Reimer, 1989), we utilized sodium arsenate dibasic heptahydrate (Na_2HAsO_4 $7H_2O$; hereon, arsenic), obtained from MilliporeSigma (A6756). C57BL/6J females were exposed to arsenic (0, 50 ppb, 36 ppm) in their drinking water (distilled deionized H_2O) starting at six weeks of age. Breeding began at two months of age and arsenic exposure continued after parturition to simulate protracted human exposure conditions. The juvenile offspring were then used for experiments prior to weaning (P17–P23), such that the pups are still nursing-dependent. Both male and female pups were used for experiments. At least two mice from two litters per exposure group were used for each experiment. Mice were maintained with a 12:12 h light:dark cycle, constant temperature of 23 °C and ad libitum feeding. An overview of arsenic exposure is shown in the Graphical Abstract. Fresh arsenic solutions were prepared and exchanged every 2–3 days to avoid oxidation.

2.2. Electrophysiology

Acute hippocampal slices were prepared from male and female juvenile mice (P17–P23) prior to weaning. After decapitation and rapid extraction of the brains into ice-cold artificial cerebrospinal fluid (ACSF), 400-micrometer-thick hippocampal slices were prepared. Slices were then allowed to recover in room temperature (RT) ACSF for at least one hour prior to experiments. Field recordings were conducted at Schaffer collateral-CA1 synapses in RT

ACSF at a flow rate of 2–3 mL/min. A borosilicate recording electrode (1–3 MΩ) filled with 1 M NaCl was placed in CA1 stratum radiatum and a monopolar stimulating electrode was placed on Schaffer collaterals between CA3 and CA1. The ACSF solution consisted of, in mM: 120.0 NaCl; 2.5 KCl; 2.5 CaCl2; 1.3 MgSO4; 1.0 NaH2PO4; 26.0 NaHCO3; and 11.0 D-glucose. ACSF was aerated with carbogen (95% O2, 5% CO2) throughout slice preparation, incubation, and recordings. Responses were elicited every 15 s. Basal synaptic transmission was assessed using input–output (IO) curves, comparing the fiber volley, a more direct measure of axonal stimulation, to the field excitatory post-synaptic potential (fEPSP) slope. Short-term pre-synaptic plasticity was assayed using paired-pulse facilitation (PPF) of varying inter-pulse intervals. Finally, LTP was induced by a single tetanus of one second, 100 Hz stimulation. The following sample sizes were used for each experiment (Group: mice, slices): IO (Control: 5, 10; 36 ppm: 6, 13; 50 ppb: 7, 15); PPF (Control: 5, 12; 36 ppm: 6, 14; 50 ppb: 8, 18); LTP (Control: 4, 7; 36 ppm: 6, 13; 50 ppb: 6, 7). Electrophysiology recordings were collected with a MultiClamp 700A amplifier (Axon Instruments, San Jose, CA, USA), PCI-6221 data acquisition device (National Instruments, Austin, TX, USA), and Igor Pro 7 (Wavemetrics, Portland, OR, USA) with a customized software package (Recording Artist, http://github.com/rgerkin/recording-artist, last commit 31 October 2019).

2.3. Analysis

Electrophysiology data was analyzed using R (version 4.0.2) and GraphPad Prism (version 9.2.0). To generate IO curves, fiber volley amplitudes were binned at ±0.05 mV, with the exception of 0.025 (0,0.025), 0.05 (0.025,0.05), and 0.1 mV (0.075,1.5). To measure the magnitude of LTP, we normalized the last five minutes of fEPSPs (25–30 min after LTP induction) to the 10 min baseline. Males and females were pooled for all analyses, with no sub-analysis of the effect of sex. Statistical significance between means was calculated using t-tests or two-way ANOVAs and Dunnett post hoc comparisons, with arsenic exposure and stimulation (fiber volley; inter-pulse interval) as factors.

3. Results

Juvenile mice (P17–P23) exposed to either a low (50 ppb) or high level (36 ppm) of arsenic in utero exhibit a significant decrease in basal synaptic transmission in the hippocampus at the Schaffer collateral-CA1 synapse (Figure 1A; two-way ANOVA, arsenic exposure: $F_{(2,216)} = 22.31$, $p < 0.0001$; fiber volley: $F_{(7,216)} = 99.87$, $p < 0.0001$). Interestingly, low and high arsenic exposure levels result in a similar decrease, such that high exposure does not reduce transmission beyond the deficit seen with low-level exposure. However, the two arsenic exposure levels differ in their effects on short-term pre-synaptic plasticity, as assessed by paired-pulse facilitation (PPF) (Figure 1D). High arsenic exposure reduced the PPF at short inter-pulse intervals (both 25 ms and 15 ms IPI, two-way ANOVA, arsenic exposure: $F_{(2,242)} = 11.93$, $p < 0.0001$; IPI: $F_{(5,242)} = 49.22$, $p < 0.0001$; Dunnett's post hoc, $p < 0.05$), whereas there is no significant change from the control with low-level arsenic exposure. There was no significant difference between the groups at longer inter-pulse intervals.

To assess the long-term changes in synaptic plasticity, we gave high-frequency stimulation to induce LTP, a cellular model for neural circuit development as well as learning and memory. Surprisingly, high arsenic exposure levels increased LTP by about 11% (Figure 2; 36 ppm: 29% ± 0.03, $n = 13$; control: 18% ± 0.04, $n = 7$, $p < 0.05$). Developmental exposure to low-level arsenic led to an 8% increase in LTP compared to the control (50 ppb: 26% ± 0.06, $n = 7$), falling between the control and high arsenic exposure, but did not reach statistical significance.

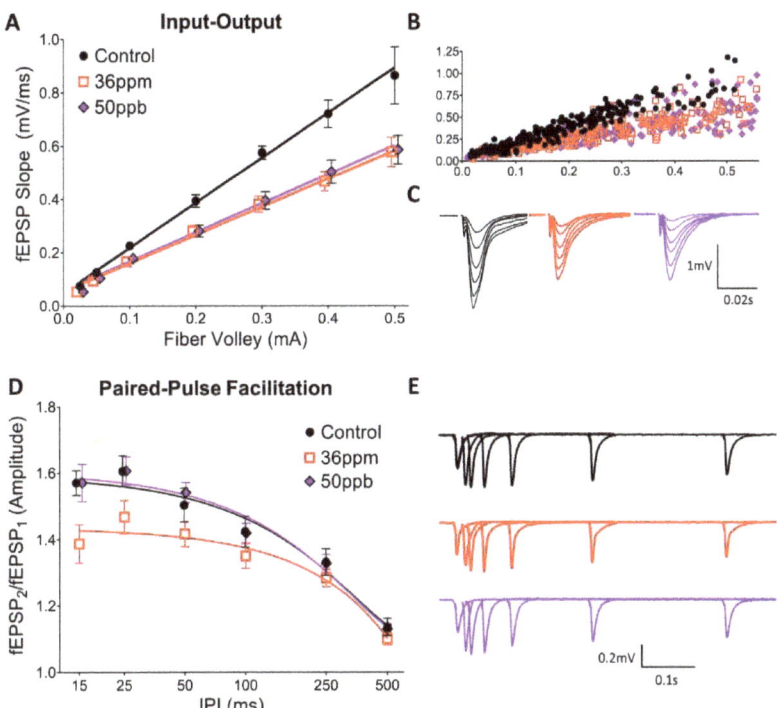

Figure 1. Effect of arsenic on hippocampal basal synaptic transmission (**A–C**) and short-term presynaptic plasticity (**D,E**), as assessed using input–output curves and paired-pulse facilitation, respectively. (**A**) Averaged responses for input-output experiments, (**B**) pre-binned, raw values from all experiments, and (**C**) waveforms averaged by fiber volley bin from representative experiments. (**D**) Averaged responses for paired-pulse facilitation experiments and (**E**) waveforms averaged by inter-pulse interval from representative experiments.

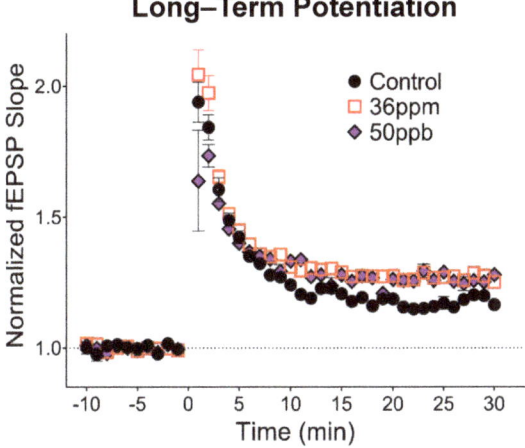

Figure 2. Effect of arsenic on hippocampal synaptic plasticity. Long-term potentiation induced by 1×100 Hz stimulation at time 0. Responses have been binned by one-minute intervals to aid visualization (average of four responses per minute).

4. Discussion

Our findings show that early developmental arsenic exposure results in significant changes to hippocampal synaptic transmission and plasticity. Most strikingly, even a relatively low dose of arsenic exposure decreases synaptic transmission. This is accompanied by no change in PPF, an indicator of glutamate release in presynaptic neurons; in our case, CA3 neurons. Therefore, our studies suggest that a low dose of arsenic led to a postsynaptic change in the CA1 neurons contributing to the decrease in synaptic transmission. This is consistent with the decrease in the neurite number and complexity observed in a cell culture model of arsenic exposure [31] as well as alterations in the hippocampal synaptic structure in arsenic-exposed juvenile mice [32]. On the other hand, a high level of arsenic decreased PPF, suggesting that glutamate release is increased in presynaptic CA3 neurons. However, there is no overall synaptic transmission change between two doses of arsenic (50 and 36 ppm). This is consistent with the notion that the higher concentration of arsenic caused a progression of changes to the hippocampal circuitry, such that there was a compensatory increase in glutamate release from presynaptic CA3 neurons in response to the changes in postsynaptic CA1 neurons for synaptic transmission.

Our data also indicate that both low and high levels of arsenic led to a trend of increased LTP expression, with an 8 and 11% increase, respectively. Our studies build upon and extend previous studies that utilize acute in vitro exposure and chronic in vivo adulthood exposure. First, our findings replicate the decrease in the hippocampal basal synaptic transmission observed both in acute, high-concentration in vitro exposure of arsenite metabolites in young rat hippocampal slices [27,28,30] as well as chronic, high-concentration (20 ppm) in vivo exposure in adult mice [20]. Second, we found that the decrease in PPF previously observed with a high-concentration arsenic exposure in adult mice [20] was also observed in our high-concentration exposure in juvenile mice (Figure 1D). Importantly, in our model, changes in basal synaptic transmission were observed even with low-concentration (50 ppb) gestational exposure. Third, our findings reveal differences between juvenile and adult mice exposed to arsenic. Whereas a previous study found a decrease in the degree of LTP in mice exposed to high-concentration arsenic from gestation through adulthood [20], we observed an increase in LTP in our juvenile mice. Together, these findings suggest a progression of changes induced by arsenic exposure, where LTP is facilitated in juvenile mice but attenuated with prolonged exposure, consistent with differential effects depending upon the timing of exposure, i.e., the critical window of exposure.

Given the epidemiological evidence [11–18], and the memory deficits observed in rodent models [11,12,20,33], we expected to see a decrease in LTP in the arsenic-exposure groups. The observed increase in LTP is difficult to reconcile, since the facilitation of LTP generally correlates with improvements in learning and memory in adults [34]. However, enhanced LTP has been observed in other rodent models of neurodevelopmental conditions, such as prenatal exposure to valproic acid, an insult-based animal model of autism [35]. In addition to an increase in LTP, valproic acid-exposed rodents show abnormal fear conditioning, decreased social interactions, and deficits in sensorimotor gating, which reflects an impairment in information processing or attention [36,37]. Therefore, while an increase in LTP represents an enhancement of learning and memory machinery at the cellular level, there may still be deficits at the behavioral level, especially when attention and information gating are impaired.

It is also important for the LTP data to be interpreted in the context of the changes in basal synaptic transmission and short-term pre-synaptic plasticity. Basal transmission is highly regulated, and alterations can lead to hippocampal circuit dysfunction. The decrease in basal transmission for both arsenic exposure groups could, therefore, lead to neurobehavioral impairments. Given that arsenic decreases neurite outgrowth in a neuronal cell culture model [31], the decrease in basal synaptic transmission in our juvenile mice may represent a decrease in hippocampus connectivity, i.e., a decrease in the number of functional excitatory dendritic spines. With fewer functional synapses, it is possible that LTP is preserved, whereas the capacity for the longer-term storage of information is greatly diminished: while short-term memory and LTP primarily depend on strengthening pre-existing synaptic connections, long-term storage is believed to require structural reorganization including the formation of new synapses [38–42]. A reduction in basal connectivity can, therefore, reduce the dynamic range of the structural reorganization necessary for the long-term storage of information. Relevant to this possibility, in two-month-old rats exposed to 3 ppm arsenic during gestation and after weaning, an impairment in contextual fear conditioning was only observed at 72 h after conditioning, but not at 1, 6, or 24 h [43].

Most strikingly, the pups in our experiments had little direct exposure to arsenic, rather they were exposed via the mother. Previous research suggests that the transmission of arsenic through the placenta is higher than the transmission into breast milk [29,44]. Therefore, it is likely that our results are primarily due to the gestational exposure in our paradigm. It is important to highlight the possible differences between the direct consumption of inorganic arsenic and gestational exposure. Inorganic arsenic undergoes biomethylation within the body, generating mono- and di-methylated organic arsenic metabolites. The metabolites possess different properties, toxicities, as well as rates of excretion [45,46]. Overall, methylated arsenicals are historically believed to be less acutely toxic than inorganic arsenic, though still biochemically active. The arsenic species that constitute gestational exposure will include both inorganic and methylated arsenicals, but primarily the methylated arsenicals according to studies of maternal and fetal cord blood [29,47]. While there is a strong correlation between the total arsenic concentration and the individual metabolite concentration in maternal and cord blood, one study in mice found that the accumulation of arsenicals in the newborn mouse brain was significantly higher than that of the exposed mother [48]. Therefore, although gestational exposure may decrease inorganic arsenic exposure compared to adult exposure, it is possible that arsenicals are more likely to pass the blood–brain barrier in the fetus.

One caveat of animal models is that biomethylation and excretion kinetics differ between species, even between mice and rats [49,50]. Indeed, even within species, it is possible that the age-related effects of arsenic could in part be due to differences in metabolism and excretion. Interestingly, among the different properties arsenic metabolites can possess, they can also have opposing effects on AMPA receptors and NMDA receptors, the major excitatory receptors underlying synaptic transmission and plasticity, respectively [30,51]. Relevant to this study, acute, ex vivo exposure to 1 μM monomethylarsonous acid (MMA(III)) is the only metabolite and concentration that has been demonstrated to significantly increase LTP in young rats [30].

Overall, we found that early developmental arsenic exposure alters hippocampal synaptic transmission and plasticity in juvenile mice. Our findings generate intriguing questions regarding the age-related effects of gestational and chronic arsenic exposure, and how they might be altered. The 50 ppb findings are of particular interest for the following several reasons: (1) 50 ppb was the effective arsenic MCL in the United States for decades prior to 2006; (2) the permissible level of arsenic continues to be above 10 ppb in many countries; and (3) mean arsenic levels exceed 10 ppb in many common beverages and 50 ppb in many common foods [52]. The current results suggest that indirect, ecologically relevant arsenic exposure in early development impacts hippocampal synaptic transmission.

Author Contributions: Conceptualization, D.A.C.-S. and H.X.; experiments, K.F.W.F. and D.B.; analysis, K.F.W.F.; writing—original draft preparation, K.F.W.F.; writing—review and editing, D.B., D.A.C.-S., and H.X.; supervision, D.A.C.-S. and H.X.; funding acquisition, D.A.C.-S. and H.X. All authors have read and agreed to the published version of the manuscript.

Funding: This study was funded by a Pilot Grant from the University of Rochester Environmental Health Sciences Center (P30 ES00247). Research in the lab of H.X. is supported by NSF (IOS 1457336) and NIH (R01MH109719). K.F.W.F. is supported by NIMH (F30MH122046).

Institutional Review Board Statement: All experimental protocols were approved by the Institutional Animal Care and Use Committee of the University of Rochester and carried out in compliance with ARRIVE guidelines.

Informed Consent Statement: Not applicable.

Acknowledgments: The graphical abstract was created in part using Servier Medical Art images (https://smart.servier.com, accessed on 1 June 2021), licensed under a Creative Commons Attribution 3.0 Unported License.

Conflicts of Interest: The authors declare no conflict of interest. The funders had no role in the design of the study; in the collection, analyses, or interpretation of data; in the writing of the manuscript, or in the decision to publish the results.

References

1. Santucci, D.; Rankin, J.; Laviola, G.; Aloe, L.; Alleva, E. Early exposure to aluminium affects eight-arm maze performance and hippocampal nerve growth factor levels in adult mice. *Neurosci. Lett.* **1994**, *166*, 89–92. [CrossRef]
2. Liu, J.; Liu, X.; Wang, W.; McCauley, L.; Pinto-Martin, J.; Wang, Y.; Li, L.; Yan, C.; Rogan, W.J. Blood lead concentrations and children's behavioral and emotional problems: A cohort study. *JAMA Pediatr.* **2014**, *168*, 737–745. [CrossRef] [PubMed]
3. Cory-Slechta, D.A.; Allen, J.L.; Conrad, K.; Marvin, E.; Sobolewski, M. Developmental exposure to low level ambient ultrafine particle air pollution and cognitive dysfunction. *Neurotoxicology* **2018**, *69*, 217–231. [CrossRef] [PubMed]
4. Naujokas, M.F.; Anderson, B.; Ahsan, H.; Aposhian, H.V.; Graziano, J.H.; Thompson, C.; Suk, W.A. The broad scope of health effects from chronic arsenic exposure: Update on a worldwide public health problem. *Environ. Health Perspect.* **2013**, *121*, 295–302. [CrossRef]
5. DeSimone, L.A.; McMahon, P.B.; Rosen, M.R. *The Quality of Our Nation's Waters: Water Quality in Principal Aquifers of the United States, 1991–2010*; US Geological Survey: Reston, VA, USA, 2015; pp. 2330–5703.
6. Guo, X.; Chen, X.; Wang, J.; Liu, Z.; Gaile, D.; Wu, H.; Yu, G.; Mao, G.; Yang, Z.; Di, Z.; et al. Multi-generational impacts of arsenic exposure on genome-wide DNA methylation and the implications for arsenic-induced skin lesions. *Environ. Int.* **2018**, *119*, 250–263. [CrossRef]
7. Biswas, S.; Banna, H.U.; Jahan, M.; Anjum, A.; Siddique, A.E.; Roy, A.; Nikkon, F.; Salam, K.A.; Haque, A.; Himeno, S.; et al. In vivo evaluation of arsenic-associated behavioral and biochemical alterations in F0 and F1 mice. *Chemosphere* **2020**, *245*, 125619. [CrossRef]
8. Htway, S.M.; Suzuki, T.; Kyaw, S.; Nohara, K.; Win-Shwe, T.T. Effects of maternal exposure to arsenic on social behavior and related gene expression in F2 male mice. *Environ. Health Prev. Med.* **2021**, *26*, 34. [CrossRef] [PubMed]
9. Environmental Protection Agency. National primary drinking water regulations: Arsenic and clarifications to compliance and new source contaminants monitoring. *Fed. Regist.* **2001**, *66*, 69–76.
10. NRC. *Arsenic in Drinking Water*; National Academy Press: Washington, DC, USA, 1999.
11. Tolins, M.; Ruchirawat, M.; Landrigan, P. The developmental neurotoxicity of arsenic: Cognitive and behavioral consequences of early life exposure. *Ann. Glob. Health* **2014**, *80*, 303–314. [CrossRef]
12. Tyler, C.R.; Allan, A.M. The Effects of Arsenic Exposure on Neurological and Cognitive Dysfunction in Human and Rodent Studies: A Review. *Curr. Environ. Health Rep.* **2014**, *1*, 132–147. [CrossRef] [PubMed]
13. O'Bryant, S.E.; Edwards, M.; Menon, C.V.; Gong, G.; Barber, R. Long-term low-level arsenic exposure is associated with poorer neuropsychological functioning: A Project FRONTIER study. *Int J. Environ. Res. Public Health* **2011**, *8*, 861–874. [CrossRef]
14. Rosado, J.L.; Ronquillo, D.; Kordas, K.; Rojas, O.; Alatorre, J.; Lopez, P.; Garcia-Vargas, G.; Del Carmen Caamano, M.; Cebrian, M.E.; Stoltzfus, R.J. Arsenic exposure and cognitive performance in Mexican schoolchildren. *Environ. Health Perspect.* **2007**, *115*, 1371–1375. [CrossRef]
15. Wasserman, G.A.; Liu, X.; Parvez, F.; Ahsan, H.; Factor-Litvak, P.; van Geen, A.; Slavkovich, V.; LoIacono, N.J.; Cheng, Z.; Hussain, I.; et al. Water arsenic exposure and children's intellectual function in Araihazar, Bangladesh. *Environ. Health Perspect.* **2004**, *112*, 1329–1333. [CrossRef] [PubMed]
16. Calderon, J.; Navarro, M.E.; Jimenez-Capdeville, M.E.; Santos-Diaz, M.A.; Golden, A.; Rodriguez-Leyva, I.; Borja-Aburto, V.; Diaz-Barriga, F. Exposure to arsenic and lead and neuropsychological development in Mexican children. *Environ. Res.* **2001**, *85*, 69–76. [CrossRef]
17. Wang, X.; Huang, X.; Zhou, L.; Chen, J.; Zhang, X.; Xu, K.; Huang, Z.; He, M.; Shen, M.; Chen, X.; et al. Association of arsenic exposure and cognitive impairment: A population-based cross-sectional study in China. *Neurotoxicology* **2020**, *82*, 100–107. [CrossRef]
18. Tsai, S.-Y.; Chou, H.-Y.; The, H.-W.; Chen, C.-M.; Chen, C.-J. The Effects of Chronic Arsenic Exposure from Drinking Water on the Neurobehavioral Development in Adolescence. *NeuroToxicology* **2003**, *24*, 747–753. [CrossRef]
19. Nigra, A.E.; Chen, Q.; Chillrud, S.N.; Wang, L.; Harvey, D.; Mailloux, B.; Factor-Litvak, P.; Navas-Acien, A. Inequalities in Public Water Arsenic Concentrations in Counties and Community Water Systems across the United States, 2006–2011. *Environ. Health Perspect.* **2020**, *128*, 127001. [CrossRef] [PubMed]

20. Nelson-Mora, J.; Escobar, M.L.; Rodriguez-Duran, L.; Massieu, L.; Montiel, T.; Rodriguez, V.M.; Hernandez-Mercado, K.; Gonsebatt, M.E. Gestational exposure to inorganic arsenic (iAs^{3+}) alters glutamate disposition in the mouse hippocampus and ionotropic glutamate receptor expression leading to memory impairment. *Arch. Toxicol.* **2018**, *92*, 1037–1048. [CrossRef]
21. Siddoway, B.H.H.; Xia, H. Glutamatergic Synapses: Molecular Organisation. *eLS* **2011**. Available online: https://doi.org/10.1002/9780470015902.a0000235.pub2 (accessed on 1 June 2021).
22. Luo, J.H.; Qiu, Z.Q.; Shu, W.Q.; Zhang, Y.Y.; Zhang, L.; Chen, J.A. Effects of arsenic exposure from drinking water on spatial memory, ultra-structures and NMDAR gene expression of hippocampus in rats. *Toxicol. Lett.* **2009**, *184*, 121–125. [CrossRef]
23. Zhang, C.; Li, S.; Sun, Y.; Dong, W.; Piao, F.; Piao, Y.; Liu, S.; Guan, H.; Yu, S. Arsenic downregulates gene expression at the postsynaptic density in mouse cerebellum, including genes responsible for long-term potentiation and depression. *Toxicol. Lett.* **2014**, *228*, 260–269. [CrossRef] [PubMed]
24. Luo, J.H.; Qiu, Z.Q.; Zhang, L.; Shu, W.Q. Arsenite exposure altered the expression of NMDA receptor and postsynaptic signaling proteins in rat hippocampus. *Toxicol. Lett.* **2012**, *211*, 39–44. [CrossRef] [PubMed]
25. Tyler, C.R.; Allan, A.M. Adult hippocampal neurogenesis and mRNA expression are altered by perinatal arsenic exposure in mice and restored by brief exposure to enrichment. *PLoS ONE* **2013**, *8*, e73720. [CrossRef] [PubMed]
26. Htway, S.M.; Sein, M.T.; Nohara, K.; Win-Shwe, T.T. Effects of Developmental Arsenic Exposure on the Social Behavior and Related Gene Expression in C3H Adult Male Mice. *Int. J. Environ. Res. Public Health* **2019**, *16*, 174. [CrossRef]
27. Kruger, K.; Binding, N.; Straub, H.; Musshoff, U. Effects of arsenite on long-term potentiation in hippocampal slices from young and adult rats. *Toxicol. Lett.* **2006**, *165*, 167–173. [CrossRef] [PubMed]
28. Kruger, K.; Repges, H.; Hippler, J.; Hartmann, L.M.; Hirner, A.V.; Straub, H.; Binding, N.; Musshoff, U. Effects of dimethylarsinic and dimethylarsinous acid on evoked synaptic potentials in hippocampal slices of young and adult rats. *Toxicol. Appl. Pharm.* **2007**, *225*, 40–46. [CrossRef]
29. Concha, G.; Vogler, G.; Lezcano, D.; Nermell, B.; Vahter, M. Exposure to Inorganic Arsenic Metabolites during Early Human Development. *Toxicol. Sci.* **1998**, *44*, 185–190. [CrossRef]
30. Kruger, K.; Straub, H.; Hirner, A.V.; Hippler, J.; Binding, N.; Musshoff, U. Effects of monomethylarsonic and monomethylarsonous acid on evoked synaptic potentials in hippocampal slices of adult and young rats. *Toxicol. Appl. Pharm.* **2009**, *236*, 115–123. [CrossRef] [PubMed]
31. Frankel, S.; Concannon, J.; Brusky, K.; Pietrowicz, E.; Giorgianni, S.; Thompson, W.D.; Currie, D.A. Arsenic exposure disrupts neurite growth and complexity in vitro. *Neurotoxicology* **2009**, *30*, 529–537. [CrossRef]
32. Zhao, F.; Liao, Y.; Tang, H.; Piao, J.; Wang, G.; Jin, Y. Effects of developmental arsenite exposure on hippocampal synapses in mouse offspring. *Metallomics* **2017**, *9*, 1394–1412. [CrossRef]
33. Jing, J.; Zheng, G.; Liu, M.; Shen, X.; Zhao, F.; Wang, J.; Zhang, J.; Huang, G.; Dai, P.; Chen, Y.; et al. Changes in the synaptic structure of hippocampal neurons and impairment of spatial memory in a rat model caused by chronic arsenite exposure. *Neurotoxicology* **2012**, *33*, 1230–1238. [CrossRef]
34. Barnes, C.A. Memory deficits associated with senescence: A neurophysiological and behavioral study in the rat. *J. Comp. Physiol Psychol* **1979**, *93*, 74–104. [CrossRef]
35. Rinaldi, T.; Kulangara, K.; Antoniello, K.; Markram, H. Elevated NMDA receptor levels and enhanced postsynaptic long-term potentiation induced by prenatal exposure to valproic acid. *Proc. Natl. Acad. Sci. USA* **2007**, *104*, 13501–13506. [CrossRef] [PubMed]
36. Markram, K.; Rinaldi, T.; La Mendola, D.; Sandi, C.; Markram, H. Abnormal fear conditioning and amygdala processing in an animal model of autism. *Neuropsychopharmacology* **2008**, *33*, 901–912. [CrossRef]
37. Schneider, T.; Przewlocki, R. Behavioral alterations in rats prenatally exposed to valproic acid: Animal model of autism. *Neuropsychopharmacology* **2005**, *30*, 80–89. [CrossRef] [PubMed]
38. Engert, F.; Bonhoeffer, T. Dendritic spine changes associated with hippocampal long-term synaptic plasticity. *Nature* **1999**, *399*, 66–70. [CrossRef]
39. Hofer, S.B.; Bonhoeffer, T. Dendritic spines: The stuff that memories are made of? *Curr. Biol.* **2010**, *20*, R157–R159. [CrossRef]
40. Yuste, R.; Bonhoeffer, T. Morphological changes in dendritic spines associated with long-term synaptic plasticity. *Annu. Rev. Neurosci.* **2001**, *24*, 1071–1089. [CrossRef]
41. Foley, K.; McKee, C.; Nairn, A.C.; Xia, H. Regulation of Synaptic Transmission and Plasticity by Protein Phosphatase 1. *J. Neurosci.* **2021**, *41*, 3040–3050. [CrossRef] [PubMed]
42. Gao, J.; Hu, X.D.; Yang, H.; Xia, H. Distinct Roles of Protein Phosphatase 1 Bound on Neurabin and Spinophilin and Its Regulation in AMPA Receptor Trafficking and LTD Induction. *Mol. Neurobiol.* **2018**, *55*, 7179–7186. [CrossRef] [PubMed]
43. Martinez, L.; Jimenez, V.; Garcia-Sepulveda, C.; Ceballos, F.; Delgado, J.M.; Nino-Moreno, P.; Doniz, L.; Saavedra-Alanis, V.; Castillo, C.G.; Santoyo, M.E.; et al. Impact of early developmental arsenic exposure on promotor CpG-island methylation of genes involved in neuronal plasticity. *Neurochem. Int.* **2011**, *58*, 574–581. [CrossRef] [PubMed]
44. Carignan, C.C.; Cottingham, K.L.; Jackson, B.P.; Farzan, S.F.; Gandolfi, A.J.; Punshon, T.; Folt, C.L.; Karagas, M.R. Estimated exposure to arsenic in breastfed and formula-fed infants in a United States cohort. *Environ. Health Perspect.* **2015**, *123*, 500–506. [CrossRef] [PubMed]
45. Thomas, D.J. Arsenic methylation—Lessons from three decades of research. *Toxicology* **2021**, *457*, 152800. [CrossRef]
46. Watanabe, T.; Hirano, S. Metabolism of arsenic and its toxicological relevance. *Arch. Toxicol.* **2013**, *87*, 969–979. [CrossRef]

47. Hall, M.; Gamble, M.; Slavkovich, V.; Liu, X.; Levy, D.; Cheng, Z.; van Geen, A.; Yunus, M.; Rahman, M.; Pilsner, J.R.; et al. Determinants of arsenic metabolism: Blood arsenic metabolites, plasma folate, cobalamin, and homocysteine concentrations in maternal-newborn pairs. *Environ. Health Perspect.* **2007**, *115*, 1503–1509. [CrossRef]
48. Jin, Y.; Xi, S.; Li, X.; Lu, C.; Li, G.; Xu, Y.; Qu, C.; Niu, Y.; Sun, G. Arsenic speciation transported through the placenta from mother mice to their newborn pups. *Environ. Res.* **2006**, *101*, 349–355. [CrossRef]
49. Odanaka, Y.; Matano, O.; Goto, S. Biomethylation of Inorganic Arsenic by the Rat and Some Laboratory-Animals. *Bull. Environ. Contam Tox.* **1980**, *24*, 452–459. [CrossRef]
50. Vahter, M. Species-Differences in the Metabolism of Arsenic Compounds. *Appl. Organomet. Chem.* **1994**, *8*, 175–182. [CrossRef]
51. Kruger, K.; Gruner, J.; Madeja, M.; Hartmann, L.M.; Hirner, A.V.; Binding, N.; Musshoff, U. Blockade and enhancement of glutamate receptor responses in Xenopus oocytes by methylated arsenicals. *Arch. Toxicol.* **2006**, *80*, 492–501. [CrossRef]
52. Wilson, D. Arsenic Consumption in the United States. *J. Environ. Health* **2015**, *78*, 8–14.

MDPI
St. Alban-Anlage 66
4052 Basel
Switzerland
Tel. +41 61 683 77 34
Fax +41 61 302 89 18
www.mdpi.com

Toxics Editorial Office
E-mail: toxics@mdpi.com
www.mdpi.com/journal/toxics

www.ingramcontent.com/pod-product-compliance
Lightning Source LLC
LaVergne TN
LVHW070446100526
838202LV00014B/1674